SIGNALLING PATHWAYS IN APOPTOSIS

Modern Genetics

A series of books covering new developments across the entire field of genetics. Edited by Richard Lathe, Centre for Genome Research, University of Edinburgh, Kings Buildings, Edinburgh, EH9 3JQ, UK

Volume 1
Embryonal Stem Cells: Introducing Planned Changes into the Animal Germline
Martin L. Hooper

Volume 2
Molecular Genetics of Inherited Eye Disorders
edited by *Alan F. Wright* and *Barrie Jay*

Volume 3
Molecular Genetics of Drug Resistance
edited by *John D. Hayes* and *C. Roland Wolf*

Volume 4
Animal Breeding: Technology for the 21st Century
edited by *A. John Clark*

Volume 5
Signalling Pathways in Apoptosis
edited by *Dianne Watters* and *Martin Lavin*

Additional volumes in preparation

Genetics of Steroid Biosynthesis and Function
edited by *Ian Mason*

This book is part of a series. The publisher will accept continuation orders which may be cancelled at any time and which provide for automatic billing and shipping of each title in the series upon publication. Please write for details.

SIGNALLING PATHWAYS IN APOPTOSIS

Edited by

Dianne Watters
and
Martin Lavin

The Queensland Institute of Medical Research
Herston
Australia

harwood academic publishers
Australia • Canada • China • France • Germany • India • Japan
Luxembourg • Malaysia • The Netherlands • Russia • Singapore
Switzerland

Amsteldijk 166
1st Floor
1079 LH Amsterdam
The Netherlands

British Library Cataloguing in Publication Data

Signalling pathways in apoptosis. – (Modern genetics; v. 5)
1. Apoptosis
I. Watters, Diane, 1952– II. Lavin, Martin
571.9′36

ISBN: 90-5702-392-X
ISSN: 1056-4497

Front cover electron micrograph taken by Wen Yu and Deborah Stenzel of the Analytical Electron Microscope Facility, Queensland University of Technology, Gardens Point, Brisbane 4001, Australia.

CONTENTS

PREFACE TO THE SERIES

The Modern Genetics series, established under the editorship of Professor H. John Evans, was intended to cover new developments across the entire field of genetics of plants and animals, including man, and at all levels from the molecule to the population. This aim will be sustained and built upon, with increasing emphasis on the practical applications of the new genetics, be they in agriculture, medicine or biotechnology.

The present volume underlines the contribution that molecular genetics has made to the understanding of basic cellular mechanisms. Indeed, the characteristic pattern of chromosomal fragmentation accompanying programmed cell death, and first described in 1980 by Andrew Wyllie here in Edinburgh, relied on the (then) relatively new technique of displaying DNA fragments by agarose gel electrophoresis. From that simple beginning has emerged the current concept that apoptosis is of fundamental importance to all aspects of development and differentiation. An ever-expanding field, the present state of the art is concisely compiled here by the editors of this volume, both acknowledged experts in the area.

R. Lathe

PREFACE

In the five years since the publication of our first volume *Programmed Cell Death: The cellular and molecular biology of apoptosis* interest in apoptosis research has increased exponentially. At that time our understanding of the signalling pathways involved was in its infancy, some information was available on signalling through the Fas/Apo1 receptor but caspases had not yet come of age and the nuclease involved in DNA fragmentation remained elusive. It is now evident that there exist families of death receptors and their downstream effectors, the caspases, which cleave a set of crucial proteins with roles in cellular homeostasis, nuclear and cytoskeletal structure, and sensing and repairing DNA damage. While there are a large number of agents (radiation, chemotherapeutic agents, toxins and receptor ligation) which induce apoptosis by damaging or altering cellular functions in different ways, the central mechanisms remain highly conserved. In the case of death receptors the initial stimulus involves receptor ligation followed by recruitment of caspases and other signalling molecules prior to activation of a cascade of downstream caspases. These caspases cleave a number of molecules to activate them, including DNA fragmentation factor and/or I-CAD resulting in activation of a nuclease which fragments chromatin. Caspase cleavage also leads to inactivation of several proteins including DNA-PK, PARP and lamin which are involved in DNA damage repair and maintenance of nuclear structure respectively.

The end result of this process is the packaging of cellular contents into apoptotic bodies in preparation for phagocytosis. While signalling through the death receptors has been relatively well described, the initiation of events resulting from cellular damage remains largely undefined. The aim of this volume is to describe the process of apoptosis at the induction, regulation, and execution phases. Emphasis is placed on the various death receptors, the pathways that are used either to induce or prevent apoptosis, and the steps involved in the cellular stress response. The pattern of activation of kinases that determines whether a cell will live or die is discussed. A greater understanding of the role of cytoplasmic events has been achieved by the identification of the mitochondrion as an apoptotic effector. Release of two apoptosis inducing factors (AIF and cytochrome c) from the mitochrondrion as well as the role of Bcl-2 in preventing this release are discussed.

Convergence of the pathways activated by different agents appears to occur at the execution phase of apoptosis, where activated caspases cleave a series of molecules ultimately leading to the morphological changes characteristic of apoptosis.

An understanding of all the intermediates involved in the process of apoptosis will provide a means of manipulating the system for the activation of apoptosis in cells resilient to death and the prevention of apoptosis in neurodegenerative and immune disorders, such as AIDS. While this treatment of apoptosis is not designed to be all embracing we trust that it will put into perspective the major events that occur in apoptosis from the initial stimulus to the execution process for at least some systems.

CONTRIBUTORS

Abrams, J.M.
Department of Cell Biology and Neuroscience
The University of Texas Southwestern Medical Center
5323 Harry Hinds Boulevard
Dallas, TX 75235–9039
USA

Brenner, C.
Centre National de la Recherche Scientifique
Unité Propre de Recherche 420
19 rue Guy Môquet
F-94801 Villejuif
France

Butt, A.J.
The Hanson Centre for Cancer Research
Frome Road
Adelaide, SA 5000
Australia

Chahal, H.
Department of Immunology
Birmingham University Medical School
Birmingham, B15 2TT
UK

Chen, P.
Department of Cell Biology and Neuroscience
The University of Texas Southwestern Medical Center
5323 Harry Hinds Boulevard
Dallas, TX 75235–9039
USA

Cotter, T.G.
Tumor Biology Laboratory
Biochemistry Department
University College Cork
Ireland

Coulson, E.J.
Walter and Eliza Hall Institute of Medical Research
PO Royal Melbourne Hospital
Parkville, VIC 3050
Australia

Deacon, E.M.
Department of Immunology
Birmingham University Medical School
Birmingham, B15 2TT
UK

Degenhardt, K.
Center for Advanced Biotechnology and Medicine
Rutgers University
679 Hoes Lane
Piscataway, NJ 08854
USA

Finucane, D.M.
Division of Cellular Immunology
La Jolla Institute for Allergy and Immunology
10355 Science Center Drive
San Diego, CA 92121
USA

Green, D.R.
Division of Cellular Immunology
La Jolla Institute for Allergy and Immunology
10355 Center Drive
San Diego, CA 92121
USA

Griffiths, G.
Department of Immunology
Birmingham University Medical School
Birmingham, B15 2TT
UK

Hancock, J.F.
Department of Pathology
University of Queensland Medical School
Herston Road
Brisbane, QLD 4006
Australia

Hannun, Y.A.
Division of Oncology
Department of Medicine
Duke University Medical Center
Durham, NC 27710
USA

Hawkins, C.J.
Division of Biology 156–29
California Institute of Technology
1201 East California Boulevard
Pasadena, CA 91125
USA

Kasof, G.
Center for Advanced Biotechnology and Medicine
Rutgers University
679 Hoes Lane
Piscataway, NJ 08854
USA

Kroemer, G.
Centre National de la Recherche Scientifique
Unité Propre de Recherche 420
19 rue Guy Môquet
F-94801 Villejuif
France

Kumar, S.
The Hanson Centre for Cancer Research
Frome Road
Adelaide, SA 5000
Australia

Lavin, M.F.
The Queensland Cancer Fund Research Unit
The Queensland Institute of Medical Research
PO Royal Brisbane Hospital
Herston
Brisbane, QLD 4029
Australia

Long, S.D.
Department of Medicine
Duke University Medical Center
Durham, NC 27710
USA

Lord, J.M.
Department of Immunology
Birmingham University Medical School
Birmingham, B15 2TT
UK

McMillan, L.
Department of Immunology
Birmingham University Medical School
Birmingham, B15 2TT
UK

Nordstrom, W.
Department of Cell Biology and Neuroscience
The University of Texas Southwestern Medical Center
5323 Harry Hinds Boulevard
Dallas, TX 75235–9039
USA

Perez, D.
Center for Advanced Biotechnology and Medicine
Rutgers University
679 Hoes Lane
Piscataway, NJ 08854
USA

Peter, M.E.
Tumor Immunology Program
German Cancer Research Center
Heidelberg
Germany

Pongracz, J.
Department of Immunology
Birmingham University Medical School
Birmingham, B15 2TT
UK

Schulze-Osthoff, K.
Department of Internal Medicine I
Medical Clinics
Eberhard-Karls University
Otfried-Müller Str. 10
D-72076 Tübingen
Germany

Smyth, M.J.
Austin Research Institute
Studley Road
Heidelberg, VIC 3084
Australia

Sutton, V.R.
Austin Research Institute
Studley Road
Heidelberg, VIC 3084
Australia

Thomas, A.
Center for Advanced Biotechnology and Medicine
Rutgers University
679 Hoes Lane
Piscataway, NJ 08854
USA

Trapani, J.A.
Austin Research Institute
Studley Road
Heidelberg, VIC 3084
Australia

Varkey, J.
Department of Cell Biology and Neuroscience
The University of Texas Southwestern Medical Center
5323 Harry Hinds Boulevard
Dallas, TX 75235–9039
USA

Vaux, D.L.
Walter and Eliza Hall Institute of Medical Research
PO Royal Melbourne Hospital
Parkville, VIC 3050
Australia

Waterhouse, N.
The Queensland Cancer Fund Research Unit
The Queensland Institute of Medical Research
PO Royal Brisbane Hospital, Herston
Brisbane, QLD 4029
Australia

Watters, D.
The Queensland Cancer Fund Research Unit
The Queensland Institute of Medical Research
PO Royal Brisbane Hospital, Herston
Brisbane, QLD 4029
Australia

White, E.
Center for Advanced Biotechnology and Medicine
Rutgers University
679 Hoes Lane
Piscataway, NJ 08854
USA

OVERVIEW: A MATTER OF LIFE AND DEATH

GARY KASOF[*], KURT DEGENHARDT[*], DENISE PEREZ[*], ANJU THOMAS[*], AND EILEEN WHITE[*,,†]**

[*]*Center for Advanced Biotechnology and Medicine*
[**]*Department of Molecular Biology and Biochemistry, and the Cancer Institute of New Jersey; Rutgers University, Piscataway, New Jersey 08854*

KEY WORDS: apoptosis, Bcl-2, capase, TNF-α, Fas, p53.

Apoptosis or programmed cell death (PCD) is a genetically controlled process whereby cells die in response to environmental or developmental cues. The morphological characteristics of apoptosis include cytoplasmic blebbing, chromatin condensation and nucleosomal fragmentation (Wyllie, 1980). Dead cells are rapidly phagocytized to prevent damage to neighboring cells. Regulation of apoptosis is critical for normal development and tissue homeostasis and disruption of this process can have severe consequences (Jacobson *et al.*, 1997). Too much cell death may produce neurodegenerative diseases and impaired development, while insufficient cell death can lead to increased susceptibility to cancer and sustained viral infection. Progress has been made in the past decade to identify many of the basic components that contribute to apoptosis, including transcriptional mediators, membrane-bound receptors (e.g. TNF-α receptor and Fas), Bcl-2 family members, kinases/phosphatases, and cysteine proteases (White, 1996). Some of these proteins have been evolutionarily conserved from nematodes to mammals (Steller, 1995). *bcl-2* was one of the first genes shown to regulate apoptosis (Hockenbery *et al.*, 1990) and can inhibit apoptosis in a wide variety of systems (White, 1996). *bcl-2* belongs to a growing family of genes that can either positively or negatively regulate apoptosis. One of these gene products, Bax, binds to Bcl-2 and antagonizes its ability to block apoptosis (Oltvai *et al.*, 1993). Another critical element of the apoptotic process is the activation of cysteine proteases, which are currently referred to as caspases (Alnemri *et al.*, 1996; Rao and White, 1997). In general, the caspases act downstream of Bcl-2-like proteins to induce apoptosis. Thus, the regulation of apoptosis appears to be a precarious balance between factors that promote survival and those responsible for initiating and executing cellular suicide. In this review, we present an overview of the basic components of apoptosis, like Bcl-2 and caspases, and how they are believed to function in this process.

TRANSCRIPTIONAL REGULATION OF APOPTOSIS

Gene expression has been shown to be required for both apoptosis and survival depending on the cell type and stimulus. Indeed, inhibition of RNA and protein

† *Corresponding Author*: Center for Advanced Biotechnology and Medicine, 679 Hoes Lane, Piscataway, New Jersey 08854. Tel.: (732)235–5329. Fax: (732)235–5795. e-mail: ewhite@cabm.rutgers.edu

synthesis can block apoptosis induced by a number of circumstances, including glucocorticoid treatment (Cohen and Duke, 1983; Wyllie *et al.*, 1984), growth factor deprivation (Martin *et al.*, 1988; Scott and Davies, 1990), treatment with some chemotherapeutic drugs (Barry *et al.*, 1990; Walker *et al.*, 1991; Mizumoto *et al.*, 1994), ischemia (Goto *et al.*, 1990; Shigeno *et al.*, 1990) and seizure (Schreiber *et al.*, 1993). In contrast, others have shown that in some cases RNA and protein synthesis inhibitors either do not block or actually promote cellular demise (Rubin *et al.*, 1988; Itoh *et al.*, 1991; Gong *et al.*, 1993; Vaux and Weissman, 1993). Some agents, such as the protein kinase inhibitor, staurosporine (Jacobson *et al.*, 1994), and Fas (Schulze-Ostoff *et al.*, 1994), can induce apoptotic-like events in enucleated cells. These data suggest that the basic machinery for apoptosis is constitutively expressed but can be modulated by changes in gene expression in order to trigger the death program. A number of transcription factors have been associated with induction of apoptosis, most notably p53; whereas others, such as NF-κB, have been implicated in promoting survival.

p53 was first detected in rodent cells transformed by simian virus SV40 in a complex with the transforming protein SV40 T antigen, suggesting that it plays a role in growth control (Lane and Crawford, 1979; Linzer and Levine, 1979). Subsequently, it was noted that p53 mutations occur in a wide variety of tumors including lung, breast, colon, esophagus, liver, bladder, ovary, brain, and haemopoetic tissues (Hollstein *et al.*, 1991; Levine *et al.*, 1991). In fact, p53 loss-of-function mutations are the most common genetic alteration found in human cancer. Disruption of p53 in "knock out" mice results in spontaneous neoplasms by 6 months of age (Donehower *et al.*, 1992). Furthermore, overexpression of wild-type p53 can suppress tumor formation in culture (Eliyahu *et al.*, 1989; Finlay *et al.*, 1989; Baker *et al.*, 1990; Diller *et al.*, 1990; Mercer *et al.*, 1990). These data suggest that p53 functions as a tumor suppressor gene and has sparked extensive research into understanding its mechanism of action.

p53 is a transcriptional regulatory protein capable of transactivating and repressing cellular genes (Ko and Prives, 1996; Levine, 1997). Transcriptional activation by p53 requires direct interaction with a sequence-specific DNA motif (PuPuPuC(A/T)(T/A)GPyPyPy) (El-Deiry *et al.*, 1992; Farmer *et al.*, 1992). p53-mediated transrepression, on the other hand, occurs on genes lacking the p53 DNA binding site and is probably dependent on interactions with components of the basal transcriptional machinery (Seto *et al.*, 1992; Mack *et al.*, 1993). Mutational analysis of p53 has revealed distinct domains that contribute to its gene regulatory activity (Prives, 1994). The N-terminus of p53 contains the activation domain, as well as binding sites for the cellular protein Mdm-2 (Momand *et al.*, 1992) and the adenovirus E1B 55K protein (Kao *et al.*, 1990). Complexes with both of these proteins interfere with p53 activity (Yew and Berk, 1992; Chen, Wu *et al.*, 1996; Haupt *et al.*, 1996). The central region of p53 has been shown by X-ray crystallography to contain sites for sequence-specific DNA binding (Cho *et al.*, 1994). The C-terminus of p53 is critical for oligimerization (Stürzbecher *et al.*, 1992; Clore *et al.*, 1994). Stable complexes of p53 with DNA suggest that p53 exists as a tetramer. A truncated protein containing just the C-terminal region can act as a dominant negative to inhibit p53 activity (Shaulian *et al.*, 1992; Eizenberg *et al.*, 1996; Sabbatini *et al.*, 1997).

There also exists a proline-rich region between the activation and DNA binding domains that is thought to be capable of binding to SH3-containing proteins and could provide a link between p53 and signal transduction pathways (Walker and Levine, 1996; Sakamuro *et al.*, 1997). The presence of these multiple domains underscores the functional, as well as the regulational, complexity of p53.

Consistent with its role as a tumor suppressor gene, expression of p53 can induce either cell cycle arrest in G1 or apoptosis (Kastan *et al.*, 1991; Yonish-Rouach *et al.*, 1991). Functional p53 is required for apoptosis induced by ionizing radiation and chemotherapeutic drugs (Clarke *et al.*, 1993; Lowe *et al.*, 1993), as well as transforming oncogenes like c-*myc* (Hermeking and Eick, 1994; Wagner *et al.*, 1994; Sakamuro *et al.*, 1995) and adenovirus *E1A* (Debbas and White, 1993). However, p53 is clearly not responsible for all modes of apoptosis, since its disruption does not appear to effect cell death triggered by glucocorticoids (Clarke *et al.*, 1993; Lowe *et al.*, 1993) or, for the most part, cell death observed during normal development (Donehower *et al.*, 1992). The function of p53 is likely to depend on the cell type and/or physiological circumstances. Indeed, cytokines (Yonish-Rouach *et al.*, 1991; Lin and Benchimol, 1995) and growth factors (Canman *et al.*, 1995) can affect the ability of p53 to induce apoptosis or growth arrest. These results have led to the hypothesis that genotoxic induction of p53 acts as a checkpoint control to stop further progression in the cell cycle and thereby maintain genomic integrity (White, 1994). However, higher accumulation of p53, or induction accompanied by either a lack of growth factors/cytokines or the presence of a proliferative signal such as c-*myc* or *E1A*, results in cell suicide (Chen *et al.*, 1996).

p53-mediated growth arrest may at least in part be explained by its ability to transcriptionally induce the cdk inhibitor *p21/WAF1/CIP1* (El-Deiry *et al.*, 1993; Harper *et al.*, 1993; Xiong *et al.*, 1993; Brugarolas *et al.*, 1995), as well as other possible growth inhibitory genes such as *GADD45* (Kastan *et al.*, 1992). The mechanism by which p53 induces apoptosis, however, is somewhat unclear. Generally, p53-mediated apoptosis is dependent on its gene regulatory capability. Point mutations in residues 22 and 23 of p53, which render p53 defective in both transcriptional activation and repression (Lin *et al.*, 1994), block E1A-induced apoptosis (Sabbatini, Lin *et al.*, 1995). However, Harris and colleagues have shown that the C-terminal region of p53, and not the transactivation domain, is required for apoptosis (Wang, Vermeulen *et al.*, 1996). Another critical issue that remains is whether p53-dependent cell death involves activation or repression of cellular genes. Studies have indicated that at least in some cases p53-induced apoptosis is not affected by the presence of RNA and protein synthesis inhibitors, suggesting that repression rather than activation is the primary mechanism (Caelles *et al.*, 1994). Several p53-repressible genes have been identified that could affect cell survival, including *bcl-2* (Miyashita *et al.*, 1994), *MAP4* (Murphy *et al.*, 1996), interleukin-6 (Santhanam *et al.*, 1991), cyclin A (Desdouets *et al.*, 1996), DNA topoisomerase IIα (Wang *et al.*, 1997), c-*fos* (Kley *et al.*, 1992), the TR2 steroid receptor (Lin and Chang, 1996), and c-*myc* (Moberg *et al.*, 1992). The anti-apoptotic gene *bcl-2* and its adenoviral homologue *E1B 19K*, which inhibit apoptosis (but not growth arrest) triggered by p53 (Debbas and White, 1993; Chiou, Rao *et al.*, 1994), abrogate p53-mediated repression suggesting one mechanism by which these genes function (Shen and Shenk, 1994; Sabbatini, Chiou *et al.*, 1995). Alternatively, p53 can transactivate several genes that could contribute to apoptosis, including *bax* (Miyashita and Reed, 1995; Han, Sabbatini, Perez *et al.*, 1996), *fas* (Owen-Schaub *et al.*, 1995), and insulin-like growth factor-binding protein-3 (*IGF-BP3*) (Buckbinder *et al.*, 1995). Bax and Fas are bonified apoptotic inducers as will be discussed in the following sections. IGF-BP3, which is capable of inhibiting mitogenic signalling by the insulin-like growth factor IGF-1, could potentiate apoptosis via suppression of growth factor signalling. Thus, it is conceivable that both p53-mediated repression and transactivation play a role in apoptosis.

In addition to p53, several other transcription factors have been associated with apoptosis regulation, including Fos/Jun, c-Myc, Nur77 (NGFI-B), and NF-κB (Soares *et al.*, 1994). The correlation between expression of these genes with cell death and survival suggests that they may be involved either directly, or indirectly, in the apoptotic program. For example, blocking Fos and Jun with antisense oligonucleotides inhibits apoptosis in lymphocytes following growth factor withdrawal (Colotta *et al.*, 1992). Application of functional blocking antibodies against Fos and Jun (Estus *et al.*, 1994), or dominant interfering Jun (Ham *et al.*, 1995), inhibits apoptosis in neuronal cells following growth factor withdrawal. Overexpression of c-*myc* in myeloid cells (Askew *et al.*, 1991) or fibroblasts (Evan *et al.*, 1992) can also accelerate apoptosis, and antisense oligonucleotides against c-*myc* inhibit apoptosis in lymphocytes (Shi *et al.*, 1992; Thulasi *et al.*, 1993). Transdominant mutations of Nur77 (Woronicz *et al.*, 1994), as well as antisense oligonucleotides (Liu *et al.*, 1994), suppress apoptosis suggesting that this steroid receptor-like transcription factor may also have a positive role in cell death. Some transcription factors, like NF-κB, may actually protect cells from apoptosis. NF-κB-mediated inhibition of apoptosis plays a critical role in TNF-α signalling and will be discussed later in more detail (Beg and Baltimore, 1996; Van Antwerp *et al.*, 1996; Wang, Mayo *et al.*, 1996). These data suggest that cell death can be regulated by a number of transcriptional mediators which can either induce or suppress apoptosis. A critical issue that remains is to identify relevant downstream targets of these transcription factors.

THE BCL-2 FAMILY

One of the first mammalian genes discovered to regulate cell death was the anti-apoptotic gene *bcl-2* (Vaux *et al.*, 1988; Hockenbery *et al.*, 1990). The role of *bcl-2* as an inhibitor of apoptosis has since been established in many circumstances, including treatment with TNF-α, Fas, UV radiation, chemotherapeutic drugs, growth factor/hormone withdrawal, viral infection and tumor formation (White, 1996). The *bcl-2* proto-oncogene was originally discovered as a common translocation in non-Hodgkins B-cell lymphoma (Bakhshi, 1985; Tsujimoto *et al.*, 1985; Cleary *et al.*, 1986). This chromosomal translocation event places the *bcl-2* gene under the transcriptional control of the powerful enhancer elements of the immunoglobulin heavy chain resulting in high levels of Bcl-2 expression and the abrogation of normal programmed cell death and promotion of malignant potential. Disruption of *bcl-2* in "knockout" mice leads to impaired kidney function manifested by polycystic disease and postnatal immune failure due to dramatic cell loss through apoptosis (Nakayama *et al.*, 1993; Veis *et al.*, 1993; Kamada *et al.*, 1995). Thus, gain of Bcl-2 function is associated with tumor development (Bakhshi, 1985; Tsujimoto *et al.*, 1985; Cleary *et al.*, 1986; McDonnell *et al.*, 1989; McDonnell and Korsmeyer, 1991), while loss of Bcl-2 has only restricted consequences to normal development (Nakayama *et al.*, 1993; Veis *et al.*, 1993; Kamada *et al.*, 1995). This suggests that there may be some redundancy in the Bcl-2 family or that other members have a more critical role (Table 1).

Table 1 Bcl-2 family.

Survival proteins	*Death proteins*
Mammalian	*Mammalian*
Bcl-2	Bax
Bcl-x_L	Bak
Mcl-2	Bad
Bfl-1	Nbk/Bik*
Bcl-w	Bid*
A1	Hrk*
	Bcl-x_S*
C. elegans	
Ced-9	
Viral	
E1B 19K	
BHRF1	
LMW5-HL	
HVS ORF16	
KSbcl-2	

Bcl-2 homologues are catagorized according to apoptotic function.
* Proteins which only contain BH3.

One such gene, *bcl-*x_L, encodes a death repressing protein that may play a more general role in the regulation of apoptosis since it is widely expressed during mouse development (Boise *et al.*, 1993). Bcl-x_L is one of two products of the *bcl-x* gene (Boise *et al.*, 1993). A shorter spliced variant of *bcl-x, bcl-*x_S, is an inducer of apoptosis and antagonizes the protective activity of *bcl-*x_L (Boise *et al.*, 1993). In contrast to *bcl-2*, disruption of the *bcl-x* gene leads to embryonic lethality with massive cell death in haemopoetic cells as well as the developing nervous system (Motoyama *et al.*, 1995). In addition to *bcl-*x_L, other *bcl-2*-like genes exist (e.g. Mcl-1, Bfl-1, Bcl-w, and A1), some of which may be tissue specific and thereby participate in unique modes of apoptosis (Table 1). Mcl-1 is an apoptotic inhibitor whose expression increases early in the differentiation of a myeloblastic leukemia cell line (Reynolds *et al.*, 1994). Expression of another Bcl-2 homologue, Bfl-1, is correlated with development of stomach cancer (Choi *et al.*, 1995). The gene *bcl-w* was identified in both murine peripheral blood cells and brain cDNA libraries as an inhibitor of apoptosis in FDC-P1 cells following interleukin-3 withdrawal (Gibson *et al.*, 1996). Human endothelial cells, which have low levels of Bcl-2, express a homologue, A1, that is induced in response to TNF-α treatment (Karsan *et al.*, 1996). Thus, inhibition of apoptosis is likely to be mediated by a complex set of Bcl-2-like proteins.

It is now realized that *bcl-2* has also been evolutionarily conserved (Table 1). Homologues exist in the nematode *C.elegans* (*ced-9* (Hengartner *et al.*, 1992)) as well as many viruses (adenovirus, E1B 19K (Rao *et al.*, 1992; White *et al.*, 1992; Boyd *et al.*, 1994; Chiou, Tseng *et al.*, 1994); Epstein-Barr virus, BHRF1 (Henderson *et al.*, 1993); African swine fever virus, LMW5-HL (Neilan *et al.*, 1993); human herpesvirus 8, KSbcl-2 (Cheng *et al.*, 1997); and herpesvirus saimiri, ORF16 (Nava *et al.*, 1997)). Like *bcl-2*, these worm and viral counterparts can inhibit apoptosis. These findings suggest that the apoptotic program has been at least partly conserved throughout evolution.

Several *bcl-2*-like genes have been identified that can promote, rather than inhibit, apoptosis. The best characterized and prototypic gene of this class is *bax* (Oltvai *et al.*, 1993). While overexpression of *bax* can induce apoptosis, disruption of this gene in "knock-out" mice leads to lymphoid hyperplasia (Knudson *et al.*, 1995). Recent data also demonstrates that expression of Bax can suppress tumorigenesis *in vivo* (Yin *et al.*, 1997). These data suggest that Bax plays a critical role during apoptosis. Biochemical studies have indicated that Bax directly interacts with itself as well as several anti-apoptotic proteins such as Bcl-2, Bcl-x_L, and E1B 19K (Oltvai *et al.*, 1993; Sedlak *et al.*, 1995; Han, Sabbatini, Perez *et al.*, 1996). Interaction with these inhibitors of cell death, functionally antagonize the death-promoting activity of Bax. Other *bcl-2* homologues have also been identified which may function similarly to *bax*, including *bad* and *bak* (Table 1). The death-promoting gene, *bad*, was isolated from a yeast two-hybrid and expression cloning screens against *bcl-2* (Yang *et al.*, 1995). Bad strongly interacts with Bcl-x_L and reverses its suppression of apoptosis (Yang *et al.*, 1995). Another gene, *bak*, was identified simultaneously by a two-hybrid screen as well as a PCR based strategy aimed at looking for *bcl-2*-related genes (Chittenden *et al.*, 1995; Farrow *et al.*, 1995; Kiefer *et al.*, 1995). Bak interacts with Bcl-x_L and E1B 19K and in most cases accelerates cell death.

Interaction between death-promoting and suppressing Bcl-2-like proteins has led to a rheostat model for explaining how these proteins function (Oltavi and Korsmeyer, 1994). According to this model, the ratio between survival factors, such as Bcl-2, and death promoters, such as Bax, controls the fate of the cell. However, the biochemical mechanism of these proteins is still unclear. What is also unclear is which set of *bcl-2* proteins act as effector molecules. For example, do the death promoting genes simply inhibit a required survival factor (e.g. Bcl-2, Bcl-x_L) or do they actually trigger apoptosis?

Bcl-2 and its family members bind to a number of unrelated proteins that has provided insight into their apoptotic activity. For example, Bcl-2 interacts with the small molecular weight G-protein, R-ras (Fernandez-Sarabia and Bischoff, 1994), and expression of R-ras can antagonize Bcl-2 function (Wang, Milan *et al.*, 1995). Although related proteins, such as H-ras, N-ras, and K-ras, contribute to oncogenic transformation, R-ras does not exhibit this property. However, R-ras can promote apoptosis following growth factor deprivation (Wang, Milan *et al.*, 1995). The role of R-ras in signal transduction is unclear although it is known that it interacts with the serine/threonine kinase Raf-1 (Spaargaren *et al.*, 1994). Bcl-2 cooperates with Raf-1 to inhibit apoptosis and targets Raf-1 to the mitochondrial membrane (Wang, Rapp *et al.*, 1996). Interestingly, other Bcl-2 binding proteins can regulate Raf-1 activity. Bag-1, a protein that interacts and cooperates another Bcl-2 family member with Bcl-2, also binds to Raf-1 (Takayama *et al.*, 1995). Serine phosphorylation of Bad results in binding to the protein, 14–3–3, which is an activator of Raf-1 (Li *et al.*, 1995; Zha, Harada *et al.*, 1996). Phosporylation of Bad is triggered by the kinase Akt in response to cytokines and inhibits Bad-induced cell death (Datta *et al.*, 1997; del Peso *et al.*, 1997). Thus, the Bcl-2 family and associated proteins R-ras, Bag-1, and 14–3–3 may function by regulating the activity of kinases and in turn may be regulated by kinases themselves.

E1B 19K, but not Bcl-2, interacts with the nuclear structural protein, lamin A/C (White *et al.*, 1984; White and Cipriani, 1989; Rao *et al.*, 1996; Rao *et al.*, 1997). The lamins are cleaved by caspases during apoptosis and lamin proteolysis is required for the nuclear events of apoptosis (Lazebnik *et al.*, 1995; Rao *et al.*, 1996; Takahashi *et al.*, 1996). Binding to lamin A/C, however, does not prevent lamin cleavage (Rao *et al.*, 1997). Rather, lamin A/C may serve to target E1B 19K to the nuclear envelope.

53Bp2 was identified in three separate two-hybrid screens using Bcl-2, p53, and phosphorylase phosphatase-1 (PP-1) as baits (Helps *et al.*, 1995; Gorina and Pavletich, 1996; Naumovski and Cleary, 1996). 53Bp2 inhibits p53 DNA-binding and induces G2/M growth arrest (Naumovski and Cleary, 1996). Phosphorylation of p53, such as by cyclin-dependent kinases, is associated with p53 transcriptional activity (Wang and Prives, 1995). This supports a model whereby 53Bp2 binding to a phosphatase may inhibit p53 activity, thereby abrogating p53-mediated apoptosis and inducing G2/M arrest. It is intriguing that Bcl-2 can also bind to 53Bp2, suggesting that the Bcl-2 family may also have a role in controlling p53 function.

Recent structural experiments on Bcl-2-like proteins have also helped to elucidate their biochemical mechanism. The Bcl-2 family members contain four conserved regions (BH1, BH2, BH3, and BH4) that are important for protein-protein interactions and apoptotic regulation (Reed *et al.*, 1996). Point mutations in Bcl-2 within the BH1 (gly145) and BH2 (trp188) domains eliminate its association with Bax and compromise its death repressing activity but has no effect on homodimerization (Yin *et al.*, 1994). In contrast, a deletion mutant of Bcl-2 which lacks the BH4 domain is unable to block apoptosis and homodimerization, but is still capable of binding to Bax (Reed *et al.*, 1996). Thus far, the BH4 domain has only been found in the survival-promoting proteins. The BH3 domain was originally identified as a region within Bax that is essential for homodimeration as well as interaction with antagonist proteins such as E1B 19K and Bcl-2 (Han, Sabbatini, Perez *et al.*, 1996; Zha, Aimé-Sempé *et al.*, 1996). A chimeric protein of Bcl-2 containing the BH3 domain of Bax provides death promoting activity to Bcl-2 (Hunter and Parslow, 1996). Interestingly, a number of proteins exist that contain just the BH3 domain, including Nbk/Bik (Boyd *et al.*, 1995; Han, Sabbatini and White, 1996), Hrk (Inohara *et al.*, 1997), Bid (Wang, Yin *et al.*, 1996), and Bcl-x_S (Boise *et al.*, 1993) (Table 1). These proteins can bind to and antagonize Bcl-2-like survival proteins. While these data suggest that the BH3 domain plays a more critical role in triggering apoptosis, some Bcl-2 like proteins such as Bad can induce cell death but do not a contain a BH3 domain (Yang *et al.*, 1995). Thus, it is likely that while these conserved regions have some role in protein function, their exact significance may be protein dependent and will probably require a further understanding of their structure.

Recently, the structure of Bcl-x_L was solved using NMR and X-ray crystallographic techniques (Muchmore *et al.*, 1996). Bcl-x_L contains 5 amphipathic α-helices surrounding 2 central hydrophobic helices. Although spread throughout the primary amino acid sequence, the functionally important BH1, BH2, and BH3 domains are in close spatial proximity forming a hydrophobic cleft that is involved in mediating protein-protein interactions important for its anti-apoptotic function (Sattler *et al.*, 1997). An interesting observation from these studies is that the structure of Bcl-x_L resembles the membrane insertion domain found in bacterial toxins such as colicins and diphtheria toxin. These proteins insert into cellular membranes and multimerize to form pores (London, 1992), suggesting that Bcl-x_L may regulate apoptosis through a pore-forming activity. Indeed, Bcl-2 related proteins are generally found associated with membrane structures, particularly the mitochondria, endoplasmic reticulum, and the nuclear envelope (White *et al.*, 1984; Monagan *et al.*, 1992; Gonzalez-Garcia *et al.*, 1994). Furthermore, *in vitro* experiments demonstrate that Bcl-x_L can facilitate ion transport across a lipid bilayer (Minn *et al.*, 1997). It will be of interest to determine if the Bcl-2 family members actually function as channels *in vivo* and if this activity is responsible for apoptosis regulation.

These and other studies have led to several hypotheses regarding the biochemical mechanism of this family of proteins. One possibility that has emerged is that they form a channel capable of regulating ion flux, such as calcium. Calcium is a critical component of signalling pathways and high levels of intracellular calcium is often associated with cell death (Trump and Berezesky, 1995). Bcl-2 can suppress calcium release from the endoplasmic reticulum, although it is possible that this is simply a downstream consequence of its ability to inhibit apoptosis (Lam *et al.*, 1994). Nevertheless, the experiments indicating that the Bcl-2-like proteins can form ion channels adds some credence to this model.

Others have speculated that Bcl-2 regulates the generation of reactive oxygen radicals since Bcl-2 is able to attenuate cell death induced by oxidative damaging agents (Hockenbery *et al.*, 1993; Zhong *et al.*, 1993; Satoh *et al.*, 1996). Free radicals can contribute to cell death by disrupting DNA, proteins, and lipids (Olanow, 1993). The localization of Bcl-2 at mitochondrial membranes makes it plausible that the protein is directly involved in oxidative phosphorylation (Hockenbery *et al.*, 1990). However, further evidence demonstrates that Bcl-2 can inhibit apoptosis in cells that lack mitochondrial DNA and thereby respiration (Jacobson *et al.*, 1993).

An alternative mechanism for Bcl-2 action at the mitochondria may be the control of cytochrome c release. Release of cytochrome c from the mitochondria into the cytosol is required for caspase activation and DNA fragmentation in cell-free extracts (Liu, Kim *et al.*, 1996), and Bcl-2 can act to suppress cytochrome c efflux from mitochondria (Kluck *et al.*, 1997; Yang *et al.*, 1997). It will certainly be important to determine if efflux of proteins, such as cytochrome c, from the mitochondria actually contributes to apoptosis *in vivo* or if this is simply a downstream phenomenon. Additionally, there is tremendous interest in elucidating the downstream events of apoptosis, most notably, caspase activation.

CASPASES

Evidence suggesting that caspases play an integral part in the final executionary steps of the apoptotic pathway was first elucidated with the discovery that the *ced-3* gene product, which is required for programmed cell death in the nematode *C. elegans*, has a high sequence homology to ICE (caspase-1) (Yuan *et al.*, 1993). Caspase-1, is a mammalian cysteine protease that specifically cleaves pro-IL-1β into the mature IL-1β cytokine in response to inflammation (Thornberry *et al.*, 1992). Activation of caspase-1 or related proteases has been shown to be a common downstream component of the apoptotic program (Rao and White, 1997). In general, nascent caspases consist of a regulatory prodomain followed by two domains that are cleaved into two subunits which bind to form a heterotetrameric active protease. ICE consists of a long N-terminal prodomain that is cleaved at a conserved aspartic acid residue as one of the steps in protease activation. The C-terminal portion is cleaved after a putative aspartic acid residue within the consensus pentapeptide sequence QACRG to produce two subunits (p10 and p20). The 20 kDa catalytic subunit contains a conserved cysteine residue that is important for substrate recognition and the 10 kDa subunit confers substrate specificity. The subunits are then capable of interacting with one another to form an active tetrameric protease $(p20{:}p10)_2$ that is able to bind and cleave substrates for subsequent steps of apoptosis. Thus far, 10 additional mammalian proteases have been discovered that are homologous to caspase-1 (Table 2). Based upon sequence homology, these proteases have been classified into three major categories, ICE, CPP32 (Ced-3-like), or Ich-1 (Alnemri *et al.*, 1996).

The role of these proteases in programmed cell death has been aided by the identification of specific caspase inhibitors, such as CrmA. CrmA, a gene product of cowpox virus, acts as a strong competitive inhibitor of caspase-1 (Ray *et al.*, 1992; Miura *et al.*, 1993; Komiyama *et al.*, 1994) as well as a weak inhibitor of caspase-3 (Nicholson *et al.*, 1995), and blocks numerous apoptotic signals, including TNF-α, Fas, and serum withdrawal (Enari *et al.*, 1995; Los *et al.*, 1995; Tewari and Dixit, 1995). Similarly, p35, a protein product of the baculovirus, acts as an irreversible inhibitor by blocking autoactivation of caspase-1 -2, -3, -4 and Ced-3 (Bump *et al.*, 1995; Xue and Horvitz, 1995). Synthetic chemical inhibitors have recently been developed that mimic the substrate pentapeptide cleavage sequence with the residue at the P4 position determining specificity (Nicholson, 1996; Rotonda *et al.*, 1996). The specific caspase-1 inhibitor acetyl-Tyr-Val-Ala-Asp-chloromethylketone (Ac-Y-VAD-CHO) has a hydrophobic residue at the P4 position that confers specificity within the protease's binding pocket. Acetyl-Asp-Glu-Val-Asp-chloromethylketone (Ac-DEVD-CHO), however, is a potent inhibitor of caspase-3 and other Ced-3-like proteases due to the aspartic acid residue at the P4 position. These synthetic inhibitors allow one to differentiate between the different classes of proteases in kinetic inhibitory studies and thereby develop a protease activation cascade model. Application of either synthetic inhibitors or CrmA blocks Fas-and TNF-α-induced apoptosis. However, the caspase-1 specific inhibitor Ac-Y-VAD-CHO blocks caspase-1-like and caspase-3-like activation as opposed to Ac-DEVD-CHO which only prevents activation of caspase-3-like proteases (Enari *et al.*, 1996). Thus, one can conclude that a sequential cascade-like activation of both proteases is involved in mediating this apoptotic pathway.

Table 2 Human caspase family.

Protease	*Alternative Name*	*Reference*	*Active Site*	*Substrate*
ICE sub-family				
caspase-1	ICE	Yuan *et al.*, 1993	QACRG	pro-Il-1β, caspase-1, -3, and -4
caspase-4	ICErel-III, TY	Faucheu *et al.*, 1995 Kamens *et al.*,1995 Munday *et al.*, 1995	QACRG	
caspase-5	TX, ICH-2, ICErel-II	Faucheu *et al.*, 1995 Munday *et al.*, 1995 Nicholson *et al.*, 1995	QACRG	caspase-1 and -5
ICH-1 sub-family				
caspase-2	ICH-1	Wang *et al.*, 1994 Kumar *et al.*, 1994	QACRG	PARP
caspase-9	ICE-LAP6, Mch-6	Duan *et al.*, 1996	QACGG	PARP
Ced-3 sub-family				
caspase-3	CPP-32, Yama, apopain	Fernandes-Alnemri *et al.*, 1994 Nicholson *et al.*, 1995 Tewari *et al.*, 1995	QACRG	PARP, caspase-6, -7, and -9, NEDD2, DNA-PK, SREBP, rho-GDI, DFF,
caspase-7	Mch-3, ICE-LAP3, CMH-1	Fernandes-Alnemri, Takahashi *et al.*, 1995 Duan *et al.*, 1996 Lippke *et al.*, 1996	QACRG	PARP, caspase-6
caspase-6	Mch2	Fernandes-Alnemri, Litwack *et al.*, 1995	QACRG	lamin A/C, caspase-3, PARP
caspase-8	MACH, FLICE, Mch5	Muzio *et al.*, 1996 Boldin *et al.*, 1996 Fernandes-Alnemri *et al.*, 1996	QACQG	All caspases
caspase-10	Mch4	Fernandes-Alnemri *et al.*, 1996	QACQG	All caspases

Caspase-3, one of the most highly studied downstream proteases, is activated by several proteases including caspases-1, -6, -8, and -10 (Tewari *et al.*, 1995; Fernandes-Alnemri *et al.*, 1996; Srinivasula *et al.*, 1996), as well as the serine protease granzyme B (Darmon *et al.*, 1995). Its activation is inhibited by CrmA (Nicholson *et al.*, 1995; Tewari *et al.*, 1995) and also more potently by Z-VAD, DEVD-CHO, p35 (Bump *et al.*, 1995), and E1B 19K and Bcl-2 (Boulakia *et al.*, 1996; Sabbatini *et al.*, 1997) in several apoptotic pathways and in various cell types. In contrast to caspase-1 knockout mice whose phenotype is relatively normal (although they do have defects in their inflammatory response), caspase-3 null mice die at 1–3 weeks of age, and show brain and neuronal defects early in embryonic development with fewer pyknotic clusters indicative of a decreased rate of apoptosis (Kuida *et al.*, 1996). Fewer thymocytes are also present but are able to undergo apoptosis by Fas antibody, dexamethasone and staurosporine, possibly due to the presence of redundant proteases such as caspases-6 or -7.

Several groups are currently seeking the downstream targets of caspases that contribute to apoptosis. Among the first substrates identified was poly(ADP-ribose) polymerase (PARP), a DNA damage sensing and repair protein that is cleaved to an 85 kDa protein very early in the apoptotic process (Kaufmann *et al.*, 1993). This protein is also one of the components cleaved by an ICE-like enzyme in a cell free system (Lazebnik *et al.*, 1994). PARP, however, appears not to be essential in the apoptotic process since mice containing a mutation in (ADP-ribosyl) transferase (ADPRT), which compromises poly(ADP-ribosyl)ation, are able to develop relatively normally (Wang, Auer *et al.*, 1995).

Several other proteins are cleaved by caspases and their role in apoptosis is being evaluated. Lamins A/C and B are substrates for proteases in a cell free system with lamin A/C being specifically cleaved by caspase-6 but not caspase-3 or -9 (Lazebnik *et al.*, 1995; Orth *et al.*, 1996; Takahashi *et al.*, 1996). Y-VAD-cmk and TLCK inhibits lamin cleavage but not PARP cleavage (Lazebnik *et al.*, 1995). Thus, one can conclude there are multiple proteases that cleave specific substrates. The role of lamins as relevant substrates in apoptosis has been determined through studies demonstrating that cells that express mutant lamins with an altered caspase cleavage site have attenuated onset of apoptosis, perhaps by maintaining the nuclear structural integrity (Rao *et al.*, 1996). The disassembly of the structural components of the cell may facilitate cell death. Another protein, DNA fragmentation factor (DFF), is cleaved by caspase-3 generating an active protein capable of inducing DNA fragmentation (Liu *et al.*, 1997). Other possible protease targets include Rb (Bing and Dou, 1996), SREBP, U1-70 Kd ribonucleoprotein, Gas 2, DNA dependent kinases and actin, but the relevance of the cleavage of these proteins by caspases in the apoptotic pathway remains to be determined (Fraser and Evan, 1996).

Caspases are an evolutionarily conserved family of proteins that not only execute the final steps of apoptosis but also, in some cases, regulate upstream induction of cell death. There appears to be a family of regulatory proteases upstream in the protease cascade that sequentially activates a second set of proteases. Lengthy prodomain proteases tend to be activated upstream in the apoptotic process by interacting directly with the receptor complex (or possibly other apoptotic genes), and then these caspases activate a second set of smaller prodomain proteases which contribute to the final steps of apoptosis. Evidence suggests these secondary proteases cleave the structural substrates that are necessary for the morphological changes associated with apoptosis. Within this cascade of events there may be redundancy amongst the proteases. Thus, determining which proteases are pertinent for each cell type and apoptotic pathway and the substrates they cleave remains to be elucidated.

Much effort has been made recently to elucidate the mechanism by which the Bcl-2 family regulates caspases (Rao and White, 1997). In general, it appears that the Bcl-2-like proteins are upstream of the cysteine proteases in the apoptotic pathway. Thus, inhibition of cell death by Bcl-2, E1B 19K, or Bcl-x_L abrogates activation of the known proteases (Boulakia *et al.*, 1996; Chinnaiyan, Orth *et al.*, 1996). A significant development was achieved by the discovery in *C. elegans* that

Ced-4 acts as a bridge to join Ced-9 (Bcl-2-like) and Ced-3 (caspase) (Chinnaiyan *et al.*, 1997; Spector *et al.*, 1997; Wu *et al.*, 1997). An analogous scenario may be present in mammals. Indeed, Ced-4 binds to Bcl-x_L as well as the prodomains of caspase-1 and -8 (Chinnaiyan *et al.*, 1997; Wu *et al.*, 1997). In addition, a mammalian protein, Apaf-1, has recently been identified that contains homology to Ced-4 (Zou *et al.*, 1997). Thus far, Apaf-1 has only been reported to interact with Bcl-x_L, although future studies may reveal associations with other Bcl-2 family connection between the Bcl-2 family and caspase activation.

THE TNF-A/FAS PATHWAY OF APOPTOSIS

TNF-α and FasL, as previously mentioned, are potent cytokines that are capable of inducing cell death by apoptosis (Tartaglia *et al.*, 1993; Nagata, 1997). TNF-α is secreted predominantly by activated macrophages and has pleiotropic activities including cytotoxicity during inflammation, immunoregulation, and antiviral responses (Beutler and Cerami, 1986). Signalling by TNF-α is mediated by two distinct and widely expressed receptor subtypes, TNFR1 (55 kDa) and TNFR2 (75 kDa) (Tartaglia *et al.*, 1991). In most cases TNFR1 is responsible for TNF-α-induced apoptosis and activation of the transcription factor NF-κB (Tartaglia *et al.*, 1991). Signalling through TNFR2 occurs less frequently and appears to be restricted to cells of the immune system. FasL binding to its receptor Fas (also known as CD95 and Apo-1) induces apoptosis and plays a critical role in apoptosis triggered by activated T and B cells (Singer and Abbas, 1994; Lagresle *et al.*, 1995; Rothstein *et al.*, 1995) and in maintaining immune privilege sites (Griffith *et al.*, 1995). The expression of Fas and FasL is greatly reduced in *lpr* (lymphoproliferation) and *gld* (general lymphoproliferative disease) mutant mice, respectively (Watanabe-Fukunaga *et al.*, 1992; Takahashi *et al.*, 1994). *Lpr* and *gld* mice develop autoimmune diseases as well as lymphadenopathy and splenomegaly by accumulating CD4-CD8-thymocytes. Humans with mutations in the Fas pathway, develop autoimmune disorders (Fisher *et al.*, 1995; Rieux-Laucat *et al.*, 1995; Giordano *et al.*, 1997). A new mechanism has recently been proposed for cells of the immune privilege sites such as the eye and the testes to evade inflammatory responses by activated inflammatory cells (Bellgrau *et al.*, 1995; Griffith *et al.*, 1995). Activated immune cells expressing Fas enter the immune privilege sites but are thought to be rapidly killed by FasL expressed by cells in these sites. It has also been demonstrated that cancer cells express FasL to kill immune effector cells in order to evade the immune system (Hahne *et al.*, 1996). Infiltrating T cells expressing Fas were rapidly killed by melanoma cells expressing FasL *in vitro*. In contrast, Fas-deficient *lpr* mutant mice immune effector cells were not killed by FasL expressing cells. Thus, the phenotypes in mice and humans suggest that Fas induces signals responsible for cell death and deletion of lymphocytes as well as other cell types.

The biochemical mechanism of TNF-α and FasL signalling have been well studied and it has been noted that they share many similarities (Nagata, 1997). The receptors for both of these cytokines, TNFR1/2 and Fas, are members of a growing receptor family that includes CD27, CD30, CD40, OX40, 4-1BB, lymphotoxin-β receptor, NGF receptor (Smith *et al.*, 1994), DR-3/Wsl-1 (Chinnaiyan, O'Rourke *et al.*, 1996; Kitson *et al.*, 1996), CAR1 (Brojatsch *et al.*, 1996), and DR-4 (Pan *et al.*, 1997). This family is defined by an extracellular group of cysteine-rich repeats, and they generally play important roles in regulating cell survival. In addition, Fas and TNFR1 both contain a death domain in the cytoplasmic region of about 80 amino acids, that is required for initiation of cell death (Itoh *et al.*, 1991; Tartaglia *et al.*, 1993). Death domains are also present in other family members including DR-3/Wsl-1, DR-4 and CAR1 (Brojatsch *et al.*, 1996; Chinnaiyan, O'Rourke *et al.*, 1996; Kitson *et al.*, 1996; Pan *et al.*, 1997). Overexpression of these receptors has been associated with apoptosis. Deletion of the death domain in either TNFR1 (Tartaglia *et al.*, 1993), Fas (Cascino *et al.*, 1996), DR-3/Wsl-1 (Chinnaiyan, O'Rourke *et al.*, 1996; Kitson *et al.*, 1996), or DR-4 (Pan *et al.*, 1997) abrogates their apoptotic response. NMR and mutational analysis of the Fas death domain has revealed that it consists of six antiparallel, amphipathic α-helices and that there are specific regions that allow for self association as well as binding to downstream partners (Huang *et al.*, 1996). It is therefore hypothesized that the mechanism of these receptor molecules is to recruit other death-associated adaptor proteins via death domain interactions.

A novel 34 kDa protein termed TNF-associated death domain protein (TRADD) was identified by specific interaction with the death domain of TNFR1 using the yeast two-hybrid system (Hsu *et al.*, 1995). Overexpression of TRADD in a wide variety of cell lines can trigger apoptosis (Hsu *et al.*, 1995). Interestingly, TRADD does not interact with Fas even though Fas and TNFR1 contain homologous death domains in their intracellular region (Hsu *et al.*, 1995). However, this disparity between TRADD recruitment to TNFR1 but not Fas has been resolved by the identification of another death domain protein, Fas-associated death domain protein (FADD/MORT 1), which binds TRADD as well as the cytoplasmic region of Fas (Boldin *et al.*, 1995; Chinnaiyan *et al.*, 1995; Chinnaiyan, Tepper *et al.*, 1996). Overexpression of FADD can also induce apoptosis in several cell types (Boldin *et al.*, 1995; Chinnaiyan *et al.*, 1995). The C-terminal region of FADD contains a death domain homologous to the cytoplasmic regions of Fas and TNFR1 (Chinnaiyan, Tepper *et al.*, 1996). The death domain allows interaction of FADD with Fas and TRADD, but not with TNFR1 (Chinnaiyan *et al.*, 1996). The N-terminal portion of FADD contains a death effector domain (DED) that is responsible for binding downstream effector proteins (Chinnaiyan, Tepper *et al.*, 1996). Deletion mutants of FADD which lack this N-terminal region, act in a dominant negative fashion to block TNF-α- and FasL-induced apoptosis, suggesting that effector molecule interactions are required for cell death signalling (Chinnaiyan, Tepper *et al.*, 1996).

Recently, an effector protein was identified termed FLICE/MACH (caspase-8) that is capable of binding to FADD in the yeast two-hybrid assay and *in vivo* and can induce apoptosis when overexpressed (Boldin *et al.*, 1996; Muzio *et al.*, 1996). The N-terminal prodomain of caspase-8 contains two DEDs of approximately 60 amino acids that is homologous to the DED of FADD, while the C-terminal region of caspase-8 shows homology with the ICE family of proteases (Boldin *et al.*, 1996; Muzio *et al.*, 1996). Caspase-8 directly interacts with the DED of FADD leading to activation of the protease and apoptosis (Boldin *et al.*, 1996; Muzio *et al.*, 1996). Indeed, TNF-α-and Fas-mediated apoptosis are both inhibited by expression of the caspase inhibitor CrmA (Enari *et al.*, 1995; Los *et al.*, 1995; Tewari and Dixit, 1995; Muzio *et al.*, 1996). This has led to the following model for TNF-α and FasL-induced cell death (Figure 1). Ligand binding leads to oligimerization of the respective receptors and facilitates binding of TRADD and FADD. In TNF-α-induced apoptosis, TRADD then binds FADD so that there is recruitment of FADD in both pathways. FADD subsequently binds to and recruits caspase-8 to the DISC allowing the caspase to be activated. Caspase-8 is capable of autocatalyzing the cleavage of its prodomain releasing the active C-terminal subunits. This is believed to begin a cascade of protease activation leading toward cell death (Srinivasula *et al.*, 1996). Multimerized FADD produces filaments throughout the cell (Perez and White, 1998; Siegel *et al.*, 1998). E1B 19K, which blocks FADD-induced caspase activation and cell death, may work in part by disrupting FADD filaments causing FADD to relocalize with regions normally associated with 19K (Perez and White, 1998). It will be interesting to determine whether other Bcl-2 family members have a similar function.

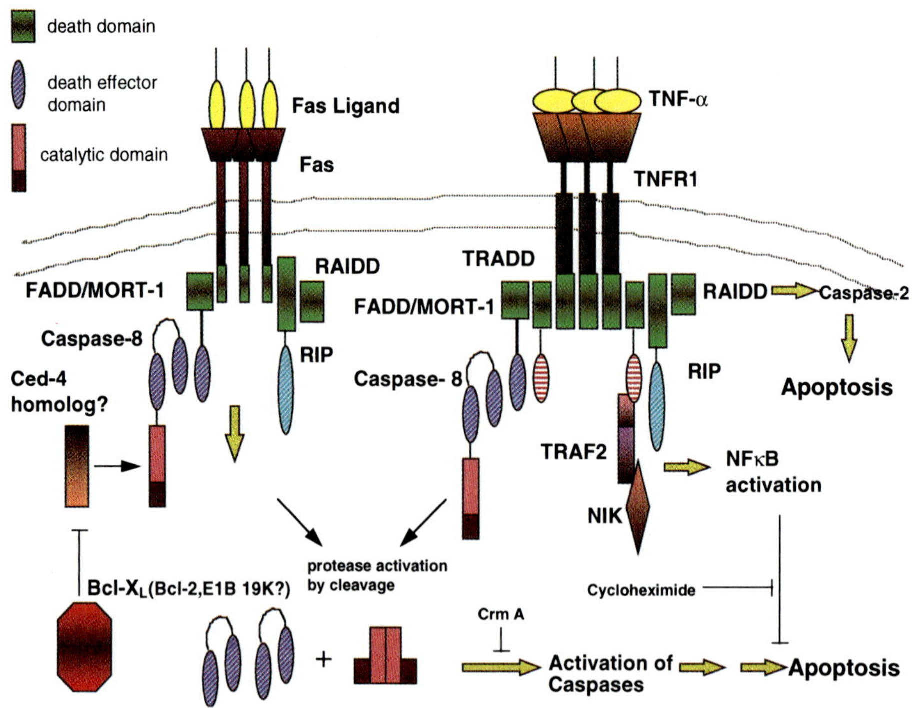

Figure 1 TNF-α and Fas signalling pathways.

In addition to FADD, a number of other Fas-interacting proteins have been identified by two hybrid screening using the cytoplasmic region of Fas as the bait.

One of these proteins, RIP, contains a C-terminal death domain as well as an N-terminal region that is strongly homologous to serine/threonine kinases (Stanger *et al.*, 1995). Transient expression of RIP results in both apoptosis and NF-κB activation, although the mechanism for these activities appear to be distinct (Stanger *et al.*, 1995; Ting *et al.*, 1996). For example, deletion of the death domain of RIP abrogates its apoptotic ability but does not effect its capacity to activate NF-κB (Stanger *et al.*, 1995; Liu, Hsu *et al.*, 1996). Death domains may play a role in binding to another death domain containing protein, RAIDD, which is capable of recruiting the cysteine protease caspase-2 and therefore may initiate a protease cascade similar to FLICE (Duan and Dixit, 1997). Another novel Fas binding protein, FAF-1 (Fas-associated protein factor) can also induce apoptosis following transient transfection (Chu *et al.*, 1995). FAP-1 (Fas-associated protein), is a tyrosine protein phosphatase that appears to suppress cell death (Sato *et al.*, 1995). Overexpression of FAP-1 in a T cell line attenuates Fas-induced apoptosis (Sato *et al.*, 1995). FAP-1 expression is highest in cell lines resistant to Fas-induced apoptosis (Sato *et al.*, 1995). The regulation of apoptosis by FAF-1 and FAP-1 are thus far unknown.

TNF-α and Fas signalling also promotes a kinase cascade leading to the activation of the stress-responsive, mitogen-activated kinases, p38 (MAP kinase) and JNK (Jun kinase) which have been implicated as potential mediators of apoptosis (Liu, Hsu *et al.*, 1996; Juo *et al.*, 1997; Nishina *et al.*, 1997). These kinases trigger changes in transcriptional regulation particularly by AP-1 which have been reported to play a role in numerous physiological processes including cell death (Soares *et al.*, 1994). The role of these kinases in TNF-α and Fas mediated apoptosis is poorly understood. The MAP kinase, ASK1, is activated upon TNF-α treatment (Ichijo *et al.*, 1997). Overexpression of ASK1 induces apoptosis and a catalytic inactive form of ASK1 blocks TNF-α-induced cell death (Ichijo *et al.*, 1997). ASK1 phosphorylates other MAP kinases such as SEK1 and MKK3/MAPKK6 which in turn can activate the JNK and p38 respectively, suggesting that the MAP kinases can trigger apoptosis (Ichijo *et al.*, 1997). However, others have suggested that the stress kinase pathway does not contribute to cell death induced by TNF-α and Fas (Liu, Hsu *et al.*, 1996; Lenczowski *et al.*, 1997). Expression of a deletion mutant of RIP lacking the death domain does not induce apoptosis but is still able to activate JNK (Liu, Hsu *et al.*, 1996). Furthermore, the specific p38 inhibitor SB 203580 completely blocks Fas-induced p38 activation but has no effect on inhibiting cell death (Stanger *et al.*, 1995; Liu, Hsu *et al.*, 1996). Disruption of SEK1, a direct activator of JNK, enhanced susceptibility to Fas-induced apoptosis, suggesting that the kinase pathway may have a protective role (Nishina *et al.*, 1997). Thus, it appears that the kinase cascade initiated by TNF-α and Fas may have a role in either inducing or protecting from cell death; however, the mechanism and significance of this process remains controversial.

In addition to the apoptotic response elicited by TNF-α, the cytokine also induces a distinct pathway leading to the activation of the transcription factor NF-κB (Tartaglia *et al.*, 1991). Transcriptional regulation by NF-κB can suppress TNF-α-induced apoptosis and may be the dominant pathway following TNF-α treatment

(Beg and Baltimore, 1996; Hsu, Sho *et al.*, 1996; Liu, Hsu *et al.*, 1996; Van Antwerp *et al.*, 1996; Wang, Mayo *et al.*, 1996). In fact, many cells require the presence of RNA or protein synthesis inhibitors to elicit TNF-α-mediated cell death (Rubin *et al.*, 1988; White *et al.*, 1992). Pretreatment of cells with interleukin-1, which induces NF-κB activation, protects from TNF-α-induced apoptosis (Wang, Mayo *et al.*, 1996). Furthermore, overexpression of Rel A, one of the common components of NF-κB, protects cells from TNF-α (Beg and Baltimore, 1996). NF-κB is normally sequestered in the cytoplasm by binding to the inhibitory protein IκB (Baeurle and Baltimore, 1996). Phosphorylation of IκB leads to its degradation freeing NF-κB so that it can be translocated to the nucleus and regulate gene expression (Baeurle and Baltimore, 1996). Inhibition of NF-κB, either by disrupting one of its subunits (e.g. Rel A) (Beg and Baltimore, 1996) or by expression of an IκB mutant that is resistant to degradation (Van Antwerp *et al.*, 1996; Wang, Mayo *et al.*, 1996), facilitates TNF-α-induced apoptosis. The biochemical mechanism in which TNF-α activates NF-κB is somewhat unclear. NF-κB activation is not blocked by dominant negative mutants of FADD or by CrmA (Chinnaiyan, Tepper *et al.*, 1996), suggesting that the pathway is different than that used for apoptosis. It is, however, triggered by overexpression of downstream proteins RIP or TRAF2 which are both recruited by TRADD to TNFR1 in a TNF-α dependent process (Hsu, Huang *et al.*, 1996; Hsu, Shu *et al.*, 1996). The binding of TRAF2 to TRADD, which occurs through its RING finger domain, is facilitated by interaction with the kinase domain of RIP (Hsu, Huang *et al.*, 1996) Interestingly, it is the kinase domain of RIP that appears to be mediate activation of NF-κB (Ting *et al.*, 1996). Dominant negative mutants of TRAF2 which lack the RING finger motif block TNF-α-induced NF-κB, but do not block apoptosis (Hsu, Shu *et al.*, 1996). In fact, these mutants actually promote TNF-α-mediated cell death, further supporting the role of the NF-κB pathway in suppressing apoptosis (Hsu, Shu *et al.*, 1996). This pathway was further characterized by the identification of a MAP (mitogen activated protein)-like kinase, NIK (NF-κB-inducing kinase), that binds to TRAF2 and promotes NF-κB activation and inhibition of cell death (Malinin *et al.*, 1997). However, the downstream effectors of NIK as well as the transcriptional targets for NF-κB remain unknown.

Conclusion

We have described several of the basic components of the apoptotic program including specific pathways initiated by TNF-α and Fas. In just the past year, we have begun to understand the structure of Bcl-2 family members and the possibility that they function as membrane spanning ion channels. A biochemical connection has been made between the Bcl-2 family and caspases. Although this junction is via the *C. elegans* protein Ced-4, it will probably not be too long before we realize its mammalian functional homologues. Finally, the recruitment of several proteins to the TNFR1/Fas complex has resulted in the identification of an effector protein, FLICE, which links this pathway to other apoptotic pathways which require the activation of cysteine proteases. The following chapters in this book will further explore these and other aspects of the apoptotic process.

REFERENCES

Alnemri, E.S., Livingston, D.J., Nicholson, D.W., Salvesen, G., Thornberry, N.A., Wong, W.W., *et al.* (1996) Human ICE/CED-3 protease nomenclature. *Cell*, **87**, 171.

Askew, D.S., Ashmun, R.A., Simmons, B.C. and Cleveland, J.L. (1991) Constitutive c-*myc* expression in an IL-3-dependent myeloid cell line suppresses cell cycle arrest and accelerates apoptosis. *Oncogene*, **6**, 1915–1922.

Baeurle, P.A. and Baltimore, D. (1996) NF-kappa B: ten years after. *Cell*, **87**, 13–20.

Baker, S.J., Markowitz, S., Fearon, E.R., Willson, J.K.V. and Vogelstein, B. (1990) Suppression of human colorectal carcinoma cell growth by wild-type p53. *Science*, **249**, 912–915.

Bakhshi, A., Jensen, J.P., Goldman, P., Wright, J.J. , McBride, O.W., Epstein, A.L. and Korsmeyer S.J. (1985) Cloning the chromosomal break-point of the t(14:18) human lymphomas: clustering around JH on Chromosome 14 and near a transcriptional unit on 18. *Cell*, **41**, 889–906.

Barry, M.A., Behnke, C.A. and Eastman, A. (1990) Activation of programmed cell death (apoptosis) by cisplatin, other anticancer drugs, toxins and hyperthermia. *Biochem. Pharmacol.*, **40**, 2353–2362.

Beg, A.A. and Baltimore, D. (1996) An essential role for NF-κB in preventing TNF-α-induced cell death. *Science*, **274**, 782–784.

Bellgrau, D., Gold, D., Selawry, H., Moore, J., Franzusoff, A. and Duke, R.C. (1995) A role for CD95 ligand in preventing graft rejection. *Nature*, **377**, 630–632.

Beutler, B. and Cerami, A. (1986) Cachectin and tumor necrosis factor as two sides of the same biological coin. *Nature*, **320**, 584–688.

Bing, A. and Dou, Q.P. (1996) Cleavage of retinoblastoma protein during apoptosis: an interleukin 1 beta-converting enzyme-like protease as candidate. *Cancer Res.*, **56**, 438–442.

Boise, L.H., Gonzalez-Garcia, M., Postema, C.E., Ding, L., Lindsten, T., Turka, L.A., *et al.* (1993) *bcl-x*, a *bcl-2*-related gene that functions as a dominant regulator of apoptotic death. *Cell*, **74**, 597–608.

Boldin, M.P., Goncharov, T.M., Goltsev, Y.V. and Wallach, D. (1996) Involvement of MACH, a novel MORT1/FADD-interacting protease, in Fas/APO-1-and TNF receptor-induced cell death. *Cell*, **85**, 803–815.

Boldin, M.P., Varfolomeev, E.E., Pancer, Z., Mett, I.L., Camonis, J.H. and Wallach, D. (1995) A novel protein that interacts with the death domain of Fas/APO1 contains a sequence motif related to the death domain. *J. Biol. Chem.*, **270**, 7795–7798.

Boulakia, C.A., Chen, G., Ng, F.W.H., Teodoro, J.G., Branton, P.E., Nicholson, D.W., *et al.* (1996) Bcl-2 and adenovirus E1B 19 kDa protein prevent E1A-induced processing of CPP32 and cleavage of poly(ADP-ribose) polymerase. *Oncogene*, **12**, 529–535.

Boyd, J., Malstrom, S., Subramanian, T., Venkatesh, L., Schaeper, U., Elangovan, B., *et al.* (1994) Adenovirus E1B 19kDa and bcl-2 proteins interact with a common set of cellular proteins. *Cell*, **79**, 341–351.

Boyd, J.M., Gallo, G.J., Elangovan, B., Houghton, A.B., Malstrom, S., Avery, B.J., *et al.* (1995) Bik1, a novel death-inducing protein shares a distinct sequence motif with Bcl-2 family proteins and interacts with viral and cellular survival-promoting proteins. *Oncogene*, **11**, 1921–1928.

Brojatsch, J., Naughton, J., Rolls, M.M., Zingler, K. and Young, J.A.T. (1996) CAR1, a TNFR-related protein, is a cellular receptor for cytopathic avian leukosis-sarcoma viruses and mediates apoptosis. *Cell*, **87**, 845–855.

Brugarolas, J., Chandrasekaran, C., Gordon, J.I., Beach, D., Jacks, T. and Hannon, G.J. (1995) Radiation-induced cell cycle arrest compromised by p21 deficiency. *Nature*, **377**, 552–557.

Buckbinder, L., Talbott, R., Velasco-Miguel, S., Takenaka, I., Faha, B., Seizinger, B.R., *et al.* (1995) Induction of the growth inhibitor IGF-binding protein 3 by p53. *Nature*, **377**, 646–649.

Bump, N.J., Hackett, M., Hugunin, M., Seshagiri, S., Brady, K., Chen, P., *et al.* (1995) Inhibition of ICE family proteases by baculovirus anti-apoptotic protein p35. *Science*, **269**, 1885–1888.

Caelles, C., Helmberg, A. and Karin, M. (1994) p53-dependent apoptosis in the absence of transcriptional activation of p53-target genes. *Nature*, **370**, 220–223.

Canman, C., Gilmer, T.M., Coutts, S.B. and Kastan, M.B. (1995) Growth factor modulation of p53-mediated growth arrest versus apoptosis. *Genes Dev.*, **9**, 600–611.

Cascino, I., Dapoff, G., DeMaria, R., Testi, R. and Ruberti, G. (1996) Fas/Apo-1 (CD95) receptor lacking the intracytoplasmic signalling domain protects tumor cells from Fas-mediated apoptosis. *J. Immunol.*, **156**, 13–17.

Chen, J., Wu, X., Lin, J. and Levine, A.J. (1996) mdm-2 inhibits the G1 arrest and apoptosis functions of the p53 tumor suppressor protein. *Mol. Cell. Biol.*, **16**, 2445–2452.

Chen, X., Ko, L.J., Jayaraman, L. and Prives, C. (1996) p53 levels, functional domains, and DNA damage determine the extent of the apoptotic response of tumor cells. *Genes Dev.*, **10**, 2438–2451.

Cheng, E.H., Nicholas, J., Bellows, D.S., Hayward, G.S., Guo, H.G., Reitz, M.S., *et al.* (1997) A Bcl-2 homolog encoded by Kaposi sarcoma-associated virus, human herpes virus 8, inhibits apoptosis but does not heterodimerize with Bax or Bak. *Proc. Natl. Acad. Sci.*, **94**, 690–694.

Chinnaiyan, A.M., O'Rourke, K., Lane, B.R. and Dixit, V.M. (1997) Interaction of CED-4 with CED-3 and CED-9: a molecular framework for cell death. *Science*, **275**, 1122–1126.

Chinnaiyan, A.M., O'Rourke, K., Tewari, M. and Dixit, V.M. (1995) FADD, a novel death domain-containing protein, interacts with the death domain of Fas and initiates apoptosis. *Cell*, **81**, 505–512.

Chinnaiyan, A.M., O'Rourke, K., Yu, G.L., Lyons, R.H., Garg, M., Duan, D.R., *et al.* (1996) Signal transduction by DR3, a death domain-containing receptor related to TNFR-1 and CD95. *Science*, **274**, 990–992.

Chinnaiyan, A.M., Orth, K., O'Rourke, K., Duan, H., Poirier, G.G. and Dixit, V.M. (1996) Molecular ordering of the cell death pathway. *J. Biol. Chem.*, **271**, 4573–4576.

Chinnaiyan, A.M., Tepper, C.G., Seldin, M.F., O'Rourke, K., Kischkel, F.C., Hellbardt, S., *et al.* (1996) FADD/MORT1 Is a common mediator of CD95 (Fas/APO-1) and tumor necrosis factor receptor-induced apoptosis. *J. Biol. Chem.*, **271**, 4961–4965.

Chiou, S.-K., Rao, L. and White, E. (1994) Bcl-2 blocks p53-dependent apoptosis. *Mol. Cell. Biol.*, **14**, 2556–2563.

Chiou, S.-K., Tseng, C.C., Rao, L. and White, E. (1994) Functional complementation of the adenovirus E1B 19K protein with Bcl-2 in the inhibition of apoptosis in infected cells. *J. Virol.*, **68**, 6553–6566.

Chittenden, T., Harrington, E.A., O'Connor, R., Flemington, C., Lutz, R.J., Evan, G.I., *et al.* (1995) Induction of apoptosis by the Bcl-2 homologue Bak. *Nature (London)*, **374**, 733–736.

Cho, Y., Gorina, S., Jeffrey, P.D. and Pavletich, N.P. (1994) Crystal structure of a p53 tumor suppressor-DNA complex: understanding tumorigenic mutations. *Science*, **265**, 346–355.

Choi, S.S., Park, I., Yun, J.W., Sung, Y.C., Hong, S. and Shin, H. (1995) A novel Bcl-2 related gene, Bfl-1, is overexpressed in stomach cancer and preferentially expressed in bone marrow. *Oncogene*, **11**, 1693–1698.

Chu, K., Niu, X.H. and Williams, L.T. (1995) A novel protein, FAF-1, potentiates Fas-mediate apoptosis. *Proc. Natl. Acad. Sci.*, **92**, 11894–11898.

Clarke, A.R., Purdie, C.A., Harrison, D.J., Morris, R.G., Bird, C.C., Hooper, M.L., *et al.* (1993) Thymocyte apoptosis induced by p53-dependent and independent pathways. *Nature*, **362**, 849–852.

Cleary, M.L., Smith, S.D. and Sklar, J. (1986) Cloning and structural analysis of cDNAs for bcl-2 and a hybrid bcl-2/immunoglobulin transcript resulting from the t(14; 18) translocation. *Cell*, **47**, 19–28.

Clore, G.M., Omichinski, J.G., Sakaguchi, K., Zambrano, N., Sakamoto, H., Appella, E., *et al.* (1994) High-resolution structure of the oligomerization domain of p53 by multidimensional NMR. *Science*, **265**, 386–391.

Cohen, J.J. and Duke, R.C. (1983) Glucocorticoid activation of a calcium-dependent endonuclease in thymocyte nuclei leads to cell death. *J. Immunol.*, **132**, 38–42.

Colotta, F., Polentarutti, N., Sironi, M. and Mantovani, A. (1992) Expression and involvement of c-fos and c-jun protooncogenes in programmed cell death induced by growth factor deprivation in lymphoid cell lines. *J. Biol. Chem.*, **267**, 18278–18283.

Datta, S.R., Dudek, H., Tao, X., Masters, S., Fu, H., Gotoh, Y. and Greenberg, M.E. (1997) Akt phosphorylation of BAD couples survival signals to the cell-intrinsic death machinery. *Cell*, **91**, 231–241.

Debbas, M. and White, E. (1993) Wild-type p53 mediates apoptosis by E1A which is inhibited by E1B. *Genes Dev.*, **7**, 546–554.

del Peso, L., González-Garcia, M., Page, C., Herrera, R. and Nuñez, G. (1997) Interleukin-3-induced phosphorylation of BAD through the protein kinase Akt. *Science*, **278**, 687–689.

Desdouets, C., Ory, C., Matesic, G., Soussi, T., Brechot, C. and Sobczak-Thepot, J. (1996) ATF/CREB site mediated transcriptional activation and p53 dependent repression of the cyclin A promoter. *FEBS Lett.*, **385**, 34–38.

Donehower, L.A., Harvey, M., Slagle, B.L., McArthur, M.J., Montgomery, C.A., Butel, J.S., *et al.* (1992) Mice deficient for p53 are developmentally normal but susceptible to spontaneous tumors. *Nature*, **356**, 215–221.

Darmon, A.J., Nicholson, D.W. and Bleackley, R.C. (1995) Activation of the apoptotic protease CPP32 by cytotoxic T-cell-derived granzyme B. *Nature*, **377**, 446–448.

Diller, L., Kassel, J., Nelson, C.E., Gryka, M.A., Litwak, G., Geghardt, M., *et al.* (1990) p53 functions as a cell cycle control protein in osteosarcomas. *Mol. Cell. Biol.*, **10**, 5772–5781.

Duan, H., Chinnaiyan, A.M., Hudson, P.L., Wing, J.P., He, W. and Dixit, V.M. (1996) ICE-LAP3, a novel mammalian homologue of the *Caenorhabditis elegans* cell death protein Ced-3 is activated during Fas- and tumor necrosis factor-induced apoptosis. *J. Biol. Chem.*, **271**, 1621–1625.

Duan, H. and Dixit, V.M. (1997) RAIDD is a new 'death' adaptor molecule. *Nature*, **385**, 86–89.

Eizenberg, O., Faber-Elman, A., Gottlieb, E., Oren, M., Rotter, V. and Schwartz, M. (1996) p53 plays a regulatory role in differentiation and apoptosis of central nervous system-associated cells. *Mol. Cell. Biol.*, **16**, 5178–5185.

El-Deiry, W.S., Kern, S.E., Pietenpol, J.A., Kinzler, K.W. and Vogelstein, B. (1992) Definition of a consensus binding site for p53. *Nature Gen.*, **1**, 45–49.

El-Deiry, W.S., Tokino, T., Velculescu, V.E., Levy, D.B., Parsons, R., Trent, J.M., *et al.* (1993) WAF1, a potential mediator of p53 tumor suppression. *Cell*, **75**, 817–825.

Eliyahu, D., Michalovitz, D., Eliyahu, S., Pinhasi-Kimhi, O. and Oren, M. (1989) Wild-type p53 can inhibit oncogene-mediated focus formation. *Proc. Natl. Acad Sci.*, **86**, 8763–8767.

Enari, M., Hug, H. and Nagata, S. (1995) Involvement of an ICE-like protease in Fas-mediated apoptosis. *Nature*, **375**, 78–81.

Enari, M., Talanian, R.V., Wong, W.W. and Nagata, S. (1996) Sequential activation of ICE-like and CPP32-like proteases during Fas-mediated apoptosis. *Nature*, **380**, 723–726.

Estus, S., Zaks, W.J., Freeman, R.S., Gruda, M., Bravo, R. and Johnson, E.M., Jr. (1994) Altered gene expression in neurons during programmed cell death: identification of *c-jun* as necessary for neuronal apoptosis. *J. Cell Biol.*, **127**, 1717–1727.

Evan, G.I., Wyllie, A.H., Gilbert, C.S., Littlewood, T.D., Land, H., Brooks, M., *et al.* (1992) Induction of apoptosis in fibroblasts by c-*myc* protein. *Cell*, **69**, 119–128.

Farmer, G., Bargonetti, J., Zhu, H., Friedman, P., Prywes, R. and Prives, C. (1992) Wild-type p53 activates transcription *in vitro*. *Nature*, **358**, 83–86.

Farrow, S.N., White, J.H.M., Martinou, I., Raven, T., Pun, K.-T., Grinham, C.J., *et al,*. (1995) Cloning of a bcl-2 homologue by interaction with adenovirus E1B 19K. *Nature*, **374**, 731–733.

Faucheu, C., Diu, A., Chan, A.-M., Miossec, C., Herve, F., Collard-Dutilleul, V., *et al.* (1995) A novel human protease similar to the interleukin-1β converting enzyme induces apoptosis in transfected cells. *EMBO J.*, **14**, 1914–1922.

Fernandes-Alnemri, T., Armstrong, R.C., Krebs, J., Srinivasula, S.M., Wang, L., Bullrich, F., *et al.* (1996) In vitro activation of CPP32 and Mch3 by Mch4, a novel human apoptotic cysteine protease containing two FADD-like domains. *Proc. Natl. Acad. Sci.*, **93**, 7464–7469.

Fernandes-Alnemri, T., Litwack, G. and Alnemri, E.S. (1995) *Mch2*, a new member of the apoptotic *Ced-3/ICE* cysteine protease gene family. *Cancer Res.*, **55**, 2737–2742.

Fernandes-Alnemri, T., Litwack, G. and Alnemri, E. (1994) CPP32, a novel human apoptotic protein with homology to *Caenorhabditis elegans* cell death protein Ced-3 and mammalian interleukin-1β-converting enzyme. *J. Biol. Chem.*, **269**, 30761–30764.

Fernandes-Alnemri, T., Takahashi, A., Armstrong, R., Krebs, J., Fritz, L., Tomaselli, K.J., *et al.* (1995) Mch3, a novel human apoptotic cysteine protease highly related to CPP32. *Cancer Res.*, **55**, 6045–6052.

Fernandez-Sarabia, M.J. and Bischoff, J.R. (1994) Bcl-2 associates with the ras-related protein R-ras p23. *Nature*, **366**, 274–275.

Finlay, C.A., Hinds, P.W. and Levine, A.J. (1989) The p53 proto-oncogene can act as a suppressor of transformation. *Cell*, **57**, 1083–1093.

Fisher, G.H., Rosenberg, F.J., Straus, S.E., Dale, J.K., Middelton, L.A., Lin, A.Y., *et al.* (1995) Dominant interfering Fas gene mutations impair apoptosis in a human autoimmune lymphoproliferative syndrome. *Cell*, **81**, 935–946.

Fraser, A. and Evan, G. (1996) A license to kill. *Cell*, **85**, 781–784.

Gibson, L., Holmgreen, S.P., Huang, D.C., Bernard, O., Copeland, N.G., Jenkins, N.A., *et al.* (1996) bcl-w, a novel member of the bcl-2 family, promotes cell survival. *Oncogene*, **13**, 665–675.

Giordano, C., Stassi, G., De Maria, R., Todaro, M., Richiusa, P., Papoff, G., *et al.* (1997) Potential involvement of Fas and its ligand in the pathogenesis of Hashimoto's thyroiditis. *Science*, **275**, 960–963.

Gong, J., Li, X. and Darzynkiewicz, Z. (1993) Different patterns of apoptosis of HL-60 cells induced by cycloheximide and camptothecin. *J. Cell. Physiol.*, **157**, 263–270.

Gonzalez-Garcia, M., Perez-Ballestero, R., Ding, L., Duan, L., Boise, L.H., Thompson, C.B., *et al.* (1994) bcl-XL is the major bcl-x mRNA form expressed during murine development and its product localizes to mitochondria. *Development*, **120**, 3033–3042.

Gorina, S. and Pavletich, N.P. (1996) Structure of the p53 tumor suppressor bound to the ankyrin and SH3 domains of 53BP2. *Science*, **274**, 1001–1005.

Goto, K., Ishige, A., Sekiguchi, K., Lizuka, S., Sugimoto, A., Yuzurihara, M., *et al.* (1990) Effects of cycloheximide on delayed neuronal death in rat hippocampus. *Brain Res.*, **534**, 299–302.

Griffith, T.S., Brunner, T., Fletcher, S.M., Green, D.R. and Ferguson, T.A. (1995) Fas ligand-induced apoptosis as a mechanism of immune privilege. *Science*, **270**, 1189–1192.

Hahne, M., Rimoldi, D., Schröter, M., Romero, P., Schreier, M., French, L.E., *et al.* (1996) Melanoma cell expression of Fas(Apo-1/CD95) ligand: implications for tumor immune escape. *Science*, **274**, 1363–1366.

Ham, J., Babij, C., Whitfield, J., Pfarr, C.M., Lallemand, D., Yaniv, M., *et al.* (1995) A c-Jun dominant negative mutant protects sympathetic neurons against programmed cell death. *Neuron*, **14**, 927–939.

Han, J., Sabbatini, P., Perez, D., Rao, L., Mohda, D. and White, E. (1996) The E1B 19K protein blocks apoptosis by interacting with and inhibiting the p53-inducible and death-promoting Bax protein. *Genes Dev.*, **10**, 461–477.

Han, J., Sabbatini, P. and White, E. (1996) Induction of apoptosis by human Nbk/Bik, a BH3 containing E1B 19K interacting protein. *Mol. Cell. Biol.*, **16**, 5857–5864.

Harper, J.W., Adami, G.R., Wei, N., Keyomarsi, K. and Elledge, S.J. (1993) The p21 cdk-interacting protein cip 1 is a potent inhibitor of G1 cyclin-dependent kinases. *Cell*, **75**, 805–816.

Haupt, Y., Barak, Y. and Oren, M. (1996) Cell type-specific inhibition of p53-mediated apoptosis by mdm2. *EMBO J.*, **15**, 1596–1606.

Helps, N.R., Barker, H.M., Elledge, S.J. and Cohen, P.T.W. (1995) Protein phosphatase 1 interacts with p53BP2, a protein which binds to the tumour suppressor p53. *FEBS Lett.*, **377**, 295–300.

Henderson, S., Huen, D., Rowe, M., Dawson, C., Johnson, G. and Rickinson, A. (1993) Epstein-Barr virus-coded BHRF1 protein, a viral homologue of Bcl-2, protects human B cells from programmed cell death. *Proc. Natl. Acad. Sci.*, **90**, 8479–8483.

Hengartner, M.O., Ellis, R.E. and Hovitz, H.R. (1992) *Caenorhabditis elegans* gene *ced-9* protects cells from programmed cell death. *Nature*, **356**, 494–499.

Hermeking, H. and Eick, D. (1994) Mediation of c-Myc-induced apoptosis by p53. *Science*, **265**, 2091–2093.

Hockenbery, D., Nuñez, G., Milliman, C., Schreiber, R.D. and Korsmeyer, S. (1990) Bcl-2 is an inner mitochondrial membrane protein that blocks programmed cell death. *Nature*, **348**, 334–336.

Hockenbery, D.M., Oltvai, Z.N., Yin, X.-M., Milliman, C.L. and Korsmeyer, S.J. (1993) Bcl-2 functions in an antioxidant pathway to prevent apoptosis. *Cell*, **75**, 241–251.

Hollstein, M., Sidransky, D., Vogelstein, B. and Harris, C. (1991) p53 mutations in human cancers. *Science*, **253**, 49–53.

Hsu, H., Huang, J., Shu, H.B., Baichwal, V. and Goedell, D.V. (1996) TNF-dependent recruitment of the protein kinase RIP to the TNF receptor-1 signalling complex. *Immunity*, **4**, 387–396.

Hsu, H., Shu, H.-B., Pan, M.G. and Goeddel, D.V. (1996) TRADD-TRAF2 and TRADD-FADD interaction define two distinct TNF receptor 1 signal transduction pathways. *Cell*, **84**, 299–308.

Hsu, H., Xiong, J. and Goeddel, D.V. (1995) The TNF receptor 1-associated protein TRADD signals cell death and NFκB activation. *Cell*, **81**, 495–504.

Hu, Y., Benedict, M.A., Wu, D., Inohara, N. and Nuñez, G. (1998) Bcl-x_L interacts with Apaf-1 and inhibits Apaf-1-dependent caspase-9 activation. *Proc. Natl. Acad. Sci.*, **95**, 4386–4391.

Huang, B., Eberstodt, M., Olejniczak, E.T., Meadows, R.P. and Fesik, S.W. (1996) NMR structure and mutagenesis of Fas (Apo-1/CD95) death domain. *Nature*, **384**, 638–641.

Hunter, J.J. and Parslow, T.G. (1996) A peptide sequence from Bax that converts Bcl-2 into an activator of apoptosis. *J. Biol. Chem.*, **271**, 8521–8524.

Ichijo, H., Nishida, E., Irie, K., ten Dijke, P., Saitoh, M., Moriguchi, T., *et al.* (1997) Induction of apoptosis by ASK1, a mammalian MAPKKK that activates SAPK/JNK and p38 signalling pathways. *Science*, **275**, 90–94.

Inohara, N., Ding, L., Chen, S. and Nuñez, G. (1997) *harakiri*, a novel regulator of cell death, encodes a protein that activates apoptosis and interacts selectively with survival-promoting proteins Bcl-2 and Bcl-X_L. *EMBO J.*, **16**, 101–109.

Itoh, N., Yonehara, S., Ishii, A., Yonehara, M., Mizushima, S., Sameshima, M., *et al.* (1991) The polypeptides encoded by the cDNA for human cell surface antigen Fas can mediate apoptosis. *Cell*, **66**, 233–243.

Jacobson, M.D., Burne, J.F., King, M.P., Miyashita, T., Reed, J.C. and Raff, M.C. (1993) Bcl-2 blocks apoptosis in cells lacking mitochondrial DNA. *Nature*, **361**, 365–369.

Jacobson, M.D., Burne, J.F. and Raff, M.C. (1994) Programmed cell death and Bcl-2 protection in the absence of a nucleus. *EMBO J.*, **13**, 1899–1910.

Jacobson, M.D., Weil, M. and Raff, M.C. (1997) Programmed cell death in animal development. *Cell*, **88**, 347–354.

Juo, P., Kuo, C.J., Reynolds, S.E., Konz, R.F., Raingeaud, J., Davis, R.J., *et al.* (1997) Fas activation of the p38 mitogen-activated protein kinase signalling pathway require ICE/CED-3 family proteases. *Mol. Cell. Biol.*, **17**, 24–35.

Kamada, S., Shinto, A.A., Tsujimura, Y., Takahashi, T., Noda, T., Kitamura, Y., *et al.* (1995) Bcl-2 deficiency in mice leads to pleiotropic abnormalities: accelerated lymphoid cell death in the thymus and spleen, polycystic kidney, hair hypopigmentation, and distorted small intestine. *Can. Res.*, **55**, 354–359.

Kamens, J., Paskind, M., Huguinin, M., Talanian, R.V., Allen, H., Banach, D., *et al.* (1995) Identification and characterization of Ich-2, a novel member of the interleukin-1β converting enzyme family of cysteine proteases. *J. Biol. Chem.*, **270**, 15250–15256.

Kao, C.C., Yew, P.R. and Berk, A.J. (1990) Domains required for in vitro association between the cellular p53 and the adenovirus 2 E1B 55K proteins. *Virology*, **179**, 806–814.

Karsan, A., Yee, E., Kaushansky, K. and Harlan, J.M. (1996) Cloning of human Bcl-2 homologue: inflammatory cytokines induce human A1 in cultured endothelial cells. *Blood*, **87**, 3089–3096.

Kastan, M.B., Onyekwere, O., Sidransky, D., Vogelstein, B. and Craig, R.W. (1991) Participation of p53 protein in the cellular response to DNA damage. *Cancer Res.*, **51**, 6304–6311.

Kastan, M.B., Zhan, Q., El-Deiry, W.S., Carrier, F., Jacks, T., Walsh, W.V., *et al.* (1992) A mammalian cell cycle checkpoint pathway utilizing p53 and GADD45 is defective in ataxia-telangiectasia. *Cell*, **13**, 587–597.

Kaufmann, S.H., Desnoyers, S., Ottaviano, Y., Davidson, N.E. and Poirier, G.G. (1993) Specific proteolytic cleavage of poly(ADP-ribose) polymerase: An early marker of chemotherapy-induced apoptosis. *Cancer Res.*, **53**, 3976–3985.

Kiefer, M.C., Brauer, M.J., Powers, V.C., Wu, J.J., Umansky, S.R., Tomei, L.D., *et al.* (1995) Modulation of apoptosis by the widely distributed Bcl-2 homolgue Bak. *Nature*, **374**, 736–739.

Kitson, J., Raven, T., Jiang, Y.-P., Goeddel, D.V., Giles, K.M., Pun, K.-T., *et al.* (1996) A death-domain-containing receptor that mediates apoptosis. *Nature*, **384**, 372–375.

Kley, N., Chung, R.Y., Fay, S., Loeffler, J.P. and Seizinger, B.R. (1992) Repression of the basal c-fos promoter by wild-type p53. *Nucl. Acids Res.*, **20**, 4083–4087.

Kluck, R.M., Bossy-Wetzel, E., Green, D.R. and Newmeyer, D.D. (1997) The release of cytochrome c from mitochrondria: A primary site for Bcl-2 regulation of apoptosis. *Science*, **275**, 1132–1136.

Knudson, C.M., Tung, K., Brown, G. and Korsmeyer, S.J. (1995) Bax deficient mice demonstrate lympoid hyperplasia but male germ cell death. *Science*, **270**, 96–99.

Ko, L.J. and Prives, C. (1996) p53: puzzle and paradigm. *Genes Dev.*, **10**, 1054–1072.

Komiyama, T., Ray, C.A., Pickup, D.J., Howard, A.D., Thornberry, N.A., Peterson, E.P., *et al.* (1994) Inhibition of interleukin-1β converting enzyme by the cowpox virus serpin CrmA. *J. Biol. Chem.*, **269**, 19331–19337.

Kuida, K., Zheng, T.S., Na, S., Kuan, C.-Y., Yang, D., Karasuyama, H., *et al.* (1996) Decreased apoptosis in the brain and premature lethality in CPP32-deficient mice. *Nature*, **384**, 368–372.

Kumar, S., Kinoshita, M., Noda, M., Copeland, N.G. and Jenkins, N.A. (1994) Induction of apoptosis by the mouse *Nedd2* gene, which encodes a protein similar to the product of the *Caenorhabditis elegans* cell death gene *ced-3* and the mammalian IL-1β converting enzyme. *Genes Dev.*, **8**, 1613–1626.

Lagresle, C., Bella, C., Daniel, P.T., Krammer, P.H. and Defrance, T. (1995) Regulation of germinal center B cell differentiation. Role of the human Apo-1/Fas (CD95) molecule. *J. Immunol.*, **154**, 5746–5756.

Lam, M., Dubyak, G., Chen, L., Nuñez, G., Miesfeld, R.L. and Distelhorst, C.W. (1994) Evidence that BCL-2 represses apoptosis by regulating endoplasmic reticulum-associated Ca2+ fluxes. *Proc. Natl. Acad. Sci.*, **91**, 6569–6573.

Lane, D.P. and Crawford, L.V. (1979) T antigen is bound to a host protein in SV40-transformed cells. *Nature*, **278**, 261–263.

Lazebnik, Y., Takahashi, A., Moir, R., Goldman, R., Poirier, G., Kaufmann, S., *et al.* (1995) Studies of the lamin proteinase reveal multiple parallel biochemical pathways during apoptotic execution. *Proc. Natl. Acad. Sci.*, **92**, 9042–9046.

Lazebnik, Y.A., Kaufmann, S.H., Desnoyers, S., Poirier, G.G. and Earnshaw, W.C. (1994) Cleavage of poly(ADP-ribose) polymerase by a proteinase with properties like ICE. *Nature*, **371**, 346–347.

Lenczowski, J.M., Dominguez, L., Eder, A., King, L.B., Zacharachuk, C.M. and Aswell, J.D. (1997) Lack of a role for Jun Kinase and AP-1 in fas-induced apoptosis. *Mol. Cell. Biol.*, **17**, 170–181.

Levine, A., Momand, J. and Finlay, C.A. (1991) The p53 tumour suppressor gene. *Nature*, **351**, 453–456.

Levine, A.J. (1997) p53, the cellular gatekeeper for growth and division. *Cell*, **88**, 323–331.

Li, S., Janosch, P., Tanji, M., Rosenfeld, G.C., Waymire, J.C., Mischak, H., *et al.* (1995) Regulation of Raf-1 kinase activity by the 14–3–3 family of proteins. *EMBO J.*, **14**, 685–696.

Lin, D.L. and Chang, C. (1996) p53 is a mediator for radiation-repressed human TR2 orphan receptor expression in MCF-7 cells, a new pathway from tumor suppressor to member of the steroid receptor superfamily. *J. Biol. Chem.*, **271**, 14649–14652.

Lin, J., Chen, J., Elenbaas, B. and Levine, A.J. (1994) Several hydrophobic amino acids in the p53 amino-terminal domain are required for transcriptional activation, binding to mdm-2 and the adenovirus 5 E1B 55-kD protein. *Genes Dev.*, **8**, 1235–1246.

Lin, Y. and Benchimol, S. (1995) Cytokines inhibit p53-mediated apoptosis but not p53-mediated G_1 arrest. *Mol. Cell. Biol.*, **15**, 6045–6054.

Linzer, D.I.H. and Levine, A.J. (1979) Characterization of a 54K Dalton cellular SV40 tumor antigen present in SV40-transformed cells and uninfected embryonal carcinoma cells. *Cell*, **17**, 43–52.

Lippke, J.A., Gu, Y., Sarnecki, C., Caron, P.R. and Su, M.S. (1996) Identification and characterization of CPP32/Mch2 homolog 1, a novel cysteine protease similar to CPP32. *J. Biol. Chem.*, **271**, 1825–1828.

Liu, X., Kim, C.N., Yang, J., Jemmerson, R. and Wang, X. (1996) Induction of apoptotic program in cell-free extracts: requirement for dATP and cytochrome c. *Cell*, **86**, 147–157.

Liu, X., Zou, H., Slaughter, C. and Wang, X. (1997) DFF, a heterodimeric protein that functions downstream of caspase-3 to trigger DNA fragmentation during apoptosis. *Cell*, **89**, 175–184.

Liu, Z.-g., Hsu, H., Goeddel, D.V. and Karin, M. (1996) Dissection of TNF receptor 1 effector functions: JNK activation is not linked to apoptosis while NF-κB activation prevents cell death. *Cell*, **87**, 565–576.

Liu, Z.-G., Smith, S.W., McLaughlin, K.A., Schwartz, L.M. and Osborne, B.A. (1994) Apoptotic signals delivered through the T-cell receptor of a T-cell hybrid require the immediate-early gene nur77. *Nature*, **367**, 281–284.

London, E. (1992) Diphtheria toxin: membrane interaction and membrane translocation. *Biochim. Biophys. Acta.*, **1113**, 25–51.

Los, M., Van de Craen, M., Penning, L.C., Schenk, H., Westendorp, M., Baeuerle, P., *et al.* (1995) Requirement of an ICE/CED-3 protease for Fas/APO-1-mediated apoptosis. *Nature*, **375**, 81–83.

Lowe, S.W., Schmitt, E.M., Smith, S.W., Osborne, B.A. and Jacks, T. (1993) p53 is required for radiation-induced apoptosis in mouse thymocytes. *Nature*, **362**, 847–849.

Mack, D.H., Vartikar, J., Pipas, J.M. and Laimins, L. (1993) Specific repression of TATA-mediated but not initiator-mediated transcription by wild-type p53. *Nature*, **363**, 281–283.

Malinin, N.L., Boldin, M.P., Kovalenko, A.V. and Wallach, D. (1997) MAP3K-related kinase involved in NF-kappaB induction by TNF, CD95 and IL-1. *Nature*, **385**, 540–544.

Martin, D.P., Schmidt, R.E., DiStefano, P.S., Lowry, O.H., Carter, J.G. and Johnson, E.M., Jr. (1988) Inhibitors of protein synthesis and RNA synthesis prevent neuronal death caused by nerve growth factor deprivation. *J. Cell Biol.*, **106**, 829–844.

McDonnell, T.J., Deane, N., Platt, F.M., Nunez, G., Jaeger, U., McKearn, J.P., *et al.* (1989) bcl-2-immunoglobulin transgenic mice demonstrate extended B cell survival and follicular lymphoproliferation. *Cell*, **57**, 79–88.

McDonnell, T.J. and Korsmeyer, S.J. (1991) Progression from lymphoid hyperplasia to high-grade malignant lymphoma in mice transgenic for the t(14:18). *Nature*, **349**, 254–256.

Mercer, W.E., Shields, M.T., Amin, M., Suave, G.J., Appella, E., Romano, J.W., *et al.* (1990) Negative growth regulation in a glioblastoma tumor cell line that conditionally expresses human wild-type p53. *Proc. Natl. Acad. Sci.*, **87**, 6166–6170.

Minn, A.J., Vélez, P., Schendel, S.L., Liang, H., Muchmore, S.W., Fesik, S.W., *et al.* (1997) Bcl-X_L forms an ion channel in synthetic lipid membranes. *Nature*, **385**, 353–357.

Miura, M., Zhu, H., Rotello, R., Hartwieg, E.A. and Yuan, J. (1993) Induction of apoptosis in fibroblasts by IL-1β converting enzyme, a mammalian homolog of the *C. elegans* cell death gene *ced-3*. *Cell*, **75**, 653–660.

Miyashita, T., Krajewski, S., Krajewska, M., Wang, H.G., Lin, H.K., Liebermann, D.A., *et al.* (1994) Tumor suppressor p53 is a regulator of *bcl-2* and *bax* gene expression in vitro and in vivo. *Oncogene*, **9**, 1799–1805.

Miyashita, T. and Reed, J.C. (1995) Tumor suppressor p53 is a direct transcriptional activator of the human *bax* gene. *Cell*, **80**, 293–299.

Mizumoto, K., Rothman, R.J. and Farber, J.L. (1994) Programmed cell death (apoptosis) of mouse fibroblasts is induced by the topoisomerase II inhibitor etoposide. *Mol. Pharmacol.*, **46**, 890–895.

Moberg, K.H., Tyndall, W.A. and Hall, D.J. (1992) Wild-type murine p53 represses transcription from the murine c-myc promoter in a human glial cell line. *J. Cell Biochem.*, **49**, 208–215.

Momand, J., Zambetti, G.P., Olson, D.C., George, D. and Levine, A.J. (1992) The mdm-2 oncogene product forms a complex with the p53 protein and inhibits p53-mediated transactivation. *Cell*, **69**, 1237–1245.

Monagan, P., Robertson, D., Amos, T.A., Dyer, M.J., Mason, D.Y. and Greaves, M.F. (1992) Ultrastructural localization of Bcl-2 protein. *J. Histochem. Cytochem.*, **40**, 1819–1825.

Motoyama, N., Wang, F., Roth, K.A., Sawa, H., Nakayama, K., Nakayama, K., *et al.* (1995) Massive cell death of immanture hematopoeitic cells and neurons in Bcl-x-deficient mice. *Science*, **267**, 1506–1510.

Muchmore, S.W., Sattler, M., Liang, H., Meadows, R.P., Harlan, J.E., Yoon, H.S., *et al.* (1996) X-ray and NMR structure of human Bcl-x_L, an inhibitor of programmed cell death. *Nature*, **381**, 335–341.

Munday, N.A., Vaillancourt, J.P., Ali, A., Casano, F.J., Miller, D.K., Molineaux, S.M., *et al.* (1995) Molecular cloning and pro-apoptotic activity of ICErel-II and ICErel-III, members of the ICE/CED-3 family of cysteine proteases. *J. Biol. Chem.*, **270**, 15870–15876.

Murphy, M., Hinman, A. and Levine, A.J. (1996) Wild-type p53 negatively regulates the expression of a microtubule-associated protein. *Genes Dev.*, **10**, 2971–2980.

Muzio, M., Chinnaiyan, A.M., Kischkel, F.C., O'Rourke, K., Shevchenko, A., Ni, J., *et al.* (1996) FLICE, a novel FADD-homologous ICE/CED-3-like protease, is recruited to the CD95 (Fas/APO-1) death-inducing signalling complex. *Cell*, **85**, 817–827.

Nagata, S. (1997) Apoptosis by death factor. *Cell*, **88**, 355–365.

Nakayama, K.-I., Nakayama, K., Negishi, I., Kuida, K., Shinkai, Y., Louie, M.C., *et al.* (1993) Disappearance of the lymphoid system in *bcl-2* homozygous mutant chimeric mice. *Science*, **261**, 1584–1588.

Naumovski, L. and Cleary, M.L. (1996) The p53-binding protein 53BP2 also interacts with Bcl2 and impedes cell cycle progression at G_2/M. *Mol. Cell. Biol.*, **16**, 3884–3892.

Nava, V.E., Cheng, E.H.-Y., Veliuona, M., Zou, S., Clem, R.J., Mayer, M.L., *et al.* (1997) Herpesvirus Saimiri encodes a functional homolog of the human bcl-2 oncogene. *J. Virol.*, **71**, 4118–4122.

Neilan, J.G., Lu, Z., Afonzo, C.L., Kutish, G.F., Sussman, M.D. and Rock, D.L. (1993) An african swine fever virus gene with similarity to the proto-oncogene *bcl-2* and the Epstein-Barr virus gene BHRF1. *J. Virol.*, **67**, 4391–4394.

Nicholson, D.W. (1996) ICE/CED3-like proteases as therapeutic targets for the control of inappropriate apoptosis. *Nature Biotech.*, **14**, 297–301.

Nicholson, D.W., Ali, A., Thornberry, N.A., Vaillancourt, J.P., Ding, C.K., Gallant, M., *et al.* (1995) Identification and inhibition of the ICE/CED-3 protease necessary for mammalian apoptosis. *Nature*, **376**, 37–43.

Nishina, H., Fischer, K.D., Radvanyi, L., Shahinian, A., Hakem, R., Rubie, E.A., *et al.* (1997) Stress-signalling kinase Sek 1 protects thymocytes from apoptosis mediated by CD95 and CD3. *Nature*, **385**, 350–353.

Olanow, C.W. (1993) A radical hypothesis for neurodegeneration, *TINS*, **16**, 439–444.

Oltavi, Z.N. and Korsmeyer, S.J. (1994) Checkpoints of dueling dimers foil death wish. *Cell*, **79**, 189–192.

Oltvai, Z.N., Millman, C.L. and Korsmeyer, S.J. (1993) Bcl-2 heterodimerizes in vivo with a conserved homolog, Bax, that accelerates programmed cell death. *Cell*, **74**, 609–619.

Orth, K., Chinnaiyan, A.M., Garg, M., Froelich, C.J. and Dixit, V.M. (1996) The CED-3/ICE-like protease Mch2 is activated during apoptosis and cleaves the death substrate lamin A. *J. Biol. Chem.*, **271**, 16443–16446.

Owen-Schaub, L., Zhang, W., Cusack, J.C., Angelo, L.S., Santee, S.M., Fujiwara, T., *et al.* (1995) Wild-type human p53 and a temerature-sensitive mutant induce Fas/Apo-1 expression. *Mol. Cell. Biol.*, **15**, 3032–3040.

Pan, G., O'Rourke, K., Chinnaiyan, A.M., Gentz, R., Ebner, R., Ni, J., *et al.* (1997) The receptor for the cytotoxic ligand TRAIL. *Science*, **276**, 111–113.

Pan, G., O'Rourke, K. and Dixit, V.M. (1998) Caspase-9, Bcl-x_L, and Apaf-1 form a ternary complex. *J. Biol. Chem.*, **273**, 5841–5845.

Perez, D. and White, E. (1998) E1B 19K inhibits Fas-mediated apoptosis through FADD-dependent sequestration of FLICE. *J. Cell Biol.*, **141**, 1255–1266.

Prives, C. (1994) How loops, β sheets, and α helices help us to understand p53. *Cell*, **78**, 543–546.

Rao, L., Debbas, M., Sabbatini, P., Hockenberry, D., Korsmeyer, S. and White, E. (1992) The adenovirus E1A proteins induce apoptosis which is inhibited by the E1B 19K and Bcl-2 proteins. *Proc. Natl. Acad. Sci.*, **89**, 7742–7746.

Rao, L., Modha, D. and White, E. (1997) The E1B 19K protein associates with lamins *in vivo* and its proper localization is required for inhibition of apoptosis, *Oncogene*, **15**, 1587–1597.

Rao, L., Perez, D. and White, E. (1996) Lamin proteolysis facilitates nuclear events during apoptosis. *J. Cell Biol.*, **135**, 1441–1455.

Rao, L. and White, E. (1997) Bcl-2 and ICE family of apoptotic regulators: Making a connection. *Curr. Opin. Genet. Dev.*, **7**, 52–58.

Ray, C.A., Black, R.A., Kronheim, S.R., Greenstreet, T.A., Sleath, P.R., Salvesen, G.S., *et al.* (1992) Viral inhibition of inflammation: cowpox virus encodes an inhibitor of the interleukin-1β converting enzyme. *Cell*, **69**, 597–604.

Reed, J.C., Zha, H., Aime-Sempe, C., Takayama, S. and Wang, H.G. (1996) Structure-function analysis of Bcl-2 family proteins. Regulators of programmed cell death. *Adv. Exp. Med. Biol.*, **406**, 96–112.

Reynolds, J.E., Yang, T., Qian, L.P., Jenkinson, J.D., Zhou, P., Eastman, A., *et al.* (1994) MCL-1, a member of the Bcl-2 family, delays apoptosis induced by c-myc overexpression in Chinese hamster ovary cells. *Can. Res.*, **54**, 6348–6352.

Rieux-Laucat, F., Deist, T.L., Hivroz, C., Roberts, I.A.G., Debatin, K.M., Fisher, A., *et al.* (1995) Mutations in Fas associated with human lymphoproliferative syndroms and autoimmunity. *Science*, **268**, 1347–1349.

Rothstein, T.L., Wang, J.K.M., Panka, D.J., Foote, L.C., Wang, Z., Stanger, B., *et al.* (1995) Protection against Fas-dependent Th-1 mediated apoptosis by antigen receptor engagement in B cells. *Nature*, **374**, 163–165.

Rotonda, J., Nicholson, D.W., Fazil, K.M., Gallant, M., Gareau, Y., Labelle, M., *et al.* (1996) The three-dimensional structure of apopain/CPP32, a key mediator of apoptosis. *Nature Struct. Biol.*, **3**, 619–625.

Rubin, B.Y., Smith, L.J., Hellermann, G.R., Lunn, R.M., Richardson, N.R. and Anderson, S.L. (1988) Correlation between the anticellular and DNA fragmenting activities of tumor necrosis factor. *Cancer Res.*, **48**, 6006–6010.

Sabbatini, P., Chiou, S.-K., Rao, L. and White, E. (1995) Modulation of p53-mediated transcription and apoptosis by the adenovirus E1B 19K protein. *Mol. Cell. Biol.*, **15**, 1060–1070.

Sabbatini, P., Han, J.H., Chiou, S.-K., Nicholson, D. and White, E. (1997) Interleukin 1β converting enyzme-like proteases are essential for p53-mediated transcriptionally dependent apoptosis. *Cell Growth Diff.*, **8**, 643–653.

Sabbatini, P., Lin, J., Levine, A.J. and White, E. (1995) Essential role for p53-mediated transcription in apoptosis but not growth suppression. *Genes Dev.*, **9**, 2184–2192.

Sakamuro, D., Eviner, V., Elliot, K.J., Showe, L., White, E. and Prendergast, G.C. (1995) c-Myc induces apoptosis in epithelial cells by both p53-dependent and p53-independent mechanisms. *Oncogene*, **11**, 2411–2418.

Sakamuro, D., Sabbatini, P., White, E. and Prendergast, G.C. (1997) The polyproline region of p53 is required to activate apoptosis but not growth arrest. *Oncogene*, **15**, 887–898.

Santhanam, U., Ray, A. and Sehgal, P.B. (1991) Repression of the interleukin 6 gene promoter by p53 and the retinoblastoma susceptibility gene product. *Proc. Natl. Acad. Sci.*, **88**, 7605–7609.

Sato, T., Irie, S., Kitada, S. and Reed, J.C. (1995) FAP-1: A protein tyrosine phosphatase that associates with Fas. *Science*, **268**, 411–415.

Satoh, T., Sakai, N., Enokido, Y., Uchiyama, Y. and Hatanaka, H. (1996) Free radical-independent protection by nerve growth factor and Bcl-2 of PC12 cells from hydrogen peroxide-triggered apoptosis. *J. Biochem.*, **120**, 540–546.

Sattler, M., Liang, H., Nettesheim, D., Meadows, R.P., Harlan, J.E., Eberstadt, M., *et al.* (1997) Structure of Bcl-x_L-Bak peptide complex: recognition between regulators of apoptosis. *Science*, **275**, 983–986.

Schreiber, S.S., Tocco, G., Najm, I., Thompson, R.F. and Baudry, M. (1993) Cycloheximide prevents kainate-induced neuronal death and c-Fos expression in adult rat brain. *J. Mol. Neurosci.*, **4**, 149–159.

Schulze-Ostoff, K., Walczak, H., Droge, W. and Krammer, P.H. (1994) Cell nucleus and DNA fragmentation are not required for apoptosis. *J. Cell Biol.*, **127**, 15–20.

Scott, S.A. and Davies, A.M. (1990) Inhibition of protein synthesis prevents cell death in sensory and parasympathetic neurons deprived of neurotrophic in vitro. *J. Neurobiol.*, **21**, 630–638.

Sedlak, T.W., Oltvai, Z.N., Yang, E., Wang, K., Boise, L.H., Thompson, C.B., *et al.* (1995) Multiple Bcl-2 family members demonstrate selective dimerizations with Bax. *Proc. Natl. Acad. Sci.*, **92**, 7834–7838.

Seto, E., Usheva, A., Zambetti, G.P., Momand, J., Horikoshi, N., Weinmann, R., *et al.* (1992) Wild-type p53 binds to the TATA-binding protein and represses transcription. *Proc. Natl. Acad. Sci.*, **89**, 12028–12032.

Shaulian, E., Zauberman, A., Ginsberg, D. and Oren, M. (1992) Identification of a minimal transforming domain of p53: negative dominance through abrogation of sequence-specific DNA binding. *Mol. Cell. Biol.*, **12**, 5581–5592.

Shen, Y. and Shenk, T. (1994) Relief of p53 mediated transcriptional repression by the adenovirus E1B 19-kDa protein or the cellular Bcl-2 protein. *Proc. Natl. Acad. Sci.*, **91**, 8940–8944.

Shi, Y., Glynn, J.M., Guilbert, L., Cotter, T.G., Bissonnette, R.P. and Green, D.R. (1992) Role for c-myc in activation-induced apoptotic cell death in T cell hybridomas. *Science*, **257**, 212–214.

Shigeno, T., Yamasaki, Y., Kato, G., Kusaka, K., Mima, T., Takakura, K., *et al.* (1990) Reduction of delayed neuronal death by inhibition of protein synthesis. *Neurosci. Lett.*, **120**, 117–119.

Siegel, R.M., Martin, D.A., Zheng, L., Ng, S.Y., Bertin, J., Cohen, J. and Lenardo, M.J. (1998) Death-effector filaments: novel cytoplasmic structures that recruit caspases and trigger apoptosis. *J. Cell Biol.*, **141**, 1243–1253.

Singer, G.G. and Abbas, A.K. (1994) The Fas antigen is involved in peripheral but not thymic deletion of T lymphocytes in T cell receptor transgenic mice. *Immunity*, **1**, 365–371.

Smith, C.A., Farrah, T. and Goodwin, R.G. (1994) The TNF receptor superfamily of cellular and viral proteins: Activation, costimulation, and death. *Cell*, **76**, 959–962.

Soares, H.D., Curran, T. and Morgan, J.I. (1994) Transcription factors as molecular mediators in cell death. *Ann. NY Acad. Sci.*, **747**, 172–182.

Spaargaren, M., Martin, G.A., McCormick, F., Fernandez-Sarabia, M.J. and Bischoff, J.R. (1994) The Ras-related protein R-ras interacts directly with Raf-1 in a GTP-dependent manner. *J. Biochem.*, **300**, 303–307.

Spector, M.S., Desnoyers, S., Hoeppner, D.J. and Hengartner, M.O. (1997) Interaction between the *C. elegans* cell-death regulators CED-9 and CED-4. *Nature*, **385**, 653–656.

Srinivasula, S.M., Ahmad, M., Fernandes-Alnemri, T., Litwack, G. and Alnemri, E.S. (1996) Molecular ordering of the Fas-apoptotic pathway: The Fas/APO-1 protease Mch5 is a CrmA-inhibitable protease that activates multiple Ced-3/ICE-like cysteine proteases. *Proc. Natl. Acad. Sci.*, **93**, 14486–14491.

Stanger, B.Z., Leder, P., Lee, T.-H., Kim, E. and Seed, B. (1995) RIP: A novel protein containing a death domain that interacts with FAS/APO-1 (CD95) in yeast and causes cell death. *Cell*, **81**, 513–523.

Steller, H. (1995) Mechanisms and genes of cellular suicide. *Science*, **267**, 1445–1449.

Stürzbecher, H.-W., Brain, R., Addison, C., Rudge, K., Remm, M., Grimaldi, M., *et al.* (1992) A C-terminal α-helix plus basic region motif is the major structural determinant of p53 tetramerization. *Oncogene*, **7**, 1513–1523.

Takahashi, A., Alnemri, E.S., Lazebnik, Y.A., Fernandes-Alnemri, T., Litwack, G., Moir, R.D., *et al.* (1996) Cleavage of lamin A by Mch2α but not CPP32: Multiple ICE-related proteases with distinct substrate recognition properties are active in apoptosis. *Proc. Natl. Acad. Sci.*, **93**, 8395–8400.

Takahashi, T., Tanaka, M., Brannan, C.I., Jenkins, N.A., Copeland, N.G., Suda, T., *et al.* (1994) Generalized lymphoproliferative disease in mice, caused by a point mutation in the Fas ligand. *Cell*, **76**, 969–976.

Takayama, S., Sato, T., Krajewski, S., Kochel, K., Irie, S., Millan, J.A., *et al.* (1995) Cloning and functional analysis of Bag-1: a novel Bcl-2-binding protein with anti-cell death activity. *Cell*, **80**, 279–284.

Tartaglia, L.A., Ayres, T.M., Wong, G.H. and Goeddel, D.V. (1993) A novel domain within the 55 KD TNF receptor signals cell death. *Cell*, **74**, 845–853.

Tartaglia, L.A., Weber, R.F., Figari, I.S., Reynolds, C., Palladino, M.A., Jr. and Goeddel, D.V. (1991) The two different receptors for tumor necrosis factor mediate distinct cellular responses. *Proc. Nat. Acad. Sci.*, **88**, 9292–9296.

Tewari, M. and Dixit, V.M. (1995) Fas- and TNF-induced apoptosis is inhibited by the poxvirus *crmA* gene product. *J. Biol. Chem.*, **270**, 3255–3260.

Tewari, M., Quan, L.T., O'Rourke, K., Desnoyers, S., Zeng, Z., Beidler, D.R., *et al.* (1995) Yama/CPP32β, a mammalian homolog of CED-3, is a CrmA-inhibitable protease that cleaves the death substrate poly(ADP-ribose) polymerase. *Cell*, **81**, 801–809.

Thornberry, N.A., Bull, H.G., Calaycay, J.R., Chapman, K.T., Howard, A.D., Kostura, M.J., *et al.* (1992) A novel heterodimeric cysteine protease is required for interleukin-1β processing in monocytes. *Nature*, **356**, 768–774.

Thulasi, R., Harbour, D.V. and Thompson, E.B. (1993) Suppression of c-myc is a critical step in glucocorticoid-induced human leukemic cell lysis. *J. Biol. Chem.*, **268**, 18306–18312.

Ting, A.T., Pimentel-Muinos, F.X. and Seed, B. (1996) RIP mediates tumor necrosis factor receptor 1 activation of NF-kappaB but not Fas/Apo-1-initiated apoptosis. *EMBO J.*, **15**, 6189–6196.

Trump, B.F. and Berezesky, I.K. (1995) Calcium-mediated cell injury and cell death. *FASEB J.*, **9**, 219–228.

Tsujimoto, Y., Gorham, J., Cossman, J., Jaffe, E. and Croce, C.M. (1985) The t(14;18) chromosome translocations involved in B cell neoplasms result from mistakes in VDJ joining. *Science*, **229**, 1390–1393.

Van Antwerp, D.J., Martin, S.J., Kafri, T., Green, D.R. and Verma, I.M. (1996) Suppression of TNF-α-induced apoptosis by NF-κB. *Science*, **274**, 787–789.

Vaux, D.L., Cory, S. and Adams, T.M. (1988) Bcl-2 promotes the survival of haemopoietic cells and cooperates with *c-myc* to immortalize pre-B cells. *Nature*, **335**, 440–442.

Vaux, D.L. and Weissman, I.L. (1993) Neither macromolecular synthesis nor myc is required for cell death via the mechanism that can be controlled by Bcl-2. *Mol. Cell. Biol.*, **13**, 7000–7005.

Veis, D.J., Sorenson, C.M., Shutter, J.R. and Korsmeyer, S.J. (1993) Bcl-2-deficient mice demonstrate fulminant lymphoid apoptosis, polycyctic kidneys, and hypopigmented hair. *Cell*, **75**, 229–240.

Wagner, A.J., Kokontis, J.M. and Hay, N. (1994) Myc-mediated apoptosis requires wild-type p53 in a manner independent of cell cycle arrest and the ability of p53 to induce *p21 waf1/cip1*. *Genes Dev.*, **8**, 2817–2830.

Walker, K.K. and Levine, A.J. (1996) Identification of a novel p53 functional domain that is necessary for efficient growth suppression. *Proc. Natl. Acad. Sci.*, **93**, 15335–15340.

Walker, P.R., Smith, C., Youdale, T., Leblanc, J., Whitfield, J.F. and Sikorska, M. (1991) Topoisomerase II-reactive chemotherapeutic drugs induce apoptosis in thymocytes. *Cancer Res.*, **51**, 1078–1085.

Wang, C.-Y., Mayo, M.W. and Baldwin, A.S., Jr. (1996) TNF- and cancer therapy-induced apoptosis: potentiation by inhibition of NF-κB. *Science*, **274**, 784–787.

Wang, H.-G., Milan, J.A., Cox, A.D., Der, C.J., Rapp, U.R., Beck, T., *et al.* (1995) R-ras promotes apoptosis caused by growth factor deprivation via a Bcl-2 suppressable mechanism. *J. Cell Biol.*, **129**, 1103–1114.

Wang, H.G., Rapp, U.R. and Reed, J.C. (1996) Bcl-2 targets the protein kinase Raf-1 to mitochondria. *Cell*, **87**, 629–638.

Wang, K., Yin, X.-M., Chao, D.T., Milliman, C.L. and Korsmeyer, S.J. (1996) BID: a novel BH3 domain-only death agonist. *Genes Dev.*, **10**, 2859–2869.

Wang, L., Miura, M., Bergeron, L., Zhu, H. and Yuan, J. (1994) *Ich-1*, an *ICE/ced-3*-related gene, encodes both positive and negative regulators of programmed cell death. *Cell*, **78**, 739–750.

Wang, Q., Zambetti, G.P. and Suttle, D.P. (1997) Inhibition of DNA topoisomerase II alpha gene expression by the p53 tumor suppressor. *Mol. Cell. Biol.*, **17**, 389–397.

Wang, X.W., Vermeulen, W., Coursen, J.D., Gibson, M., Lupold, S.E., Forrester, K., *et al.* (1996) The XPB and XPD DNA helicases are components of the p53-mediated apoptosis pathway. *Genes Dev.*, **10**, 1219–1232.

Wang, Y. and Prives, C. (1995) Increased and altered DNA binding of human p53 by S and G2/M but not G1 cyclin-dependent kinases. *Nature*, **376**, 88–91.

Wang, Z.-Q., Auer, B., Stingl, L., Berghammer, H., Haidacher, D., Schweiger, M., *et al.* (1995) Mice lacking ADPRT and poly(ADP-ribosyl)ation develop normally but are susceptible to skin disease. *Genes Dev.*, **9**, 509–520.

Watanabe-Fukunaga, R., Brannan, C.I., Copeland, N.G., Jenkins, N.A. and Nagata, S. (1992) Lymphoproliferation disorder in mice explained by defects in Fas antigen that mediates apoptosis. *Nature*, **356**, 314–317.

White, E. (1994) p53, guardian of Rb. *Nature*, **371**, 21–22.

White, E. (1996) Life, death, and the pursuit of apoptosis. *Genes Dev.*, **10**, 1–15.

White, E., Blose, S.H. and Stillman, B. (1984) Nuclear envelope localization of an adenovirus tumor antigen maintains the integrity of cellular DNA. *Mol. Cell. Biol.*, **4**, 2865–2875.

White, E. and Cipriani, R. (1989) Specific disruption of intermediate filaments and the nuclear lamina by the 19-kDa product of the adenovirus E1B oncogene. *Proc. Natl. Acad. Sci. USA*, **86**, 9886–9890.

White, E., Sabbatini, P., Debbas, M., Wold, W.S.M., Kusher, D.I. and Gooding, L. (1992) The 19-kilodalton adenovirus E1B transforming protein inhibits programmed cell death and prevents cytolysis by tumor necrosis factor α. *Mol. Cell. Biol.*, **12**, 2570–2580.

Woronicz, J.D., Calnan, B., Ngo, V. and Winoto, A. (1994) Requirement for the orphan steroid receptor Nur77 in apoptosis of T-cell hybridomas. *Nature*, **367**, 277–281.

Wu, D., Wallen, H.D. and Nuñez, G. (1997) Interaction and regulation of subcellular localization of CED-4 by CED-9. *Science*, **275**, 1126–1128.

Wyllie, A.H. (1980) Cell death: The significance of apoptosis. *Int. Rev. Cytol.*, **68**, 251–306.

Wyllie, A.H., Morris, R.G., Smith, A.L. and Dunlop, D. (1984) Chromatin cleavage in apoptosis: association with condensed chromatin morphology and dependence on macromolecular synthesis. *J. Pathol.*, **142**, 67–77.

Xiong, Y., Hannon, G., Zhang, H., Casso, D., Kobayashi, R. and Beach, D. (1993) p21 is a universal inhibitor of cyclin kinases. *Nature*, **366**, 701–704.

Xue, D. and Horvitz, H.R. (1995) Inhibition of the *Ceanorhabditis elegans* cell-death protease CED-3 by a CED-3 cleavage site in baculovirus p35 protein. *Nature*, **377**, 248–251.

Yang, E., Zha, J., Jockel, J., Boise, L.H., Thompson, C.B. and Korsmeyer, S.J. (1995) Bad, a heterodimeric partner for Bcl-xL and Bcl-2, displaces Bax and promotes cell death. *Cell*, **80**, 285–291.

Yang, J., Liu, X., Bhalla, K., Kim, C.N., Ibrado, A.M., Cai, J., *et al.* (1997) Prevention of apoptosis by Bcl-2: release of cytochrome c from mitochondria blocked. *Science*, **275**, 1129–1132.

Yew, P.R. and Berk, A.J. (1992) Inhibition of p53 transactivation required for transformation by adenovirus early 1B protein. *Nature*, **357**, 82–85.

Yin, C., Knudson, C.M., Korsmeyer, S.J. and Van Dyke, T. (1997) Bax suppresses tumorigenesis and stimulates apoptosis *in vivo*. *Nature*, **385**, 637–640.

Yin, X.-M., Oltvai, Z. and Korsmeyer, S. (1994) BH1 and BH2 domains of Bcl-2 are required for inhibition of apoptosis and heterodimerization with Bax. *Nature*, **369**, 321–323.

Yonish-Rouach, E., Resnitzky, D., Lotem, J., Sachs, L., Kimchi, A. and Oren, M. (1991) Wild-type p53 induces apoptosis of myeloid leukaemic cells that is inhibited by interleukin-6. *Nature*, **352**, 345–347.

Yuan, J., Shaham, S., Ledoux, S., Ellis, H.M. and Horvitz, H.R. (1993) The *C. elegans* cell death gene ced-3 encodes a protein similar to mammalian interleukin-1β converting enzyme. *Cell*, **75**, 641–652.

Zha, H., Aimé-Sempé, C., Sato, T. and Reed, J.C. (1996) Proapoptotic protein Bax heterodimerizes with Bcl-2 and homodimerizes with Bax via a novel domain (BH3) distinct from BH1 and BH2. *J. Biol. Chem.*, **21**, 7440–7444.

Zha, J., Harada, H., Yang, E., Jockel, J. and Korsmeyer, S.J. (1996) Serine phosphorylation of death agonist BAD in response to survival factor results in binding to 14–3–3 not BCL-X_L. *Cell*, **87**, 619–628.

Zhong, L.T., Sarafian, T., Kane, D.J., Charles, A.C., Mah, S.P., Edwards, R.H., *et al.* (1993) bcl-2 inhibits death of central neural cells induced by multiple agents. *Proc. Natl. Acad. Sci.*, **90**, 4533–4537.

Zou, H., Henzel, W.J., Liu, X., Lutschg, A. and Wang, X. (1997) Apaf-1, a human protein homologous to *C. elegans* CED-4, participates in cytochrome c-dependent activation of caspase-3. *Cell*, **90**, 405–413.

Part 1
INDUCERS OF APOPTOSIS

1. THE DEATH RECEPTORS

KLAUS SCHULZE-OSTHOFF[*,†] AND MARCUS E. PETER[**]

[*]*Department of Internal Medicine I, Medical Clinics, University of Tübingen*
[**]*Tumor Immunology Program, German Cancer Research Center, Heidelberg, Germany*

INTRODUCTION

Higher organisms have developed elaborate mechanisms to rapidly and selectively eliminate unwanted cells by programmed cell death. Exposure or depletion of certain steroid hormones, incubation with noxious agents, loss of cell adhesion to the extracellular matrix, or dysregulated expression of oncogenes are only some of the conditions that can lead to cell death. A fine-tuned mechanism to regulate life and death of a cell is the interaction of surface receptors with their cognate ligands, which may either trigger a survival or, oppositely, an apoptogenic signal. Several receptors are able to transmit cytotoxic signals into the cytoplasm, but in most cases they have a wide range of other functions unrelated to cell death. The T and B cell receptors, CD2, CD4 and cytokine receptors, such as those for interferons, TGF-ß or TNF-related ligands, are examples of molecules that induce diverse signals resulting in cell activation, differentiation, proliferation, or induction of apoptosis. Whether the signals induced by a given receptor lead to cell activation or death is highly cell-type specific and tightly regulated during differentiation of cells. For example, TNF receptors can exert costimulatory signals for proliferation of naive lymphocytes as well as can induce death signals required for deletion of activated lymphocytes.

Many receptors with important functions in differentiation, survival and cell death belong to an emerging family of structurally related molecules, called the TNF/NGF receptor superfamily (Figure 1.1). For some members of the family an apoptosis-inducing activity has been reported. However, most of them also have other functions such as induction of proliferation, differentiation, immune regulation and gene expression. Receptors with pleiotropic functions include TNF-R1 (CD120a) (Loetscher *et al.*, 1990; Schall *et al.*, 1990; Smith *et al.*, 1990), TNF-R2 (CD120b) (Dembic *et al.*, 1990), CD40 (Stamenkovic *et al.*, 1989), CD30 (Durkop *et al.*, 1992), CD27 (Camerini *et al.*, 1991), OX-40 (Mallett *et al.*, 1990), 4-1BB (Kwon and Weissman, 1989), NGF-R (Radeke *et al.*, 1987), TRAMP (DR3/wsl-1/APO-3/LARD) (Chinnaiyan *et al.*, 1996a; Bodmer *et al.*, 1997; Kitson *et al.*, 1996; Marsters *et al.*, 1996a; Screaton *et al.*, 1997), HVEM (ATAR/TR2) (Montgomery *et al.*, 1995; Hsu *et al.*, 1997a; Kwon *et al.*, 1997), GITR (Nocentini *et al.*, 1997) and RANK (Anderson *et al.*, 1997). The receptor molecules of this family are type I membrane proteins and are structurally homologous. Each contains in its extracellular

† *Corresponding Author*: Department of Internal Medicine I, Medical Clinics, Eberhard-Karls-University, Otfried-Müller-Str.10 D-72076, Tübingen, Germany. Tel.: +49–7071–29 84113. Fax:+49–7071–29 5865.

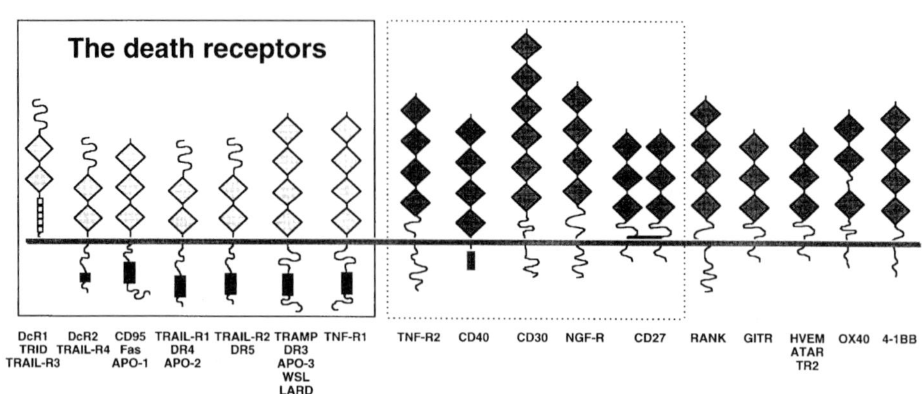

Figure 1.1 Members of the tumor necrosis factor receptor superfamily. The subfamily of the death receptors is boxed. The tinted box in the cytoplasmic regions of the receptors represents the death domain (DD). Members of the family that have been shown to induce apoptosis without having a DD are grouped in the stippled box.

domain two to six imperfect repeats of about 40 amino acids, each containing approximately six cysteine residues. Their cytoplasmic domains generally lack sequence homology.

APO-1/Fas, now called CD95, was the first member of the TNF/NGF receptor superfamily described in terms of its function in apoptosis (Itoh *et al.*, 1991; Watanabe-Fukunaga *et al.*, 1992a; Oehm *et al.*, 1992). Sequence comparison of the intracellular domain of CD95 with TNF-R1 revealed that both receptors contained a homologous stretch of about 80 amino acids. This region has been designated the 'death domain' (DD) since it enables transmission of a cytotoxic signal by both molecules (Tartaglia *et al.*, 1993a; Itoh and Nagata, 1993). Recent homology searches in EST databases led to the cloning of a number of novel membrane receptors that contain such a death domain and are therefore referred to as the death receptors (DRs) (Figure 1.1). TRAMP (DR3/wsl-1/APO-3/LARD) is both structurally and functionally similar to TNF-R1 and is abundantly expressed in T lymphocytes (Chinnaiyan *et al.*, 1996a; Bodmer *et al.*, 1997; Kitson *et al.*, 1996; Marsters *et al.*, 1996a; Screaton *et al.*, 1997). TRAIL-R1 (DR4, APO-2) and TRAIL-R2 (DR5) have been found as receptors binding to a novel cytokine, called TRAIL (TNF-related apoptosis-inducing ligand). The two TRAIL receptors are functionally similar to CD95 as their main function seems to be to induce apoptosis (Pan *et al.*, 1997a; 1997b; Sheridan *et al.*, 1997; Walczak *et al.*, 1997; MacFarlane *et al.*, 1997). They are the only TNF-R-like molecules possessing only two cysteine-rich domains. The TRAIL system, in addition, consists of two neutralizing decoy receptors, called DcR1 (TRAIL-R3, TRID) and DcR2 (Pan *et al.*, 1997b; Sheridan *et al.*, 1997; Degli-Eposti *et al.*, 1997; MacFarlane *et al.*, 1997; Marsters *et al.*, 1997). The sequence of DcR1 encodes a protein that contains the external TRAIL-binding region as well as a glycosyl-phosphatidylinositol residue that anchors the receptor to the membrane. But unlike the other receptors, DcR1 lacks an intracellular tail needed to spark the death pathway. DcR2 is also able to bind TRAIL but contains a truncated death domain. Thus, both decoy receptors will prevent TRAIL from engaging functional TRAIL receptors and thereby may render cells resistant to apoptosis. Collectively, this underlines that the death domain is required to induce apoptosis triggered by the different surface receptors. In this review, we will survey the function of the death receptors and their respective ligands with special emphasis on the signal transduction pathways used by these receptors.

THE DEATH LIGANDS

For most members of the TNF-R superfamily their cognate ligands have been identified. Five of them, CD95L, TNF, lymphotoxin-α (LTα, TNFß), TRAIL and APO-3L (TWEAK) bind to death receptors (Suda *et al.*, 1993; Beutler and van Huffel, 1994; Wiley *et al.*, 1995 Marsters *et al.*, 1998). It was not surprising to find that, in addition to the receptors, also the ligands display striking structural homologies, which are reflected by similar mechanisms of receptor recognition and triggering. With the exception of NGF, all ligands recognize their receptors through a shared structure composed of anti-parallel ß-sheets arranged in a jelly roll structure. The crystal structure of TNF and LTα alone or LTα in complex with the extracellular domain of the TNF-R1 revealed a trimeric structure (Eck and Sprang, 1989; Eck *et al.*, 1992; Jones *et al.*, 1992; Banner *et al.*, 1993). Together with a number of biochemical data (Karpusas *et al.*, 1995; Pitti *et al.*, 1996), it is believed that all active ligands have a trimeric structure in solution and activate their cognate receptors by trimerization. Data on the CD95 receptor demonstrating that dimerization of CD95 was not sufficient to trigger apoptosis supported this notion (Dhein *et al.*, 1992; Kischkel *et al.*, 1995). Another common feature of the ligands is that almost all of them are synthesized as type II transmembrane proteins. The only exception is LTα which, although formed as a soluble protein, can bind as a subunit to another membrane-bound ligand of the family, LTß, and also act as a cell-bound form. Lymphotoxins can be found as homo- ($LT\alpha_3$) or heterotrimers ($LT\alpha_1/ß_2$ or $LT\alpha_2/ß_1$). The LTα homotrimer binds the TNF receptors, whereas the heterotrimers bind to the LTß receptor which does not contain a death domain.

Although TNF-related ligands are synthesized as membrane-bound molecules, for most of them soluble forms have been identified. The secreted forms of the ligands are generated by the activity of rather specific metalloproteinases. This was suggested for CD95L (Kayagaki *et al.*, 1995; Mariani *et al.*, 1995). For TNF, a zinc-dependent metalloprotease, called TACE (TNFα-converting enzyme) was recently cloned and shown to specifically cleave TNF (Black *et al.*, 1997; Moss *et al.*, 1997). Activation of a death receptor pathway is in many cases controlled by the inducible *de novo* expression of the respective death ligands such as TNF, CD95L or TRAIL. As described below, several apparently different death-inducing conditions, such as the exposure to chemotherapeutic agents or irradiation, can induce the expression of death ligands and may therefore, at least partially, mediate cell death via receptor-ligand interaction.

BIOLOGICAL FUNCTIONS OF DEATH RECEPTORS

TNF and LTα were isolated more than 15 years ago, on the basis of their ability to kill tumor cells *in vitro* and to cause hemorrhagic necrosis of transplantable tumors in mice. Because TNF proved to be highly toxic in animals and humans, it did not fulfill initial expectations that would be useful in the treatment of cancer. Considerable evidence suggests that overproduction or inappropriate expression of TNF plays a role in various chronic inflammatory disorders, modifies anticoagulant properties of endothelial cells, activates neutrophils and induces cytokine expression. Bone resorption, fever, anemia and wasting may all, in some measure, be attributable to TNF (reviewed in Tracey and Cerami, 1993; Bazzoni and Beutler, 1996; Fiers *et al.*, 1996).

The CD95 (APO-1, Fas) molecule has been identified much later as a cell surface receptor that could mediate apoptotic cell death of transformed cells and cause regression of experimental tumors growing in nude mice (Trauth *et al.*, 1989; Yonehara *et al.*, 1989). Although due to the high systemic side effects, application of CD95-mediated apoptosis has also not hold its promise as a potential cancer therapy, several evidences have now documented a pivotal role of CD95 in various physiological and pathological forms of cell death (reviewed in Schulze-Osthoff, 1994; Krammer *et al.*, 1994; Nagata and Golstein 1995; Nagata, 1997).

Finally, TRAIL and its different receptors have been identified as the last, but probably not least, receptor-mediated apoptosis system. TRAIL is able to induce apoptosis in many transformed cells. In contrast to TNF and CD95L, no side effects of TRAIL on normal primary cells have yet been reported. Thus, whether the TRAIL pathway represents the long-sought-after means to selectively kill tumor cells remains to be shown.

In the following, we will describe the relevance and some biological functions of these receptors and their ligands within the organism. Although most functions have been described in the immune system, death receptors are involved in a number of very different apoptotic settings ranging from cell homeostasis, organ development, immune privilege and anticancer treatment. The apoptotic machinery utilizing death receptor/ligand systems is very powerful and requires tight regulation. Disturbance of these systems can cause severe disease.

The CD95 system

The importance of the CD95 system has mainly been addressed in the immune system. Our understanding of the function of the CD95 receptor and its ligand has greatly been enhanced by the finding that both molecules are mutated in mouse strains suffering from severe autoimmune diseases. *Lpr* (for lymphoproliferation) mice which lack a functional CD95 receptor (Watanabe *et al.*, 1992b) as well as *gld* (for generalized lymphoproliferative disease) mice which bear a mutant CD95 ligand (Takahashi *et al.*, 1994a) exhibit various autoimmune phenomena resembling systemic lupus erythematosus in men. Both mouse strains produce autoantibodies and accumulate $CD4^-CD8^-$ T cells leading to lymphadenopathy, splenomegaly and other signs of autoimmune disorders. The *lpr* mutation is caused by the insertion of a transposable element into intron 2 of the gene encoding CD95, thereby preventing full-length transcription (Adachi *et al.*, 1993). The *gld* defect arises from a point mutation within the CD95L gene, changing an amino acid critical for CD95 binding. Recently, in men a similar disease with a dysfunction of the CD95 system was reported (Fisher *et al.*, 1995; Rieux-Laucat *et al.*, 1995; Drappa *et al.*, 1996). Children with "autoimmune lymphoproliferative syndrome" (ALPS), also called Canale-Smith syndrome, have massive nonmalignant lymphadenopathy, hepatosplenomegaly, altered and enlarged T cell populations and other manifestations of systemic autoimmunity. The loss-of-function phenotype in mice and men indicates that CD95 plays an important role in the regulation of a normal immune response and the maintenance of self-tolerance.

Molecular studies provided evidence that CD95 is the mediator of activation-induced cell death (AICD), a form of apoptosis important for the downsizing of the immune response, as well as an effector component of cytotoxic T cell activity (Dhein *et al.*, 1995; Brunner *et al.*, 1995; Ju *et al.*, 1995; Rouvier *et al.*, 1993; Hanabuchi *et al.*, 1994; Alderson *et al.*, 1995). In contrast, there is no convincing evidence so far to believe that CD95 is involved in negative selection. This assumption is consistent with a relatively normal thymic architecture and proper thymic deletion of superantigen-activated T cells in *lpr* mice (Herron *et al.*, 1993; Singer and Abbas, 1994). *lpr* and *gld* mice, in addition to abnormalities in the T cell compartment, show B cell hyperreactivity associated with the production of autoantibodies, suggesting that CD95 also controls the expansion of the B cell compartment. Indeed, it has been found that CD40 ligand sensitizes B cells to CD95-mediated apoptosis, whereas CD40 ligation plus engagement of surface immunoglobulin protects cells. This indicates that the CD95L-mediated demise may represent a mechanism to prevent nonspecific B cell activation and confer antigen specificity to the interaction of helper T cells with B cells (reviewed in Krammer *et al.*, 1994; Cornall *et al.*, 1995).

Just as a defect of the CD95 system is intimately linked to autoimmune diseases caused by the impaired removal of autoreactive lymphocytes, so may inappropriate induction of apoptosis lead to various pathological conditions. Accumulating evidence exists that CD95 is critically involved in the progression of viral diseases, such as HIV-1 or hepatitis virus infections where massive apoptosis occurs. It could be shown that indirect mechanisms lead to a sensitization of noninfected T cells towards AICD after HIV-1 infection (Westendorp *et al.*, 1995; Li *et al.*, 1995a; Szawlowski *et al.*, 1993; Zagury *et al.*, 1993). T lymphocytes from HIV-1 infected patients exhibit an elevated expression of CD95 and sensitivity towards CD95-mediated apoptosis (Debatin *et al.*, 1994; Katsikis *et al.*, 1995). Two HIV-1 derived soluble proteins, gp120 and Tat, have been found to activate the inducible expression of CD95L in T lymphocytes (Westendorp *et al.*, 1995). This event may then cause a fratricide or suicide death of uninfected T lymphocytes and result in the continuous depletion of $CD4^+$ T cells during AIDS disease.

Although CD95L has been originally found on activated T lymphocytes, various other non-lymphoid cells can express CD95L. A high constitutive expression can be detected in Sertoli cells of the testis and epithelial cells of the anterior eye chamber (Griffith *et al.*, 1995; Bellgrau *et al.*, 1995). This finding led to the proposal that CD95L may be responsible for the maintenance of immune privilege, which characterizes the ability of certain organs to suppress graft rejection, even when transplanted in non-matched individuals. After viral inoculation into the anterior eye chamber, infiltrating lymphocytes and granulocytes undergo apoptosis probably due to high expression of CD95L on epithelial cells (Griffith *et al.*, 1995; 1996). This apoptosis is not observed in eyes of animals with defective CD95L (*gld* mice), and the resulting uncontrolled inflammation destroys the tissue. Thus, CD95L is necessary for the maintenance of the privileged status of the eye by killing infiltrating lymphocytes of the host. It is interesting to note that *gld* and *lpr* mice have no apparent ocular abnormalities and no increased lymphocytic infiltration. Presumably, even with aberrant CD95L expression, organs such as the eye maintain their function through other additional mechanisms that insure immune privilege.

The exciting novel function of CD95 in immune privilege has presumably enormous practical implications for future transplantation strategies aimed to avoid allograft rejection. It was recently shown that human corneas express functional CD95L (Stuart *et al.*, 1997), raising the possibility that this molecule could act to protect cornea grafts. Examination of corneal transplants in mice supported this idea; while approximately 45% of allogeneic cornea transplants survived for an extended period, no graft survival was seen with corneas expressing defective CD95 receptor or ligand. Thus, the protection of allogeneic grafts was dependent upon the presence of a functional CD95 system.

Other studies on the CD95L involvement in graft acceptance are less clear and currently very controversially discussed. A protective effect of CD95L expressed in the testis was observed after transplantation of allogeneic testis under the kidney capsule (Bellgrau *et al.*, 1995), but this result could not be confirmed by others (Allison *et al.*, 1997). In another example, syngeneic myoblasts expressing ectopic CD95L protected allogeneic pancreatic islets co-implanted under the kidney capsule of animals with streptozotocin-induced diabetes (Lau *et al.*, 1996). These grafts, which were quickly rejected if myoblasts did not express CD95L, maintained their function for an extended period of time. Consistent with this was the observation that allogeneic islets showed delayed rejection when co-implanted with CD95L-expressing testis tissue (Selawry and Cameron, 1993). However, in other studies no protective effect of CD95L was observed. Allison *et al.* (1997) reported that transgenic expression of CD95L in pancreatic islets failed to protect these from allogeneic transplant rejection when placed under the kidney capsule. The presence of CD95L rather induced a granulocytic infiltrate in the animals, which damaged but did not destroy the islets. The finding is related to a report showing that CD95L on tumor cells can induce a granulocyte-mediated rejection reaction (Seino *et al.*, 1997).

CD95L-mediated depletion of cytotoxic T lymphocytes may not only be beneficial but may also play a role for tumor cells to escape the host's immunosurveillance. Recently, high constitutive CD95L expression has been found in distinct lineages of tumors, such as colon, lung, renal carcinoma, melanoma, hepatocellular carcinoma, astrocytoma and T- and B-cell derived neoplasms (O'Connell *et al.*, 1996; Hahne *et al.*, 1996; Strand *et al.*, 1996; Niehans *et al.*, 1997; Saas *et al.*, 1997; Shiraki *et al.*, 1997; Tanaka *et al.*, 1996; Villunga *et al.*, 1997; unpublished results). This suggested that the same mechanisms responsible for protecting tissues from autoimmune destruction during inflammation and graft rejection may be also used by tumors in establishing immunologically privileged environments.

An increasing body of data implies a role of CD95 in inflammatory situations and autoimmune diseases that are associated with tissue destruction. Most likely CD95L expression causes tissue damage directly, or indirectly by recruitment of granulocytes. In graft-versus-host disease, the ability of the graft effector cells to express functional CD95L contributes to the destructive assault (Baker *et al.*, 1996; Braun *et al.*, 1996). The observation that anti-CD95 antibody induces apoptosis in hepatocytes *in vivo* led to the idea that CD95L-induced apoptosis may be involved in some forms of hepatitis (Ogasawara *et al.*, 1993). This was supported at least in patients with alcoholic liver damage, in which the hepatocytes express CD95L (Galle *et al.*, 1995). There is also some evidence for a role of CD95L-induced apoptosis in autoimmune diseases, such as diabetes and encephalomyelitis (Chervonsky *et al.*, 1997; Sabelko *et al.*, 1997; Waldner *et al.*, 1997).

A particular situation was found in the thyroid gland. Normal thyrocytes constitutively express functional CD95 ligand, but do not express the receptor. However, in Hashimoto's thyroiditis patients, thyrocytes do express CD95, and these cells undergo apoptosis (Giordano *et al.*, 1997). *In vitro*, normal thyrocytes express CD95 after exposure to IL-1, and the resulting apoptosis can be blocked by antibodies that disrupt CD95/CD95L interactions. Hence, in Hashimoto's thyroiditis the normally protective function of CD95L on thyrocytes leads to the destruction of the thyroid gland. The cause for this dysfunction is unclear, but it is likely that the CD95 system contributes to the disease process.

Although CD95 has a predominant function in inducing apoptosis, there are some conditions where CD95L may obviously trigger an inflammatory reaction. In this respect, it has been found that CD95 ligation can result in secretion of the chemokine IL-8 (Abreu-Martin *et al.*, 1995). In conjunction with T-cell receptor activation or other signals in T cells, CD95 may act as a costimulatory molecule, enhancing gene expression of IL-2 and other cytokines (Alderson *et al.*, 1993). In addition, some cell types can respond to CD95 ligation via proliferation and not cell death (Alderson *et al.*, 1993; Mapara *et al.*, 1993; Aggarwal *et al.*, 1995; Freiberg *et al.*, 1997). However, the overall *in vitro* and *in vivo* data suggest that CD95 is a receptor which mainly mediates apoptosis. This is also reflected by the observation that, in contrast to TNF, CD95 does generally not induce the activation of proinflammatory transcription factors, including NF-κB, AP-1 or NF-AT (Schulze-Osthoff *et al.*, 1994).

An exciting finding was that several unrelated death-inducing agents and conditions obviously utilize physiological means of induction of apoptosis. Apoptosis mediated by p53 may involve the CD95 system, as the gene encoding CD95 has been found as a putative target of this transcription factor (Owen-Schaub *et al.*, 1995). Also overexpression of the c-Myc proto-oncogene, which induces cell death under growth-limiting conditions, appears to mediate death, at least partially, by a mechanism requiring CD95/CD95L interaction (Huber *et al.*, 1997). An apoptosis-inducing effect of c-Myc was not observed in *lpr* and *gld* mice. Furthermore, AICD in mature and immature T lymphocytes is blocked by c-myc antisense oligonucleotides (Bissonnette *et al.*, 1994). The mechanism how c-Myc sensitizes cells for CD95-mediated apoptosis remains to be elucidated.

Recent data demonstrate that anticancer drug-induced cell death may involve the CD95 system. Several different drugs, widely used in chemotherapy of cancers induce CD95L expression in leukemic, hepatocellular and neuroblastoma cells (Friesen *et al.*, 1996; Müller *et al.*, 1997; Fulda *et al.*, 1997a). Binding of CD95L to the receptor then triggers the apoptosis cascade in chemosensitive tumor cells. In support of these data, anticancer drug-induced cell death has been found in most cases to be inhibitable by CD95L neutralizing reagents. The upregulation of CD95L may therefore provide an new clue to the mechanism of action of chemotherapeutic drugs.

Still another situation has been found in cell death induced by UV irradiation. UV-induced apoptosis is strongly attenuated in CD95-resistant cells (Rehemtulla *et al.*, 1997; Aragane *et al.*, 1998). It has been shown that UV irradiation directly oligomerizes and thereby activates death receptors, such as CD95 and TNF-R1. This is presumably mediated by energy transfer which may allow for a subsequent conformational change of the receptors (Rosette and Karin, 1996; Aragane *et al.*, 1998). UV-induced apoptosis therefore does not require CD95L *de novo* expression, but directly engages the CD95 signalling pathway. Altogether, these findings demonstrate that CD95 and presumably other death receptors play a role in very diverse apoptosis settings.

The TNF system

In contrast to the CD95 system, the biological function of the TNF/TNF-R system is much more complex. In addition of being cytotoxic for mainly transformed cells, TNF exerts a number of other activities related to proinflammatory processes on almost all cell types (reviewed in Fiers *et al.*, 1996; Bazzoni and Beutler, 1996). The function of TNF is also complicated by the fact that two different TNF receptors, TNF-R1 and TNF-R2, exist which can be occupied by two different ligands. TNF and the LTα homotrimer bind to the TNF receptors, while LTα/LTß complexes selectively ligate the LTß receptor.

TNF was originally found as a serum factor in endotoxin-primed mice which caused hemorrhagic necrosis of transplanted tumors. Almost concurrently, the factor was identified as a catabolic substance that suppressed the expression of lipoprotein lipase and other anabolizing enzymes in fat and was therefore termed cachectin. The gene encoding TNF was cloned by several groups and found to cluster with the genes for LTα and LTß in the major histocompatibility complex. A major cellular source of TNF are activated macrophages, but also other cell types such as lymphoid cells, NK cells, neutrophils, keratinocytes, fibroblasts and smooth muscle cells produce the cytokine in response to inflammatory and environmental challenges. The inducible expression of TNF is regulated at the transcriptional but also translational level. Whereas TNF can be secreted by a variety of cell types, the lymphotoxins are mostly produced by activated lymphocytes and NK cells.

Though the principal interest in TNF arose from its antitumor activity, it soon became clear that TNF has a wide range of other biological effects and is a mediator of endotoxic shock. An important cellular target of TNF action is the endothelium where TNF induces the release of platelet activating factor (PAF), the secretion of various cytokines, such as IL-1, IL-6, IL-8, GM-CSF, and the expression of adhesion molecules including ICAM-1, VCAM-1 and E-selectin. These responses together with the activation of arachidonic acid metabolism commonly result in increased vascular permeability, anticoagulant activity and leukocyte adhesion. Because TNF receptors are ubiquitously expressed with the exception of erythrocytes, it is not suprising that almost all cell types respond to TNF. In neutrophils, TNF activates respiratory burst and degranulation leading to the release of reactive oxygen intermediates, elastase, lysozymes and other granular enzymes. Macrophages respond to TNF with enhanced cytotoxic activity and cytokine synthesis. In hepatocytes, TNF is, together with IL-1 and IL-6, an important mediator of the synthesis of acute-phase proteins. In addition, a multitude of biological effects have been described for several cell types including osteoblasts, fibroblasts, smooth muscle cells and others. Many of the proinflammatory activities of TNF are regulated by the transcription factor NF-κB.

Most of the biological activities of TNF including programmed cell death, antiviral activity, and activation of transcription factor NF-κB, are mediated by TNF-R1, while an involvement of TNF-R2 has been demonstrated particularly in T lymphocytes (Engelmann *et al.*, 1990; Espevik *et al.*, 1990; Tartaglia *et al.*, 1991; 1993b; Wong *et al.*, 1992; Vandenabeele *et al.*, 1995). Membrane-bound TNF (mTNF) and soluble TNF (sTNF) have different affinities to the two receptors, with TNF-R2 preferentially binding mTNF (Grell *et al.*, 1995). TNF-R2 appears to play an auxiliary role in cellular responses to sTNF. It has been suggested that TNF-R2–bound ligand may be passed over to TNF-R1 to enhance TNF-R1 signalling. This process, termed ligand passing, is favored by the distinct kinetics of ligand association and dissociation at the two receptors. TNF binding to TNF-R2 has a fast off-rate that creates a locally high TNF concentration at the cell surface, which in turn facilitates binding to TNF-R1 which has a slow dissociation rate (Tartaglia *et al.*, 1993b). The prime physiological activator of TNF-R2 seems to be mTNF, since TNF-R2 can be more strongly stimulated by mTNF rather than by sTNF. As mTNF also signals via TNF-R1, the resulting cooperativity of both receptors leads to cellular responses much stronger than those achievable with sTNF alone. Moreover, it was shown that upon appropriate activation of TNF-R2, a switch of the cellular response pattern to TNF occurred, such that, as an example, cells fully resistant to the cytotoxic action of sTNF become highly susceptible upon contact with mTNF (Grell *et al.*, 1995).

Gene targeting and transgene technologies have been used to unravel the *in vivo* role of the TNF system and to establish genetically defined models of human diseases. $TNF^{(-/-)}$ mice show an almost normal phenotype histologically, but have reduced sensitivity to LPS-mediated toxicity and increased sensitivity to intracellular pathogens such as *Listeria* and *Candida*, due to severely impaired macrophage functions (Pasparakis *et al.*, 1996; Marino *et al.*, 1997). Apart from their deficiency in effector functions, $TNF^{(-/-)}$ mice have defects in lymphoid organogenesis and formation of germinal centers. Gene targeting of LTα also results in a lack of primary and secondary lymphoid follicles, but moreover these mice have a defective lymphnode development (reviewed in von Boehmer *et al.*, 1997).

The importance of mTNF *in vivo* has been elegantly demonstrated in several transgenic models. For example, it was reported that the deficiencies of $TNF^{(-/-)}$ mice are reconstituted by ectopical expression of noncleavable mTNF (Korner and Sedgwick, 1996). In a different transgenic model it was shown that mTNF induces multiple sclerosis-like disease with paralysis and a histopathology resembling experimental autoimmune encephalomyelitis when expressed in microglia, but not neuronal cells.

In contrast to $TNF\text{-}R1^{(-/-)}$ mice (Rothe *et al.*, 1993), deletion of TNF-R2 has no apparent influence on lymphoid organ development. TNF-R2 is critically involved in mediating pathogenicity during cerebral malaria, is essential for LPS-induced leukostasis and downregulates TNF-R1 dependent neutrophil influx in a lung inflammation model (Garcia *et al.*, 1995). A dominant role of TNF-R1 in mediating pathogenic activities was evident early on from models of septic shock and arthritis (Espevik and Waage, 1988; Shimamoto *et al.*, 1988; Hayward and Fiedler-Nagy, 1987). The growing knowledge about the pathophysiological role of TNF in acute and especially in chronic diseases calls for strategies to intervene with the deleterious effects of TNF. Clinical trials employing anti-TNF reagents have been impressingly successful in diseases such as rheumatoid arthritis, septic shock and inflammatory bowel diseases (Maini, 1996; Glauser, 1996; Stokkers *et al.*, 1995).

The TRAMP and TRAIL Systems

The biological functions of the new death receptors are largely undefined at present. TRAMP (TNF receptor-related apoptosis-mediating protein, DR3) is abundantly expressed in thymocytes and lymphocytes and may therefore play a role in lymphocyte development (Chinnaiyan *et al.*, 1996a; Bodmer *et al.*, 1997). The chromosomal localization of the TRAMP gene has been assigned to the long arm of chromosome 11 where other related receptors (CD30R, TNFR2, OX40R) have been mapped. TRAMP is both structurally and functionally related to the TNF receptors, because its overexpression or stimulation by its recently identified ligand APO-3L (TWEAK) (Chicheportiche *et al.*, 1997; Marsters *et al.*, 1998) leads to NF-κB activation and apoptosis.

The recent cloning of TRAIL and its three receptors revealed a new apoptosis system with apparently high complexity. Among the ligands of the TNF family, TRAIL is most closely related to CD95L. However, in contrast to the restricted expression of CD95L, TRAIL is more abundantly expressed in several tissues. Two of the TRAIL receptors, TRAIL-R1 and TRAIL-R2, can induce apoptosis in various cancer cells, whereas the decoy receptors DcR1 and DcR2, which are mostly expressed on normal cells, do not contain a functional death domain and therefore confer resistance against TRAIL action (Pan *et al.*, 1997a; 1997b; Sheridan *et al.*, 1997; Walczak *et al.*, 1997; Degli-Eposti *et al.*, 1997; MacFarlane *et al.*, 1997; Masters *et al.*, 1997). Whether TRAIL might be used to kill more selectively tumor cells, awaits further *in vivo* experiments. Because TRAIL only weakly triggers NF-κB activation, it can be expected that *in vivo* administration of TRAIL will not lead to severe side effects related to proinflammatory gene expression. At the moment, it is unclear why there are two death-signalling TRAIL receptors. It is possible that either the two TRAIL receptors are redundant or provide an additional and versatile means to regulate apoptosis. So far, there is only limited information on target cells of TRAIL-induced apoptosis. It was reported that TRAIL can induce AICD in activated T cells (Marsters *et al.*, 1996a; 1996b). Furthermore, it has been shown that TRAIL-induced AICD may contribute to cell death of T lymphocytes during HIV infection (Katsikis *et al.*, 1997).

DEATH RECEPTOR-ASSOCIATING MOLECULES

A major advancement in our understanding of death receptor signalling was the definition of the so-called death domain (DD) within the death receptors, which is required for the transmission of a cytotoxic signal. The DD has been characterized in detail in TNF-R1 and CD95 and consists of a stretch of about 80 amino acids that are essential for triggering cell death. Its importance is also demonstrated by *lpr* ^{cg} mice which carry a point mutation in the DD of CD95. Delineation of the DD was not only a major aid for the identification of new death receptors, when used as a screen in EST databases, but also allowed for the identification of new adaptor molecules, when used as a bait in interactive cloning approaches.

The DD exerts its effects via interactive properties. As part of some proteins, it was found to self-associate and to be capable of binding to the DD in other proteins. These associations between DDs occur as a consequence of receptor-ligand binding and seem to involve electrostatic interactions. NMR spectroscopy of the DD of CD95 confirms that this region, which comprises a series of antiparallel amphipathic α-helices, has many exposed charged residues (Huang *et al.*, 1996). The tendency of the DD to self-associate apparently fortifies the interactions of the receptors imposed by ligand binding. Following self-association, the DD of the receptors recruits and binds other DD-containing proteins which then serve as adaptors in the signalling cascades (Figure 1.2).

The first DD-containing adaptor proteins identified were FADD (MORT1) (Chinnaiyan *et al.*, 1995; Boldin *et al.*, 1995), RIP (Stanger *et al.*, 1995) and TRADD (Hsu *et al.*, 1995). TRADD is most effectively bound following ligation of TNF-R1, where it then likely serves to recruit the DD proteins FADD and RIP as well as the RING domain adaptor protein TRAF2. FADD, in contrast, is preferentially recruited to CD95. Thus, the DD of FADD can bind to the DD of TRADD and the DD of RIP to the DDs of both TRADD and FADD. These mutual interactions may therefore account for a potential crosstalk of the different death receptor signalling pathways.

Overexpression of DD proteins causes cell death indicating that these molecules are involved in apoptosis signalling. In the case of FADD, transient expression of the N-terminal region was found to be sufficient to cause apoptosis (Chinnaiyan *et al.*, 1995). This part of FADD was therefore termed the death effector domain (DED) (Chinnaiyan *et al.*, 1996b). In contrast, overexpression of the C-terminal DD-containing part lacking the DED (FADD-DN) protected cells from CD95-mediated apoptosis and functioned as a dominant-negative mutant. This suggested

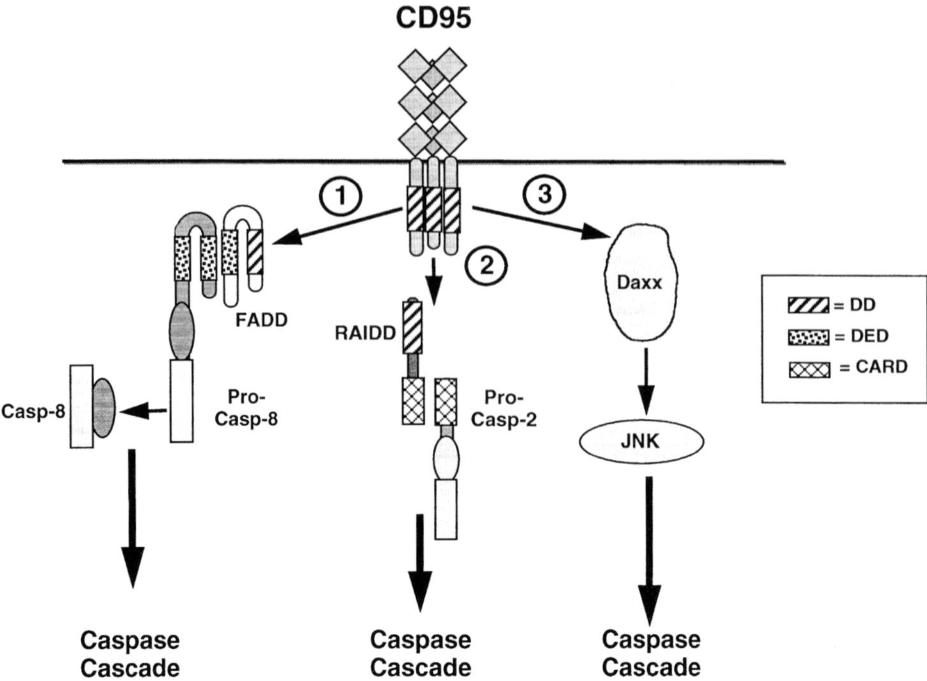

Figure 1.2 Proposed signalling pathways of CD95-mediated apoptosis. One important pathway includes recruitment of the adapter protein FADD through interaction between the death domains (DD) of FADD and CD95. The death effector domain (DED) of FADD in turn recruits procaspase-8 which is cleaved and activated at the DISC. An alternative pathway may involve activation of caspase-2 through the RIP-interacting protein RAIDD. RAIDD contains DD and a caspase recruitment domain (CARD) that is also present in procaspase-2. A third pathway may include recruitment of Daxx to the cytoplasmic domain of CD95. This pathway involves JNK activation and further downstream activation of caspases. It should be mentioned that the physiological relevance of the latter two pathways is rather unknown.

that the N-terminus of FADD is coupled to the cytotoxic machinery. Both TRADD and RIP induce apoptosis but can also activate NF-κB, which is a typical feature of TNF-induced signalling (Hsu *et al.*, 1995; 1996a; Park *et al.*, 1996; Ting *et al.*, 1996). Similar to FADD, the C-terminus of TRADD contains a DD enabling self-association and association with the DD of other signalling molecules including TNF-R1 and FADD. TRADD, however, lacks the typical DED present in FADD.

RIP (receptor-interacting protein) contains an N-terminal kinase domain and a C-terminal DD. It was originally identified as a molecule binding to the cytoplasmic domain of CD95 in a two-hybrid system (Stanger *et al.*, 1995). Therefore, RIP was suggested to play a role in CD95-signalling. Later however, it was demonstrated that RIP does not directly bind to CD95 or TNF-R1 but is recruited through the TNF-R adaptor protein TRAF2. RIP was identified to be crucial for TNF-R1-mediated NF-κB activation. In a mutant cell line that had lost expression of RIP, CD95 signalling was not affected, whereas TNF-R1-mediated NF-κB activation was blocked

(Ting *et al.*, 1996). After reconstitution with RIP, NF-κB activation in this cell line was restored. The intermediate region of RIP was found to be responsible for activating NF-κB which was potentiated by the presence of the kinase region and DD.

While most of the information regarding death pathways has been obtained from yeast two-hybrid assays or supra-physiological overexpression of DD proteins in mammalian cells, for CD95 the signalling complexes have been also identified *in vivo* using classical biochemical methods (Kischkel *et al.*, 1995). Treatment of cells with the agonistic mAb anti-APO-1 and subsequent co-immunoprecipitation of CD95 resulted in the identification of four cytotoxicity-dependent APO-1-associated proteins (CAP1-4) on two-dimensional gels within seconds after receptor triggering. Together with the receptor these proteins formed the death-inducing signalling complex (DISC). Using specific antisera two of the proteins (CAP1 and 2) were identified as two different serine phosphorylated species of FADD and it was demonstrated that FADD bound to CD95 in a stimulation-dependent fashion *in vivo*.

Sequencing of the other immunoprecipitated proteins, CAP3 and CAP4, resulted in the identification of a further downstream DED-containing protein. Using nanoelectrospray tandem mass spectrometry, sequence information of CAP3 and CAP4 was obtained that led to the retrieval of a full-length clone from a cDNA data base (Muzio *et al.*, 1996). This protein contained two DEDs at its N-terminus, which associate with the DED of FADD. At its C-terminus it had the typical domain structure of an ICE-like protease and was therefore termed FLICE (FADD-like ICE). FLICE was also cloned by two other groups and named MACH and Mch5 (Boldin *et al.*, 1996; Fernandes-Alnemri *et al.*, 1996a). It belongs to cysteine proteases of the caspase family and is therefore now referred to as caspase-8 (Alnemri *et al.*, 1996). Identification of caspase-8 as part of the DISC connected two different levels in apoptosis pathways, the receptor level with the level of the apoptosis executioners, the caspases.

Another possible route for death receptor signalling was recently suggested by the discovery of a new death adaptor protein, called RAIDD or CRADD (Figure 1.2) (Duan and Dixit 1997; Ahmad *et al.*, 1997). This protein contains a carboxy-terminal DD that binds to the DD of RIP. At its amino terminus RAIDD has a domain that is responsible for the recruitment and binding to the prodomain of caspases, in particular caspase-2. The contribution of this pathway to TNF and CD95 cytotoxicity is still unknown.

Furthermore, it was demonstrated that death receptors can directly trigger signalling pathways other than caspases. MADD was cloned as DD-containing protein that binds to TNF-R1 and activates ERK2, a MAP kinase member (Schievella *et al.*, 1997). Similarly for CD95, a protein, called Daxx, was identified that specifically associates with the DD of CD95 (Yang *et al.*, 1997). Overexpression of Daxx stimulates stress-activated protein kinases of the MAP kinase family (JNK/SAPK) and enhances apoptosis (Figure 1.2). Thus, a single receptor is able to trigger multiple pathways which, in addition to the FADD/caspase cascade, participate in induction of cell death. Despite the key role of the DD, the possibility that proteins associated with other intracellular regions of the receptors contribute to the overall pattern of apoptosis cannot be excluded. As will be described below, there are some evidences for such proteins, which either utilize other signalling mechanisms or which may suppress a death signal elicited at the DD.

THE DEATH-INDUCING SIGNALLING COMPLEX (DISC)

As described previously, caspase-8 was identified *in vivo* as a part of the CD95 DISC, suggesting that caspase-8 activation occurred at the DISC level. Indeed, it was recently confirmed that the entire cellular amount of cytoplasmic caspase-8 can be converted into active caspase-8 subunits at the DISC (Medema *et al.*, 1997a). After receptor engagement, FADD and caspase-8 are recruited to CD95 within seconds. Binding of caspase-8 to receptor-associated FADD then presumably causes a structural change, resulting in autoproteolytic activation of the protease. The active subunits p10 and p18 are released into the cytoplasm whereas part of the prodomain remains bound to the DISC. Presently, it is assumed that active caspase-8 subunits cleave various death substrates including other caspases, such as caspase-3, which then leads to the execution of apoptosis.

Using anti-caspase-8 antibodies it became clear that from all of the eight published caspase-8 isoforms two were predominantly expressed at the protein level (Scaffidi *et al.*, 1997). These isoforms, termed caspase-8/a and caspase-8/b, are recruited and processed with similar kinetics at the DISC. Recently, it was demonstrated that recombinant caspase-8 lacking the prodomain could cleave caspase-8 *in vitro* suggesting an amplification step with caspase-8 at the top of a caspase cascade (Srinivasula *et al.*, 1996a; Muzio *et al.*, 1997). However, using the *in vivo* DISC assay, this could not be confirmed in intact cells (Medema *et al.*, 1997a). It is possible that recombinant caspase-8 lacking the prodomain displays different substrate specificity in comparison to full-length caspase-8 *in vivo*.

Overexpression of functionally inactive FADD and caspase-8 did not only block CD95, but also inhibited TNF-R1-induced signalling (Chinnaiyan *et al.*, 1996b; Boldin *et al.*, 1996). This suggested that both receptors are coupled to a similar signalling complex following cell activation. However, FADD does not directly bind to TNF-R1 but becomes associated upon binding of the DD-containing protein TRADD. Indeed, in cells stimulated with TNFα caspase-8 is processed, suggesting that TNF-R1 uses a similar signalling pathway for induction of apoptosis (unpublished data). So far a direct biochemical association has not been shown at the DISC level in intact cells. It remains therefore unclear whether caspase-8 is activated by association with TNF-R1 or whether its activation is the result of a secondary event.

In contrast to CD95, TNF-R1 signalling seems to be more complex (see below). It was demonstrated that RAIDD/CRADD, another DD-containing protein, can bind to TNF-R1 *in vitro* or when overexpressed in 293 cells (Duan and Dixit, 1997; Ahmad *et al.*, 1997). RAIDD/CRADD carries the DD at its C-terminus and at its N-terminus it displays homologies with the prodomain of caspase-2. It binds to TNF-R1 more efficiently in the presence of RIP. It was therefore suggested that RAIDD would engage TNF-R1 via RIP and induce activation of caspase-2. However, no *in vivo* data are available and the mechanism of caspase-2 activation remains to be determined.

TRAMP has been reported to bind FADD, TRADD, TRAF2, and caspase-8 (Chinnaiyan *et al.*, 1996a). Due to its TNF-R1-like structure it is expected to have a signalling function analogous to TNF-R1. Similarly to TNF-R1, TRAMP can induce apoptosis and activate NF-κB (Chinnaiyan *et al.*, 1996a; Bodmer *et al.*, 1997; Kitson *et al.*, 1996; Marsters *et al.*, 1996b; Screaton *et al.*, 1997).

TRAIL is known to bind to two receptors, TRAIL-R1 and TRAIL-R2, both of which can signal for cell death. As with the other death receptors, TRAIL-R1 and TRAIL-R2-mediated apoptosis involves caspases, because caspase inhibitors as well as overexpression of CrmA or p35 blocked TRAIL-induced cell death (Pan *et al.*, 1997a; 1997b; Sheridan *et al.*, 1997; MacFarlane *et al.*, 1997; Mariani *et al.*, 1997). In fact, also caspase-8 activation was detected upon treatment sensitive BJAB cells with TRAIL (unpublished results). The data support the view that caspase-8 among other caspases is involved in the TRAIL-R signalling pathway. Moreover, it was recently suggested that caspase-10 (FLICE-2), which has a structure similar to caspase-8, is preferentially activated by TRAIL, whereas CD95 more strongly activates caspase-8 (Pan *et al.*, 1997b).

A number of reports indicate that the proximal signalling pathways of the TRAIL receptors are similar but distinct from CD95. For instance, Ag8 mouse myeloma cells have been found to be sensitive to TRAIL-induced, but not CD95L-mediated apoptosis, although CD95 was detected on the surface of these cells (Mariani *et al.*, 1997). In addition, it was demonstrated that HeLa cells transfected with dominant-negative FADD became resistant to CD95 but remained sensitive to TRAIL (Marsters *et al.*, 1996a). This finding indicates that FADD does obviously not play an obligatory role in TRAIL-induced apoptosis. It is now generally established that TRAIL-R1 does not couple to FADD. However, whether TRAIL-R2-mediated cell death requires association with FADD remains controversial. Most reports point out that, similarly to TRAIL-R1, also TRAIL-2 signalling is independent of FADD (Pan *et al.*, 1997b; Sheridan *et al.*, 1997; MacFarlane *et al.*, 1997). In contrast, Walczak *et al.*, (1997) showed that overexpression of dominant-negative FADD strongly attenuated TRAIL-R2-mediated apoptosis in CV-1 cells. Testing BJAB cells expressing FADD-DN, we also observed a complete block of TRAIL-induced signalling (unpublished data). Thus, whether FADD is directly or indirectly coupled to TRAIL-R2 signalling requires further investigation. It is possible that TRAIL-R2 recruits an unknown adaptor protein that triggers the caspase cascade or that, like TRADD for TNF-R1 and TRAMP, another molecule mediates binding of FADD to TRAIL-R2.

EFFECTOR CASPASES

Caspase-8 (FLICE/MACH/Mch5) belongs to a growing family of proteases, which were recently named caspases (Alnemri *et al.*, 1996). Caspases are aspartate-directed cysteine proteases that seem to be required for most apoptotic pathways. A detailed survey on the biochemistry of caspases will be provided in other chapters in this book. The following section briefly summarizes the involvement of individual caspases in death receptor-mediated apoptosis.

First evidence for the involvement of caspases in CD95-mediated apoptosis came from pharmacological experiments that employed selective caspase inhibitors.

YVAD-CMK, DEVD-CHO and similar peptides that mimic the P1 aspartate residue of caspase substrates, strongly suppressed CD95 and TNF-R1-mediated killing (Los *et al.*, 1995; Enari *et al.*, 1995; 1996; Miura *et al.*, 1995; Hsu *et al.*, 1995). Another clue came from studies with the poxvirus-derived CrmA protein, designated for cytokine response modifier A. *CrmA* encodes a highly specific serpin-like inhibitor of a number of caspases, including caspase-1 and caspase-8 (Ray *et al.*, 1992; Komiyama *et al.*, 1994; Pickup, 1994). Several groups have shown that overexpression of *crmA* efficiently suppressed CD95-mediated apoptosis in a variety of cell types (Los *et al.*, 1995; Enari *et al.*, 1995; Tewari and Dixit, 1995; Heinkelein *et al.*, 1996). It was demonstrated that expression of microinjected or liposome-transfected CrmA protected cells from CD95-mediated killing. In addition, enzymatic measurements revealed that triggering of CD95 readily induces caspase proteolytic activity. Maximal caspase activity appeared within 15 to 20 min after CD95 engagement (Los *et al.*, 1995).

In contrast to *Caenorhabditis elegans*, so far more than ten mammalian members of the caspase family have been identified. Based on phylogenetic analysis they can be divided into three subfamilies. The ICE-like protease family includes ICE (caspase-1) (Thornberry *et al.*, 1992; Ceretti *et al.*, 1992), TX/ICH-2/ICE_{rel}II (caspase-4) (Faucheu *et al.*, 1995; Munday *et al.*, 1995; Kamens *et al.*, 1995), TY/ICE_{rel}III (caspase-5) (Munday *et al.*, 1995; Faucheu *et al.*, 1996) and ICH-3 (caspase-11) (Wang *et al.*, 1996a). The CED-3-like family includes CPP32/YAMA/apopain (caspase-3) (Fernandes-Alnemri *et al.*, 1994; Tewari *et al.*, 1995; Nicholson *et al.*, 1995), Mch2 (caspase-6) (Fernandes-Alnemri *et al.*, 1996b), Mch3/ICE-LAP3/CMH-1 (caspase-7) (Fernandes-Alnemri *et al.*, 1995; Lippke *et al.*, 1996; Duan *et al.*, 1996), caspase-8 (FLICE/MACH/Mch5) (Muzio *et al.*, 1996; Boldin *et al.*, 1996; Fernandes-Alnemri *et al.*, 1996a), Mch6/ICE-LAP6 (caspase-9) (Duan *et al.*, 1996; Srinivasula *et al.*, 1996b), and Mch4/FLICE2 (caspase-10) (Fernandes-Alnemri *et al.*, 1996a; Vincenz and Dixit, 1997). The third subfamily consists only of Nedd2/ICH-1 (caspase-2) (Wang *et al.*, 1994; Kumar *et al.*, 1994). All caspases are synthesized as zymogens that need to be activated by proteolytic cleavage. The active enzyme is composed of a heterotetrameric complex of two large subunits, containing the active center, and two small subunits, as can be deduced from the crystal structure of both caspase-1 and caspase-3 (Wilson *et al.*, 1994; Walker *et al.*, 1994; Mittl *et al.*, 1997).

Based on their structure and order of action in the death pathway caspases can be divided into initiators and executioners. It is known that at least for CD95-mediated apoptosis signalling is transmitted by sequential caspase activation. However, the exact order of caspase activation during execution of the death pathways is still obscure. A direct link between death receptor triggering and caspase activation was established by cloning of caspase-8 as part of the CD95 DISC (Muzio *et al.*, 1996). The proform of caspase-8 is recruited to the multimerized receptor and then likely activated by autoproteolytic cleavage at the DISC (Medema *et al.*, 1997a). Thus, caspase-8 is regarded as the most upstream caspase in the CD95 pathway. As discussed previously, either caspase-8 itself or simultaneously a caspase-8-like caspase, such as caspase-10, may be involved in a similar fashion in the signal transduction of the other death receptors.

It is assumed that an apical initiator caspase cleaves and activates downstream executioner caspases, though it is unknown which and how many caspases are needed for the final demise of the cell. Caspase-8 has been shown to directly cleave caspase-3, caspase-4, caspase-7, caspase-9 and caspase-10 *in vitro*, while caspase-2 and caspase-6 were cleaved indirectly by other caspase-8-activated caspases present in cellular extracts (Muzio *et al.*, 1997). The order of other caspases in this cascade is not clear so far. Orth *et al.* (1996) place caspase-6 upstream of caspase-3 and caspase-7. It has also been demonstrated that caspase-3 can cleave and activate caspase-6, caspase-7 and caspase-9 (Fernandes-Alnemri *et al.*, 1995, 1996b; Srinivasula *et al.*, 1996b).

The reason for such a great variability of caspases in the mammalian system, in comparison to *C. elegans*, is at present unclear. So far, there is no report demonstrating that a single caspase is crucial for apoptosis signalling by death receptors. The most intensively studied caspase member is caspase-3, which is activated by multiple apoptotic signals including serum withdrawal, activation of death receptors, treatment with granzyme B, ionizing radiation, or staurosporine. Depletion of caspase-3 due to homologous recombination results in excessive accumulation of neuronal cells, due to a lack of apoptosis in the brain, whereas it has no effect in other tissues. This indicates that caspase-3 may be redundant in many cell types (Kuida *et al.*, 1996).

As it is the case with other caspases, the role of caspase-1 in apoptosis is also controversial. It has been suggested that caspase-1 is involved in CD95-mediated apoptosis of thymocytes, in apoptosis of mammary cells following matrix loss, and in DNA damage-induced interferon regulatory factor-1 (JRF-1)-dependent lymphocyte apoptosis (Kuida *et al.*, 1995; Boudreau *et al.*, 1995; Tamura *et al.*, 1995). However, others could not find an impairment of apoptosis in caspase-$1^{(-/-)}$ mice (Li *et al.*, 1995b; Smith *et al.*, 1997) or failed to demonstrate activation of caspase-1 upon CD95 triggering (Muzio *et al.*, 1997). Therefore, either caspase-1 does not play a role in apoptosis signalling through the death receptors, or another caspase-1-like caspase substitutes for its function in different cellular contexts.

In addition to the growing number of caspases, different splice variants of numerous caspases have been reported (Wang *et al.*, 1994; Alnemri *et al.*, 1995, Fernandes-Alnemri *et al.*, 1995; 1996b; Boldin *et al.*, 1996; Wang *et al.*, 1996a; Vincenz and Dixit, 1997). Such splice variants were shown to either function as activators or inhibitors of caspase activation. Some of them might also represent nonfunctional protease species. Interestingly, most of the reported splice variants are only detected at the mRNA level and do not undergo translation. Therefore, the number of isoforms expressed as proteins is limited (Scaffidi *et al.*, 1997).

An increasing number of proteins have been found to be cleaved by caspases, yet the critical apoptosis-relevant substrates are still unknown (reviewed in Cohen, 1997; Nicholson and Thornberry, 1997). Sometimes cleavage results in the activation of a protein, sometimes in its inactivation. Substrates include enzymes involved in genome function, such as the DNA repair enzyme poly(ADP-ribose) polymerase (PARP), DNA-PK, 70 kDa U1, heteronuclear ribonucleoproteins C, and the 140 kDa component of the DNA replication complex. Regulators of the cell-cycle progression including the retinoblastoma protein, the p53 regulator MDM-2 and protein kinase C-δ are also cleaved. Structural proteins of the nucleus and cytoskeleton that are cleaved by caspases include lamins, Gas2, gelsolin and fodrin, a non-erythroid spectrin. Furthermore, it has been found that endonucleolytic DNA cleavage is triggered upon caspase-mediated degradation of the 45 kDa subunit of DNA fragmentation factor (DFF) (Liu *et al.*, 1997).

One of the first death substrates found to be cleaved by caspases, particularly caspase-3 and -7, was PARP. Already before the discovery of caspases, PARP had been implicated in a variety of apoptotic events (Kaufmann *et al.*, 1993). The enzyme catalyzes the transfer of ADP-ribose moieties from NAD to nuclear proteins which may result in protein modification and, following excessive PARP activation, NAD depletion. As DNA strand breaks activate the enzyme, PARP has been proposed to trigger DNA damage-induced apoptosis by depleting intracellular NAD stores. On the other hand, due to its DNA repair activity, PARP may exert a protective function. It has been proposed that during CD95-induced apoptosis proteolytic cleavage of PARP inhibits most of its DNA repair activity, and thus may contribute to the demise of the cell (Tewari *et al.*, 1995). To analyze whether PARP cleavage is a prerequisite for cell death, we have recently investigated CD95 and TNF-R1-mediated apoptosis in $PARP^{(-/-)}$ mice. In a variety of cells and tissues, no significant differences between the apoptosis sensitivity of $PARP^{(-/-)}$ and parental mice were detected (Wang *et al.*, 1997a; Los *et al.*, unpublished results). Therefore, although $PARP^{(-/-)}$ mice have defects in maintaining genomic stability, cleavage of PARP is dispensable for death receptor signal transduction.

Recently, cleavage of the p21-activated kinase, PAK2, during CD95 and TNF-mediated apoptosis has been reported (Rudel and Bokoch, 1997). This cleavage results in a constitutively active kinase. Since PAK2 activates the stress-activated protein kinase pathway, it may provide the link between caspases and JNK/SAPK activation in cells undergoing apoptosis. Interestingly, blocking the activity of PAK2 by a dominant-negative mutant prevents the formation of apoptotic bodies during CD95-mediated apoptosis, whereas nuclear apoptosis as well as phosphatidylserine externalization remain unaffected (Rudel and Bokoch, 1997). This illustrates how different features of apoptosis might be discriminated at the level of caspase targets.

A direct link between caspase-3 activation and DNA fragmentation was found by cloning of a heterodimeric factor, called DFF. It was demonstrated that DFF, when activated after cleavage of its 45 kDa subunit by caspase-3, induces DNA fragmentation on isolated nuclei (Liu *et al.*, 1997). Therefore theoretically, a death receptor signalling pathway may involve only caspases, i.e. CD95 could activate caspase-8 that cleaves caspase-3 which in turn activates DFF.

Despite compelling evidence for a key role of FADD-mediated recruitment of caspase-8 to CD95 and TNF-R1, it should be stressed that at present the precise scenario of receptor-mediated caspase activation is still fragmentary and most experiments have been only performed in a limited number of cell lines. The molecular modifications that control the recruitment process of caspases are not fully understood.

A novel molecular clue was provided by the cloning of Apaf-1 (apoptotic protease-activating factor-1), which represents a mammalian homologue of the *C. elegans* upstream death regulator protein Ced-4 (Zou *et al.*, 1997). In *C. elegans*, Ced-4 physically interacts with both the Bcl-2 homologue Ced-9 and the caspase Ced-3, thus linking the upstream inhibitor and the downstream effector to a multicomponent death complex (Chinnaiyan *et al.*, 1997; Wu *et al.*, 1997). In addition, a direct binding between Ced-4 and mammalian caspase-1 and caspase-8, but not caspase-3 and caspase-8 has been observed. Apaf-1 is a 130 kDa protein that possesses three distinct domains. The C-terminal part of Apaf-1 is composed of 12 putative WD40 repeats, a motif known to mediate protein-protein interactions. This region is followed by a stretch of 320 amino acids that are homologous to Ced-4. The N-terminal region of Apaf-1 shares sequence similarity with the N-terminal domain of Ced-3 and some other mammalian caspases. This domain serves as a so-called caspase-recruitment domain (CARD) by binding to caspases that have a similar CARD motif at their N-terminus (Hofmann *et al.*, 1997). In particular, caspase-9 (Mch6, ICE-Lap6) is recruited to Apaf-1 (Li *et al.*, 1997). In cells not undergoing apoptosis, the CARD is not exposed and therefore not bound to caspase-9. However, binding of ATP and cytochrome c, that is released from mitochondria during early cell death (Liu *et al.*, 1996a), presumably induces a conformational change and unmasks the CARD in Apaf-1. This event finally culminates in the recruitment and activation of caspase-9 at the apoptosome complex.

Similar to caspase-9, also caspase-1 and caspase-2 contain a CARD region. Thus, it is possible that these initiator caspases are recruited to Apaf-1 and may therefore act independently of FADD and other DED-containing proteins. This idea is supported by observations that some anticancer drugs induce caspase activation by a mechanism not requiring CD95/CD95L interaction (Fulda *et al.*, 1997b, Wesselborg *et al.*, submitted). In summary, further studies are necessary to unravel the caspase cascade induced by the different death receptors and to identify crucial targets for caspases that constitute the link between caspase activation and more downstream events in apoptosis.

ALTERNATIVE DEATH SIGNALLING PATHWAYS

Role of non-caspase proteases

During apoptosis various nuclear and cytoplasmic proteins, which are not necessarily substrates for caspases, may be subjected to proteolytic breakdown. It is likely that caspases, once being activated, stimulate a cascade of other proteases resulting in the final demise of the cell. Indeed, the partially protective effect of certain protease inhibitors suggests that, in addition to caspases, other classes of proteases with different target specificity may contribute to the various morphological alterations of apoptosis (Chow *et al.*, 1995; Schlegel *et al.*, 1995).

There are some hints that activation of caspases may be required but not sufficient to cause apoptosis in certain systems. Caspase activation may not necessary lead to apoptosis, as transient activation of caspase-3 during T-cell stimulation with PHA, which is not linked to apoptosis, can be observed (Miossec *et al.*, 1997). Based on the protective effect of certain inhibitors, an involvement of serine proteases has been reported in some forms of apoptosis. In U937 cells, a 24 kDa elastase-like serine protease, called AP24, has been purified that can induce the formation of a DNA ladder when added to isolated nuclei (Wright *et al.*, 1994). The activity of the protease is rapidly activated during apoptosis triggered by UV irradiation and TNF treatment. Interestingly, there are some inhibitors of AP24 activation that do not affect caspase-3, but fully prevent DNA fragmentation and apoptosis (Wright *et al.*, 1997). In addition, bFGF was found to indirectly prevent AP24 activation and cell death, but not PARP cleavage or caspase-3 activity in U937 cells.

The involvement of serine proteases in death receptor pathways has been already suggested in early investigations on the mechanism of TNF-mediated cytotoxicity (Ruggiero *et al.*, 1987; Suffys *et al.*, 1988). In support of the participation of serine proteases in cell death execution, it was demonstrated that overexpression of plasminogen activator inhibitor-2 (PAI-2) prevented apoptosis in HT-180 and HeLa cells (Kumar and Bagglioni, 1991; Dickinson *et al.*, 1995). Interestingly, in TNF-sensitive L929 cells overexpressing CD95, certain inhibitors of serine proteases such as TLCK abolish TNF-, but not CD95-mediated cell death (Vercammen *et al.*, 1997). As will be described in a later section, these observations indicate that in some cell types distinct effector molecules may participate in both pathways.

Based on data obtained in a cloning approach to isolate positive regulators of apoptosis, cathepsin D, a lysosomal cysteine protease, was found and suggested to play a role in cell death mediated by IFNγ, TNF and CD95 (Deiss *et al.*, 1996). In HeLa cells, cell death was inhibited following overexpression of an antisense cathepsin D construct or inactivation of the protease with the inhibitor pepstatin A. Since HeLa cells are normally completely protected against TNF-R1 and CD95-mediated cell death by caspase inhibitors, the relationship of cathepsin D and activation of caspases remains to be demonstrated.

It should be noted that sometimes proteases are not involved in the execution but rather in the initial activation phase of apoptosis. For instance in T lymphocytes, inhibitors of proteasome function block T cell receptor (TCR)-mediated cell death but do not directly interfere with CD95-mediated apoptosis (Grimm *et al.*, 1996a; Cui *et al.*, 1997). It is established that TCR-mediated apoptosis is CD95-dependent and requires the inducible expression of CD95L (Dhein *et al.*, 1995; Brunner *et al.*, 1995; Ju *et al.*, 1995). Some evidence suggests that CD95L expression in response to TCR ligation is controlled by a proteolytic step which involves the transcription factor NF-κB (Ivanow *et al.*, 1997; Cui *et al.*, 1997). Activation of NF-κB requires the proteolytic degradation of its inhibitory subunit IκB at the proteasome which allows active NF-κB to translocate into the nucleus (Schulze-Osthoff *et al.*, 1995). We recently found that the proteasome-specific inhibitor PSI prevents TCR-induced cell death by inhibiting NF-κB-controlled CD95L expression (unpublished results). Thus, proteases may not only control the execution of cell death, but may be involved in signalling events required for the sensitization of cells to a specific apoptotic pathway.

The sphingomyelin pathway

Another apoptotic pathway implicated in death receptor-mediated apoptosis involves the generation of ceramide by the hydrolysis of the phospholipid sphingomyelin. Ceramide is a second messenger produced upon activation of sphingomyelinases (SMases) or via *de novo* synthesis by ceramide synthetase. Two forms of SMases can be distinguished based on their pH optima. Neutral SMase has a pH optimum of 7.4, requires Mg^{2+} ions and is found at the plasma membrane. Acidic SMase has the highest enzymatic activity at pH 5.0, is activated by diacylglycerol and mainly present in endosomes and lysosomes. A multitude of non-apoptotic and apoptotic stimuli can activate sphingomyelin turnover including ionizing irradiation, oxidative stress, treatment with doxorubicin or ligation of TNF-R1 and CD95 (Haimovitz-Friedman *et al.*, 1994, Bose *et al.*, 1995; Jaffrezou *et al.*, 1996; Wiegmann *et al.*, 1994; Cifone *et al.*, 1995; Gulbins *et al.*, 1995; Tepper *et al.*, 1995). Ceramide generated as a result of sphingomyelin turnover in turn can stimulate various target molecules, such as ceramide-activated protein kinase (CAP kinase), ceramide-activated protein phosphatase (identical to PP2A), the protein kinase C isoform ζ, and Raf-1. A specific role for ceramide in mediating apoptotic signals was also suggested by the apoptotic effect of exogenous short-chain ceramides or the treatment of cells with bacterial SMase.

TNF-R1 has been shown to activate neutral SMase through FAN (factor-associated neutral SMase), a protein that interacts with a stretch of 9 amino acids upstream of the DD. A dominant-negative mutant of FAN is able to block TNF-R1-mediated neutral SMase activation completely without affecting cell death (Adam-Klages *et al.*, 1996). Recent data from two different groups, however, indicate that all functions exerted by TNF-R1 require a functional DD. First, overexpression of a trimerized TNF-R1 DD was sufficient to induce apoptosis and to activate NF-κB (Vandevoorde *et al.*, 1997). Second, knock-out mice expressing a TNF-R1 transgene lacking the 30 terminal amino acids have the same phenotype as TNF-R1$^{-/-}$ mice. They are resistant to endotoxic shock and susceptible to *Listeria* infection (Pfeffer *et al.*, 1993; T. Plitz and K. Pfeffer, personal communication). All these data suggest that the juxtamembrane region of TNF-R1 is not required for the main functions of the receptor. Hence, the relevance of FAN binding and activation of neutral SMase needs to be shown. Similar to TNF-R1, mutant CD95, which is defective in death signalling, is still able to activate neutral SMase (Cifone *et al.*, 1995). Therefore, neutral SMase mediated ceramide production is independent of cell death signalling by CD95 and TNF-R1.

Ceramide production by acidic SMase is mediated through the prior activation of the phosphatidylcholine-specific phospholipase C (PC-PLC). The region of TNF-R1 which initiates the PC-PLC/acidic SMase pathway corresponds to the DD of TNF-R1. The xanthogenate compound D609 inhibits this pathway and is able to prevent TNF-induced cell death in various cell types (Machleidt *et al.*, 1996). However, it has been found that cells from patients with Niemann-Pick disease type A, which lack functional acidic SMase, are resistant to ionizing irradiation, but not to CD95- or TNF-R1-induced apoptosis (Santana *et al.*, 1996). Therefore, although both neutral and acidic SMase have been implicated in ceramide production and death signalling through CD95 and TNF-R1, neither of them seems to be essential or sufficient for apoptosis induction by these receptors.

Another metabolizing pathway for ceramide was recently proposed by Testi and coworkers to play a role in apoptosis (De Maria *et al.*, 1997). Ceramide can be shuttled to the Golgi complex where it is converted to gangliosides. In myeloid and lymphoid cells it was found that CD95 ligation or treatment with ceramide resulted in the accumulation of the ganglioside GD3, an event, which was inhibited by caspase inhibitors. Moreover, antisense oligonucleotides to GD3 synthetase, which resides in the Golgi complex, attenuated apoptosis, whereas overexpression of wild-type enzyme was associated with massive cell death. The authors suggested that during CD95-mediated apoptosis, GD3 ganglioside may be targeted to mitochondria where it alters mitochondrial function and causes cell death.

It should be stressed that there is currently much confusion about the role of endogenous ceramide in apoptosis. Whereas some publications place ceramide production upstream of caspases, others suggest that it acts downstream of caspases, as it can be blocked by caspase inhibitors, such as CrmA, zVAD or DEVD (Dbaibo *et al.*, 1997; Sillence and Allan, 1997; Gamen *et al.*, 1996). Moreover, ceramide production may be not necessarily linked to apoptosis, as it is also observed after Ca^{2+} ionophore treatment (Sillence and Allan, 1997) without being associated with cell death. A possible reason for the discrepancy on the functional role of ceramides may lie in methodological problems. Ceramide production is generally determined in assays using bacterial DAG kinase. In a recent investigation in T lymphocytes, no ceramide production in response to CD95 ligation could be detected by mass spectroscopy, whereas under the same conditions an apparent increase of ceramides was measured by the classical DAG kinase assay (Watts *et al.*, 1997). It was suggested that lysates from apoptotic cells may stimulate DAG kinase activity directly, which then may falsely reflect an increase in ceramide production. Thus, whether sphingomyelin hydrolysis is functionally involved in the propagation of death signals or rather represents a secondary modulatory pathway has to await careful reexamination.

Stress-activated protein kinases

A multitude of noxious stimuli lead to the activation of two related signalling pathways which center on two MAP kinase homologs, called the stress-activated protein kinase (SAPKs), also known as Jun N-terminal kinase JNK, and p38 which is the mammalian counterpart of yeast HOG1 (reviewed in Cahill *et al.*, 1996a; Su and Karin, 1996). Known targets of these kinases include mostly transcription factors such as ATF-2, c-Jun or JunD which become activated after cell toxin exposure. Because many of the stress stimuli activating these kinases such as UV irradiation, heat shock and protein synthesis inhibitors, often also induce apoptosis, this hinted at the possibility that SAPK/JNK and p38 may be involved in the transmission of the death signal (Derijard *et al.*, 1994; Hibi *et al.*, 1993; Zanke *et al.*, 1996; Meier *et al.*, 1996).

There are several reports linking apoptosis induced by CD95 and TNF-R1 ligation or other stress stimuli to the activation of SAPK/JNK and/or p38. Both CD95 and TNF-R1 are able to increase the activity of the kinases, although compared to many other stress stimuli activation by CD95 is usually delayed. Inhibition of p38 activity was found to be unable to prevent TNF-induced cell death of fibroblasts (Beyaert *et al.*, 1996). Moreover, when the relationship between SAPK/JNK activation in response to TNF-R1 ligation, activation of NF-κB and induction of apoptosis was studied, TRAF2 and RIP were found to be involved in SAPK/JNK and NF-κB activation but not apoptosis, whereas dominant-negative FADD inhibited apoptosis but not kinase activation (Natoli *et al.*, 1997; Liu *et al.*, 1996b). Thus, dominant-negative TRAF2 and FADD mutants clearly dissociate SAPK/JNK activation from induction of apoptosis. The cytotoxic signal of TNF-R1, mediated through TRADD/FADD, and the activation of SAPK/JNK, mediated through TRAF2 and RIP, are therefore two separate pathways that bifurcate at the level of receptor-associated molecules.

CD95 can activate SAPK/JNK and p38, although TRAF2 is not associated with this receptor (Cahill *et al.*, 1996b; Latinis *et al.*, 1996; Goillot *et al.*, 1997). SAPK/JNK activation has been located downstream of caspases in the CD95 pathway, since it can be blocked by the caspase inhibitors zVAD and CrmA (Cahill *et al.*, 1996b; Juo *et al.*, 1997). In addition, SEK1, an upstream activator kinase of SAPK/JNK, is able to inhibit SAPK/JNK activation when expressed as a dominant-negative mutant without affecting CD95-mediated apoptosis (Lenczowski *et al.*, 1997). This again suggests that the pathway of SAPK/JNK activation is independent from apoptosis induction.

At present, there are contradictory results placing SAPK/JNK activation right into the apoptotic pathway. A recent study identified a novel upstream activator kinase of the MAP kinase pathway, termed ASK1 (apoptosis signalling kinase-1) (Ichijo *et al.*, 1997). The enzyme is assumed to contribute to TNF-mediated cytotoxicity, because a kinase-dead ASK-1 mutant inhibited TNF-induced apoptosis. Some studies further revealed that certain upstream elements of the different MAP kinase cascades are targeted and cleaved by caspases. As mentioned above, p21-activated kinase, PAK2, is cleaved during CD95 and TNF-mediated apoptosis leading to a constitutively active kinase. Overexpression of a dominant-negative PAK2 mutant resulted in inhibition of the formation of apoptotic bodies, whereas other signs of apoptosis remained unaffected (Rudel and Bokoch, 1997). In addition, MAP kinase kinase-6b (MKK6b), an upstream mediator of p38 and SAPK/JNK activation, was found to be activated in a caspase-dependent manner and to be necessary for CD95-mediated apoptosis in Jurkat cells (Huang *et al.*, 1997). In contrast, also a protecting effect of upstream components of MAP kinase cascades has been reported. Thymocytes deficient in SEK1 were found to be more sensitive towards CD95 and anti-CD3-induced apoptosis, whereas apoptosis induced by other environmental stresses was unaffected (Nishina *et al.*, 1997). Finally, a direct link between death receptor signalling and activation of SAPK/JNK pathway was identified through the cloning of Daxx, a protein that interacts with the DD of CD95 and leads to caspase-independent SAPK/JNK activation. It was found that a dominant-negative SEK1 mutant was able to block both SAPK/JNK activation and cell death in certain cells (Yang *et al.*, 1997a). Therefore, a secondary apoptotic pathway may exist in certain cells which is dependent on activation of MAP kinases and may cooperate with the caspase cascade. However, the relevance of this pathway in biological systems needs to be established.

Reactive oxygen intermediates: necrosis versus apoptosis

Cells die by one of the two mechanisms, necrosis or apoptosis, that can be distinguished by biochemical and morphological criteria (Farber, 1994). While triggering of a given death receptor will lead to apoptosis in most cells, there are some conditions and cell types where death receptors clearly mediate necrotic cell death. Necrosis is often referred to as accidental cell death and is induced when the plasma membrane of a cell is irreversibly damaged. Biochemically, these alterations seem to be less regulated than apoptosis, and a number of pathways have been implicated in necrosis, including generation of reactive oxygen intermediates (ROIs), activation of phospholipases, perturbation of calcium homeostasis, and unspecific DNA and protein damage (reviewed in Beyaert and Fiers, 1994).

In both necrosis and apoptosis, mitochondria play obviously a critical role, as in both forms of cell death a rapid and dramatic decrease in the mitochondrial membrane potential ($\Delta\Psi m$) is observed (reviewed in Kroemer *et al.*, 1997a). The drop in $\Delta\Psi m$ is due to permeability transition and allows molecules to leak out from the mitochondrial matrix. The exact molecular events causing permeability transition are not known, but it is believed that the decrease in $\Delta\Psi m$ represents the "point of no return" in a death pathway. In cells treated with apoptogenic agents, anti-apoptotic members of the Bcl-2 family that are localized at the mitochondrial membrane prevent permeability transition and the release of cytochrome c, thereby keeping the Ced-4 homologue Apaf-1 in an inactive state. During necrotic cell death, membrane permeability transition may lead to increased radical production, which in turn will cause cell damage through the oxidation of lipids, proteins and other components. This common occurrence of mitochondrial alterations, such as permeability transition, in necrosis and apoptosis indicates that some signalling processes might be shared between the two forms of cell death. Although it is not entirely clear which event decides whether a cell undergoes apoptosis or necrosis, the supply with ATP and other energy equivalents are likely determinants in the process (Tsujimoto, 1997; Leist *et al.*, 1997).

Among the death ligands, at least TNF has been reported to be able to induce apoptosis and necrosis (Laster *et al.*, 1988). A necrotic cell death is exemplified for instance by TNF-treated L929 fibroblasts, which are often used as the prototype of TNF-sensitive cells. There is substantial evidence that during TNF-induced necrosis of fibroblasts mitochondria-derived ROIs are the critical mediators of cell death. Already early studies showed that TNF treatment caused ultrastructural abnormalities of mitochondria, as they appeared swollen and contained fewer cristae (Matthews and Neale, 1987; Schulze-Osthoff *et al.*, 1992). Furthermore, when cells were treated with certain antioxidants or kept under anaerobic conditions, where no or less ROIs are produced, TNF cytotoxicity was strongly reduced (Schulze-Osthoff *et al.*, 1992). Pharmacological experiments revealed that the mitochondrial respiratory chain was the major source of TNF-induced ROI formation. TNF cytotoxicity was strongly inhibited by rotenone and amytal, two drugs which inhibit the electron transfer at the level of complex I, thereby preventing ROI formation at the distally located ubiquinone (Schulze-Osthoff *et al.*, 1992,

1993). In addition, it was observed that L929 cell clones, which had been selected for the depletion of mitochondrial DNA (mt-DNA) and therefore lacked mitochondrial respiration, were almost completely resistant to TNF-induced cytotoxicity (Schulze-Osthoff *et al.*, 1993). Collectively, these evidences suggest that ROIs generated in the mitochondrial electron transport chain play an important role for TNF-induced necrosis.

Interestingly, cells that are devoid of mt-DNA and hence of a functional respiratory chain can still undergo apoptosis, for instance induced by staurosporine treatment or CD95 ligation (Jacobson *et al.*, 1993). This can be explained by the fact that mt-DNA-deficient cells still maintain a mitochondrial membrane potential and thus may release cytochrome c or other factors that can engage the apoptotic machinery. Thus, although in some cells TNF-R1 and CD95-mediated cell death appears to involve similar signalling pathways, there are also examples demonstrating that death induction by the two death receptors must be different.

Differences between TNF-R1 and CD95 signalling are sometimes evident even within the same cell. L929 fibroblasts overexpressing human CD95 exhibit typical alterations of apoptosis when stimulated with anti-CD95, such as membrane blebbing, cytoplasmic shrinkage and internucleosomal DNA fragmentation. In contrast, in the same cells, TNF induces necrosis as evident by changes in the mitochondrial ultrastructure and lack of nuclear apoptotic alterations (Schulze-Osthoff *et al.*, 1994; Vercammen *et al.*, 1997). Differences between both receptors are also observed when the effects of pharmacological inhibitors are analyzed. While mitochondrial inhibitors or antioxidants almost completely block TNF-R1-mediated cell death, inhibition of ROI production does not affect CD95-mediated cytotoxicity (Schulze-Osthoff *et al.*, 1994; Hug *et al.*, 1994). The different drug effects may be related to the distinct morphological forms of cell death that are induced upon CD95 and TNF-R1 in certain cell types.

There are other differences between TNF-R1 and CD95-mediated signal transduction. As described above, TNF is a potent inducer of transcription factor NF-κB and pro-inflammatory gene expression, whereas the biological function of CD95 is largely restricted to apoptosis. As ROIs have been proposed as key second messengers of NF-κB activation (reviewed in Schulze-Osthoff *et al.*, 1995), the lack of NF-κB activation by CD95 concurs with the notion that CD95 signal transduction is ROI-independent. Another difference between CD95 and TNF-R1 relies in the time course of killing induced by both death receptors. In most cell lines, CD95 triggers death very rapidly, in line with the instant recruitment of caspase-8 to the death receptor complex. In contrast, TNF-R1-mediated cell death proceeds slower, as there is mostly a delay of several hours between receptor ligation and the initial signs of cell death. The reasons for the different kinetics are unknown, but they are inconsistent with the idea of similar death inducing complexes of the two receptors. It has been further observed that a number of cell lines are only sensitive to either CD95 or TNF-R1, although both receptors are expressed at similar amounts (Wong and Goeddel, 1994; Grell *et al.*, 1994).

While ROIs may be selectively involved in death pathways leading to necrosis, caspases may be restricted to apoptotic alterations of cell death. Under certain circumstances, such as depletion of ATP by the drug oligomycin, apoptosis can be shifted to necrosis (Leist *et al.*, 1997). Furthermore, in some cells, caspase inhibitors potently block the appearance of apoptotic alterations, but the final cell death is not stopped. For instance, we recently observed that in the presence of caspase inhibitors CD95 ligation induces necrosis instead of apoptosis in some fibroblast cell lines (Los *et al.*, submitted). This may be explained by the fact that under these conditions cytochrome c is still released, $\Delta\Psi m$ drops, ROIs are produced, and consequently cells die by a caspase-independent mechanism. It should be noted that these findings may also constrain the potential therapeutic use of caspase inhibitors under certain conditions of cell death.

CELLULAR ANTI-APOPTOTIC MECHANISMS

Whether triggering of a death pathway results in cell death, not only depends on the expression level of a death receptor and effector molecule, but also on resistance mechanisms that counteract an apoptogenic signal. Often apoptosis is enhanced by inhibitors of protein synthesis indicating that cells produce short-lived proteins that would normally prevent cell death. Thus, the balance between destructive apoptotic signals and protective mechanisms determines the outcome of a given death stimulus. Among many other functions, apoptosis serves as a important defense mechanism of the organism to combat viral infections and to prevent virus spreading (Collins, 1995; Teodoro and Branton, 1997). It is therefore not suprising to find that viruses have developed their own or adopted the host's mechanisms to suppress apoptosis. The identification of viral anti-apoptotic genes led in many cases to the subsequent discovery of their cellular homologues that act to prevent cell death. In the following section, we will describe various mechanisms that interfere with cell death at very distinct steps in the signal transduction pathway of death receptors.

Receptor-associated mechanisms

The most proximal step to suppress a death receptor pathway is of course the inhibition of ligand binding. Members of the death receptor family can sometimes be found as truncated, soluble forms of the extracellular domain, which are either derived from alternative gene splicing or from proteolytic shedding of the receptor molecule. It has been proposed that expression of the soluble extracellular part of CD95 is enhanced and may account for defective apoptosis and development of systemic lupus erythematosus (SLE) (Cheng *et al.*, 1994), though this finding could not be confirmed by other investigators (Mysler *et al.*, 1994; Knipping *et al.*, 1995). A very interesting mechanism of neutralizing a death ligand is found in the TRAIL system. The membrane expression of the specialized decoy receptors, DcR1 and DcR2, which binds TRAIL without signalling for cell death, is held responsible for the resistance of normal cells to TRAIL cytotoxicity (Pan *et al.*, 1997b; Sheridan *et al.*, 1997; Degli-Eposti *et al.*, 1997; MacFarlane *et al.*, 1997).

A further downstream level, at which induction of apoptosis can be prevented, is the signalling activity of a death receptor. For CD95, a negative regulatory role for the C-terminus of the receptor has been suggested, since deletion of its last 15 amino acids increases the sensitivity towards CD95-induced apoptosis (Itoh and Nagata, 1993). By interaction cloning and *in vitro* binding studies, this region of CD95 has been found to interact with a protein tyrosine phosphatase, called Fas-associated phosphatase-1 (FAP-1) (Sato *et al.*, 1995). Overexpression of FAP-1 partially inhibits CD95-induced apoptosis, while its expression inversely correlates with the sensitivity in T helper cell subsets (Zhang *et al.*, 1997). The region of CD95 that is required for interaction with FAP-1 has recently been narrowed down to the last 3 amino acids, and microinjection of this tripeptide into cells, which blocks binding of FAP-1 to CD95 *in vitro*, facilitates CD95 signalling (Yanagisawa *et al.*, 1997). These results are consistent with a negative regulatory role for FAP-1 in CD95 signalling. Yet, so far conclusions are mainly based on correlations and only an association of FAP-1 with human but not mouse CD95 has been detected (Cuppen *et al.*, 1997). Together with the markedly different expression patterns of FAP-1 and CD95, it is unlikely that FAP-1 plays a key role in CD95 signal transduction.

Recently, it was suggested that signal transduction of TNF-R1 and CD95 may be modulated by a pathway related to ubiquitination. A novel protein, called sentrin, was isolated in a two-hybrid screen and found to bind to the DD of CD95 and TNF-R1 (Okura *et al.*, 1996). Sentrin exhibits homology to ubiquitin but itself does not contain a DD. When overexpressed, sentrin provides protection against TNF and CD95-mediated apoptosis. Sentrin associates strongly with the ubiquitin-conjugating enzyme UBC9 (Gong *et al.*, 1997) that was found to be also associated with CD95 (Becker *et al.*, 1997). It is believed that UBC9 catalyzes the conjugation of ubiquitin or sentrin to other molecules. Whether this process, called sentrinization, inhibits cell death by blocking binding of DD adaptor proteins or other molecules remains to be elucidated.

FLICE (caspase-8) inhibitory proteins

When it became evident that caspase-8 contains two regions at its N-terminus that shared sequence homology with the DED of FADD, this information was used to search for other regulatory proteins containing DED motifs. In screening EST databases, a family of proteins encoded by different viral genomes with homology to the DEDs of FADD and caspase-8 was identified. Members of this family are the E8 protein of equine herpesvirus 2, the ORF71 of human herpesvirus 8 and herpesvirus saimiri, a protein encoded by the bovine herpesvirus 4 and two genes (ORF159L and ORF160L) of human moluscipoxvirus (reviewed by Peter *et al.*, 1997a). These proteins were called v-FLIPs (viral FLICE inhibitory proteins), because, following overexpression in mammalian cells, they bind to the death receptor-FADD complex and prevent caspase-8 recruitment and DISC formation (Thome *et al.*, 1997; Hu *et al.*, 1997a; Bertin *et al.*, 1997). v-FLIPs have a unique structure, as they consist of two DEDs without bearing a caspase-like catalytic domain. v-FLIPS can thereby act as dominant-negative proteins and compete for caspase-8 recruitment to the DISC. Consequently, overexpression of v-FLIPs results in protection against CD95-, TNF-R1-, and TRAIL receptor-induced apoptosis (Thome *et al.*, 1997; Hu *et al.*, 1997a; Bertin *et al.*, 1997). v-FLIPs are the first known anti-apoptotic viral proteins that interfere with the most proximal signalling events of death receptors. For the herpesvirus saimiri (ORF71)-FLIP, it was shown that the protein is expressed late in the lytic cycle and thereby renders the cells resistant to CD95. Therefore, v-FLIPs may protect infected cells from premature apoptosis induced by viral overload (Thome *et al.*, 1997).

Shortly after the discovery of v-FLIPs, a cellular homologue, called c-FLIP, was found that presumably acts as a principal death regulator in mammalian cells (Irmler *et al.*, 1997). c-FLIP was independently identified by several groups and is also termed Casper, I-FLICE, FLAME-1, CASH, CLARP, and MRIT (Shu *et al.*, 1997; Hu *et al.*, 1997b, Srinivasula *et al.*, 1997, Goltsev *et al.*, 1997; Inohara *et al.*, 1997; Han *et al.*, 1997; reviewed in Wallach, 1997). Its gene was localized to the long arm of chromosome 2 (2q33-34) where also the caspase-8 and caspase-10 genes are clustered. c-FLIP is expressed in a variety of tissues and occurs in two splice variants. Similar to v-FLIP, the short form c-$FLIP_S$ contains two DED motifs at the N-terminus and through them it can bind to other DED-containing proteins. The long form $FLIP_L$, in addition, contains at its C-terminus a region that resembles the proteolytic region in caspase-8 and -10. The most remarkable feature of c-FLIP is that the active-site cysteine is absent and substituted by a tyrosine residue. Therefore c-FLIP can be suggested to be proteolytically inactive.

It was found that c-FLIP can associate with different proteins including FADD, caspase-8 and caspase-3. By some investigators, also an independent association with the TNF-R1-bound protein TRAF2 and with Bcl-x_L was noted (Shu *et al.*, 1997; Han *et al.*, 1997). Transfection of c-FLIP cDNA into mammalian cells had profound effects on cell death. Curiously, quite opposite effects were reported by different groups. Mostly, a protective effect of c-FLIP on TNF-R1, CD95, TRAMP or TRAIL-induced apoptosis was detected (Irmler *et al.*, 1997; Hu *et al.*, 1997; Srinivasula *et al.*, 1997). This concurs with observations made for the viral homologues that FLIPs act as dominant-negative inhibitors that prevent caspase-8 activation. In contrast, other studies reported a marked cytotoxic effect of c-FLIP which was inhibitable by caspase inhibitors (Shu *et al.*, 1997; Han *et al.*, 1997b; Inohara *et al.*, 1997). The pro-apoptotic effect could result from the ability of c-FLIP to activate caspase-8 via homophilic interaction through its corresponding DED. This mechanism would be similar to that previously suggested for the FADD/caspase-8 interaction. Alternatively, since c-FLIP can associate with TRAF2, it was suggested that c-FLIP overexpression might displace the anti-apoptotic proteins c-IAP1 and c-IAP2 from the TNF-R2/TRAF2 complex (Shu *et al.*, 1997). This discrepancy in the observed effects may be also due to the cell line used, and the artificial system of the transfection experiments. It underscores that *in vitro* effects may not necessarily correspond to the function of a protein when expressed in the physiological range within the organism. Because a pro-apoptotic effect of c-FLIP was seen especially following massive overexpression of the protein, the true function of c-FLIP may be restricted to the inhibition of apoptosis. This view also correlates with the amount of c-FLIP expressed in T lymphocytes. c-FLIP is largely expressed during the early stage of T cell activation. In these cells it was shown that caspase-8 is not recruited to the CD95/FADD complex (Peter *et al.*,

1997b). c-FLIP disappears when T cells become susceptible to CD95-induced apoptosis and a functional CD95 DISC is formed. High levels of c-FLIP were also found in melanoma cells and malignant melanoma tumors (Irmler *et al.*, 1997).

The family of IAPs

Inhibitor of apoptosis proteins (IAPs) constitute a family of molecules that are conserved throughout evolution and prevent cell death in several systems. Originally two IAPs, Cp-IAP and Op-IAP, were discovered in baculovirus and found to functionally complement the death inhibitor p35 (Crook *et al.*, 1993; Birnbaum *et al.*, 1994). Recently, Drosophila IAP-like proteins, designated DIAP1 and DIAP2 (DILP/DIAP) were cloned that inhibit cell death in insects (Hayand *et al*; 1995; Liston *et al.*, 1996; Duckett *et al.*, 1996). The first human IAP to be identified was the neuronal apoptosis inhibitory protein (NIAP) which was isolated based on its contribution to the neurodegenerative disorder, spinal muscular atrophy (Roy *et al.*, 1995). Subsequently, four other human IAPs, called c-IAP1, c-IAP2, X-IAP and survivin, have been isolated and demonstrated to counteract cell death (Rothe *et al.*, 1995; Duckett *et al.*, 1996; Liston *et al.*, 1996; Ambrosini *et al.*, 1997). The c-IAP1 and c-IAP2 proteins were originally identified as molecules that are recruited to the cytosolic domain of TNF-R2 via their association with TRAF1 and TRAF2 (Rothe *et al.*, 1995). c-IAP1, in addition, has been shown to be a component of the TNF-R1 complex via its association with TRAF2 (Shu *et al.*, 1996).

The common structural feature of all IAP family members is a Cys/His-rich motif termed the baculovirus IAP repeat (BIR) that is present in one, two or three copies. With the exception of NIAP and survivin, all other IAP family members also contain a zinc finger-like RING domain at their carboxy terminus. The fact that the BIR motif is shared by all members suggest a central role for this domain in mediating cellular protection. The function of the RING domain, however, is elusive. Recently, it was shown that some IAP members, including X-IAP, c-IAP1 and c-IAP2, can bind to and potently inhibit certain caspases, such as caspase-3 and caspase-7 (Deveraux *et al.*, 1997; Roy *et al.*, 1997). Inhibition of these caspases was also found in *in vitro* assays, whereas the activity of the proximal proteases caspase-1, caspase-6 and caspase-8 was not affected. Mutational analysis revealed that the BIR domains were sufficient for the inhibitory effect, though proteins that retained the RING finger domain were more efficient. In contrast to the other human IAPs, NIAP did neither bind nor inhibit caspases (Roy *et al.*, 1997), suggesting that this and perhaps other IAP members may additionally have alternative targets of apoptosis inhibition. A detailed analysis of the mechanism of inhibition revealed that IAPs bind but are not cleaved by the caspases (Roy *et al.*, 1997). This is in contrast to the poxvirus inhibitor CrmA and baculovirus protein p35 that are suicide inactivators which require peptide bond hydrolysis as part of their inhibitory mechanism (Bertin *et al.*, 1996; Xue and Horvitz, 1995; Bump *et al.*, 1995; Komiyama *et al.*, 1994).

At present, the role of c-IAP1/2 recruitment to the TNF receptors is unknown. Inhibition of caspases is independent of TNF-R binding function (Roy *et al.*, 1997), though it is possible that recruitment of IAPs to the TNF-R may promote their interaction with caspases. Alternatively, it has been also suggested that the function of c-IAP2 may be associated with the activation of NF-κB, which is known to play a role of an anti-apoptotic transcription factor (see below). Overexpression of c-IAP2 directly stimulated NF-κB activation and interfered with TNF-induced cell death (Chu *et al.*, 1997). This effect required the RING domain of c-IAP2 indicating that, in addition to caspase inhibition, IAPs may prevent apoptosis by alternative mechanisms.

The Bcl-2 family of proteins

Bcl-2-related proteins constitute another important decisional point of cell death. The Bcl-2 family consists of two functional classes of proteins including anti-apoptotic members, such as Bcl-2, Bcl-xL, Mcl-1, Bcl-w, Bfl-1, Brag-1 and A1, as well as pro-apoptotic molecules, such as Bax, Bad, Bak, Bid, Bik and Hrk (reviewed by Otvai and Korsmeyer, 1994; Reed, 1997; Kroemer, 1997b). Members of the Bcl-2 family are characterized by up to four conserved regions, termed the Bcl-2 homology (BH) domains. Many of these proteins interact with each other through a complex network of homo- and heterodimers. Originally, it has been proposed that the relative ratio of pro- and anti-apoptotic members dictates whether a cell will respond to a proximal apoptotic stimulus or not. Indeed, overexpression of Bcl-2/Bcl-x_L abrogates cell death induced by a multitude of apoptotic stimuli (Cory, 1995, Boise *et al.*, 1993), while enforced expression of Bax promotes apoptosis (Oltvai *et al.*, 1993).

Recent evidence, however, has challenged and at least partially invalidated the model of heterodimerization of Bcl-2 proteins. It has been proposed that heterodimerization may be due to the presence of non-ionic detergents in the dimerization assays and that the family members might not interact *in vivo* (Hsu *et al.*, 1997b). In accordance, selected Bcl-x_L mutants which fail to heterodimerize with Bax can still inhibit apoptosis (Cheng *et al.*, 1996; Clair *et al.*, 1997). *Vice versa*, mutants of Bax and Bak which do not bind Bcl-x_L still exert their pro-apoptotic activity (Simonian *et al.*, 1996, 1997). It therefore seems that these Bcl-2 family members instead of heterodimerizing compete for a common binding partner.

A variety of models have been proposed to explain how Bcl-2-related proteins inhibit apoptosis. Bcl-2 seems do prevent all alterations of apoptosis that occur at the level of mitochondria. Thus, Bcl-2-inhibitory proteins have been shown to prevent the formation of mitochondrial ROIs, alterations in calcium homeostasis, and mitochondrial membrane permeability transition. Interesting findings also came from cell-free experiments. *In vitro*, Bcl-2 prevents the release of an apoptogenic protease, called apoptosis-inducing factor (AIF), by isolated mitochondria (Susin *et al.*, 1996). In addition, Bcl-2 blocks the redistribution of cytochrome c which released in early stages of apoptosis (Kluck *et al.*, 1997; Yang *et al.*, 1997b). These observations therefore suggest that the outer mitochondrial membrane is the principal site of action of Bcl-2.

Recent studies demonstrated that after transient transfection Bcl-x_L can bind to caspase-8, suggesting that this interaction would inhibit activation of caspase-8

(Chinnaiyan *et al.*, 1997). In contrast, in MCF7-Fas cells, which were selected for CD95-resistance by overexpression of Bcl-x_L, no association between caspase-8 and Bcl-x_L was detected (Medema *et al.*, 1997b). Likewise, activation of caspase-8 in response to CD95 ligation was unaffected. However, although DISC-bound caspase-8 may not be targeted by Bcl-2, the processing of downstream caspases including caspase-3 is prevented by Bcl-2 or Bcl-x_L in a number of apoptotic systems (Monney *et al.*, 1996; Boulakia *et al.*, 1996). Thus Bcl-2-related proteins may inhibit cell death by acting downstream of initiator but upstream of executioner caspases. Experiments *in vitro* demonstrated that Bcl-2 can interact with Apaf-1, a mammalian Ced-4 homologue. It is possible that Bcl-2 thereby sequesters the Apaf-1 complex at the outer mitochondrial membrane. Studies with isolated mitochondria revealed that Bcl-2 inhibits the release of the mitochondrial protease AIF and cytochrome c which otherwise, upon binding to Apaf-1, would activate the caspase cascade (Kluck *et al.*, 1997; Yang *et al.*, 1997a; Susin *et al.*, 1996).

An interesting clue to the mechanism of action of Bcl-2 proteins may be provided by the three-dimensional structure of Bcl-x_L. Its X-ray and NMR structure reveals a similarity to the pore-forming domains of bacterial toxins such as diphteria toxin and colicins (Muchmore *et al.*, 1996). Indeed, both Bcl-2 and Bcl-x_L have been shown to form channels in artificial lipid bilayers with a selectivity for K^+ (Schendel *et al.*, 1997; Minn *et al.*, 1997). Different Bcl-2 family members might form channels in mitochondria with distinct selectivity which could regulate permeability of the outer mitochondrial membrane. In this context, it has been shown that Bcl-2 inhibits mitochondrial swelling during apoptosis that is presumably caused by a damage of the outer mitochondrial membrane (Vander Heiden *et al.*, 1997). This osmotic swelling precedes redistribution of cytochrome c and membrane permeability transition.

It is still very controversial whether Bcl-2 affects apoptosis induced by CD95 or other death receptors. From several reports it appears that there are certain cell types where Bcl-2 or related proteins provide potent protection against CD95-induced apoptosis (Itoh *et al.*, 1993; Mandal *et al.*, 1996), while there are other cells in which Bcl-2 is completely ineffective (Vanhaesebroeck *et al.*, 1993; Memon *et al.*, 1995). Since CD95-mediated caspase-8 activation at the DISC level is not inhibited by Bcl-x_L, execution of apoptosis could be independent of the release of cytochrome c and other pro-apoptotic mitochondrial components. In contrast, Bcl-2 may inhibit cell death in such cells, where execution of apoptosis depends on Apaf-1 and cytochrome c-induced activation of executioner caspases. It will be interesting to investigate, how the FADD/caspase-8 pathway is connected to mitochondria and whether caspase-8 can directly cause mitochondrial swelling, for instance by cleavage of proteins at the outer mitochondrial membrane.

Important progress has been obtained how extracellular signals prevent cell death at the level of Bcl-2 proteins. Survival factors such as the cytokine IL-3 induce the phosphorylation of the pro-apoptotic protein Bad (Zha *et al.*, 1996). This phosphorylation of Bad occurs at two serine residues which are embedded in the binding site of the phosphoserine-binding protein 14-3-3. As a consequence, phosphorylated Bad is bound to 14-3-3 and retained in the cytosol, but cannot interact with Bcl-x_L and exert its pro-apoptotic function at the mitochondrial membrane. Two protein kinases, Raf-1 and Akt/PKB, have been implicated in phosphorylating Bad. Sequence comparison of Bcl-2 and other anti-apoptotic members reveals a domain (BH4) which is not present in the pro-apoptotic members. This domain interacts with Raf-1 and thereby targets the kinase to outer mitochondrial membrane (Wang *et al.*, 1996b). Raf-1 in turn can phosphorylate the pro-apoptotic protein Bad. Raf-1-mediated phosphorylation of Bad prevents Bad from binding to Bcl-2 and Bcl-x_L, thus relieving repression of these anti-apoptotic proteins by allowing them to homodimerize with themselves. In addition, there is strong evidence that the phosphoinositide-3 kinase (PI3K) pathway is the essential component in growth factor-mediated protection *in vivo*. The lipids produced by PI3K bind to and activate the serine/threonine kinase Akt/PKB. Activated forms of Akt/PKB have been shown to protect cells from apoptosis, for instance induced by growth factor withdrawal or by detachment of adherent cells from their extracellular matrix (anoikis) (reviewed in Marte and Downward 1997). It has been demonstrated that Akt/PKB phosphorylates Bad (Datta *et al.*, 1997; del Peso *et al.*, 1997). Upon phosphorylation, Bad dissociates from Bcl-x_L which is then free to resume its activity as a suppressor of cell death. Very recently, it has been reported that CD95-mediated apoptosis is inhibited by overexpression of Akt/PKB (Häusler *et al.*, 1998).

Other anti-apoptotic mechanisms

Cells contain a variety of other mechanisms providing protection against cell death. Heat shock proteins (hsps) constitute a group of proteins that based on their molecular size are divided into small hsps (shsp), hsp60, hsp70, and hsp90 subfamilies (reviewed in Jäättelä and Wissing, 1992). All hsps are rapidly induced by heat shock treatment and other stress stimuli. Their principal function seems to be to act as molecular chaperones in the process of protein folding. Overexpression of hsps was shown to enhance survival of cells exposed to numerous injuries that lead to death including heat shock, oxidative stress, treatment with anti-cancer drugs and other apoptosis-inducing agents. However, the molecular mechanism how hsps protect against apoptosis, is largely undefined at present. It has been proposed that hsps may counteract stress-induced disruption of the microfilament network which often occurs during apoptosis (Lavoie *et al.*, 1993). Overexpression of hsp70 protects against TNF-induced apoptosis (Jäättelä, 1993). Whether this effect is associated with the attenuation of TNF-induced phospholipase A_2 activation remains to be shown. Likewise, overexpression of shsps confers enhanced resistance to apoptosis induced by oxidative stress, TNF, CD95, staurosporine and anticancer drugs (Mehlen *et al.*, 1996a; 1996b). It has been hypothesized that shsps protect cells by inhibiting the action of ROIs. Though no *in vitro* detoxificant activity was found to be associated with shsps, overexpression led to an increased cellular content of the antioxidant glutathione and decrease in lipid peroxidation.

The Zn-finger protein A20, another anti-apoptotic protein, was first identified by differential screening of a cDNA library from TNF-activated endothelial cells (Opipari *et al.*, 1992). Gene expression of A20 is tightly regulated by transcription factor NF-κB that binds at two κB-binding sites within the A20 promoter (Krikos *et al.*, 1992; Laherty *et al.*, 1992). Overexpression of A20 strongly attenuates TNF-induced NF-κB activation and cytotoxicity. However, the ability to protect against apoptosis seems to be largely restricted to the kind of apoptotic stimulus. While protection against TNF-induced apoptosis has been shown in several cell types, no protection could be demonstrated against cell death triggered by serum depletion, oxidative stress or CD95 (Opipari *et al.*, 1992; Jäättelä *et al.*, 1996). It is likely that this selectivity is conferred by the specific molecular mechanism of A20 action. Though the biochemical mechanism of action of A20 is unknown, it has been found that A20 interacts with the TNF-receptor associated proteins TRAF1 and TRAF2 (Song *et al.*, 1996). How A20 inhibits apoptosis, which, as mentioned above, is not mediated by TRAF1/TRAF2 but by TRADD, remains to be determined.

NF-κB as an anti-apoptotic transcription factor

Many cell types are resistant to apoptosis induced by death receptors, but become sensitive in the presence of inhibitors of RNA and protein synthesis. This observation can be explained by the fact that triggering of a death receptor not only mounts an apoptotic response, but simultaneously initiates a genetic programme that serves to block cell death and so sets up a delicate life-death balance. Transcription factor NF-κB, which was previously known mainly as a central mediator of inflammatory gene expression, has been recently implicated in protecting cells against apoptosis, most likely by inducing the expression of anti-apoptotic genes (Figure 1.3) (reviewed in Baichwal and Baeuerle, 1997). NF-κB consists of a heterodimeric complex which is often composed of the p50 and the p65 kDa RelA subunit (reviewed in Baeuerle and Baltimore, 1996). In its inactive state, the complex is sequestered in the cytosol by a third inhibitory subunit, called IκB. Upon stimulation of cells with TNF or a variety of other inflammatory stimuli, IκB is phosphorylated at its N-terminal serine residues and rapidly degraded at the proteasome. This event in turn releases the active form of NF-κB and allows for nuclear translocation, subsequent DNA binding and activation of NF-κB target genes.

A first hint for a connection between NF-κB and cell death came from studies on knockout mice that lacked the RelA subunit of NF-κB (Beg *et al.*, 1995). These mice died before birth and showed massive degeneration of liver cells caused by apoptosis. It was further demonstrated that fibroblasts and macrophages from RelA-deficient mice are hypersensitive to TNF-induced cytotoxicity, whereas wildtype cells survive this treatment (Beg and Baltimore, 1996). The susceptibility of RelA-deficient cells was reversed following transfection of cells with the wild-type relA gene.

In another approach, it was shown that cells expressing a dominant-negative mutant of IκB were sensitized to TNF and even killed in the absence of protein synthesis inhibitors (Wang *et al.*, 1996c; van Antwerp *et al.*, 1996). Conversely, pretreatment of cells with IL-1ß, a cytokine that activates NF-κB but is not pro-apoptotic, protected cells from TNF-induced apoptosis. An increased resistance conferred by NF-κB was also observed with proapoptotic stimuli other than TNF. The chemotherapeutic drug daunorubicin and ionizing irradiation which also induce NF-κB are more toxic, when NF-κB activation was blocked by a dominant-negative IκB mutant. In addition, a role of NF-κB in preventing apoptosis is evident in

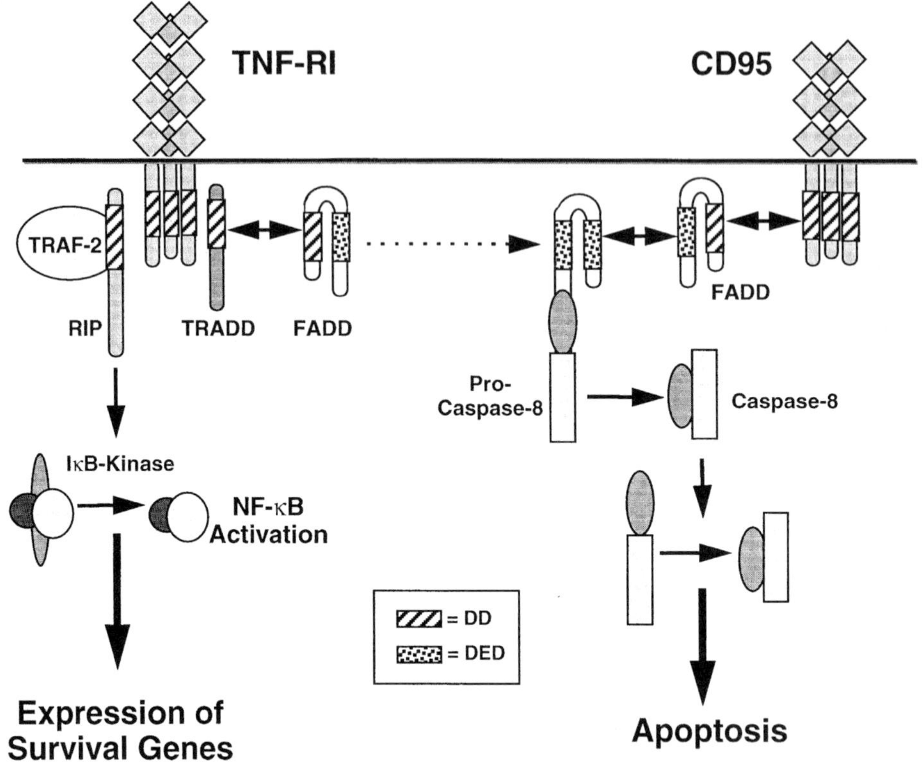

Figure 1.3 Model of TNF-R1 induced signal transduction. TNF-R1 ligation results in two major cellular responses, activation of transcription factor NF-κB and induction of cell death. NF-κB activation by TNF-R1 involves recruitment of the adapter proteins TRAF-2 and RIP. This event finally results in activation of the multisubunit IκB kinase which phosphorylates the NF-κB inhibitor IκB. Phosphorylated IκB dissociates from the inactive NF-κB heterodimer and allows NF-κB to translocate to the nucleus and to activate gene expression of survival genes. The cytotoxic response of TNF-R1 ligation is mediated by the DD-containing protein TRADD which can interact with FADD and therefore allow for a potential crosstalk between the TNF-R1 and CD95 death receptor pathways.

B-lymphocyte cell lines. Such cell lines express constitutively NF-κB, but inactivation of NF-κB by various means induces apoptosis (Wu *et al.*, 1996). Thus, in B cells constitutive NF-κB has a role in ensuring cell survival.

A general role for NF-κB as a transcription factor that prevents cell death is, however, far from being established. Apoptosis triggered by CD95 is not countered by NF-κB (van Antwerp *et al.*, 1996; unpublished results), presumably because CD95 does not lead to detectable NF-κB activation in most cell types. This may be due to the rapid induction of cell death, as NF-κB activation by CD95 can be restored when apoptosis is inhibited (Ponton *et al.*, 1996). In addition, there are some reports pointing to a pro-apoptotic role of NF-κB. For instance, glutamate-induced toxicity in neurons has been found to be accompanied by NF-κB induction (Grilli *et al.*, 1996). In this case, NF-κB seems to cause cell death, as blocking its activation with aspirin and sodium salicylate protected cells from excitotoxic damage. In another study, apoptotic death induced by serum withdrawal was demonstrated to induce NF-κB activation which was prevented by overexpression of Bcl-2 (Grimm *et al.*, 1996b). Furthermore, Sindbis virus-induced apoptosis in a carcinoma cell line was shown to require NF-κB activation, as antisense NF-κB oligonucleotides prevented this type of cell death (Lin *et al.*, 1995).

The putative anti-apoptotic genes that are activated by NF-κB remain to be identified. Likely candidates are the genes encoding for manganese superoxide dismutase (MnSOD), c-IAP2 and the zinc finger protein A20. Expression of all three genes is induced by TNF. MnSOD, an enzyme that converts superoxide anion to hydrogen peroxide inside mitochondria, may be involved by providing protection against TNF cytotoxicity, in particular necrosis which is largely dependent on ROI formation (Wong *et al.*, 1989). c-IAP2 is an NF-κB-controlled target gene product, which, as described in a previous section, may inhibit apoptosis at the level of either caspases or proximal TNF receptor signal transduction. Also A20 expression is tightly regulated by NF-κB and provides protection against TNF in number of cell types (Opipari *et al.*, 1992; Krikos *et al.*, 1992). Yet, transfection of A20 in RelA$^{(-/-)}$ 3T3 cells does not rescue these cells from TNF-induced apoptosis (Beg and Baltimore, 1996), suggesting that other resistance genes must be also involved. Certainly, the function of NF-κB as a regulatory loop that operates to provide protection against apoptosis, has important implications for tumor therapy. If tumor cell sensitivity is controlled by NF-κB, a combination of NF-κB inhibitors and chemotherapeutic drugs may result in improved efficacy of anticancer therapy.

CONCLUSIONS

This review describes the different death receptors and the pathways that are used either to induce or to counteract apoptosis. Certainly, the rapid and very recent discovery of a great variety of different death receptors with their multitude of control points to either execute or to inhibit cell death proceeds well ahead of our understanding of their role in multicellular systems. Our present view of the different functions of the various death systems is still fairly restricted and it can be anticipated that more death receptors and elements of their signalling pathways will be identified. A major objective of future research will be to study the biology of death systems in the whole organism and to address questions such as why there are so many different death systems, when and how do they operate and what ensures their specificity. The other major challenge is to apply the knowledge of death receptor networks to therapy. Therapeutic targeting of certain death pathways would be beneficial for very diverse diseases including AIDS, hepatitis, neurodegeneration, multiple sclerosis, stroke, myocardial ischemia and others. While the goal of treating acute degenerative diseases is to prevent unwanted cell death, the major challenge in treating cancer is to kill cells that have become resistant to available chemotherapy. Since the failure to undergo apoptosis is often associated with drug resistance, direct activation of physiological death pathways may be an effective strategy to kill tumor cells.

REFERENCES

Abreu-Martin, M.T., Vidrich, A., Lynch, D.H., Targan, S.R. (1995) Divergent induction of apoptosis and IL-8 secretion in HT-29 cells in response to TNF-alpha and ligation of Fas antigen. *J. Immunol.*, **155**, 4147–4154.

Adachi, M., Watanabe-Fukunaga, R., Nagata, S. (1993) Aberrant transcription caused by the insertion of an early transposable element in an intron of the Fas antigen gene of lpr mice. *Proc. Natl. Acad. Sci. USA*, **90**, 1756–1760.

Adam-Klages, S., Adam, D., Wiegmann, K., Struve, S., Kolanus, W., Schneider-Mergener, J., Krönke, M. (1996) FAN, a novel WD-repeat protein, couples the p55 TNF-receptor to neutral sphingomyelinase. *Cell*, **86**, 937–947.

Aggarwal, B.B., Singh, S., LaPushin, R., Totpal, K. (1995) Fas antigen signals proliferation of normal human diploid fibroblast and its mechanism is different from tumor necrosis factor receptor. *FEBS Lett.*, **364**, 5–8.

Ahmad, M., Srinivasula, S.M., Wang, L., Talanian, R.V., Litwack, G., Fernandes-Alnemri, T., Alnemri E.S. (1997) CRADD, a novel human apoptotic adaptor molecule for caspase-2, and FasL/tumor necrosis factor receptor-interacting protein RIP. *Cancer Res.*, **57**, 615–619.

Alderson, M.R., Armitage, R.J., Maraskovsky, E., Tough, T.W., Roux, E., Schooley, K., Ramsdell, F., Lynch, D.H. (1993) Fas transduces activation signals in normal human T lymphocytes. *J. Exp. Med.*, **178**, 2231–2235.

Alderson, M.R., Tough, T.W., Davis Smith, T., Braddy, S., Falk, B., Schooley, K.A., Goodwin, R.G., Smith, C.A., Ramsdell, F., Lynch, D.H. (1995) Fas ligand mediates activation-induced cell death in human T lymphocytes. *J. Exp. Med.*, **181**, 71–77.

Allison, J., Georgiou, H.M., Strasser, A., Vaux, D.L. (1997) Transgenic expression of CD95 ligand on islet ß-cells induces a granulocytic infiltration but does not confer immune privilege upon islet allografts. *Proc. Natl. Acad. Sci. USA*, **94**, 3943–3947.

Alnemri, E.S., Fernandes-Alnemri, T., Litwack, G. (1995) Cloning and expression of four novel isoforms of human interleukin-1ß converting enzyme with different apoptotic activities. *J. Biol. Chem.*, **270**, 4312–4317.

Alnemri, E.S., Livingston, D.J., Nicholson, D.W., Salvesen, G., Thornberry, N.A., Wong, W.W., Yuan, J. (1996) Human ICE/CED-3 protease nomenclature. *Cell*, **87**, 171.

Ambrosini, G., Adida, C., Altieri, D. (1997) A novel anti-apoptosis gene, survivin, expressed in cancer and lymphoma. *Nature Med.*, **3**, 917–921.

Anderson, D.M., Maraskovsky, E., Billingsley, W.L., Dougall, W.C., Tometsko, M.E., Roux, E.R., Teepe, M.C., DuBose, R.F., Cosman, D., Galibert, L. (1997) A homologue of the TNF receptor and its ligand enhance T-cell growth and dendritic cell function. *Nature*, **390**, 175–179.

Aragane, Y., Kulms, D., Metze, D., Wilkes, G., Pöppelmann, B., Luger, T.A., Schwarz, T. (1998) Ultraviolet light induces apoptosis via activation of CD95 (Fas/APO-1) independently from its ligand CD95L. *J. Cell Biol.*, **27**, 557–562.

Baeuerle, P.A., Baltimore, D. (1996). NF-κB: ten years after. *Cell*, **87**, 13–20.

Baichwal, V.R., Baeuerle, P.A. (1997) Activate NF-κB or die? *Curr. Biol.*, **7**, R94–96.

Baker, M.B., Altman, N.H., Podack, E.R., Levy, R.B. (1996) The role of cell-mediated cytotoxicity in acute GVHD after MHC-matched allogeneic bone marrow transplantation in mice. *J. Exp. Med.*, **183**, 2645–2656.

Banner, D.W., D'Arcy, A., Janes, W., Gentz, R., Schoenfeld, H.J., Broger C., Loetscher H., Lesslauer W. (1993) Crystal structure of the soluble human 55 kd TNF receptor-human TNFß complex: implications for TNF receptor activation. *Cell*, **73**, 431–445.

Bazzoni, F and Beutler, B. (1996) The tumor necrosis factor ligand and receptor families. *New Engl. J. Med.*, **334**, 1717–1725.

Becker, K., Schneider, P., Hofmann, K., Mattmann, C., Tschopp, J. (1997) Interaction of Fas (APO-1/CD95) with proteins implicated in the ubiquitination pathway. *FEBS Lett.*, **412**, 102–106.

Beg, A.A., Baltimore, D. (1996) An essential role for NF-κB in preventing TNF-α-induced cell death. *Science*, **274**, 782–784.

Beg, A.A., Sha, W.C., Bronson, R.T., Ghosh, S., Baltimore, D. (1995) Embryonic lethality and liver degeneration in mice lacking the RelA component of NF-κB. *Nature*, **376**, 167–170.

Bellgrau, D., Gold, D., Selawry, H., Moore, J., Franzusoff, A., Duke, R.C. (1995) A role for CD95 ligand in preventing graft rejection. *Nature*, **377**, 630–632.

Bertin, J., Mendrysa, S.M., LaCount, D.J., Gaur, S., Krebs, J.F., Armstrong, R.C., Tomaselli, K.J., Friesen, P.D. (1996) Apoptotic suppression by baculovirus P35 involves cleavage by and inhibition of a virus-induced CED-3/ICE-like protease. *J. Virol.*, **70**, 6251–6259.

Bertin, J., Armstrong, R.C., Ottilie, S., Martin, D.A., Wang, Y., Banks, S., Wang, G.H., Senkevich, T.G., Alnemri, E.S., Moss, B., Lenardo, M.J., Tomaselli, K.J., Cohen, J.I. (1997) Death effector domain-containing herpesvirus and poxvirus proteins inhibit both Fas- and TNFR1-induced apoptosis. *Proc. Natl. Acad. Sci. USA*, **94**, 1172–1176.

Beutler, B., van Huffel, C. (1994) Unraveling function in the TNF ligand and receptor families. *Science*, **264**, 667–668.

Beyaert, R., Fiers, W. (1994) Molecular mechanisms of tumor necrosis factor-induced cytotoxicity. What we do understand and what we do not. *FEBS Lett.*, **340**, 9–16.

Beyaert, R., Cuenda, A., Vanden, Berghe, W., Plaisance, S., Lee, J.C., Haegeman, G., Cohen, P., Fiers, W. (1996) The p38/RK mitogen activated protein kinase pathway regulates interleukin-6 synthesis response to tumor necrosis factor. *EMBO J.*, **15**, 1914–1923.

Birnbaum, M.J., Clem, R.J., Miller. L.K. (1994) An apoptosis-inhibiting gene from a nuclear polyhedrosis virus encoding a polypeptide with Cys/His sequence motifs. *J. Virol.*, **68**, 2521–2528.

Bissonnette, R.P., McGahon, A., Mahboubi, A., Green, D.R. (1994) Functional Myc-Max heteodimer is required for activation-induced apoptosis in T cell hybridomas. *J. Exp. Med.*, **180**, 2413–2418.

Black, R.A., Rauch, C.T., Kozlosky, C.J., Peschon, J.J., Slack, J.L., Wolfson, M.F., Castner, B.J., Stocking, K.L., Reddy, P., Srinivasan, S., Nelson, N., Boiani, N., Schooley, K.A., Gerhart, M., Davis, R., Fitzner, J.N., Johnson, R.S., Paxton, R.J., March, C.J., Cerretti, D.P. (1997) A metalloproteinase disintegrin that releases tumour-necrosis factor-α from cells. *Nature*, **385**, 729–733.

Bodmer, J.L., Burns, K., Schneider, P., Hofmann, K., Steiner, V., Thome, M., Bornand, T., Hahne, M., Schroter, M., Becker, K., Wilson, A., French, L.E., Browning, J.L., MacDonald, H.R., Tschopp, J. (1997) TRAMP a novel apoptosis-mediating receptor with sequence homology to tumor necrosis factor receptor 1 and Fas (Apo-1/CD95). *Immunity*, **6**, 79–88.

Boise, L.H., Gonzalez Garcia, M., Postema, C.E., Ding, L., Lindsten, T., Turka, L.A., Mao, X., Nunez, G., Thompson, C.B. (1993) Bcl-x, a bcl-2–related gene that functions as a dominant regulator of apoptotic cell death. *Cell*, **74**, 597–608.

Boldin, M.P., Varfolomeev, E.E., Pancer, Z., Mett, I.L., Camonis, J.H., Wallach, D. (1995) A novel protein that interacts with the death domain of Fas/APO1 contains a sequence motif related to the death domain. *J. Biol. Chem.*, **270**, 7795–7798.

Boldin, M.P., Goncharov, T.M., Goltsev, Y.V., Wallach, D. (1996) Involvement of MACH, a novel MORT1/FADD-interacting protease, in Fas/APO-1- and TNF receptor-induced cell death. *Cell*, **85**, 803–815.

Bose, R., Verheij, M., Haimovitz-Friedman, A., Scotto, K., Fuks, Z., Kolesnick, R. (1995) Ceramide synthase mediates daunorubicin-induced apoptosis: an alternative mechanism for generating death signals. *Cell*, **82**, 405–414.

Boudreau, N., Sympson, C., Werb, Z. and Bissell, M.J. (1995). Suppression of ICE and apoptosis in mammary epithelial cells by extracellular matrix. *Science*, **267**, 891–893.

Boulakia, C.A., Chen, G., Ng, F.W., Teodoro, J.G., Branton, P.E., Nicholson, D.W., Poirier, G.G., Shore, G.C. (1996) Bcl-2 and adenovirus E1B 19 kDA protein prevent E1A-induced processing of CPP32 and cleavage of poly(ADP-ribose) polymerase. *Oncogene*, **12**, 529–535.

Braun, M.Y., Lowin, B., French, L., Acha Orbea, H., Tschopp, J. (1996) Cytotoxic T cells deficient in both functional fas ligand and perforin show residual cytolytic activity yet lose their capacity to induce lethal acute graft-versus-host disease. *J. Exp. Med.*, **183**, 657–661.

Brunner, T., Mogil, R.J., LaFace, D., Yoo, N.J., Mahboubi, A., Echeverri, F., Martin, S.J., Force, W.R., Lynch, D.H., Ware, C.F., *et al.* (1995) Cell-autonomous Fas (CD95)/Fas-ligand interaction mediates activation-induced apoptosis in T-cell hybridomas. *Nature*, **373**, 441–444.

Bump, N.J., Hackett, M., Hugunin, M., Seshagiri, S., Brady, K., Chen, P., Ferenz, C., Franklin, S., Ghayur, T., Li P., *et al.* (1995) Inhibition of ICE family proteases by baculovirus antiapoptotic protein p35. *Science*, **269**, 1885–1888.

Cahill, M.A., Janknecht, R., Nordheim, A. (1996a) Signalling pathways: Jack of all cascades. *Curr. Biol.*, **6**, 16–19.

Cahill, M.A., Peter, M.E., Kischkel, F.C., Chinnaiyan, A.M., Dixit, V.M., Krammer, P.H., Nordheim, A. (1996b) CD95 (APO-1/Fas) induces activation of SAP kinases downstream of ICE-like proteases. *Oncogene*, **13**, 2087–2096.

Camerini, D., Walz, G., Loenen, W.A., Borst, J., Seed, B. (1991) The T cell activation antigen CD27 is a member of the nerve growth factor/tumor necrosis factor receptor gene family. *J. Immunol.*, **147**, 3165–3169.

Cerretti, D.P., Kozlosky, C.J., Mosley, B., Nelson, N., Van Ness, K., Greenstreet, T.A., March, C.J., Kronheim, S.R., Druck, T., Cannizzaro, L.A., *et al.* (1992) Molecular cloning of the interleukin-1 beta converting enzyme. *Science*, **256**, 97–100.

Cheng, J., Zhou, T., Liu, C., Shapiro, J.P., Brauer, M.J., Kiefer, M.C., Barr, P.J., Mountz J.D. (1994) Protection from Fas-mediated apoptosis by a soluble form of the Fas molecule. *Science*, **263**, 1759–1762.

Cheng, E.H., Levine, B., Boise, L.H., Thompson, C.B., Hardwick, J.M. (1996) Bax-independent inhibition of apoptosis by Bcl-X_L. *Nature*, **379**, 554–556.

Chervonsky, A.V., Wang, Y., Wong, F.S., Flavell, R.A., Janeway, C.A. Jr., Matis L.A. (1997) The role of Fas in autoimmune diabetes. *Cell*, **89**, 17–24.

Chicheportiche, Y., Bourdon, P.R., Xu, H., Hsu, Y.M., Scott, H., Hession, C., Garcia, I., Browing, J.L. (1997) TWEAK, a new secreted ligand in the tumor necrosis factor family that weakly induces apoptosis. *J. Biol. Chem.*, **274**, 32401–32410.

Chinnaiyan, A.M., O'Rourke, K., Tewari, M., Dixit, V.M. (1995) FADD, a novel death domain-containing protein, interacts with the death domain of Fas and initiates apoptosis. *Cell*, **81**, 505–512.

Chinnaiyan, A.M., O'Rourke, K., Yu, G.L., Lyons, R.H., Garg, M., Duan, D.R., Xing, L., Gentz, R., Ni J., Dixit, V.M. (1996a) Signal transduction by DR3, a death domain-containing receptor related to TNFR-1 and CD95. *Science*, **274**, 990–992.

Chinnaiyan, A.M., Tepper, C.G., Seldin, M.F., O'Rourke, K., Kischkel, F.C., Hellbardt, S., Krammer, P.H., Peter, M.E., Dixit, V.M. (1996b) FADD/MORT1 is a common mediator of CD95 (Fas/APO-1) and tumor necrosis factor receptor-induced apoptosis. *J. Biol. Chem.*, **271**, 4961–4965.

Chinnaiyan, A.M., O'Rourke, K., Lane, B.R., Dixit, V.M. (1997) Interaction of CED-4 with CED-3 and CED-9: a molecular framework for cell death. *Science*, **275**, 1122–1126.

Chow, S.C., Weis, M., Kass, GEN., Holmström, T.H., Eriksson, J.E., Orrenius S. (1995) Involvement of multiple proteases during Fas mediated apoptosis in T lymphocytes. *FEBS Lett.*, **364**, 134.

Chu, Z.L., McKinsey, T.A., Liu, L., Gentry, J.J., Malim, M., Ballard, D.W. (1997) Suppression of tumor necrosis fator-induced cell death by inhibitor of apoptosis c-IAP2 is under NF-κB control. *Proc. Natl. Acad. Sci. USA*, **94**, 10657–10621.

Cifone, M.G., Roncaioli, P., De Maria, R., Camarda, G., Santoni, A., Ruberti, G., Testi, R. (1995) Multiple pathways originate at the Fas/APO-1 (CD95) receptor: sequential involvement of phosphatidylcholine-specific phospholipase C and acidic sphingomyelinase in the propagation of the apoptotic signal. *EMBO J.*, **14**, 5859–5868.

Clair, EGS., Anderson, S.J., Oltvai, Z.N. (1997) Bcl-2 counters apoptosis by Bax heterodimerization dependent and independent mechanisms in the T cell lineage. *J. Biol. Chem.*, **272**, 29347–29355.

Cohen, G.M. (1997) Caspases: the executioners of apoptosis. *Biochem. J.*, **326**, 1–16.

Collins, M. (1995) Potential roles of apoptosis in viral pathogenesis. *Am. J. Respir. Crit. Care. Med.*, **152**, S20–S24.

Cornall, R.J., Goodnow, C.C., Cyster, J.G. (1995) The regulation of self-reactive B cells. *Curr. Opin. Immunol.*, **7**, 804–811.

Cory, S. (1995) Regulation of lymphocyte survival by the BCL-2 family. *Ann. Rev. Immunol.*, **13**, 513–543.

Crook, N.E., Clem, R.J., Miller, L.K. (1993) An apoptosis-inhibiting baculovirus gene with a zinc finger-like motif. *J. Virol.*, **67**, 2168–2174.

Cui, H., Matsui, K., Omura, S., Schauer, S.L., Matulka, R.A., Sonenshein, G.E., Ju, S.T. (1997) Proteasome regulation of activation-induced cell death. *Proc. Natl. Acad. Sci. USA*, **94**, 7515–7520.

Cuppen, E., Nagata, S., Wieringa, B., Hendriks, W. (1997) No evidence for involvement of mouse proteine tyrosine phosphatase-Bas-like Fas-associated phosphatase-1 in Fas-mediated apoptosis. *J. Biol. Chem.*, **272**, 30215–30220.

Datta, S.R., Dudek, H., Tao, X., Masters, S., Fu, H., Gotoh, Y., Greenberg, M.E. (1997) Akt phosphorylation of BAD couples survival signals to the cell-intrinsic death machinery. *Cell*, **91**, 231–241.

Dbaibo, G.S., Perry, D.K., Gamard, C.J., Platt, R., Poirier, G.G., Obeid, L.M., Hannun, Y.A. (1997) Cytokine response modifier A (CrmA) inhibits ceramide formation in response to tumor necrosis factor (TNF)-α, CrmA and Bcl-2 target distinct components in the apoptotic pathway. *J. Exp. Med.*, **185**, 481–490.

Debatin, K.M., Fahrig-Faissner, A., Enenkel-Stoodt, S., Kreuz, W., Benner, A., Krammer, P.H. (1994) High expression of APO-1 (CD95) on T lymphocytes from human immunodeficiency virus-1-infected children. *Blood*, **83**, 3101–3103.

Degli-Eposti, M.A., Smolak, P.J., Walczak, H., Waugh, J., Huang, C-P., DuBose, R.F., Goodwin, R.G., Smith, C.A. (1997) Cloning and characterization of TRAIL-R3, a novel member of the emerging TRAIL receptor family. *J. Exp. Med.*, **186**, 1165–1170.

Deiss, L.P., Galinka, H., Berissi, H., Cohen, O., Kimchi, A. (1996) Cathepsin, D protease mediates programmed cell death induced by interferon-γ, Fas/APO-1 and TNF-α. *EMBO J.*, **15**, 3861–3870.

Del Peso, L., Gonzalez-García, M., Page, C., Herrera, R., Nunez, G. (1997) Interleukin 3 induced phosphorylation of Bad through the protein kinase Akt. *Science*, **278**, 687–689.

De Maria, R., Lenti, L., Malisan, F., d'Agostino, F., Tomassini, B., Zeuner, A., Rippo, M.R., Testi, R. (1997) Requirement for GD3 ganglioside in CD95- and ceramide-induced apoptosis. *Nature*, **277**, 1652–1655.

Dembic, Z., Loetscher, H., Gubler, U., Pan, Y.C., Lahm, H.W., Gentz, R., Brockhaus, M., Lesslauer, W. (1990) Two human TNF receptors have similar extracellular, but distinct intracellular, domain sequences. *Cytokine*, **2**, 231–237.

Derijard, B., Hibi, M., Wu, I.H., Barrett, T., Su, B., Deng, T., Karin, M., Davis, R.J. (1994) JNK1: a protein kinase stimulated by UV light and Ha-Ras that binds and phosphorylates the c-Jun activation domain. *Cell*, **76**, 1025–1037.

Deveraux, Q., Takahashi, R., Salvesen, G.S., Reed, J.C. (1997) X-linked IAP is a direct inhibitor of cell death proteases. *Nature*, **388**, 300–303.

Dhein, J., Daniel, P.T., Trauth, B.C., Oehm, A., Möller, P., Krammer, P.H. (1992) Induction of apoptosis by monoclonal antibody anti-APO-1 class switch variants is dependent on cross-linking of APO-1 cell surface antigens. *J. Immunol.*, **149**, 3166–3173.

Dhein, J., Walczak, H., Bäumler, C., Debatin, K.M., Krammer, P.H. (1995) Autocrine T-cell suicide mediated by APO-1/(Fas/CD95). *Nature*, **373**, 438–441.

Dickinson, J.L., Bates, E.J., Ferrante, A., Antalis, T.M. (1995) Plasminogen activator inhibitor type 2 inhibits tumor necrosis factor-α induced apoptosis. Evidence for an alternate biological function. *J. Biol. Chem.*, **270**, 27894–27904.

Drappa, J., Vaishnaw, A.K., Sullivan, K.E., Chu, J.L., Elkon, K.B. (1996) The Canale-Smith syndrome: an inherited autoimmune disorder associated with defective lymphocyte apoptosis and mutations in the Fas gene. *N. Engl. J. Med.*, **335**, 1643–1649.

Duan, H., Dixit, V.M. (1997) RAIDD is a new 'death' adaptor molecule. *Nature*, **385**, 86–89.

Duan, H., Orth, K., Chinnaiyan, A.M., Poirier, G.G., Froelich, C.J., He, W.W., Dixit, V.M. (1996) ICE-LAP6, a novel member of the ICE/Ced-3 gene family, is activated by the cytotoxic T cell protease granzyme B. *J. Biol. Chem.*, **271**, 16720–16724.

Duckett, C.S., Nava, V.E., Gedrich, R.W., Clem, R.J., Van Dongen, J.L., Gilfillan, M.C., Shiels, H., Hardwick, J.M., Thompson, C.B. (1996) A conserved family of cellular genes related to the baculovirus iap gene and encoding apoptosis inhibitors. *EMBO J.*, **15**, 2685–2694.

Durkop, H., Latza, U., Hummel, M., Eitelbach, F., Seed, B., Stein, H. (1992) Molecular cloning and expression of a new member of the nerve growth factor receptor family that is characteristic for Hodgkin's disease. *Cell*, **68**, 421–427.

Eck, M.J., Sprang, S.R. (1989) The structure of tumor necrosis factor-alpha at 2.6 A resolution. Implications for receptor binding. *J. Biol. Chem.*, **264**, 17595–17605.

Eck, M.J., Ultsch, M., Rinderknecht, E., de Vos, A.M., Sprang S.R. (1992) The structure of human lymphotoxin (tumor necrosis factor-ß) at 1.9-A resolution. *J. Biol. Chem.*, **267**, 2119–2122.

Emoto, Y., Manome, Y., Meinhardt, G., Kisaki, H., Kharbanda, S., Robertson, M., Ghayur, T., Wong, W.W., Kamen, R., Weichselbaum, R., *et al.* (1995) Proteolytic activation of protein kinase C delta by an ICE-like protease in apoptotic cells. *EMBO J.*, **14**, 6148–6156.

Enari, M., Hug, H., Nagata, S. (1995) Involvement of an ICE-like protease in Fas-mediated apoptosis. *Nature*, **375**, 78–81.

Enari, M., Talanian, R.V., Wong, W.W. and Nagata, S. (1996). Sequential activation of ICE-like and CPP-32 like proteases during Fas-mediated apoptosis. *Nature*, **380**, 723–726.

Engelmann, H., Holtmann, H., Brakebusch, C., Avni, Y.S., Sarov, I., Nophar, Y., Hadas, E., Leitner, O., Wallach, D. (1990) Antibodies to a soluble form of a tumor necrosis factor (TNF) receptor have TNF-like activity. *J. Biol. Chem.*, **265**, 14497–14504.

Espevik, T., Waage, A. (1988) The involvement of tumor necrosis factor-alpha (TNF-alpha) in immunomodulation and in septic shock. *Dev. Biol. Stand.*, **69**, 139–142.

Espevik, T., Brockhaus, M., Loetscher, H., Nonstad, U., Shalaby, R. (1990) Characterization of binding and biological effects of monoclonal antibodies against a human tumor necrosis factor receptor. *J. Exp. Med.*, **171**, 415–426.

Farber, E. (1994) Programmed cell death: necrosis versus apoptosis. *Mod. Pathol.*, **7**, 605–609.

Faucheu, C., Blanchet, A.M., Collard, Dutilleul, V., Lalanne, J.L., Diu Hercend, A. (1996) Identification of a cysteine protease closely related to interleukin-1ß-converting enzyme. *Eur. J. Biochem.*, **236**, 207–213.

Faucheu, C., Diu, A., Chan, A.W., Blanchet, A.M., Miossec, C., Herve, F., Collard Dutilleul, V., Gu, Y., Aldape, R.A., Lippke, J.A., *et al.* (1995) A novel human protease similar to the interleukin-1ß converting enzyme induces apoptosis in transfected cells. *EMBO J.*, **14**, 1914–1922.

Fernandes-Alnemri, T., Litwack, G., Alnemri, E.S. (1994) CPP32, a novel human apoptotic protein with homology to Caenorhabditis elegans cell death protein Ced-3 and mammalian interleukin-1ß-converting enzyme. J. Biol. Chem., **269**, 30761–30764.

Fernandes-Alnemri, T., Takahashi, A., Armstrong, R., Krebs, J., Fritz, L., Tomaselli, K.J., Wang, L., Yu, Z., Croce, C.M., Salveson, G., *et al.* (1995) Mch3, a novel human apoptotic cysteine protease highly related to CPP32. *Cancer Res.*, **55**, 6045–6052.

Fernandes-Alnemri, T., Armstrong, R.C., Krebs, J., Srinivasula, S.M., Wang, L., Bullrich, F., Fritz, L.C., Trapani, J.A., Tomaselli, K.J., Litwack, G., Alnemri, E.S. (1996a) In vitro activation of CPP32 and Mch3 by Mch4, a novel human apoptotic cysteine protease containing two FADD-like domains. *Proc. Natl. Acad. Sci. USA*, **93**, 7464–7469.

Fernandes-Alnemri, T., Litwack, G., Alnemri, E.S. (1996b) Mch2, a new member of the apoptotic Ced-3/Ice cysteine protease gene family. *Cancer Res.*, **55**, 2737–2742.

Fiers, W., Beyaert, R., Boone, E., Cornelis, S., Declercq, W., Decoster, E., Denecker, G., Depuydt, B., De Valck, D., De Wilde, G., Goossens, V., Grooten, J., Haegeman, G., Heyninck, K., Penning, L., Plaisance, S., Vancompernolle, K., Van Criekinge, W., Vandenabeele, P., Vanden Berghe, W., Van de Craen, M., Vandevoorde, V., Vercammen, D. (1996) TNF induced intracellular signalling leading to gene induction or to cytotoxicity by necrosis or by apoptosis. *J. Inflamm.*, **47**, 67–75.

Fisher, G.H., Rosenberg, F.J., Straus, S.E., Dale, J.K., Middleton, L.A., Lin, A.Y., Strober, W., Lenardo, M.J., Puck, J.M. (1995) Dominant interfering Fas gene mutations impair apoptosis in a human autoimmune lymphoproliferative syndrome. *Cell*, **81**, 935–946.

Freiberg, R.A., Spencer, D.M., Choate, K.A., Duh, H.J., Schreiber, S.L., Crabtree, G.R., Khavari, P.A. (1997) Fas signal transduction triggers either proliferation or apoptosis in human fibroblasts. *J. Invest. Dermatol.*, **108**, 215–219.

Friesen, C., Herr, I., Krammer, P.H., Debatin, K.M. (1996) Involvement of the CD95 (APO-1/Fas) receptor/ligand system in drug induced apoptosis in leukemia cells. *Nature Med.*, **2**, 574–577.

Fulda, S., Sieverts, H., Friesen, C., Herr, I., Debatin, K-M. (1997a) The CD95 (APO-1/Fas) system mediates drug-induced apoptosis in neuroblastoma cells. *Cancer Res.*, **57**, 3823–3829.

Fulda, S., Friesen, C., Los, M., Scaffidi, C., Mier, W., Benedict, M., Nunez, G., Krammer, P.H., Peter, M.E., Debatin, K-M. (1997b) Betulinic acid triggers CD95 (APO-1/Fas)- and p53–independent apoptosis via activation of caspases in neuroectodermal tumors. *Cancer Res.*, **57**, 4956–4964.

Galle, P.R., Hofmann, W.J., Walczak, H., Schaller, H., Otto, G., Stremmel, W., Krammer, P.H., Runkel, L. (1995) Involvement of the CD95 (APO-1/Fas) receptor and ligand in liver damage. *J. Exp. Med.*, **182**, 1223–1230.

Gamen, S., Marzo, I., Anel, A., Pineiro, A., Naval, J. (1996) CPP32 inhibition prevents Fas-induced ceramide generation and apoptosis in human cells. *FEBS Lett.*, **390**, 232–237.

Garcia, I., Miyazaki, Y., Araki, K., Araki, M., Lucas, R., Grau, G.E., Milon, G., Belkaid, Y., Montixi, C., Lesslauer, W., *et al.* (1995) Transgenic mice expressing high levels of soluble TNF-R1 fusion protein are protected from lethal septic shock and cerebral malaria, and are highly sensitive to Listeria monocytogenes and Leishmania major infections. *Eur. J. Immunol.*, **25**, 2401–2407.

Giordano, C., Stassi, G., De Maria, R., Todaro, M., Richiusa, P., Papoff, G., Ruberti, G., Bagnasco, M., Testi, R., Galluzzo, A. (1997) Potential involvement of Fas and its ligand in the pathogenesis of Hashimoto's thyroiditis. *Science*, **275**, 960–963.

Glauser, M.P. (1996) The inflammatory cytokines. New developments in the pathophysiology and treatment of septic shock. *Drugs*, **52** Suppl 2, 9–17.

Goillot ,E., Raingeaud, J., Ranger, J., Tepper, R.I., Davis, R.J., Harlow, E., Sanchez, I. (1997) Mitogen-activated protein kinase-mediated Fas apoptotic signalling pathway. *Proc. Natl. Acad. Sci. USA*, **94**, 3302–3307.

Goltsev, Y.V., Kovalenko, A.V., Arnold, E., Varfolomeev, E.E., Brodianskii, V.M., Wallach, D. (1997) CASH, a novel caspase homologue with death effector domains. *J. Biol. Chem.*, **272**, 19641–19644.

Gong, L., Kamitani, Fujise, K., Caskey, L.S., Yeh ETH. (1997) Preferential interaction of sentrin with a ubiquitin-conjugating enzyme, ubc9. *J. Biol. Chem.*, **272**, 28198–28201.

Grell, M., Krammer, P.H., Scheurich, P. (1994) Segregation of APO-1/Fas antigen and tumor necrosis factor-mediated apoptosis. *Eur. J. Immunol.*, **24**, 2563–2566.

Grell, M., Douni, E., Wajant, H., Lohden, M., Clauss, M., Maxeiner, B., Georgopoulos, S., Lesslauer, W., Kollias, G., Pfizenmaier, K., *et al.* (1995) The transmembrane form of tumor necrosis factor is the prime activating ligand of the 80 kDa tumor necrosis factor receptor. *Cell*, **83**, 793–802.

Griffith, T.S., Brunner, T., Fletcher, S.M., Green, D.R., Ferguson, T.A. (1995) Fas ligand-induced apoptosis as a mechanism of immune privilege. *Science*, **270**, 1189–1192.

Griffith, T.S., Yu, X., Herndon, J.M., Green, D.R., Ferguson, T.A. (1996) CD95-induced apoptosis of lymphocytes in an immune privileged site induces immunological tolerance. *Immunity*, **5**, 7–16.

Grilli, M., Pizzi, M., Memo, M., Spano, P. (1996) Neuroprotection by aspirin and sodium salicylate through blockade of NF-κB activation. *Science*, **274**, 1383–1385.

Grimm, L.M., Goldberg, A.L., Poirier, G.G., Schwartz L.M. and Osborne, B.A. (1996a) Proteasomes play an essential role in thymocyte apoptosis. *EMBO J.*, **15**, 3835–3844.

Grimm, S., Bauer, M.K.A., Baeuerle, P.A., Schulze-Osthoff, K. (1996b) Bcl-2 downregulates the activity of transcription factor NF-κB induced upon apoptosis. *J. Cell Biol.*, **134**, 13–23.

Gulbins, E., Bissonnette, R., Mahboubi, A., Martin, S., Nishioka, W., Brunner, T., Baier, G., Baier Bitterlich, G., Byrd, C., Lang, F., *et al.* (1995), FAS induced apoptosis is mediated via a ceramide initiated RAS signalling pathway. *Immunity*, **2**, 341–351.

Hahne, M., Rimoldi, D., Schroter, M., Romero, P., Schreier, M., French, L.E., Schneider, P., Bornand, T., Fontana, A., Lienard, D., Cerottini, J., Tschopp, J. (1996) Melanoma cell expression of Fas(Apo-1/CD95) ligand: implications for tumor immune escape. *Science*, **274**, 1363–1366.

Haimovitz-Friedman, A., Kan, C.C., Ehleiter, D., Persaud, R.S., McLoughlin, M., Fuks, Z., Kolesnick, R.N. (1994) Ionizing radiation acts on cellular membranes to generate ceramide and initiate apoptosis. *J. Exp. Med.*, **180**, 525–535.

Han, D.K.M., Chaudry, P.M., Wright, M.E., Friedman, C., Trask, B.J., Riedel, R.T., Baskin, D.G., Schwarz, S.M., Hood, L. (1997) MRIT, a novel death-effector domain-containing protein interacts with caspases and $BclX_L$ and initiates cell death. *Proc. Natl. Acad. Sci. USA*, **94**, 11333–11338.

Hanabuchi, S., Koyanagi, M., Kawasaki, A., Shinohara, N., Matsuzawa, A., Nishimura, Y., Kobayashi, Y., Yonehara, S., Yagita, H., Okumura, K. (1994) Fas and its ligand in a general mechanism of T cell mediated cytotoxicity. *Proc. Natl. Acad. Sci. USA*, **91**, 4930–4934.

Häusler, P., Papoff, G., Eramo, A., Reif, K., Cantrell, D.A., Ruberti, G. (1998) Protection of CD95-mediated apoptosis by activation of phosphatidylinositide 30kinase and protein kinase B. *Eur. J. Immunol.*, **28**, 57–69.

Hay, B.A., Wassarman, D.A., Rubin, G.M. (1995) Drosophila homologs of baculovirus inhibitor of apoptosis proteins function to block cell death. *Cell*, **83**, 1253–1262.

Hayward, M., Fiedler-Nagy, C. (1987) Mechanisms of bone loss: rheumatoid arthritis, periodontal disease and osteoporosis. *Agents Actions*, **22**, 251–254.

Heinkelein, M., Pilz, S., Jassoy, C. (1996) Inhibition of CD95 (Fas/Apo1)-mediated apoptosis by vaccinia virus W.R. *Clin. Exp. Immunol.*, **103**, 8–14.

Herron, L.R., Eisenberg, R., Roper, E., Kakkanaiah, V.N., Cohen, P.L., Kotzin, B.L. (1993) Selection of the T cell receptor repertoire in Lpr mice. *J. Immunol.*, **151**, 3450–3459.

Hibi, M., Lin, A., Smeal, T., Minden, A., Karin, M. (1993) Identification of an oncoprotein- and UV-responsive protein kinase that binds and potentiates the c-Jun activation domain. *Genes Dev.*, **7**, 2135–2148.

Hofmann, K., Bucher, P., Tschopp, J. (1997) The CARD domain: a new apoptotic signalling motif. *Trends Biochem. Sci.*, **22**, 155–156.

Hsu, H., Xiong, J., Goeddel, D.V. (1995) The TNF receptor 1-associated protein TRADD signals cell death and NF-kappa B activation. *Cell*, **81**, 495–504.

Hsu, H., Huang, J., Shu, H.B., Baichwal, V., Goeddel, D.V. (1996a) TNF-dependent recruitment of the protein kinase RIP to the TNF receptor-1 signalling complex. *Immunity*, **4**, 387–396.

Hsu, H., Shu, H.B., Pan, M.G., Goeddel, D.V. (1996b) TRADD-TRAF2 and TRADD-FADD interactions define two distinct TNF receptor 1 signal transduction pathways. *Cell*, **84**, 299–308.

Hsu, H., Solovyev, I., Colombero, A., Elliott, R., Kelley, M., Boyle, W.J. (1997a) ATAR, a novel tumor necrosis factor receptor family member, signals through TRAF2 and TRAF5. *J. Biol. Chem.*, **272**, 13471–13474.

Hsu, Y.T., Youle, R.J. (1997b) Nonionic detergents induce dimerization among members of the Bcl-2 family. *J. Biol. Chem.*, **272**, 13829–13834.

Hu, S., Vincenz, C., Buller, M., Dixit, V.M. (1997a) A novel family of viral death effector domain-containing molecules that inhibit both CD95- and tumor necrosis factor receptor-1-induced apoptosis. *J. Biol. Chem.*, **272**, 9621–9624.

Hu, S., Vincenz, C., Ni, J., Gentz, R., Dixit, V.M. (1997b) I-FLICE, a novel inhibitor of tumor necrosis factor receptor-1- and CD95-induced apoptosis. *J. Biol. Chem.*, **272**, 17255–17257.

Huang, B., Eberstadt, M., Olejniczak, E.T., Meadows, R.P., Fesik, S.W. (1996) NMR structure and mutagenesis of the Fas (APO-1/CD95) death domain. *Nature*, **384**, 638–641.

Huang, S., Jiang, Y., Li, Z., Nishida, E., Mathias, P., Lin, S., Ulevitch, R.J., Nemerow, G.R., Han, J. (1997) Apoptosis signalling pathway in T cells is composed of ICE/Ced 3 family proteases and MAP kinase kinase 6b. *Immunity*, **6**, 739–749.

Huber, A-O., Zörnig, M., Lyon, D., Suda, T., Nagata, S., Evan, G.I. (1997) Requirement for the CD95 receptor-ligand pathway in c-Myc-induced apoptosis. *Science*, **278**, 1305–1309.

Hug, H., Enari, M., Nagata, S. (1994) No requirement of reactive oxygen intermediates in Fas-mediated apoptosis. *FEBS Lett.*, **351**, 311–313.

Ichijo, H., Nishida, E., Irie, K., ten-Dijke, P., Saitoh, M., Moriguchi, T., Takagi, M., Matsumoto, K., Miyazono, K., Gotoh, Y. (1997) Induction of apoptosis by ASK1, a mammalian MAPKKK that activates SAPK/JNK and p38 signalling pathways. *Science*, **275**, 90–94.

Inohara, N., Koseki, T., Hu, Y., Chen, S., Núnez, G. (1997) CLARP, a death-effector domain-containing protein-interacts with caspase-8 and regulates apoptosis. *Proc. Natl. Acad. Sci. USA*, **94**, 10717–10722.

Irmler, M., Thome, M., Hahne, M., Schneider, P., Hofmann, K., Steiner, V., Bodme, J-L., Schröter, M., Burns, K., Mattmann, C., Rimoldi., D., French, L.E., Tschopp, J. (1997) Inhibition of death receptor signals by cellular FLIP. *Nature*, **388**, 190–195.

Itoh, N., Yonehara, S Ishii, A., Yonehara, M., Mizushima, S., Sameshima, M., Hase, A., Seto, Y., Nagata, S. (1991) The polypeptide encoded by the cDNA for human cell surface antigen Fas can mediate apoptosis. *Cell*, **66**, 233–243.

Itoh, N., Nagata, S. (1993) A novel protein domain required for apoptosis. Mutational analysis of human Fas antigen. *J. Biol. Chem.*, **268**, 10932–10937.

Itoh, N., Tsujimoto, Y., Nagata, S. (1993) Effect of Bcl-2 on Fas antigen-mediated cell death. *J. Immunol.*, **151**, 621–627.

Ivanov, V.N., Lee, R.K., Podack, E.R., Malek, T.R. (1997) Regulation of Fas-dependent activation-induced T cell apoptosis by cAMP signalling: a potential role for transcription factor NF-κB. *Oncogene*, **14**, 2455–2464.

Jäättelä, M., Wissing, D. (1992) Emerging role of heat shock proteins in biology and medicine. *Ann. Med.*, **24**, 249–258.

Jäättelä, M. (1993) Overexpression of major heat shock protein hsp70 inhibits tumor necrosis factor induced activation of phospholipase A2. *J. Immunol.*, **151**, 4286–4294.

Jäättelä, M., Mouritzen, H., Elling, F., Bastholm, L. (1996) A20 zinc finger protein inhibits TNF and IL-1 signalling. *J. Immunol.*, **156**, 1166–1173.

Jacobson, M.D., Burne, J.F., King, M.P., Miyashita, T., Reed, J.C., Raff, M.C. (1993) Bcl-2 blocks apoptosis in cells lacking mitochondrial DNA. *Nature*, **361**, 365–369.

Jones, E.Y., Stuart, D.I., Walker, N.P. (1992) Crystal structure of TNF. *Immunol. Ser.*, **56**, 93–127.

Ju, S.T., Panka, D.J., Cui, H., Ettinger, R., el Khatib, M., Sherr, D.H., Stanger, B.Z., Marshak Rothstein, A. (1995) Fas(CD95)/FasL interactions required for programmed cell death after T-cell activation. *Nature*, **373**, 444–448.

Juo, P., Kuo, C.J., Reynolds, S.E., Konz, R.F., Raingeaud, J., Davis, R.J., Biemann, H.P., Blenis, J. (1997) Fas activation of the p38 mitogen-activated protein kinase signalling pathway requires ICE/CED-3 family proteases. *Mol. Cell Biol.*, **17**, 24–35.

Kamens, J., Paskind, M., Hugunin, M., Talanian, R.V., Allen, H., Banach, D., Bump, N., Hackett, M., Johnston, C.G., Li, P., *et al.* (1995) Identification and characterization of ICH-2, a novel member of the interleukin-1 beta-converting enzyme family of cysteine proteases. *J. Biol. Chem.*, **270**, 15250–15256.

Karpusas, M., Hsu, Y.M., Wang, J.H., Thompson, J., Lederman, S., Chess, L., Thomas, D. (1995) 2 A crystal structure of an extracellular fragment of human CD40 ligand. *Structure*, **3**, 1031–1039.

Katsikis, P.D., Wunderlich, E.S., Smith, C.A., Herzenberg, L.A. (1995) Fas antigen stimulation induces marked apoptosis of T lymphocytes in human immunodeficiency virus-infected individuals. *J. Exp. Med.*, **181**, 2029–2036.

Katsikis, P.D., Garciaojeda, M.E., Torresroca, J.F., Tijoe, I.M., Smith, C.A., Herzenberg L.A., Herzenberg L.A. (1997) Interleukin-1ß converting enzyme like protease involvement in Fas induced and activation induced peripheral blood T cell apoptosis in HIV infection: TNF related apoptosis inducing ligand can mediate activation induced T cell death in HIV infection. *J. Exp. Med.*, **186**, 1365–1372.

Kaufmann, S.H., Desnoyers, S., Ottaviano, Y., Davidson, N.E., Poirier, G.G. (1993) Specific proteolytic cleavage of poly(ADP-ribose) polymerase: an early marker of chemotherapy-induced apoptosis. *Cancer Res.*, **53**, 3976–3985.

Kayagaki, N., Kawasaki, A., Ebata, T., Ohmoto, H., Ikeda, S., Inoue, S., Yoshino, K., Okumura, K., Yagita, H. (1995) Metalloproteinase-mediated release of human Fas ligand. *J. Exp. Med.*, **182**, 1777–1783.

Kischkel, F.C., Hellbardt, S., Behrmann, I., Germer, M., Pawlita, M., Krammer, P.H., Peter, M.E. (1995) Cytotoxicity-dependent APO-1 (Fas/CD95)-associated proteins form a death-inducing signalling complex (DISC) with the receptor. *EMBO J.*, **14**, 5579–5588.

Kitson, J., Raven, T., Jiang, Y.P., Goeddel, D.V., Giles, K.M., Pun, K.T., Grinham, C.J., Brown, R., Farrow, S.N. (1996) A death-domain-containing receptor that mediates apoptosis. *Nature*, **384**, 372–375.

Kluck, R.M., Bossy Wetzel, E., Green, D.R., Newmeyer, D.D. (1997) The release of cytochrome c from mitochondria: a primary site for Bcl-2 regulation of apoptosis. *Science*, **275**, 1132–1136.

Knipping, E., Krammer, P.H., Lehman, T.J., Mysler, E., Elkon, K.B. (1995) Levels of soluble Fas/APO-1/CD95 in systemic lupus erythematosus and juvenile rheumatoid arthritis. *Arthritis Rheum*, **38**, 1735–1737.

Komiyama, T., Ray, C.A., Pickup, D.J., Howard, A.D., Thornberry, N.A., Peterson, E.P., Salvesen, G. (1994) Inhibition of interleukin-1 beta converting enzyme by the cowpox virus serpin CrmA. An example of cross-class inhibition. *J. Biol. Chem.*, **269**, 19331–19337.

Korner, H., Sedgwick, J.D. (1996) Tumour necrosis factor and lymphotoxin: molecular aspects and role in tissue-specific autoimmunity. *Immunol. Cell. Biol.*, **74**, 465–472.

Krammer, P.H., Dhein, J., Walczak, H., Behrmann, I., Mariani, S., Matiba, B., Fath, M., Daniel, P.T., Knipping, E., Westendorp, M.O., *et al.* (1994) The role of APO-1-mediated apoptosis in the immune system. *Immunol. Rev.*, **142**, 175–191.

Krikos, A., Laherty, C.D., Dixit, V.M. (1992) Transcriptional activation of the tumor necrosis factor alpha-inducible zinc finger protein, A20, is mediated by kappa B elements. *J. Biol. Chem.*, **267**, 17971–17976.

Kroemer, G. (1997a) Mitochondrial control of apoptosis. *Immunol. Today*, **18**, 44–51.

Kroemer, G. (1997b) The proto-oncogene Bcl-2 and its role in regulating apoptosis. *Nature Med.*, **3**, 614–620.

Kuida, K., Lippke, J.A., Ku, G., Harding, M.W., Livingston, D.J., Su, M.S., Flavell, R.A. (1995) Altered cytokine export and apoptosis in mice deficient in interleukin-1ß converting enzyme. *Science*, **267**, 2000–2003.

Kuida, K., Zheng, T.S., Na, S., Kuan, C., Yang, D., Karasuyama, H., Rakic, P., Flavell, R.A. (1996) Decreased apoptosis in the brain and premature lethality in CPP32-deficient mice. *Nature*, **384**, 368–372.

Kumar, S., Baglioni, C. (1991) Protection from tumor necrosis factor mediated cytolysis by overexpression of plasminogen activator inhibitor type 2. *J. Biol. Chem.*, **266**, 20960–20964.

Kumar, S., Kinoshita, M., Noda, M., Copeland, N.G., Jenkins, N.A. (1994) Induction of apoptosis by the mouse Nedd2 gene, which encodes a protein similar to the product of the Caenorhabditis elegans cell death gene ced-3 and the mammalian IL-1 beta-converting enzyme. *Genes Dev.*, **8**, 1613–1626.

Kwon, B.S., Weissman, S.M. (1989) cDNA sequences of two inducible T cell genes. *Proc. Natl. Acad. Sci. USA*, **86**, 1963–1968.

Kwon, B.S., Tan, K.B., Ni, J., Lee, K.O., Kim, K.K., Kim, Y.J., Wang, S., Gentz, R., Yu, G.L., Harrop, J., Lyn, S.D., Silverman, C., Porter, T.G., Truneh, A., Young, P.R. (1997) A newly identified member of the tumor necrosis factor receptor superfamily with a wide tissue distribution and involvement in lymphocyte activation. *J. Biol. Chem.*, **272**, 14272–14276.

Laherty, C.D., Hu, H.M., Opipari, A.W., Wang, F., Dixit, V.M. (1992) The Epstein Barr virus LMP1 gene product induces A20 zinc finger protein expression by activating nuclear factor-kappa B. *J. Biol. Chem.*, **267**, 24157–24160.

Laster, S.M., Wood, J.G., Gooding, L.R. (1988) Tumor necrosis factor can both induce apoptotic and necrotic forms of cell lysis. *J. Immunol.*, **141**, 2629–2635.

Latinis, K.M., Koretzky, G.A. (1996) Fas ligation induces apoptosis and Jun kinase activation independently of CD45 and Lck in human T cells. *Blood*, **87**, 871–875.

Lau, H.T., Yu, M., Fontana, A., Stoeckert, C.J.J. (1996) Prevention of islet allograft rejection with engineered myoblasts expressing FasL in mice. *Science*, **273**, 109–112.

Lavoie, J.N., Gingras-Breton, G., Tanguay, R.M., Laundry, J. (1993) Induction of chinese hamster hsp27 gene in mouse cells confers resistance to heat shock. *J. Biol. Chem.*, **268**, 3420–3429.

Leist, M., Single, B., Castoldi, A.F., Kuhnle, S., Nicotera, P. (1997) Intracellular adenosine triphosphate (ATP) concentration: a switch in the decision between apoptosis and necrosis. *J. Exp. Med.*, **285**, 1481–1486.

Lenczowski, J.M., Dominguez, L., Eder, A.M., King, L.B., Zacharchuk, C.M., Ashwell, J.D. (1997) Lack of a role for Jun kinase and AP-1 in Fas-induced apoptosis. *Mol. Cell Biol.*, **17**, 170–181.

Li, C.J., Friedman, D.J., Wang, C., Metelev, V., Pardee, A.B. (1995a) Induction of apoptosis in uninfected lymphocytes by HIV-1 Tat protein. *Science*, **268**, 429–431.

Li, P., Allen, H., Banerjee, S., Franklin, S., Herzog, L., Johnston, C., McDowell, J., Paskind, M., Rodman, L., Salfeld, J., *et al.* (1995b) Mice deficient in IL-1 beta-converting enzyme are defective in production of mature IL-1 beta and resistant to endotoxic shock. *Cell*, **80**, 401–411.

Li, P., Nijhawan, D., Budihardjo, I., Srinivasula, S.M., Ahmad, M., Alnemri, E.S., Wang, X. (1997) Cytochrome c and dATP-dependent formation of Apaf-1/caspase-9 complex initiates an apoptotic protease cascade. *Cell*, **91**, 479–489.

Lin, K.I., Lee, S.H., Narayanan, R., Baraban, J.M., Hardwick, J.M., Ratan, R.R. (1995) Thiol agents and Bcl-2 identify an alphavirus induced apoptotic pathway that requires activation of the transcription factor NF-kappa B. *J. Cell. Biol.*, **131**, 1149–1161.

Lippke, J.A., Gu, Y., Sarnecki, C., Caron, P.R., Su, M.S. (1996) Identification and characterization of CPP32/Mch2 homolog 1, a novel cysteine protease similar to CPP32. *J. Biol. Chem.*, **271**, 1825–1828.

Liston, P., Roy, N., Tamai, K., Lefebvre, C., Baird, S., Cherton, Horvat, G., Farahani, R., McLean, M., Ikeda, J.E., MacKenzie, A., Korneluk, R.G. (1996) Suppression of apoptosis in mammalian cells by NAIP and a related family of IAP genes. *Nature*, **379**, 349–353.

Liu, X., Kim, C.N., Yang, J., Jemmerson, R., Wang, X. (1996a) Induction of apoptotic program in cell-free extracts: requirement for dATP and cytochrome c. *Cell*, **86**, 147–157.

Liu, Z.G., Hsu, H., Goeddel, D.V., Karin, M. (1996b) Dissection of TNF receptor 1 effector functions: JNK activation is not linked to apoptosis while NF-κB activation prevents cell death. *Cell*, **87**, 565–576.

Liu, X., Zou, H., Slaughter, C., Wang, X. (1997) DFF, a heterodimeric protein that functions downstream of caspase 3 to trigger DNA fragmentation during apoptosis. *Cell*, 89, 175–184.

Loetscher, H., Pan, Y.C., Lahm, H.W., Gentz, R., Brockhaus, M., Tabuchi, H., Lesslauer, W. (1990) Molecular cloning and expression of the human 55 kd tumor necrosis factor receptor. *Cell*, **61**, 351–359.

Los, M., Van de Craen, M., Penning, L.C., Schenk, H., Westendorp, M., Baeuerle, P.A., Dröge, W., Krammer, P.H., Fiers, W., Schulze-Osthoff, K. (1995) Requirement of an ICE/CED-3 protease for Fas/APO-1-mediated apoptosis. *Nature*, **375**, 81–83.

MacFarlane,, M., Ahmadi, M., Srinivasula, S.M., Fernades-Alnemri, T., Cohen, G.M., Alnemri, E.S. (1997) Identification of two novel receptors for the cytotoxic ligand TRAIL. *J. Biol. Chem.*, **272**, 25417–25420.

Machleidt, T., Kramer, B., Adam, D., Neumann, B., Schütze, S., Wiegmann, K., Krönke, M. (1997) Function of the p55 tumor necrosis factor receptor "death domain" mediated by phosphatidylcholine-specific phospholipase C. *J. Exp. Med.*, **184**, 725–733.

Maini, R.N. (1996) The role of cytokines in rheumatoid arthritis. The Croonian Lecture 1995. *J. R. Coll Physicians Lond*, **30**, 344–351.

Mallett, S., Fossum, S., Barclay, A.N. (1990) Characterization of the MRC OX40 antigen of activated CD4 positive T lymphocytes-a molecule related to nerve growth factor receptor. *EMBO J.*, **9**, 1063–1068.

Mandal, M., Maggirwar, S.B., Sharma, N., Kaufmann, S.H., Sun, S.C., Kumar, R. (1996) Bcl-2 prevents CD95 (Fas/APO-1)-induced degradation of lamin B and poly(ADP-ribose) polymerase and restores the NF-κB signalling pathway. *J. Biol. Chem.*, **271**, 30354–30359.

Mapara, M.Y., Bargou, R., Zugck, C., Dohner, H., Ustaoglu, F., Jonker, R.R., Krammer, P.H., Dörken, B. (1993) APO-1 mediated apoptosis or proliferation in human chronic B lymphocytic leukemia: correlation with bcl-2 oncogene expression. *Eur. J. Immunol.*, **23**, 702–708.

Mariani, S.M., Matiba, B., Bäumler, C., Krammer, P.H. (1995) Regulation of cell surface APO-1/Fas (CD95) ligand expression by metalloproteases. *Eur. J. Immunol.*, **25**, 2303–2307.

Mariani, S.M., Matiba, B., Armandola, E.A., Krammer, P.H. (1997) Interleukin 1 beta-converting enzyme related proteases/caspases are involved in TRAIL-induced apoptosis of myeloma and leukemia cells. *J. Cell. Biol.*, **137**, 221–229.

Marino, M.W., Dunn, A., Grail, D., Inglese, M., Noguchi, Y., Richards, E., Jungbluth, A., Wada, H., Moore, M., Williamson, B., Basu, S., Old, L. (1997) Characterization of tumor necrosis factor-deficient mice. *Proc. Natl. Acad. Sci. USA*, **94**, 8093–8098.

Marsters, S.A., Sheridan, J.P., Donahue, C.J., Pitti, R.M., Gray, C.L., Goddard, A.D., Bauer, K.D., Ashkenazi, A. (1996a) Apo-3, a new member of the tumor necrosis factor receptor family, contains a death domain and activates apoptosis and NF-κB. *Curr. Biol.*, **6**, 1669–1676.

Marsters, S.A., Pitti, R.M., Donahue, C.J., Ruppert, S., Bauer, K.D., Ashkenazi, A. (1996b) Activation of apoptosis by Apo-2 ligand is independent of FADD but blocked by CrmA. *Curr. Biol.*, **6**, 750–752.

Marsters, S.A., Sheridan, J.P., Pitti, R.M., Skubatch, M., Baldwin, D., Yuan, J., Gurney, A., Goddard, A.D., Godowski, P., Ashkenazi, A. (1997) A novel receptor for Apo2L/TRAIL contains a truncated death domain. *Curr. Biol.*, **7**, 1003–1006.

Marsters, S.A., Sheridan, J.P., Pitti, R.M., Brush, J., Goddard, A., Ashkenazi, A. (1998) Identification of a ligand for the death-domain-containing receptor apo3. *Curr. Biol.*, **8**, 525–528.

Marte, B.M., Downward, J. (1997) PKB/Akt: Connecting phosphoinositide-3-kinase to cell survival and beyond. *Trends Biochem Sci.*, **22**, 355–358.

Matthews, N., Neale, M.L. (1987) Studies on the mode of action of tumor necrosis factor on tumor cells in vitro. *Lymphokines*, **14**, 223–252.

Medema, J.P., Scaffidi, C., Kischkel, F.C., Shevchenko, A., Mann, M., Krammer, P.H., Peter, M.E. (1997a) FLICE is activated by association with the CD95 death-inducing signalling complex (DISC). *EMBO J.*, **16**, 2794–2804.

Medema, J.P., Scaffidi, C., Krammer, P.H., Peter, M.E. (1997b) Bcl-x_L acts downstream of caspase-8 activation by the death-inducing signalling complex. *J. Biol. Chem.*, **273**, 3388–3393.

Mehlen, P., Kretz Remy, C., Preville, X., Arrigo, A.P. (1996a) Human hsp27, Drosophila hsp27 and human alphaB crystallin expression mediated increase in glutathione is essential for the protective activity of these proteins against TNFα induced cell death. *EMBO J.*, **15**, 2695–26706.

Mehlen, P., Schulze-Osthoff, K., Arrigo, A.P. (1996b) Small stress proteins as novel regulators of apoptosis. Heat shock protein 27 blocks Fas/APO 1 and staurosporine induced cell death. *J. Biol. Chem.*, **271**, 16510–1654.

Meier, R., Rouse, J., Cuenda, A., Nebreda, A.R., Cohen, P. (1996) Cellular stresses and cytokines activate multiple mitogen-activated-protein kinase kinase homologues in PC12 and KB cells. *Eur. J. Biochem.*, **236**, 796–805.

Memon, S.A., Moreno, M.B., Petrak, D., Zacharchuk, C.M. (1995) Bcl-2 blocks glucocorticoid- but not Fas- or activation-induced apoptosis in a T cell hybridoma. *J. Immunol.*, **155**, 4644–4652.

Minn, A.J., Velez, P., Schendel, S.L., Liang, H., Muchmore, S.W., Fesik, S.W., Fill, M., Thompson, C.B. (1997) Bcl-x_L forms an ion channel in synthetic lipid membranes. *Nature*, **385**, 353–357.

Miossec, C., Dutilleul, V., Fassy, F., Diu-Hercend, A. (1997) Evidence for CPP32 activation in the absence of apoptosis during T lymphocyte stimulation *J. Biol. Chem.*, **272**, 13459–13462.

Mittl, P.R., Di Marco, S., Krebs, J.F., Bai, X., Karanewsky, D.S., Priestle, J.P., Tomaselli, K.J., Grutter, M.G. (1997) Structure of recombinant human CPP32 in complex with the tetrapeptide acetyl-Asp-Val-Ala-Asp fluoromethyl ketone. *J. Biol. Chem.*, **272**, 6539–6547.

Miura, M., Friedlander R.M., Yuan J. (1995) Tumor necrosis factor-induced apoptosis is mediated by a CrmA-sensitive cell death pathway. *Proc. Natl. Acad. Sci. USA*, **92**, 8318–8322.

Mogil, R.J., Radvanyi, L., Gonzalez Quintial, R., Miller, R., Mills, G., Theofilopoulos, A.N., Green, D.R. (1995) Fas (CD95) participates in peripheral T cell deletion and associated apoptosis in vivo. *Int. Immunol*, **7**, 1451–1458.

Monney, L., Otter, I., Olivier, R., Ravn, U., Mirzasaleh, H., Fellay, I., Poirier, G.G., Borner, C. (1996) Bcl-2 overexpression blocks activation of the death protease CPP32/Yama/apopain. *Biochem. Biophys. Res. Commun.*, **221**, 340–345.

Montgomery, R.I., Warner, M.S., Lum, B.J., Spear, P.G. (1996) Herpes simplex virus-1 entry into cells mediated by a novel member of the TNF/NGF receptor family. *Cell*, **87**, 427–436.

Moss, M.L., Jin, S.L., Milla, M.E., Burkhart, W., Carter, H.L., Chen, W.J., Clay, W.C., Didsbury, J.R., Hassler, D., Hoffman, C.R., Kost, T.A., Lambert, M.H., Leesnitzer, M.A., McCauley, P., McGeehan, G., Mitchell, J., Moyer, M., Pahel, G., Rocque, W., Overton, L.K., Schoenen, F., Seaton, T., Su, J.L., Warner, J., Becherer, J.D., *et al.* (1997) Cloning of a disintegrin metalloproteinase that processes precursor tumour-necrosis factor-alpha. *Nature*, **385**, 733–736.

Muchmore, S.W., Sattler, M., Liang, H., Meadows, R.P., Harlan, J.E., Yoon, H.S., Nettesheim, D., Chang, B.S., Thompson, C.B., Wong, S.L., Ng, S.L., Fesik, S.W. (1996) X-ray and NMR structure of human Bcl-x_L, an inhibitor of programmed cell death. *Nature*, **381**, 335–341.

Müller, M., Strand, S., Hug, H., Heinemann, E.M., Walczak, H., Hofmann, W.J., Stremmel, W., Krammer, P.H., Galle, P.R. (1997) Drug-induced apoptosis in hepatoma cells is mediated by the CD95 (APO-1/Fas) receptor/ligand system and involves activation of wild-type p53. *J. Clin. Invest.*, **99**, 403–413.

Munday, N.A., Vaillancourt, J.P., Ali, A., Casano, F.J., Miller, D.K., Molineaux, S.M., Yamin, T.T., Yu, V.L., Nicholson, D.W. (1995) Molecular cloning and pro-apoptotic activity of ICErelII and ICErelIII, members of the ICE/CED-3 family of cysteine proteases. *J. Biol. Chem.*, **270**, 15870–15876.

Muzio, M., Chinnaiyan, A.M., Kischkel, F.C., O'Rourke, K., Shevchenko, A., Ni, J., Scaffidi, C., Bretz, J.D., Zhang, M., Gentz, R., Mann, M., Krammer, P.H., Peter, M.E., Dixit, V.M. (1996) FLICE, a novel FADD-homologous ICE/CED-3–like protease, is recruited to the CD95 (Fas/APO-1) death-inducing signalling complex. *Cell*, **85**, 817–827.

Muzio, M., Salvesen, G.S., Dixit, V.M. (1997) FLICE induced apoptosis in a cell-free system. Cleavage of caspase zymogens. *J. Biol. Chem.*, **272**, 2952–2956.

Mysler, E., Bini, P., Drappa, J., Ramos, P., Friedman, S.M., Krammer, P.H., Elkon, K.B. (1994) The apoptosis-1/Fas protein in human systemic lupus erythematosus. *J. Clin. Invest.*, **93**, 1029–1034.

Nagata, S., Golstein, P. (1995) The Fas death factor. *Science*, **267**, 1449–1456.

Nagata, S. (1997) Apoptosis by death factor. *Cell*, **88**, 355–365.

Natoli, G., Costanzo, A., Ianni, A., Templeton, D.J., Woodgett, J.R., Balsano, C., Levrero, M. (1997) Activation of SAPK/JNK by TNF receptor 1 through a noncytotoxic TRAF2-dependent pathway. *Science*, **275**, 200–203.

Nicholson, D.N., Thornberry, N.A. (1997) Caspases: killer proteases. *Trends Biochem. Sci.*, **22**, 299–306.

Nicholson, D.W., Ali, A., Thornberry, N.A., Vaillancourt, J.P., Ding, C.K., Gallant, M., Gareau, Y., Griffin, P.R., Labelle, M., Lazebnik, Y.A., *et al.* (1995) Identification and inhibition of the ICE/CED-3 protease necessary for mammalian apoptosis. *Nature*, **376**, 37–43.

Niehans, G.A., Brunner, T., Frizelle, S.P., Liston, J.C., Salerno, C.T., Knapp, D.J., Green, D.R., Kratzke, R.A. (1997) Human lung carcinomas express Fas ligand. *Cancer Res.*, **57**, 1007–1012.

Nishina, H., Fischer, K.D., Radvanyi, L., Shahinian, A., Hakem, R., Rubie, E.A., Bernstein, A., Mak, T.W., Woodgett, J.R., Penninger, J.M. (1997) Stress-signalling kinase Sek1 protects thymocytes from apoptosis mediated by CD95 and CD3. *Nature*, **385**, 350–353.

Nocentini, G., Giunchi, L., Ronchetti, S., Krausz, L.T., Bartoli, A., Moraca, R., Migliorati, G., Riccardi, C. (1997) A new member of the tumor necrosis factor/nerve growth factor receptor family inhibits T cell receptor-induced apoptosis. *Proc. Natl. Acad. Sci. USA*, **94**, 6216–6221.

O'Connell, J., O'Sullivan, G.C., Collins, J.K., Shanahan, F. (1996) The Fas counterattack: Fas-mediated T cell killing by colon cancer cells expressing Fas ligand. *J. Exp. Med.*, **184**, 1075–1082.

Oehm, A., Behrmann, I., Falk, W., Pawlita, M., Maier, G., Klas, C., Li Weber, M., Richards, S., Dhein, J., Trauth, B.C., *et al.* (1992) Purification and molecular cloning of the APO-1 cell surface antigen, a member of the tumor necrosis factor/nerve growth factor receptor superfamily. Sequence identity with the Fas antigen. *J. Biol. Chem.*, **267**, 10709–10715.

Ogasawara, J., Watanabe-Fukunaga, R., Adachi, M., Matsuzawa, A., Kasugai, T., Kitamura, Y., Itoh, N., Suda, T., Nagata, S. (1993) Lethal effect of the anti-Fas antibody in mice. *Nature*, **364**, 806–809.

Okura, T., Gong, L., Kamitani, T., Wada, T., Okura, I., Wei, C.F., Chang, H.M., Yeh, E.T. (1996) Protection against Fas/APO-1- and tumor necrosis factor-mediated cell death by a novel protein, sentrin. *J. Immunol.*, **157**, 4277–4281.

Oltvai, Z.N., Milliman, C.L., Korsmeyer, S.J. (1993) Bcl-2 heterodimerizes in vivo with a conserved homolog, Bax, that accelerates programmed cell death. *Cell*, **74**, 609–619.

Oltvai, Z.N., Korsmeyer, S.J. (1994) Checkpoints of dueling dimers foil death wishes. *Cell*, **79**, 189–192.

Opipari, A.W., Jr., Hu, H.M., Yabkowitz, R., Dixit, V.M. (1992) The A20 zinc finger protein protects cells from tumor necrosis factor cytotoxicity. *J. Biol. Chem.*, **267**, 12424–12427.

Orth, K., O'Rourke, K., Salvesen, G.S., Dixit, V.M. (1996) Molecular ordering of apoptotic mammalian CED-3/ICE-like proteases. *J. Biol. Chem.*, **271**, 20977–20980.

Owen-Schaub, L.B., Zhang, W., Cusack, J.C., Angelo, L.S., Santee, S.M., Fujiwara, T., Roth, J.A., Deisseroth, A.B., Zhang, W.W., Kruzel, E *et al.* (1995) Wild-type human p53 and a temperature-sensitive mutant induce Fas/APO-1 expression. *Mol. Cell Biol.*, **15**, 3032–3040.

Pan, G., O'Rourke, K., Chinnaiyan, A.M., Gentz, R., Ebner, R., Ni, J., Dixit, V.M. (1997a) The receptor for the cytotoxic ligand TRAIL. *Science*, **276**, 111–113.

Pan, G., Ni, J., Wei, Y-F., Yu, G., Gentz, R., Dixit, V.M. (1997b) An antagonist decoy receptor and a death domain-containing receptor for TRAIL. *Science*, **277**, 815–818.

Park, A., Baichwal, V.R. (1996) Systematic mutational analysis of the death domain of the tumor necrosis factor receptor 1-associated protein TRADD. *J. Biol. Chem.*, **271**, 9858–9862.

Pasparakis, M., Alexopoulou, L., Episkopou, V., Kollias, G. (1996) Immune and inflammatory responses in TNF alpha-deficient mice: a critical requirement for TNF alpha in the formation of primary B cell follicles, follicular dendritic cell networks and germinal centers, and in the maturation of the humoral immune response. *J. Exp. Med.*, **184**, 1397–1411.

Peter, M.E., Medema, J.P., Krammer, P.H. (1997a) Does the Caenorhabditis elegans protein CED-4 contain a region of homology to the mammalian death effector domain? *Cell Death Diff.*, **4**, 523–525.

Peter, M.E., Kischkel, F.C., Scheuerpflug, C., Medema, J.P., Debatin, K-M and Krammer, P.H. (1997b) Resistance of cultured peripheral T cells towards activation induced cell death involves a lack of recuitment of FLICE to the death-inducing signalling complex (DISC). *Eur. J. Immunol.*, **27**, 1207–1212.

Pfeffer, K., Matsuyama, T., Kundig, T.M., Wakeham, A., Kishihara, K., Shahinian, A., Wiegmann, K., Ohashi, P.S., Kronke, M., Mak, T.W. (1993) Mice deficient for the 55 kd tumor necrosis factor receptor are resistant to endotoxic shock, yet succumb to L. monocytogenes infection. *Cell*, **73**, 457–467.

Pickup, D.J. (1994) Poxviral modifiers of cytokine responses to infection. *Infect. Agents Dis.*, **3**, 116–27.

Pitti, R.M., Marsters, S.A., Ruppert, S., Donahue, C.J., Moore, A., Ashkenazi, A. (1996) Induction of apoptosis by Apo-2 ligand, a new member of the tumor necrosis factor cytokine family. *J. Biol. Chem.*, **271**, 12687–12690.

Ponton, A., Clement, M.V., Stamenkovic, I. (1996) The CD95 (APO-1/Fas) receptor activates NF-κB independently of its cytotoxic function. *J. Biol. Chem.*, **271**, 8991–8995.

Radeke, M.J., Misko, T.P., Hsu, C., Herzenberg, L.A., Shooter, E.M. (1987) Gene transfer and molecular cloning of the rat nerve growth factor receptor. *Nature*, **325**, 593–597.

Ray, C.A., Black, R.A., Kronheim, S.R., Greenstreet, T.A., Sleath, P.R., Salvesen, G.S., Pickup, D.J. (1992) Viral inhibition of inflammation: cowpox virus encodes an inhibitor of the interleukin-1 beta converting enzyme. *Cell*, **69**, 597–604.

Reed, J.C. (1997) Double identity for proteins of the Bcl-2 family. *Nature*, **387**, 773–776.

Rehemtulla, A., Hamilton, C.A., Chinnaiyan, A.M., Dixit, V.M. (1997) Ultraviolet radiation-induced apoptosis is mediated by activation of CD-95 (Fas/APO-1). *J. Biol. Chem.*, **272**, 25783–25786.

Rieux-Laucat, F., Le Deist, F., Hivroz, C., Roberts, I.A., Debatin, K.M., Fischer, A., de Villartay, J.P. (1995) Mutations in Fas associated with human lymphoproliferative syndrome and autoimmunity. *Science*, **268**, 1347–1349.

Rosette, C., Karin, M. (1996) Ultraviolet light and osmotic stress: activation of the JNK cascade through multiple growth factor and cytokine receptors. *Science*, **274**, 1194–1197.

Rothe, J., Lesslauer, W., Lotscher, H., Lang, Y., Koebel, P., Kontgen, F., Althage, A., Zinkernagel, R., Steinmetz, M., Bluethmann, H. (1993) Mice lacking the tumour necrosis factor receptor 1 are resistant to TNF-mediated toxicity but highly susceptible to infection by Listeria monocytogenes. *Nature*, **364**, 798–802.

Rothe, M., Pan, M.G., Henzel, W.J., Ayres, T.M., Goeddel, D.V. (1995) The TNFR2-TRAF signalling complex contains two novel proteins related to baculoviral inhibitor of apoptosis proteins. *Cell*, **83**, 1243–1252.

Rouvier, E., Luciani M.F., Golstein, P. (1993) Fas involvement in Ca^{2+}-independent T cell-mediated cytotoxicity. *J. Exp. Med.*, **177**, 195–200.

Roy, N., Mahadevan, R.S., McLean, M., Shutler, G., Yaraghi, Z. (1995) The gene for neuronal apoptosis inhibitory protein is partially deleted in individuals with spinal muscular atrophy. *Cell*, **80**, 167–178.

Roy, N., Deveraux, Q.L., Takahashi, R., Salvesen, G.S., Reed, J.C. (1997) The c-IAP-1 and c-IAP-2 proteins are direct inhibitors of specific caspases. *EMBO J.*, **16**, 6914–6925.

Rudel, T., Bokoch, G.M. (1997) Membrane and morphological changes in apoptotic cells regulated by caspase-mediated activation of PAK2. *Science*, **276**, 1571–1574.

Ruggiero, V., Johnson, S.E., Baglioni, C. (1987) Protection from tumor necrosis factor cytotoxicity by protease inhibitors. *Cell Immunol.*, **107**, 317–325.

Saas, P., Walker, P.R., Hahne, M., Quiquerez, A.L., Schnuriger, V., Perrin, G., French, L., Van Meir, E.G., de Tribolet, N., Tschopp, J., Dietrich P.Y. (1997) Fas ligand expression by astrocytoma in vivo: maintaining immune privilege in the brain? *J. Clin. Invest.*, **99**, 1173–1178.

Sabelko, K.A., Kelly, K.A., Nalm, M.H., Cross, A.H., Russell, J.H. (1997) Fas and Fas ligand enhance the pathogenesis of experimental allergic encephalomyelitis but are not essential for immune privilege in the central nervous system. *J. Immunol.*, **159**, 3096–3099.

Santana, P., Pena, L.A., Haimovitz Friedman, A., Martin, S., Green, D., McLoughlin, M., Cordon Cardo, C., Schuchman, E.H., Fuks, Z., Kolesnick, R. (1996) Acid sphingomyelinase-deficient human lymphoblasts and mice are defective in radiation-induced apoptosis. *Cell*, **86**, 189–99.

Sato, T., Irie, S., Kitada, S., Reed, J.C. (1995) FAP-1: a protein tyrosine phosphatase that associates with Fas. *Science*, **268**, 411–415.

Scaffidi, C., Medema, J.P., Krammer, P.H., Peter, M.E. (1997) FLICE is predominantly expressed as two functionally active isoforms, caspase-8/a and caspase-8/b. *J. Biol. Chem.*, **272**, 26953–26958.

Schall, T.J., Lewis, M., Koller, K.J., Lee, A., Rice, G.C., Wong, G.H., Gatanaga, T., Granger, G.A., Lentz, R., Raab, H., *et al.* (1990) Molecular cloning and expression of a receptor for human tumor necrosis factor. *Cell*, **61**, 361–370.

Schendel, S.L., Xie, Z., Montal, M.O., Matsuyama, S., Montal, M., Reed, J.C. (1997) Channel formation by antiapoptotic protein Bcl-2. *Proc. Natl. Acad. Sci. USA*, **94**, 5113–5118.

Schievella, A.R., Chen, J.H., Graham, J.R., Lin, L.L. (1997) MADD, a novel death domain protein that interacts with the type 1 tumor necrosis factor receptor and activates mitogen-activated protein kinase. *J. Biol. Chem.*, **272**, 12069–12075.

Schlegel, J., Peters, I., Orrenius, S., Miller, D.K., Thornberry, N.A., Yamin, T.T., Nicholson, D.W. (1996) CPP32/apopain is a key interleukin-1ß converting enzyme-like protease involved in Fas-mediated apoptosis. *J. Biol. Chem.*, **271**, 1841–1844.

Schulze-Osthoff, K., Bakker, A.C., Vanhaesebroeck, B., Beyaert, R., Jacob, W.A., Fiers, W. (1992) Cytotoxic activity of tumor necrosis factor is mediated by early damage of mitochondrial functions. Evidence for the involvement of mitochondrial radical generation. *J. Biol. Chem.*, **267**, 5317–5323.

Schulze-Osthoff, K., Beyaert, R., Vandevoorde, V., Haegeman, G., Fiers, W. (1993) Depletion of the mitochondrial electron transport abrogates the cytotoxic and gene-inductive effects of TNF. *EMBO J.*, **12**, 3095–3104.

Schulze-Osthoff, K. (1994) The Fas/APO-1 receptor and its deadly ligand. *Trends Cell Biol.*, **4**, 421–426.

Schulze-Osthoff, K., Krammer, P.H., Dröge, W. (1994) Divergent signalling via APO-1/Fas and the TNF receptor, two homologous molecules involved in physiological cell death. *EMBO J.*, **13**, 4587–4596.

Schulze-Osthoff, K., Los, M., Baeuerle, P.A. (1995) Redox signalling by transcription factors NF-κB and AP-1 in lymphocytes. *Biochem. Pharmacol.*, **50**, 735–741.

Screaton, G.R., Xu, X.N., Olsen, A.L., Cowper, A.E., Tan, R., McMichael, A.J., Bell, J.I. (1997) LARD, a new lymphoid-specific death domain containing receptor regulated by alternative pre-mRNA splicing. *Proc. Natl. Acad. Sci. USA*, **94**, 4615–4619.

Seino, K., Kayagaki, N., Okumura, K., Yagita, H. (1997) Antitumor effect of locally produced CD95 ligand. *Nat. Med.*, **3**, 165–170.

Selawry, H.P., Cameron, D.F. (1993) Sertoli cell-enriched fractions in successful islet cell transplantation. *Cell Transplant*, **2**, 123–129.

Sheridan, J.P., Marsters, S.A., Pitti, R.M., Gurney, A., Skubatch, M., Badwin, D., Ramakrishnan, L., Gray, C.L., Baker, K., Wood, W.I., Goddard, A.D., Godowski, P., Ashkenazi, A. (1997) Control of TRAIL-induced apoptosis by a family of signalling and decoy receptors. *Science*, **277**, 818–821.

Shimamoto, Y., Chen, R.L., Bollon, A., Chang, A., Khan, A. (1988) Monoclonal antibodies against human recombinant tumor necrosis factor: prevention of endotoxic shock. *Immunol. Lett.*, **17**, 311–317.

Shiraki, K., Tsuji, N., Shioda, T., Isselbacher, K.J., Takahashi, H. (1997) Expression of Fas ligand in liver metastases of human colonic adenocarcinomas. *Proc. Natl. Acad. Sci. USA*, **94**, 6420–6425.

Shu, H-B, Takeuchi, M., Goeddel, D.V. (1996) The tumor necrosis factor receptor 2 signal transducers TRAF2 and c-IAP1 are components of the tumor necrosis factor 1 signalling complex. *Proc. Natl. Acad. Sci. USA*, **93**, 13973–13978.

Shu, H-B., Halpin, D.R., Goeddel, D.V. (1997) Casper is a FADD- and caspase-related inducer of apoptosis. *Immunity*, **6**, 751–763.

Sillence, D.J., Allan, D. (1997) Evidence against an early signalling role for ceramide in Fas-mediated apoptosis. *Biochem. J.*, **324**, 29–32.

Simonian, P.L., Grillot, D.A., Andrews, D.W., Leber, B., Nunez, G. (1996) Bax homodimerization is not required for Bax to accelerate chemotherapy-induced cell death. *J. Biol. Chem.*, **271**, 32073–32077.

Simonian, P.L., Grillot, DAM., Nunez, G. (1997) Bak can accelerate chemotherapy-induced cell death independently of its heterodimerization with Bcl-X_L and Bcl-2. *Oncogene*, **15**, 1871–1875.

Singer, G.G., Abbas, A.K. (1994) The Fas antigen is involved in peripheral but not thymic deletion of T lymphocytes in T cell receptor transgenic mice. *Immunity*, **1**, 365–371.

Smith, C.A., Williams, G.T., Kingston, R., Jenkinson, E.J., Owen, J.J. (1989) Antibodies to CD3/T-cell receptor complex induce death by apoptosis in immature T cells in thymic cultures. *Nature*, **337**, 181–184.

Smith, C.A., Davis, T., Anderson, D., Solam, L., Beckmann, M.P., Jerzy, R., Dower, S.K., Cosman, D., Goodwin, R.G. (1990) A receptor for tumor necrosis factor defines an unusual family of cellular and viral proteins. *Science*, **248**, 1019–1023.

Smith, D.J., McGuire, M.J., Tocci, M.J., Thiele, D.L. (1997) IL-1ß convertase (ICE) does not play a requisite role in apoptosis induced in T lymphoblasts by Fas-dependent or Fas-independent CTL effector mechanisms. *J. Immunol.*, **158**, 163–170.

Song, H.Y., Rothe, M., Goeddel, D.V. (1996) The tumor necrosis factor-inducible zinc finger protein A20 interacts with TRAF1/TRAF2 and inhibits NF-κB activation. *Proc. Natl. Acad. Sci. USA*, **93**, 6721–6725.

Srinivasula, S.M., Ahmad, M., Fernandes-Alnemri, T., Litwack, G., Alnemri, E.S. (1996a) Molecular ordering of the Fas-apoptotic pathway: the Fas/APO-1 protease Mch5 is a CrmA-inhibitable protease that activates multiple Ced-3/ICE-like cysteine proteases. *Proc. Natl. Acad. Sci. USA*, **93**, 14486–14491.

Srinivasula, S.M., Fernandes-Alnemri, T., Zangrilli, J., Robertson, N., Armstrong, R.C., Wang, L., Trapani, J.A., Tomaselli, K.J., Litwack, G., Alnemri, E.S. (1996b) The Ced-3/interleukin-1ß converting enzyme-like homolog Mch6 and the lamin-cleaving enzyme Mch2α are substrates for the apoptotic mediator CPP32. *J. Biol. Chem.*, **271**, 27099–27106.

Srinivasula, S.M., Ahmad, M., Ottilie, S., Bullrich, F., Banks, S., Wang, Y., Fernandes-Alnemri, T., Croce, C.M., Litwack, G., Tomaselli, K.J., Armstrong, R.C., Alnemri, E.S. (1997) FLAME-1, a novel FADD-like anti-apoptotic molecule that regulates Fas/TNFR1-induced apoptosis. *J. Biol. Chem.*, **272**, 18542–18545.

Stamenkovic, I., Clark, E.A., Seed, B. (1989) A B-lymphocyte activation molecule related to the nerve growth factor receptor and induced by cytokines in carcinomas. *EMBO J.*, **8**, 1403–1410.

Stanger, B.Z., Leder, P., Lee, T.H., Kim, E., Seed, B. (1995) RIP: a novel protein containing a death domain that interacts with Fas/APO-1 (CD95) in yeast and causes cell death. *Cell*, **81**, 513–523.

Stokkers, P.C., Camoglio, L., van Deventer, S.J. (1995) Tumor necrosis factor (TNF) in inflammatory bowel disease: gene polymorphisms, animal models, and potential for anti-TNF therapy. *J. Inflamm.*, **47**, 97–103.

Strand, S., Hofmann, W.J., Hug, H., Müller, M., Otto, G., Strand, D., Mariani, S.M., Stremmel, W., Krammer, P.H., Galle, P.R. (1996) Lymphocyte apoptosis induced by CD95 (APO-1/Fas) ligand-expressing tumor cells. A mechanism of immune evasion? *Nature Med.*, **2**, 1306–1307.

Stuart, P.M., Griffith, T.S., Usui, N., Pepose, J., Yu, X., Ferguson, T.A. (1997) CD95 ligand (FasL)-induced apoptosis is necessary for corneal allograft survival. *J. Clin. Invest.*, **99**, 396–402.

Su, B., Karin, M. (1996) Mitogen-activated protein kinase cascades and regulation of gene expression. *Curr. Opin. Immunol.*, **8**, 402–411.

Suda, T., Takahashi, T., Golstein, P., Nagata, S. (1993) Molecular cloning and expression of the Fas ligand, a novel member of the tumor necrosis factor family. *Cell*, **75**, 1169–1178.

Suffys, P., Beyaert, R., Van Roy, F., Fiers, W. (1988) Involvement of a serine protease in tumour necrosis factor-mediated cytotoxicity. *Eur. J. Biochem.*, **178**, 257–265.

Susin, S.A., Zamzami, N., Castedo, M., Hirsch, T., Marchetti, P., Macho, A., Daugas, E., Geuskens, M., Kroemer, G. (1996) Bcl-2 inhibits the mitochondrial release of an apoptogenic protease. *J. Exp. Med.*, **184**, 1331–1341.

Szawlowski, P.W., Hanke, T., Randall, R.E. (1993) Sequence homology between HIV-1 gp120 and the apoptosis mediating protein Fas. *AIDS*, **7**, 1018.

Takahashi, T., Tanaka, M., Brannan, C.I., Jenkins, N.A., Copeland, N.G., Suda, T., Nagata, S. (1994) Generalized lymphoproliferative disease in mice, caused by a point mutation in the Fas ligand. *Cell*, **76**, 969–976.

Tamura, T., Ishishara, M., Lamphier, M.S., Tanaka, N., Oishi, I., Aizawa, S., Matsuyama, T., Mak, T.W., Taki, S., Taniguchi, T. (1995). An IRF-1 dependent pathway of DNA damage-induced apoptosis in mitogen-activated T lymphocytes. *Nature*, **376**, 596–599.

Tanaka, M., Suda, T., Haze, K., Nakamura, N., Sato, K., Kimura, F., Motoyoshi, K., Mizuki, M., Tagawa, S., Ohga, S., Hatake, K., Drumond, A.H., Nagata, S. (1996) Fas ligand in human serum. *Nature Med.*, **2**, 317–322.

Tartaglia, L.A., Weber, R.F., Figari, I.S., Reynolds, C., Palladino, M.A, Jr., Goeddel, D.V. (1991) The two different receptors for tumor necrosis factor mediate distinct cellular responses. *Proc. Natl. Acad. Sci. USA*, **88**, 9292–9296.

Tartaglia, L.A., Ayres, T.M., Wong, G.H., Goeddel, D.V. (1993a) A novel domain within the 55 kd TNF receptor signals cell death. *Cell*, **74**, 845–853.

Tartaglia, L.A., Pennica, D., Goeddel, D.V. (1993b) Ligand passing: the 75-kDa tumor necrosis factor (TNF) receptor recruits TNF for signalling by the 55-kDa TNF receptor. *J. Biol. Chem.*, 268, 18542–18548.

Teodoro, J.G., Branton, P.E. (1997) Regulation of apoptosis by viral gene products. *J. Virol*, **71**, 1739–1746.

Tepper, C.G., Jayadev, S., Liu, B., Bielawska, A., Wolff, R., Yonehara, S., Hannun, Y.A., Seldin, M.F. (1995) Role for ceramide as an endogenous mediator of Fas induced cytotoxicity. *Proc. Natl. Acad. Sci. USA*, **92**, 8443–8447.

Tewari, M., Dixit, V.M. (1995) Fas- and tumor necrosis factor-induced apoptosis is inhibited by the poxvirus crmA gene product. *J. Biol. Chem.*, **270,** 3255–3260.

Tewari, M., Quan, L.T., O'Rourke, K., Desnoyers, S., Zeng, Z., Beidler, D.R., Poirier, G.G., Salvesen, G.S., Dixit, V.M. (1995) Yama/CPP32 beta, a mammalian homolog of CED-3, is a CrmA-inhibitable protease that cleaves the death substrate poly(ADP-ribose) polymerase. *Cell*, **81**, 801–809.

Thome, M., Schneider, P., Hofmann, K., Fickenscher, H., Meinl, E., Neipel, F., Mattmann, C., Burns, K., Bodmer, J.L., Schroter, M., Scaffidi, C., Krammer, P.H., Peter, M.E., Tschopp, J. (1997) Viral FLICE-inhibitory proteins (FLIPs) prevent apoptosis induced by death receptors. *Nature*, **386**, 517–521.

Thornberry, N.A., Bull, H.G., Calaycay, J.R., Chapman, K.T., Howard, A.D., Kostura, M.J., Miller, D.K., Molineaux, S.M., Weidner, J.R., Aunins, J., *et al.* (1992) A novel heterodimeric cysteine protease is required for interleukin-1 beta processing in monocytes. *Nature*, **356**, 768–774.

Ting, A.T., Pimentel, Muinos, F.X., Seed, B. (1996) RIP mediates tumor necrosis factor receptor 1 activation of NF-κB but not Fas/APO-1-initiated apoptosis. *EMBO J.*, **15**, 6189–6196.

Tracey, K.J., Cerami, A. (1993) Tumor necrosis factor, other cytokines and disease. *Annu. Rev. Cell Biol.*, **9**, 317–343.

Trauth, B.C., Klas, C., Peters, A.M., Matzku, S., Möller, P., Falk, W., Debatin, K.M., Krammer, P.H. (1989) Monoclonal antibody-mediated tumor regression by induction of apoptosis. *Science*, **245**, 301–305.

Tsujimoto, Y. (1997) Apoptosis and necrosis: Intracellular ATP level as a determinant for cell death modes. *Cell Death Differ.*, **4**, 429–434.

Van Antwerp, D.J., Martin, S.J., Kafri, T., Green, D.R., Verma, I.M. (1996) Suppression of TNF-α-induced apoptosis by NF-κB. *Science*, **274**, 787–789.

Vandenabeele, P., Declercq, W., Vanhaesebroeck, B., Grooten, J., Fiers, W. (1995) Both TNF receptors are required for TNF-mediated induction of apoptosis in PC60 cells. *J. Immunol.*, **154**, 2904–2913.

Vander Heiden, M.G., Chandel, N.S., Williamson, E.K., Schumacker, P.T., Thompson, C.G. (1997) Bcl-x_L reguates the membrane potential and volume homeostasis of mitochondria. *Cell*, **91**, 627–637.

Vandevoorde, V., Haegeman, G., Fiers, W. (1997) Induced expression of trimerized intracellular domains of the human tumor necrosis factor (TNF) p55 receptor elicits TNF effects. *J. Cell Biol.*, **137**, 1627–1638.

Vanhaesebroeck, B., Reed, J.C., De Valck, D., Grooten, J., Miyashita, T., Tanaka, S., Beyaert, R., Van Roy, F., Fiers, W. (1993) Effect of bcl-2 proto-oncogene expression on cellular sensitivity to tumor necrosis factor mediated cytotoxicity. *Oncogene*, **8**, 1075–1081.

Vercammen, D., Vandenabeele, P., Beyaert, R., Declercq, W., Fiers, W. (1997) Tumour necrosis factor-induced necrosis versus anti-Fas-induced apoptosis in L929 cells. *Cytokine*, **9**, 801–808.

Villunger, A., Egle, A., Marschitz, I., Kos, M., Böck, G., Ludwig, H., Geley, S., Kofer, R., Greil, R. (1997) Constitutive expression of Fas (APO-1/CD95) ligand on multiple myeloma cells: a potential mechanism of tumor-induced suppression of immune surveillance. *Blood*, **90**, 12–20.

Vincenz, C., Dixit, V.M. (1997) Fas-associated death domain protein interleukin-1ß-converting enzyme 2 (FLICE2), an ICE/Ced-3 homologue, is proximally involved in CD95- and p55-mediated death signalling. *J. Biol. Chem.*, **272** , 6578–6583.

von Boehmer. (1997) Lymphotoxins: from cytotoxicity to lymphoid organogenesis. *Proc. Natl. Acad. Sci. USA*, **94**, 8926–8927.

Walczak, H., Degli-Esposti, M.A., Johnson, R.S., Smolak, P.J., Waugh, J.Y., Boiani, N., Timour, M.S., Gerhart, M.J., Schooley, K.A., Smith, C.A., Goodwin, R.G., Rauch, C.T. (1997) TRAIL-R2: a novel apoptosis-mediating receptor for TRAIL. *EMBO J.*, **16**, 5386–5397.

Waldner, H., Sobel, R.A., Howard, E., Kuchroo, V.K. (1997) Fas and FasL deficient mice are resistant to induction of autoimmune encephalomyelitis. *J. Immunol.*, **159**, 3100–3103.

Walker, N.P., Talanian, R.V., Brady, K.D., Dang, L.C., Bump, N.J., Ferenz, C.R., Franklin, S., Ghayur, T., Hackett, M.C., Hammill, L.D., *et al.* (1994) Crystal structure of the cysteine protease interleukin-1 beta-converting enzyme: a $(p20/p10)_2$ homodimer. *Cell*, **78**, 343–52.

Wallach, D. (1997) Placing death under control. *Nature*, **388**, 123–126.

Wang, L., Miura, M., Bergeron, L., Zhu, H., Yuan, J. (1994) Ich-1, an Ice/ced-3-related gene, encodes both positive and negative regulators of programmed cell death. *Cell*, **78**, 739–750.

Wang, S., Miura, M., Jung, Yk., Zhu, H., Gagliardini, V., Shi, L., Greenberg, A.H., Yuan, J. (1996a) Identification and characterization of Ich-3, a member of the interleukin-1ß converting enzyme (ICE)/Ced-3 family and an upstream regulator of ICE. *J. Biol. Chem.*, **271**, 20580–20587.

Wang, H.G., Rapp, U.R., Reed, J.C. (1996b) Bcl-2 targets the protein kinase Raf-1 to mitochondria. *Cell*, **87**, 629–638.

Wang, C.Y., Mayo, M.W., Baldwin, ASJ. (1996c) TNF- and cancer therapy-induced apoptosis: potentiation by inhibition of NF-kappaB. *Science*, **274**, 784–787.

Wang, Z.Q., Stingl, L., Morrison, C., Jantsch, M., Los, M., Schulze-Osthoff, K., Wagner, E.F. (1997) PARP is important for genomic stability but dispensable in apoptosis. *Genes Dev.*, **11**, 2347–2358.

Watanabe-Fukunaga, R., Brannan, C.I., Itoh, N., Yonehara, S., Copeland, N.G., Jenkins, N.A., Nagata, S. (1992a) The cDNA structure, expression, and chromosomal assignment of the mouse Fas antigen. *J. Immunol.*, **148**, 1274–1279.

Watanabe-Fukunaga, R., Brannan, C.I., Copeland, N.G., Jenkins, N.A., Nagata, S. (1992b) Lymphoproliferation disorder in mice explained by defects in Fas antigen that mediates apoptosis. *Nature*, **356**, 314–317.

Watts, J.D., Gu, M., Polverino, A.J., Patterson, S.D., Aebersold, R. (1997) Fas-induced apoptosis of T cells occurs independently of ceramide generation. *Proc. Natl. Acad. Sci. USA*, **94**, 7292–7296.

Westendorp, M.O., Frank, R., Ochsenbauer, C., Stricker, K., Dhein, J., Walczak, H., Debatin, K.M., Krammer, P.H. (1995) Sensitization of T cells to CD95-mediated apoptosis by HIV-1 Tat and gp120. *Nature*, **375**, 497–500.

Wiegmann, K., Schütze, S., Machleidt, T., Witte, D., Krönke, M. (1994) Functional dichotomy of neutral and acidic sphingomyelinases in tumor necrosis factor signalling. *Cell*, **78**, 1005–1015.

Wiley, S.R., Schooley, K., Smolak, P.J., Din, W.S., Huang, C.P., Nicholl, J.K., Sutherland, G.R., Smith, T.D., Rauch, C., Smith, C.A. (1995) Identification and characterization of a new member of the TNF family that induces apoptosis. *Immunity*, **3**, 673–682.

Wilson, K.P., Black, J.A., Thomson, J.A., Kim, E.E., Griffith, J.P., Navia, M.A., Murcko, M.A., Chambers, S.P., Aldape, R.A., Raybuck, S.A., *et al.* (1994) Structure and mechanism of interleukin-1ß converting enzyme. *Nature*, **370**, 270–275.

Wong, G.H., Elwell, J.H., Oberley, L.W., Goeddel, D.V. (1989) Manganous superoxide dismutase is essential for cellular resistance to cytotoxicity of tumor necrosis factor. *Cell*, **58**, 923–931.

Wong, G.H., Tartaglia, L.A., Lee, M.S., Goeddel, D.V. (1992) Antiviral activity of tumor necrosis factor is signaled through the 55-kDa type I TNF receptor. *J. Immunol.*, **149**, 3350–3353.

Wong, G.H., Goeddel, D.V. (1994) Fas antigen and p55 TNF receptor signal apoptosis through distinct pathways. *J. Immunol.*, **152**, 1751–1755.

Wright, S.C., Wei, A.S., Zhong, J., Zheng, H., Kinder, D.H., Larrick, J.W. (1994) Purification of a 24-kD protease from apoptotic tumor cells that activates DNA fragmentation. *J. Exp. Med.*, **180**, 2113–2123.

Wright, S.C., Schellenberger, U., Wang, H., Kinder, D.H., Talhouk, J.W., Larrick, J.W. (1997) Activation of CPP32-like proteases is not sufficient to trigger apoptosis. Inhibition of apoptosis by agents that suppress activation of AP24, but not CPP32-like activity. *J. Exp. Med.*, **186**, 1107–1117.

Wu, M., Lee, H., Bellas, R.E., Schauer, S.L., Arsura, M., Katz, D., FitzGerald, M.J., Rothstein, T.L., Sherr, D.H., Sonenshein, G.E. (1996) Inhibition of NF-κB/Rel induces apoptosis of murine B cells. *EMBO J.*, **15**, 4682–4690.

Wu, D., Wallen, H.D., Nunez, G. (1997) Interaction and regulation of subcellular localization of CED-4 by CED-9. *Science*, **275**, 1126–1129.

Xue, D., Horvitz, H.R. (1995) Inhibition of the Caenorhabditis elegans cell-death protease CED-3 by a CED-3 cleavage site in baculovirus p35 protein. *Nature*, **377**, 248–251.

Yanagisawa, J., Takahashi, M., Kanki, H., Yano-Yanagisawa, H., Tazunoki, T., Sawa, E., Nishitoba, T., Kamishohara, M., Kobayashi, E., Kataoka, S., Sato, T. (1997) The molecular interaction of Fas and FAP-1. A tripeptide blocker of human Fas interaction with FAP-1 promotes Fas-induced apoptosis. *J. Biol. Chem.*, **272**, 8539–8545.

Yang, X., Khoravi, Far, R., Chang, H.Y., Baltimore, D. (1997a) Daxx, a novel Fas-binding protein that activates JNK and apoptosis. *Cell*, **89**, 1067–1076.

Yang, J., Liu, X., Bhalla, K., Kim, C.N., Ibrado, A.M., Cai, J., Peng, T.I., Jones, D.P., Wang, X. (1997b) Prevention of apoptosis by Bcl-2: release of cytochrome c from mitochondria blocked. *Science*, **275**, 1129–1132.

Yonehara, S., Ishii, A., Yonehara, M. (1989) A cell killing monoclonal antibody (anti-Fas) to a cell surface antigen co downregulated with the receptor of tumor necrosis factor. *J. Exp. Med.*, **169**, 1747–1756.

Zagury, J.F., Cantalloube, H., Achour, A., Cho, Y.Y., Fall, L., Lachgar, A., Chams, V., Astgen, A., Biou, D., Picard, O., *et al*; (1993) Striking similarities between HIV-1 Env protein and the apoptosis mediating cell surface antigen Fas. Role in the pathogenesis of AIDS. *Biomed. Pharmacother.*, **47**, 331–335.

Zanke, B.W., Boudreau, K., Rubie, E., Winnett, E., Tibbles, L.A., Zon, L., Kyriakis, J., Liu, F.F., Woodgett, J.R. (1996) The stress-activated protein kinase pathway mediates cell death following injury induced by cis-platinum, UV irradiation or heat. *Curr. Biol.*, **6**, 606–613.

Zha, J., Harada, H., Yang, E., Jockel, J., Korsmeyer, S.J. (1996) Serine phosphorylation of death agonist BAD in response to survival factor results in binding to 14–3–3 not BCL-X_L. *Cell*, **87**, 619–628.

Zhang, X., Brunner, T., Carter, L., Dutton, R.W., Rogers, P., Bradley, L., Sato, T., Reed, J.C., Green, D., Swain, S.L. (1997) Unequal death in T helper cell (Th)1 and Th2 effectors: Th1, but not Th2, effectors undergo rapid Fas/FasL-mediated apoptosis. *J. Exp. Med.*, **185**, 1837–1849.

Zou, H., Henzel, W.J., Liu, X., Lutschg, A., Wang, X. (1997) Apaf-1, a human protein, homologous to C. elegans Ced-4, participates in cytochrome c-dependent activation of caspase-3. *Cell*, **90**, 405–413.

2. THE ROLE OF SPHINGOLIPIDS IN STRESS RESPONSES AND APOPTOSIS IN EUKARYOTES

SHEREE D. LONG[*] AND YUSUF A. HANNUN[*†]

[*] *Department of Medicine, Duke University Medical Center, Durham, North Carolina 27710*

Sphingolipids constitute an important class of membrane lipids in eukaryotic cells whose structural complexity exceeds that of phospholipids. With the elucidation of the sphingomyelin cycle, in which intracellular ceramide is generated, sphingolipids have become recognized for their role in signal transduction and cell regulation. Investigation of sphingolipid-mediated biology has revealed ceramide as an important regulator of cellular processes including terminal differentiation, cell cycle arrest, cellular senescence, and cell death. The action of several extracellular agents and insults such as tumor necrosis factor α, Fas ligands, and chemotherapeutic agents results in the activation of an intracellular sphingomylinase which acts on membrane sphingomyelin and generates ceramide. Mechanisms for ceramide action involve regulation of protein phosphorylation via activation of a protein phosphatase and a protein kinase. Ceramide has also been shown to regulate multiple downstream targets such as activation of the proteases involved in apoptosis, stress-activated kinases, and the retinoblastoma gene product which causes cell cycle arrest. These effects appear to result in profound changes in cell growth behavior and support a role for ceramide as a pluripotent mediator of intracellular stress responses.

KEY WORDS: sphingomyelin, ceramide, CAPP, CAPK, PKCζ, NF-κB, SAPK.

INTRODUCTION

In the past, lipids were viewed largely as structural building blocks of the membrane. They are now recognized as precursors of bioactive molecules that are generated in cells following stimulation of cell-surface receptors and function as second messengers and bioeffector substrates. Sphingolipids, a diverse class of biomolecules found in all eukaryotic membranes, have recently been recognized as important mediators of membrane signal transduction pathways (Hannun *et al.*, 1986; Hannun and Bell, 1989; Hannun, 1994). They have been shown to play a role in such crucial cellular functions as regulation of cell growth, differentiation, oncogenesis, and apoptosis (Kolesnick, 1991; Dobrowsky and Hannun, 1994; Hannun and Linardic, 1994).

Since the discovery that sphingosine and lysosphingolipids are potent inhibitors of protein kinase C (PKC), products of sphingolipid hydrolysis have been demonstrated to serve as "lipid second messengers" in a number of pathways (Hannun and Bell, 1989; Kolesnick, 1991; Dbaio *et al.*, 1993; Merrill *et al.*, 1993; Chen *et al.*, 1995). Evaluation of sphingolipid turnover as a mechanism for cell regulation led to the discovery of the sphingomyelin cycle in which activation of a neutral

† *Corresponding Author*: Department of Biochemistry; Medical University of South Carolina, 171 Ashley Ave. Charleston, SC 29425. Tel.: 843 792 4321. Fax: 843 792 4322.

sphingomyelinase leads to the breakdown of sphingomyelin and the generation of phosphocholine and ceramide (Okazaki *et al.*, 1990; Kim *et al.*, 1991). The cycle is completed with the resynthesis of sphingomyelin, presumably by the transfer of a choline phosphate headgroup to ceramide. As such, the sphingomyelin cycle emerges as a sphingolipid analog of the phosphatidylinositol cycle. Various cytokines, hormones, and growth factors are known to induce the hydrolysis of membrane sphingomyelin following ligand binding of receptors. Ceramide, a potent second messenger generated by activation of this pathway, has been identified as an important mediator of growth inhibition, *c-myc* down-regulation, apoptosis, and the activation of nuclear factor κB (Kolesnick and Golde, 1994; Liscovitch and Cantley, 1994). Despite the numerous reports demonstrating biologic activity for ceramide *in vivo*, there is no clearly defined intracellular target for ceramide activity.

In addition to ceramide, the immediate product of sphingomyelin hydrolysis, attention has also focused on sphingosine, the backbone sphingoid base of all sphingolipids and on sphingosine 1-phosphate. Sphingosine has been shown to modulate numerous cellular functions, including inhibition of protein kinase C activity. Multiple biochemical targets and biological activities of sphingosine have been identified through the use of sphingosine as a pharmacologic agent. In addition to playing a role in growth suppression in PKC dependent and independent pathways, sphingosine has been demonstrated to regulate diacylglycerol and phosphatidic acid levels, calcium release, receptor tyrosine kinase activities, casein kinase II, and to modulate the activity of endogenous protein kinases (Hannun *et al.*, 1986; Hannun and Bell , 1989). On the other hand sphingosine-1-phosphate, another product of sphingosine metabolism, has been shown to induce mitogenesis and to inhibit cell motility and phagokinesis of tumor cells (Zhang *et al.*, 1991; Sadahira *et al.*, 1992). Thus, a biologic role for sphingolipids and their metabolites, including ceramide, sphingosine-1-phosphate, and sphingosine, in cell regulation is becoming increasingly apparent. It is likely that sphingolipids are components of multiple signalling cascades, generating a variety of lipid second messengers which are active participants in cellular metabolism. However, determining the biological functions and mode of action of individual sphingolipids remains difficult to elucidate because of the great variety of sphingolipids present in multicellular eukaryotes.

There is increasing evidence supporting the existence of novel lipid signal transduction pathways that may be specifically involved in stress induced apoptosis. Ceramide has been proposed to play a role in the initiation of apoptosis and other stress responses in mammalian cells. Sphingolipids are also proving to be critcal in the yeast stress responses. In the following chapter, we summarize the current understanding of the role sphingolipids play in stress responses and apoptosis.

SPHINGOLIPIDS ARE ESSENTIAL FOR CELL VIABILITY AND STRESS RESPONSES IN YEAST AND IN MAMMALIAN CELLS

Studies in yeast provided the first genetic evidence in any species that sphingolipids are necessary for survival. In contrast to the sphingolipid diversity in mammalian cells, the sphingolipids of *Saccharomyces cerevisiae* are limited to one major and two minor types of structurally related sphingolipids (Smith and Lester, 1974). The simplicity of this unicellular eukaryote serves as a model system in which to study sphingolipid biology and function.

In 1983, Lester and colleagues isolated a sphingolipid deficient mutant in a strain of *S. cerevisiae* which requires a sphingolipid long chain base, such as phytosphingosine, for growth and viability (Wells and Lester, 1983). The defect in these long chain base deficient (lcb) strains was ultimately shown to be a defect in the gene coding for serine palmitoyltransferase, the first enzyme in the sphingolipid synthetic pathway. Suppressor strains were selected which bypass the requirement for exogenous long chain bases because of a mutation in a suppressor gene termed SLC1 (sphingolipid compensation) (Dickson *et al.*, 1990). These SLC strains make a set of novel glycerolipids which contain the same polar head groups found in yeast sphingolipids; thus demonstrating that the viability function of sphingolipids in yeast resides in the head group (Lester *et al.*, 1993). These novel phospholipids structurally mimic sphingolipids and thereby may allow cell growth. The strain is, however, capable of making the normal species and levels of sphingolipids if phytosphingosine is added to the culture medium. Hence the SLC strains are of great value in understanding sphingolipid function.

Table 2.1 Systems involving sphingolipid signalling in the stress response.

Stressor	*Cell Type*	*Response*
osmostress	yeast	increased glycerol
heatshock	yeast	trehalose accumulation
pH	yeast	increased proton transport
heatshock	mammalian cells	increased heat shock proteins
TNFα	HL-60 cells	terminal differentiation
TNFα	U937 cells	apoptosis
serum starvation	Molt-4 cells	cell cycle arrest
Fas	Jurkat cells	apoptosis
hydrogen peroxide	U937 cells	apoptosis
?	fibroblast	senescence
cytotoxic chemotherapy	mammalian cells	apoptosis
UV-radiation	U937 cells	"UV response"

In addition, the study of SLC strains has revealed that sphingolipids are required for the cells to respond to certain environmental stresses (Patton *et al.*, 1992). Strains lacking sphingolipids (phytosphingosine omitted from the culture medium) were unable to grow under extremes of pH, temperature, and osmotic stress. Indeed, the product of an SLC suppressor gene permitted growth without sphingolipids, yet only in a limited range of environments. Outside this range, sphingolipids appeared to be an essential component of the cell's response and adaptive mechanisms which allow for growth under stressful conditions.

Sphingolipids have also been shown to be essential for growth in mammalian cells. Hanada *et al.* (1990) isolated a temperature-sensitive Chinese hamster ovary cell mutant (strain SPB-1) with a thermolabile serine palmitoyltransferase. When the mutant strain was cultured at non-permissive temperatures, *de novo* sphingolipid synthesis ceased and the growth rate gradually decreased suggesting that the deficiency in sphingolipids was responsible for the temperature sensitive growth of the mutant cells. Exogenous sphingosine restored the sphingomyelin and ganglioside sialyl lactosylceramide (G_{M3}) contents to normal levels and allowed the mutant cells to grow even at the non-permissive temperature (Hanada *et al.*, 1992). Similarly, exogenous sphingomyelin suppressed temperature sensitivity of the SB-1 growth and restored the sphingomyelin content 100% while G_{M3} content was restored to only 50% of the parental levels. In contrast, the addition of glucosylceramide, which restored G_{M3} but not sphingomyelin levels failed to suppress the temperature sensitivity. Thus, the results indicated that the lack of sphingomyelin is primarily responsible for the temperature sensitivity of the mutant cell growth demonstrating the essential role of sphingolipids in growth and viability. The vital importance of sphingomyelin in mammalian cells and inositol phosphorylceramides in yeast cells has become increasingly apparent and may play similar roles in generating ceramide as a potent regulator of signal transduction.

CERAMIDE, THE CARDINAL LIPID OF SPHINGOLIPID METABOLISM

The discovery of the sphingomyelin cycle led to the identification of ceramide as the primary metabolite of agonist induced sphingomyelin hydrolysis. Ceramide has been studied extensively and its role as a bioeffector is becoming increasingly apparent. Generated via the sphingomyelin pathway, ceramide emerges as a major mediator of growth suppression through induction of differentiation, initiation of apoptosis, activation of tumor suppressors, and induction of specific cell cycle arrest. These activities have been studied in a myriad of cell lines in which agonist induced ceramide production can be correlated with the effect of exogenous ceramides on cell growth and differentiation. Growth suppression was one of the earliest biologic effects attributed to ceramide. This was first demonstrated in human leukemia HL60 cells where treatment with 1α, 25-dihydroxyvitamin D_3 stimulated sphingomyelin (SM) hydrolysis within 2 to 4 hours generating choline-phosphate and ceramide (Okazaki *et al.*, 1989). Since these initial studies with HL60 cells, the list of inducers known to act via the SM cycle continues to expand and thus, a novel cell regulation pathway emerges with the lipid second messenger, ceramide, at the center.

BIOLOGIC ACTIVITY OF CERAMIDE AS A REGULATOR OF THE APOPTOTIC RESPONSE AND GROWTH SUPPRESSION

Ceramide and apoptosis

The induction of programmed cell death involves the activation of biochemical mechanisms leading to regulated cell death or apoptosis. Various studies have identified ceramide as a possible mediator of apoptosis in response to the action of multiple agents and insults including γ-interferon, the presence of hypoxia, Fas ligand, and tumor necrosis factor-α (TNFα) (Kim *et al.*, 1991; Dressler *et al.*, 1992;

Wiegmann *et al.*, 1992; Yanga and Watson, 1992; Tepper *et al.*, 1994). Mammalian cells have an intrinsic biochemical response that allows them to repair the damage caused by these agents by undergoing cell cycle arrest or, if the damage is irreversible, apoptosis is initiated. TNFα has been shown to induce activation of neutral sphingomyelinase through the activation of phospholipase A_2 and the generation of arachidonic acid as an intermediate coupling mechanism (Jayadev *et al.*, 1994). As a result of sphingomyelin hydrolysis, the elevation of intracellular ceramide levels is thought to mediate the effects of TNFα-induced apoptosis. Studies carried out by Obeid and coworkers demonstrated that in U937 monoblastic leukemia cells, TNFα-induced SM hydrolysis and ceramide generation preceded the cellular effects of TNFα-induced growth suppression and DNA fragmentation (Obeid *et al.*, 1993). In addition, exogenous addition of ceramide was found to mimic the apoptotic effects of TNFα in a time and dose dependent manner. The structurally related lipid, dihyrdroceramide, did not induce apoptosis, suggesting the action of ceramide is specific. However, the question remained whether ceramide itself was eliciting the stress responses, or if subsequent degradation to other sphingolipids such as sphingosine or sphingosine-1-phosphate was occurring. Although ceramide inhibits phospholipase D activation and sphingosine induces its activation (Venable *et al.*, 1994), in many cell systems ceramide and sphingosine induce similar activities including growth suppression and induction of Rb dephosphorylation (Dbaio *et al.*, 1995). Recent studies using compounds that result in accumulation of endogenous ceramide such as bacterial sphingomyelinase, PDMP (a cerebroside synthase inhibitor), and D-MAPP (which inhibits ceramidase) distinguished the effects of endogenous ceramide from those of sphingosine (Inokuchi *et al.*, 1989; Bielawska *et al.*, 1992a, b). The results demonstrated that an increase in ceramide alone without further metabolism could elicit the same cellular activities as exogenously added ceramide analogs. Therefore, a direct role for ceramide in mediating apoptosis and growth suppression is supported by indirect metabolic manipulation of endogenous amounts of ceramide.

Ceramide and cell cycle arrest

Among the numerous biological effects elicited by ceramide, cell cycle arrest has been recently discovered as an integral component. The progression of cell cycle is tightly regulated by multiple genes and gene products in response to proliferation and antiproliferation. Since cultured cells depend on various serum growth factors for continued growth, serum deprivation becomes a powerful mechanism by which cells can be arrested in the cell cycle and induced to undergo programmed cell death. In a study carried out by Jayadev *et al.* (1995) ceramide was found to induce a significant block in cell cycle progression in G0/G1 accompanied by apoptosis. Serum starvation of Molt-4 leukemia cells led to a remarkable increase in endogenous ceramide levels which paralleled apoptosis and cell cycle arrest. However, investigation of the mechanism of increased ceramide production led to the finding that serum withdrawal appears to stimulate a signalling cascade distinct from the TNFα stimulation of the sphingomyelin cycle. Serum deprivation resulted in substantial activation of a particulate, magnesium-dependent, neutral sphingomyelinase rather than the magnesium-independent enzyme observed with TNFα. These studies identified a novel role for ceramide in cell cycle arrest and raised the possibility that ceramide modulates the endogenous machinery regulating cell cycle progression.

The product of the retinoblastoma gene, a tumor suppressor protein known as Rb, has emerged as a possible target for serum deprivation-induced cell cycle arrest. Recent studies have also demonstrated that Rb may be an important target of ceramide regulation. Hence mechanistically, ceramide-induced cell cycle arrest could be mediated by the Rb protein since phosphorylation of Rb has been found to be regulated by ceramide. Molt-4 cells treated with C_6-ceramide showed significant Rb dephosphorylation in a time frame which precedes G0/G1 arrest and at ceramide concentrations comparable to endogenous levels which produce arrest (Dbaio *et al.*, 1995). The specificity of this effect was confirmed by the use of several other lipids, including dihydro-C_6-ceramide which failed to change the phosphorylation status of Rb. Furthermore, experiments performed in different cell lines either containing or lacking functional Rb demonstrated that the presence of functional and active Rb protein is essential for mediating the effects of ceramide on growth suppression and cell cycle arrest. Taken together, these studies show that Rb is a downstream target of ceramide and that ceramide-induced Rb dephosphorylation is a biochemical precursor to cell cycle arrest.

Ceramide and senescence

Our knowledge of lipid mediated signal transduction pathways in cellular senescence is modest. Cellular senescence is defined as the limited capacity of cells to undergo population doublings due to the inability to respond to mitogenic signals with DNA synthesis, growth, and proliferation (Goldstein, 1990; Kirkland, 1992). The mechanism by which senescent cells fail to respond to mitogenic stimuli remains, however, poorly understood. Preliminary studies have begun to indicate changes in components of the sphingomyelin cycle with respect to cellular aging. Studies by Venable and colleagues provided direct evidence that defects in the DAG/PKC pathway underlies the mitogenic defect in cell senescence (Venable *et al.*, 1994). Unlike their young counterparts, senescent human diploid fibroblasts (HDF) did not respond to serum-induced activation of phospholipase D (PLD) which resulted in an inability to generate a sustained diacylglycerol signal. Without DAG production, PKC was not translocated resulting in failure to transcribe c-fos and activate AP-1, a transcription factor required for cell replication. They also investigated the possible connection between the ceramide pathway and PLD/DAG pathway and found that the inability of senescent cells to activate PLD may be attributed to elevated levels of ceramide. During senescence, endogenous levels of ceramide increased considerably (4-fold) and specifically (compared to other lipids) (Venable *et al.*, 1995). The elevated ceramide in senescent cells was a result of increased neutral sphingomyelinase activity (8-10 fold). The observed increases in

SMase activity and ceramide appeared to be stable and prolonged as opposed to the transient signalling increases in response to inducers of apoptosis and differentiation. Moreover, the addition of C_6-ceramide to young cells induced a senescent phenotype characterized by the inability to undergo DNA synthesis and mitogenesis. In addition, exogenous ceramide at concentrations that mimic endogenous levels in senescence also induced Rb dephosphorylation and inhibited serum-induced AP-1 activation in young HDF, recapitulating the established biochemical and molecular changes of cell senescence. Based on these studies, ceramide appears to be one of the first biologically active molecules regulating a biochemical event that distinguishes senescence from cell cycle growth arrest. Important insight into the molecular mechanisms involved in cellular senescence will come with further understanding of the sphingomyelinase/ceramide pathway and its mechanism of action on cellular regulation.

Ceramide and differentiation

The role of the sphingomyelin cycle in the regulation of differentiation and cell growth was first described in HL60 leukemia cells (Okazaki *et al.*, 1989; 1990; Kim *et al.*, 1991). Treatment with the cell permeable ceramide analog, C_2-ceramide, was sufficient to induce monocytic differentiation and inhibit growth of these cells, mimicking the action of TNFα, 1α, 25-dihydroxyvitamin D_3, and γ-interferon in inducing monocytic differentiation. The effects of C_2-ceramide were probably not due to metabolic conversion to other sphingolipids since neither sphingosine or N-ethylsphingosine induced differentiation in HL60 cells. Ceramide-induced cell differentiation has been extended to other cell systems as well. In T9 glioma cells, nerve growth factor (NGF), which promotes growth inhibition and differentiation, has been shown to activate the sphingomyelin cycle resulting in ceramide generation (Dobrowsky *et al.*, 1994). The addition of C_2-ceramide mimicked the effects of NGF and resulted in a dose-dependent inhibition of proliferation and induction of differentiation.

The protooncogene, *c-myc*, which plays a central role in cell proliferation and differentiation, has been identified as a candidate downstream effector of ceramide-induced differentiation (Kim *et al.*, 1991; Wolff *et al.*, 1994). In the HL60 cell line, treament with C_2-ceramide, as well as TNFα, resulted in a rapid downregulation of *c-myc* mRNA. The mechanism by which ceramide exerted its effect was demonstrated to be a block in transcriptional elongation similar to the effect produced by TNFα. Recent work has suggested that ceramide's effects are mediated via ceramide activated protein phosphatase (CAPP) (Wolff *et al.*, 1994). A role for CAPP in the ceramide signalling pathway was supported by the observation that both CAPP activation and c-myc downregulation exhibit similar specificities for various ceramide analogs and stereoisomers. In addition, the effects of ceramide were abolished when cells were treated with okadaic acid, an inhibitor of CAPP. These studies provide significant insight for the mechanism by which ceramide regulates cell proliferation and differentiation and show that CAPP may function as an intracellular target for the action of ceramide.

A POTENTIAL ROLE FOR SPHINGOLIPIDS IN INFLAMMATION AND THE IMMUNE RESPONSE

Ceramide and the immune response

TNFα is now known to be one of the most pleiotropic cytokines and is viewed as a principal mediator of a large number of cellular responses including the immuno-inflammatory response. Most of the biological effects of TNFα are triggered by activation of a variety of genes in a multitude of target cells. Induction of many genes by TNFα is mediated at least in part, by the activation of a family of transcription factors known collectively as NF-κB (Lowenthal *et al.*, 1989; Lenardo and Baltimore, 1989; Osborn *et al.*, 1989; Molitor *et al.*, 1990); a cellular response that links cell surface receptor activation to transcriptional events in the nucleus. TNFα is one of the few cytokines known to stimulate the translocation of NF-κB from the cytosol to the nucleus where the transcription factor exerts either a positive or negative control over cellular genes. Activation of NF-κB involves the release of the inhibitory subunit Iκ-B from a cytoplasmic complex containing the DNA binding subunit p50 and the regulatory subunit Rel A. The Iκ-B proteins inhibit DNA binding and prevent nuclear uptake of NF-κB complexes (Baeuerle and Baltimore, 1994).

The role of the sphingomyelin/ceramide pathway in the activation of NF-κB remains uncertain. Exogenous ceramides have for the most part been shown not to induce nuclear translocation and activation of NF-κB (Dbaibo *et al.*, 1993). However, reports do exist of sphingomyelinases activating NF-κB in a similar manner as TNFα, suggesting a role for ceramide (Schutz *et al.*, 1992; Lozano *et al.*, 1994). Addition of acidic or neutral sphingomyelinases to cell homogenates at pH 5.0 and 7.4, respectively, activated NF-κB to an extent similar to that produced by TNFα. Further investigation revealed that, although both diacylglycerol and ceramide were produced in response to TNFα, the mechanism of hydrolysis of sphingomyelin was thought to occur via TNFα activation of PC-PLC and subsequent generation of DAG. The acid sphingomyelinase appeared to be the target since sphingomyelin hydrolysis in response to DAG occurred only at pH 5.0 *in vitro*. More recent studies have demonstrated *in vivo* that inhibition of acid sphingomyelinase by SR33557, did not block the TNFα induced activation of NF-κB. (Higuchi *et al.*, 1996). In addition, in Niemann-Pick fibroblast cells which lack acid SMase, NF-κB was still activated in response to TNFα (Kuno *et al.*, 1994).

Based on studies from Moscat and coworkers, a model has been proposed where ceramide is the second messenger that activates NF-κB by stimulation of PKCζ, which has been shown to phosphorylate and inactivate Iκ-B (Lozano *et al.*, 1994). They demonstrated that addition of exogenous sphingomyelinase to NIH-3T3 fibroblasts transactivated a κB-dependent chloramphenicol acetyltranserfase reporter plasmid, in a similar manner as TNFα, and PC-PLC. Furthermore, κB transactivation by TNFα, PC-PLC, and sphingomyelinase was dramatically inhibited by transfection of a PKCζ dominant negative mutant, suggesting a role for PKCζ in activation of NF-κB. PKCζ was also shown to be activated by ceramide *in vitro* and *in vivo* by treatment with sphingomyelinase. Taken together, these results suggest that PKCζ is a component of the SM signalling pathway leading to NF-κB activation. Although, ceramide may indeed directly activate PKCζ *in vivo*, whether this leads to induction of NF-κB translocation by ceramide remains to be seen.

An important point raised by these studies concerns the role of ceramide in modulating NF-κB activation. Although the importance of ceramide in activation of NF-κB is far from clear, it is conceivable that TNFα launches a multistep process in which numerous signals, including ceramide, are necessary but none are sufficient for activation of NF-κB.

Ceramide and inflammation

Studies in human fibroblasts showed that ceramide and sphingosine markedly enhanced the production of PGE_2 in response to IL-1 stimulation (Ballou *et al.*, 1992). Sphingomyelin hydrolysis and ceramide generation did, indeed, occur in response to IL-1 treatment. Further investigation of the mechanism of ceramide's effect on PGE_2 synthesis revealed that ceramide or sphingosine treatment resulted in a considerable increase in the expression of cyclooxygenase, the rate limiting enzyme in the synthesis of PGE_2. Interestingly, ceramide or sphingosine alone were unable to elicit PGE_2 production, implying the possibility that molecules other than ceramide may be the physiological mediators of the inflammatory response.

CERAMIDE: A KEY COMPONENT OF INTRACELLULAR STRESS RESPONSE PATHWAYS

A new subfamily of protein kinases has recently been identified as stress-activated protein kinases (SAPKs) also known as Jun nuclear kinases or JNKs (Kyriakis *et al.*, 1994; Sluss *et al.*, 1994; Westwick *et al.*, 1994). They are related to the mitogen activated protein (MAP) kinases, but activated preferentially by cellular stress and tumor necrosis factor-α (TNF). The TNFα signalling pathway which activates the SAPKs leads to the stimulation of a specific set of transcription factors. SAPK phosphorylates and activates c-Jun and ATF-2 which in turn preferentially induce genes with specific non-consensus AP-1 binding sites, such as the *c-jun* gene itself (Hibi *et al.*, 1993; Derijard *et al.*, 1994; Gupta *et al.*, 1995). TNFα-stimulated sphingomyelin hydrolysis and generation of ceramide has been suggested as an early step in the SAPK-activation pathway leading to apoptosis. Addition of exogenous ceramides or sphingomyelinases induced SAPK activity and enhanced expression of the *c-jun* gene suggesting that ceramide functions upstream of SAPK. Furthermore, a dominant negative mutant of c-jun blocked the stress-induced apoptosis but did not inhibit the generation of ceramide or SAPK activation suggesting a role for SAPK in mediating apoptosis possibly downstream of ceramide. However, another study dissociates SAPK activation from induction of apoptosis (Liu *et al.*, 1996). Therefore, whether SAPK is indeed involved in apoptosis remains to be resolved. Also, the role of ceramide in this process remains to be defined.

TNFα has been shown to have a dual function in mammalian cells, inducing both inflammation and apoptosis. Ceramide has been suggested to regulate both responses. However, the inflammatory response is thought to be induced via the MAPK cascade (Yao *et al.*, 1995) whereas the apoptotic or stress response is proposed to be activated through the SAPK pathway (Westwick *et al.*, 1995; Verheij *et al.*, 1996). Recently KSR (kinase suppressor of ras) was identified as CAPK (ceramide activated protein kinase) (Zhang *et al.*, 1997) where ceramide stimulates KSR/CAP kinase to complex with and phosphorylate Raf-1, initiating signalling down the MAPK pathway.

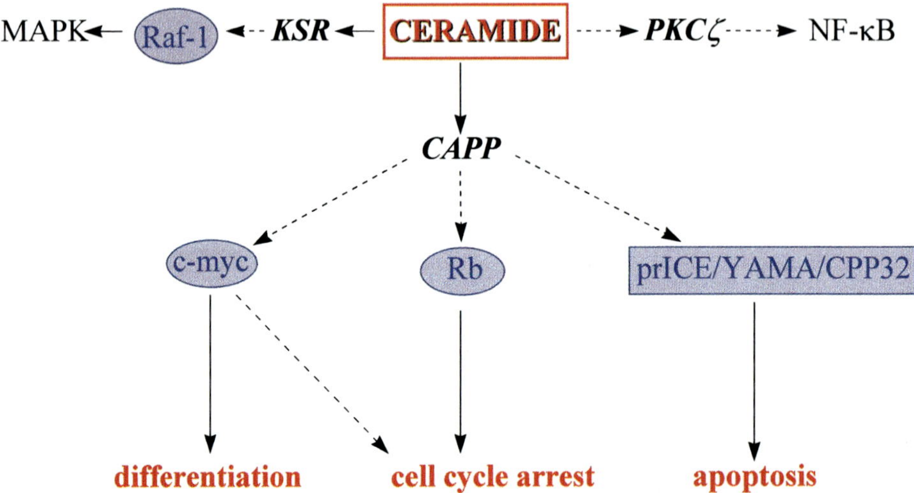

Figure 2.1 Ceramide-mediated stress responses. Ceramide has been shown to regulate differentiation, cell cycle arrest, and apoptotic cell death. The specific biologic response appears to be determined by the particular cell type. Mechanisms for ceramide action involve activation of a cytosolic serine/threonine protein phosphatase 2A (CAPP), a membrane bound serine/threonine protein kinase (CAPK/KSR), and PKC ζ. Downstream targets of ceramide include the protooncogene, c-myc, the retinoblastome gene product (Rb), and the death proteases (prICE/YAMA/CPP32). The dotted lines represent proposed pathways which are not clearly defined.

The essential signalling events linking Fas receptor to apoptosis have been the subject of intense investigation. The structural and physiological similarities between the TNFR and Fas led to the possibility that ceramide also participates in apoptosis triggered by the Fas antigen. In a recent report by Tepper and coworkers, crosslinking of the Fas antigen with agonist resulted in a coordinate increase in sphingomyelinase activity and ceramide supporting a role for ceramide in Fas induced apoptosis (Tepper *et al.*, 1995). Further delineation of the Fas signalling cascade revealed Fas activation of the SAPKs, p38 and JNK, but not activation of the mitogen activated kinases (Juo *et al.*, 1997). Fas-mediated apoptosis also requires the action of interleukin-1β converting enzyme (ICE) or ICE/CED-3 family proteases, based on the ability of the cowpox viral protein CrmA, which inhibits ICE proteases, and specific tetrapeptide ICE inhibitors to inhibit Fas-induced death (Enari *et al.*, 1995; 1996; Los *et al.*, 1995; Tewari and Dixit, 1995). In addition, the inhibitors also blocked Fas-induced activation of p38, demonstrating that Fas-dependent activation of p38 requires ICE/CED-3 family members and conversely that the proteases act as upstream regulators of p38 during Fas signalling. Interestingly, ceramide has been shown to activate the same family of ICE proteases specifically prICE/YAMA/CPP32, the protease responsible for cleavage of poly (adenosine diphosphate-ribose) polymerase (PARP) (Smyth *et al.*, 1996). Overexpression of the cell death regulator Bcl-2 resulted in inhibition of ceramide activation of prICE and induction of apoptosis (Dbaibo *et al.*, 1997) suggesting that Bcl-2 functions downstream of ceramide. Moreover, overexpression of Bcl-2 did not attenuate the increase in ceramide levels with the addition of extracellular agents. The mechanism by which ceramide directly or indirectly activates the protease remains to be determined, although potential candidates for direct targets of ceramide action have begun to emerge.

MECHANISMS OF CERAMIDE ACTION

The sphingolipid signalling pathway is thought to mediate downstream biologic effects via modulation of cellular phosphoprotein metabolism. Ceramide has been demonstrated to activate both a membrane bound kinase activity as well as a cytosolic protein phosphatase activity. More recently the protein kinase C ζ has been suggested to play a critical role in the SMase signalling pathway leading to κB-dependent promoter activation. This observation has led to the investigation of ceramide as the second messenger that activates NF-κB by stimulation of PKCζ.

Ceramide activated protein phosphatase

The search for a direct molecular target of ceramide is ongoing. Biologically relevant candidates should be activated by ceramide *in vitro* in addition to mediating the most proximal effects of ceramide *in vivo*. The identification of a serine-threonine protein phosphatase in rat T9 glioma cells led to the most promising suitor for ceramide action (Dobrowsky *et al.*, 1993). Ceramide activated protein phosphatase (CAPP) has been shown to be activated by both native (C18) and cell permeable (C2) ceramides but not by sphingosine or sphingomyelin *in vitro* (Dobrowsky and Hannun, 1992). CAPP, related to the family of PP2A phosphatases, has no cation requirement and is potently inhibited by okadaic acid. Studies in *S. cerevisiae* where ceramide-induced growth inhibition has been established, the existence of a ceramide-dependent serine/threonine phosphatase has also been proven in yeast (Fishbein *et al.*, 1993). The biochemical characteristics of the yeast CAPP are similar to that of mammalian CAPP indicating ceramide-mediated cell regulation is conserved in all eukaryotes and hence, of fundamental importance.

Recent studies have begun to implicate CAPP as a potential mediator of certain cellular activities of ceramide, including downregulation of the proto-oncogene *c-myc* and apoptosis. Primarily, the specificity of activation of CAPP *in vitro* closely matches the specificity of ceramide-induced *c-myc* gene regulation and apoptosis. CAPP is activated *in vitro* by ceramide but not by dihydroceramide which is inactive in eliciting ceramide activities in cells (Dobrowsky and Hannun, 1992). In addition, low concentrations of okadaic acid can inhibit the regulatory effect of ceramide on *c-myc* and apoptosis, providing further evidence of a role for PP2A-type phosphatase in ceramide signalling. Both ceramide and TNFα induce *c-myc* downregulation through a similar mechanism involving a block to transcription (Wolff *et al.*, 1994) which appears to involve an okadaic acid-inhibited phosphatase. Although, relevant physiological substrates for CAPP have yet to be identified, the possibility of CAPP playing a role in TNFα signal transduction via activation by ceramide is emanate.

Ceramide-activated protein kinase

Ceramide activated protein kinase (CAPK) is an exclusively membrane bound enzyme that is a member of an emerging family of proline directed serine/threonine protein kinases (Mathias *et al.*, 1991; Joseph *et al.*, 1993). The 97-kDa autophosphorylating kinase is activated by C_8-ceramide in a cell free system suggesting the possibility that CAPK may serve as a direct target for the action of ceramide. However, upon renaturation from SDS gels, the kinase is no longer ceramide responsive (Liu *et al.*, 1994). In a recent study by Zhang *et al.* (1997) the kinase suppressor of Ras (KSR) was identified genetically in *C. elegans* and Drosophilia, as CAPK. They demonstrate that recombinant KSR displays all of the previously described properties of CAPK. The designation of KSR as CAPK may allow for more precise molecular ordering of transmembrane signalling events. However, the specificity of ceramide activation of CAPK and its role in ceramide signalling awaits further investigation.

Protein kinase C ζ

Yet a third target for ceramide action has been proposed -PKC ζ. The ζ isoform of PKC is distinguished from other isoforms (except λ) with regard to its lack of activation by DAG or phorbol esters. Even more notable, is the observation by Lozano *et al.* of direct PKCζ activation by ceramide *in vitro* (1994). As discussed previously, they suggest that this activation may be important in mediating the regulation of NF-κB in response to TNFα. The kinase may be an immediate target for the lipid second messenger but its importance in regulating the various biological effects of ceramide remains to be elucidated.

Thus, there appear to be several candidates that may serve as direct targets for the action of ceramide. Indeed, multiple *in vitro* targets for ceramide may exist and may mediate distinct biologic activities of ceramide. However, determining the role of these enzymes in mediating the specific cellular activities of ceramide, emerges as the key goal for the elucidation of signal transduction pathways activated by ceramide.

LESSONS FROM YEAST

In mammalian systems sphingolipids play a part in important cellular functions and in yeast they are known to be essential for growth and viability. Studies in mutant strains of *S. cerevisiae* have identified sphingolipids as essential components of growth in high osmolarity conditions suggesting that the lipids play an important role in the signalling cascade regulating the osmostress response. The HOG1 kinase signalling cascade found in yeast is similar to the stress-activated kinase pathway described in mammalian cells. When yeast cells are confronted with high osmolarity, they induce the synthesis of glycerol to increase their internal osmolarity in an effort to re-establish osmotic equilibrium. The signalling pathway that mediates this response uses the PBS2 and HOG1 genes which code for a MEK and a MAPK homolog respectively and are required for cell growth in high osmolarity medium. By comparison, one could predict that sphingolipids activate the yeast HOG1 kinase in a similar manner to ceramide activation of SAPK in mammalian cells. Furthermore the p38 MAPK presents substantial homology to HOG1 (Kyriakis *et al.*, 1994) and like HOG1, can be activated by changes in osmolarity but the mechanism by which activation occurs remains to be determined. Studies on the yeast MAPK stress responses have the potential to identify the function of sphingolipids in regulating the osmostress response which can then be easily applied to mammalian stress activated pathways.

CONCLUSIONS AND FUTURE DIRECTIONS

All living cells elicit coordinated biochemical responses that will either enhance cell survival or lead to cell death when cells are exposed to adverse environmental conditions. This ubiquitous reaction to metabolic disturbances is designated the stress response. The intracellular stress signals generated provide the necessary network which allow the occurrence of both specific and general responses to stress challenges. The decision for cellular death or life may depend on the integration of multiple signals generated from many signal transduction pathways. Membrane lipids functioning as 'lipid second messengers' have emerged as key components in such fundamental mechanisms regulating the stress responses, growth suppression, and programmed cell death. The endeavors of many years of research are beginning to define a novel sphingolipid-mediated signal transduction pathway where several of these stress-activated biochemical pathways converge on ceramide. Although the basic blueprint of this novel signalling pathway can be outlined to illustrate sphingomyelin hydrolysis and the formation of ceramide which directly activates a number of cellular targets, the ultimate biological response depends on the context in which the signal, ceramide, is generated.

Through the definition of CAPP, CAPK, and PKZζ as intracellular targets for ceramide action, we have begun to understand the mechanism by which the proximal effects of ceramide can be mediated such as the regulation of *c-myc* expression, Rb phosphorylation, cyclooxygenase expression, and NF-κB translocation. However, we have yet to define direct endogenous substrates of these proteins.

The yeast model provides an ideal system in which to dissect the components of signal transduction pathways through the selection of mutants and cloning of involved genes. Studies in *S.cerevisiae* have demonstrated that ceramide-mediated growth regulation is conserved in all eukaryotes, thus proving its fundamental importance. Thus, yeast molecular genetics will provide valuable information in identifying proximal targets for the action of ceramide and determining their role in ceramide signalling.

Activation of the SAPKs by ceramide and sphingomyelinase provide strong evidence that ceramide, or a related metabolite, functions as the lipid second messenger in TNF signalling resulting in the activation of stress kinases. However, the mechanism that connects the upstream messengers to the activation of the cytoplasmic and nuclear kinases remains elusive. Evidence does exist suggesting that the ICE family of proteases may be involved in lipid activation of stress kinase pathways. First, ICE and CPP32 have been shown to activate JNK and p38 stress kinases leading to apoptosis. Second, ceramide has been shown to induce activation of prICE, the protease that cleaves PARP and the activation is inhibited by Bcl2. These studies support a distinct role for proteases in the mammalian stress activated pathway, placing them downstream of ceramide and upstream of the SAPKs.

Finally, the elucidation of the molecular mechanisms of sphingolipid biology will ultimately provide a basis for the development of specific therapies to control disease states related to aberations in sphingolipid signalling including cancer and inflammation.

REFERENCES

Baeuerle, P.A. and Baltimore, D. (1994) Ik-B: a specific inhibitor of the NF-kB transcription factor. *Science*, **242**, 540–546.

Ballou, L.R., Chao, C.P., Holness, M.A., Barker, S.C. and Raghow, R. (1992) Interleukin-1-mediated PGE2 production and sphingomyelin metabolism. Evidence for the regulation of cyclooxygenase gene expression by ceramide and sphingosine. *J. Biol. Chem.*, **267**, 20044–20050.

Bielawska, A., Linardic, C.M. and Hannun, Y.A. (1992a) Modulation of cell growth and differentiation by ceramide. *FEBS Lett.*, **307**, 211–214.

Bielawska, A., Linardic, C.M. and Hannun, Y.A. (1992b) Ceramide mediated biology: determination of structural and stereospecific requirements through the use of N-acyl-phenylaminoalcohol analogs. *J. Biol. Chem.*, **267**, 18493–18497.

Chen, C.S., Rosenwald, A.G., Pagano, R.E. (1995) Ceramide as a modulator of endocytosis. *J. Biol. Chem.*, **270**, 13291–13297.

Dbaibo, G.S., Obeid, L.M. and Hannun, Y.A. (1993) Tumor necrosis factor-α (TNF-α) signal transduction through ceramide. *J. Biol. Chem.*, **268**, 17762–17766.

Dbaio, G.S., Pushkareva, M.Y., Jayadev, S., Scharwz, J.K., Horowitz, J.M., Obeid, L.M. and Hannun, Y.A. (1995) Rb as a downstream target for ceramide-dependent pathway of growth arrest. *Proc. Natl. Acad. Sci. USA*, **92**, 1347–1351.

Dbaio, G.S., Perry, D.K., Gamard, C.J., Platt, R., Poirier, G.G., Obeid, L.M. and Hannun, Y.A. (1997) Cytokine response modifier A (CrmA) inhibits ceramide formation in response to tumor necrosis factor (TNF)-alpha: CrmA and Bcl-2 target distinct components in the apoptotic pathway. *J. Exp. Med.*, **185**, 481–90.

Derijard, B., Hibi, M., Wu, I., Barrett, T., Su, B., Deng, T., Karin, M. and Davis, R.J. (1994) JNK1: a protein kinase stimulated by UV light and Ha-Ras that binds and phosphorylates the c-Jun activation domain. *Cell*, **76**, 1025–1037.

Dickson, R.C., Wells, G.B., Schmidt, A. and Lester, R.L. (1990) Isolation of muntant *Saccharomyces cerevisiae* strains that survive without sphingolipids. *Mol. Cell. Biol.*, **10**, 2176–2181.

Dobrowsky, R.T. and Hannun, Y.A. (1992) Ceramide stimulates a cytosolic protein phosphatase. *J. Biol. Chem.*, **267**, 5048–5051.

Dobrowsky, R.T., Kamibayashi, C., Mumby, M.C. and Hannun, Y.A. (1993) Ceramide activates heterotrimeric protein phosphatase 2A. *J. Biol. Chem.*, **268**, 15523–15530.

Dobrowsky, R.T. and Hannun, Y.A. (1994) The sphingomyelin cycle and ceramide second messengers. In *Signal Activated phospholipases*, edited by M. Liscovitch. R.G. Landes Co., 85–99.

Dobrowsky, R.T., Werner, M.H., Castellino, A.M. Chao, M.V. and Hannun, Y.A. (1994) Activation of the sphingomyelin cycle through the low affinity neurotrophin receptor. *Science*, **265**, 1596–1599.

Dressler, K.A., Mathias, S. and Kolesnick, R.N. (1992) Tumor necrosis factor α activates the sphingomyelin signal transduction pathway in a cell free system. *Science*, **255**, 1715–1718.

Enari, M., Hug, H. and Nagata, S. Involvement of an ICE-like protease in fas-mediated apoptosis. *Nature*, **375**, 78–81.

Fishbein, J.D., Dobrowsky, R.T., Bielaska, A., Garret, S. and Hannun, Y.A.(1993) Ceramide mediated growth inhibition and CAPP are conserved in *Saccharomyces cerevisiae*. *J. Biol. Chem.*, **268**, 9255–9261.

Goldstein, S. (1990) Replicative senescence: the human fibroblast comes of age. *Science*, **249**, 1129–1133.

Gupta, S., Campbell, D., Derijard, B. and Davis, R.J. (1995) Transcription factor ATF2 regulation by the JNK signal transduction pathway. *Science*, **267**, 389–393.

Hannun, Y.A., Loomis,C.R., Merrill, A.H., Jr. and Bell, R.M. (1986) Sphingosine inhibition of protein kinase C activity and phorboldibutyrate binding in vitro and in human platlets. *J. Biol. Chem.*, **261**, 19076–19080.

Hannun, Y.A. (1994) The sphingomyelin cycle and the second messenger function of ceramide. *J.Biol. Chem.*, **269**, 3125–3128.

Hannun, Y.A. and Bell, R.M. (1989) Functions of sphingolipids and sphingolipid breakdown products in cellular regulation. *Science*, **243**, 500–507.

Hannun, Y.A. and Bell, R.M. (1989) Regulation of protein kinase C by sphingosine and lysosphingolipids. *Clin. Chim. Acta.*, **185**, 333–345.

Hannun, Y.A. and Linardic, C.M. (1994) Sphingolipid breakdown products: antiproliferative and tumor suppressor lipids. *Biochem. Biophys. Acta.*, **1154**, 223–236.

Hanada, K., Nishijama, M., Kiso, M., Hasegawa, A., Fujita, S., Ogawa, T. and Akamatsu, Y. (1992) Sphingolipids are essential for the growth of chinese hamster ovary cells. *J. Biol. Chem.* **267**, 23527–23533.

Hibi, M., Lin A., Smeal, T., Minden, A., Karin, M. (1993) Identification of an oncoprotein- and UV-responsive protein kinase that binds and potentiates the c-Jun activation domain. *Genes. Dev.*, **7**, 2135–2148.

Higuchi, M., Singh, S., Jaffrezou, J. and Aggarwal, B.B. (1996) Acidic sphingomyelinase-generated ceramide is needed but not sufficient for TNF-induced apoptosis and nuclear factor-kappa B activation. *J. Immunol.*, **157**, 297–.

Inokuchi, J., Momosaki, K., Shimeno, H., Nagamatsu, A. and Rabin, N.S. (1989) Effects od D-threo-PDMP, an inhibitor of glucosylceramide synthetase, on expression of cell surface glycolipid antigen and binding to adhesive proteins by B16 melanoma cells. *J. Cell. Physiol.*, **141**, 573–583.

Jayadev, S., Linardic, C.M. and Hannun, Y.A. (1994) Identification of arachidonic acid as a mediator of sphingomyelin hydrolysis in response to tumor necrosis factor a. *J. Biol. Chem.*, **269**, 5757–5763.

Jayadev, S., Liu, B., Bielawska, A.E., Lee, J.Y., Nazaire, F., Pushkareva, M.Y.U., Obeid, L.M. and Hannun, Y.A. (1995) Role for ceramide in cell cycle arrest. *J. Biol. Chem.*, **270**, 2047–2052.

Joseph., C.K., Byun H.S.,Bittman, R. and Kolesnick, R.N. (1993) Substrate recognition by ceramide-activated protein kinase. Evidence that kinase activity is proline-directed. *J. Biol. Chem.*, **268**, 20002–20006.

Juo, P., Kuo, C., Reynolds, S., Konz, R., Raingeaud, J., Davis, R., Biemann, H. and Blenis, J. (1997) Fas activation of the p38 mitogen-activated protein kinase signalling pathway requires ICE/CED-3 family proteases. *Mol. Cell. Biol.*, **17**, 24–35.

Kim, M.Y., Linardic, C., Obeid, L.M. and Hannun, Y.A. (1991) Identification of sphingomyelin turnover as an effector mechanism for the action of tumor necrosis factor a and gamma interferon. Specific role in cell differentiation. *J. Biol. Chem.*, **266**, 484–489.

Kirkland, J.L. (1992) The biochemistry of mammalian senescence. *Clin. Biochem.*, **25**, 61–75.

Kolesnick, R.N. (1991) Sphingomyelin and derivatives as cellular signals. *Prog. Lipid Res.*, **30**, 1–38.

Kolesnick, R.N. and Golde, D.W. (1994) The sphingomyelin pathway in tumor necrosis factor and interleukin-1 signalling. *Cell*, **77**, 325–328.

Kuno, K., Sukegawa, K., Ishikawa, Y., Orii, T. and Matsushima, K. (1994) Acid sphingomyelinase is not essential for the IL-1 and tumor necrosis factor receptor signalling pathway leading to NFkB activation. *Int. Immunol.*, **6**, 1269.

Kyriakis, J.M., Banerjee, P., Nikolakaki, E., Dai, T., Rubie, E.A., Ahmad, M.F., Avruch, J. and Woodgett, J. (1994) The stress-activated protein kinase subfamily of c-jun kinases. *Nature*, **369**, 156–160.

Lenardo, M.J. and Baltimore, D. (1989) NF-κB: a pleiotropic mediator of inducible and tissue specific gene control. *Cell*, **58**, 227–229.

Lester, R.L., Wells, G.B., Oxford, G. and Dickson, R.C. (1993) Mutant strains of *Saccharomyces cerevisiae* lacking sphingolipids synthesize novel inositol glycerophospholipids that mimic sphingolipid structures. *J. Biol. Chem.*, **268**, 845–856.

Liscovitch, M. and Cantley, L. (1994) Lipid second messengers. *Cell*, **77**, 329–334.

Merrill, A.H., Jr., Hannun, Y.A. and Bell, R.M. (1993) Introduction: sphingolipids and their metabolites in cell regulation. *Adv. Lipid Res.*, **25**, 1–24.

Liu, J., Mathias, S., Yang, Z. and Kolesnick, R. (1994) Renaturation and tumor necrosis factor α stimulation of a 97kDa ceramide activated protein kinase. *J. Biol. Chem.*, **269**, 3047–3052.

Liu, Z., Hsu, H., Goeddel, D. and Karin, M. (1996) Dissection of TNF receptor 1 effector functions: JNK activation is not linked to apoptosis while NF-kB activation prevents cell death. *Cell*, **87**, 565–576.

Los, M., de Craen, M.V., Penning, L.C., Schenk, H., Westendorp, M., Baeuerle, P.A., Droge, W., *et al.* (1995) Requirement of an ICE/CED-3 protease for Fas/APO-1 mediated apoptosis. *Nature*, **375**, 81–83.

Lowenthal, J.W., Ballard, D.W., Bogerd, H., Bohnlein, E. and Greene, W.C. (1989) Tumor necrosis factor α activation of the IL-2 receptor α gene involves the induction of κB-specific DNA binding proteins. *J. Immunol.*, **142**, 3121–3128.

Lozano, J., Berra, E., Municio, M.M., Diaz-Meco, M.T., Dominguez, I., Sanz, L. and Moscat, J. (1994) Protein kinase C ζ isoform is critical for κB-dependent promoter activation by sphingomyelinase. *J. Biol. Chem.*, **269**, 19200–19202.

Mathias, S., Dressler, K.A. and Kolesnick, R.N. (1991) Characterization of a ceramide-activated protein kinase: stimulation by tumor necrosis factor α. *Proc. Natl. Acad. Sci. USA*, **88**, 10009–10013.

Molitor, J.A., Walker, W.H., Doerre, S., Ballard, D.W. and Greene, W.C. (1990) NF-kB: a family of inducible and differentially expressed enhancer-binding proteins in human T cells. *Proc. Natl. Acad. Sci. USA*, **87**, 10028–10032.

Obeid, L. M., Linardic, C.M., Karolak, L.A. and Hannun, Y.A. (1993) Programmed cell death induced by ceramide. *Science*, **259**, 1769–1771.

Okazaki, T., Bielawska, A., Bell, R.M. and Hannun, Y.A. (1989) Sphingomyelin turnover induced by vitamin D_3 in HL-60 cells. Role in cell differentiation. *J. Biol. Chem.*, **264**, 19076–19080.

Okazaki, T., Bielawska, A., Bell, R.M. and Hannun, Y.A. (1990) Role of ceramide as a lipid mediator of 1-alpha, 25-dihydroxyvitamin D3-induced HL60 cell differentiation. *J. Biol. Chem.*, **265**, 15823–15831.

Osborn, L., Kunkel, W. and Nabel, G. (1989) Tumor necrosis factor a and interleukin 1 stimulate the human immunodeficiency virus enhancer by activation of the nuclear factor kB. *Proc. Natl. Acad. Sci. USA*, **86**, 2336–2340.

Patton, J.L., Srinivasan, B., Dickson, R.C. and Lester, R.L. (1992) Phenotypes of sphingolipi-dependent strains of *Saccharomyces cerevisiae*. *J. Bact.*, **174**, 7180–7184.

Sadahira, Y., Ruan, F., Hakomori, S. and Igarashi, Y. (1992) Sphingosine-1-phosphate, a specific endogenous signalling molecule controlling cell motility and tumor cell invasiveness. *Proc. Natl. Acad. Sci. USA*, **89**, 9686–9690.

Schutze, S., Potthoff, K., Machleidt, T., Berkovic, D., Weigmann, K. and Kronke, M. (1992) TNF activates NF-kB by phosphatidylcholine-specific phospholipase C-induced "acidic" sphingomyelin breakdown. *Cell*, **71**, 765–776.

Sluss, H.K., Barrett, T., Derijard, B. and Davis, R.J. (1994) Signal transduction by tumor necrosis factor mediated by JNK protein kinases. *Mol. Cell. Biol.*, **14**, 8376–8384.

Smith, S.W. and Lester, R.L. (1974) Inositolphosphorylceramide, a novel substance and the chief member of a major group of yeast sphingolipids containing a single inositol phosphate. *J. Biol.Chem.*, **249**, 3395–3405.

Smyth, M.J., Perry, D.K., Zhang, J., Poirier, G.G., Hannun, Y.A. and Obeid, L.M. (1996) prICE: a downstream target for ceramide-induced apoptosis and for the inhibitory action of Bcl-2. *Biochem. J.* **316**, 25–8.

Steiner, S., Smith, S. and Lester, R.L. (1969) *Biochem.*, **64**, 1042–1048.

Tepper., C.G., Jayadev, S., Liu, B., Bielawska, A., Wolff, R., Yonehara, S., Hannun, Y. and Seldin, M. (1995) Role for ceramide as endogenous mediator of Fas-induced cytotoxicity. *Proc. Natl. Acad. Sci. USA*, **92**, 8443–8447.

Tewari, M. and Dixit, V.M. (1995) Fas and tumor necrosis factor-induced apoptosis is inhibited by the poxvirus *crmA* gene product. *J. Biol.Chem.*, **270**, 3255–3260.

Venable, M.E., Blobe, G.C. and Obeid, L.M. (1994) Identification of a defect in the phospholipase D/ diacylglycerolpathway in cellular senescence. *J. Biol. Chem.*, **269**, 26040–26044.

Venable, M.E., Lee, J.Y., Smyth, M.J., Bielawska, A.E. and Obeid, L.M. (1995) Role of ceramide in cellular senescence. *J. Biol. Chem.*, **270**, 30701–30708.

Verheij, M., Bose, R., Lin, X.H., Yao, B., Jarvis, W.D., Grant, S., Birrer, M., Szabo, E., Zon, L., Kyriakis, J., Haimovitz-Friedman, A., Fuks, Z. and Kolesnick, R. (1996) Requirement for ceramide initiated SAPK/JNK signalling in stress-induced apoptosis, *Nature*, **380**, 75–79.

Wells, G.B. and Lester, R.L. (1983) The isolation and characterization of a mutant strain of *Saccharomyces cerevisiae* that requires long chain base for growth and for synthesis of phosphosphingolipids. *J. Biol Chem.*, **258**, 10200–10203.

Westwick, J.K., Weitzel, C., Minden, A., Karin, M. and Brenner, D.A. (1994) Tumor necrosis factor alpha stimulates AP-1 activity through prolonged activation of the c-Jun kinase. *J. Biol. Chem.*, **269**, 26396–26401.

Westwick, J.K., Bielawska, A., Bdaibo, G., Hannun, Y. and Brenner, D. (1995) Ceramide activates the stress-activated protein kinases. *J. Biol. Chem.*, **270**, 22689–22692.

Wiegman, K., Schutz, S., Kampen, E., Himmler, A., Machleidt, T. and Kronke, M. (1992) Human 55-kDa receptor for tumor necrosis factor coupled to signal transduction cascades. *J. Biol. Chem.*, **267**, 17997–18001.

Wolff, R.A., Dobrowsky, R.T., Bielawska, A., Obeid, L.M. and Hannun, Y.A. (1994) Role of ceramide-activated protein phosphatase in ceramide-mediated signal tranduction. *J. Biol. Chem.*, **269**, 19605–19609.

Yanaga, F. and Watson, S.P. (1994) Ceramide does not mediate the effect of tumor necrosis factor a on superoxide generation in human neutrophils. *Biochem. J.*, **298**, 733–738.

Yao, B., Zhang, Y., Delikat, S., Basu, S. and Kolesnick,R. (1995) Phosphorylation of Raf by ceramide-activated protein kinase. *Nature*, **378**, 307–310.

Zhang, H., Desai, N.N., Olivera, A., Seki, T., Brooker, G. and Spiegel, S. (1991) Sphingosine-1-phosphate, a novel sphingolipid, involved in cellular proliferation. *J. Cell. Biol.*, **114**, 155–167.

Zhang, Y., Yao, B., Delikat, S., Bayoumy, S., Lin, X.H., Basu, S., McGinley, M., Chan-Hui, P.Y., Lichenstein, H. and Kolesnick R. (1997) Kinase suppressor of ras is ceramide activated protein kinase, *Cell*, **89**, 63–72.

3. RADIATION RESPONSE PATHWAYS AND APOPTOSIS

MARTIN F. LAVIN[*†]

[*]*The Queensland Cancer Fund Research Unit, The Queensland Institute of Medical Research, Department of Surgery, University of Queensland, PO Box Royal Brisbane Hospital, Herston, Brisbane 4029, Australia*

KEY WORDS: radiation, DNA damage, nuclear and membrane signalling, cell cycle.

INTRODUCTION

Engagement of the Fas/APO-1 and TNF receptors with their respective ligands FasL and TNF leads to the initiation of apoptosis by a number of transducing molecules some of which have been identified (Stanger *et al.*, 1995; Hsu *et al.*, 1995). After ligand binding or cross-linking of the receptor, activation is initiated by association of FADD/MORT1 (Fas-associated protein with death domain) with a homologous region on the Fas receptor called the death domain (Boldin *et al.*, 1996; Chinnaiyan *et al.*, 1995). In the case of the TNF receptor, association occurs with TRADD (TNF-RI-associated death domain protein) (Boldin *et al.*, 1996), but FADD also associates with this complex at least in some cell types under conditions that lead to apoptosis. Another protein RIP also associates with both receptors and may be processed by one of several other receptor associated proteins (Kischkel *et al.*, 1995). After these initial changes in the death signalling complex, a cysteine protease MACH/FLICE (caspase 8), is activated and since it is capable of activating a series of other caspases by cleavage it has been suggested that it is at the peak of a cascade of activating reactions (Alnemri *et al.*, 1996). The end result is a series of active caspases which are thought to target specific substrates in the cell (Martin and Green, 1996). Of these caspase 3 has been shown to cleave poly (ADP-ribose) polymerase (PARP) (Lazebnik *et al.*, 1994; Fernandez-Alnemri *et al.*, 1995b) the catalytic subunit of DNA-dependent protein kinase (DNA-PKcs) (Song *et al.*, 1996) U1-70kDa (Casciola-Rosen *et al.*, 1994) and PKCδ (Emoto *et al.*, 1995). These and several other caspase substrates appear to be key targets in the nucleus, nuclear scaffold, cytoplasm and cytoskeleton during the onset of apoptosis (Nicholson, 1997). In summary, in the case of anti-Fas and TNF-mediated apoptosis, a complex series of signalling steps are initiated from the receptor which leads to protease activation, substrate degradation and ultimately apoptosis.

It seems likely that apoptosis induced by a variety of agents such as ionizing radiation, etoposide and glucocorticoids also involves some form of receptor activation at least as part of the process. The cell radiation response is more complex than

† *Corresponding Author*: Tel.: 617 3362 0341. Fax: 617 3362 0106. e-mail: martin @qimr.edu.au

simply responding to DNA damage by repairing lesions. It is now evident that a number of signalling pathways can be activated in response to radiation damage involving both the activation or inactivation of existing proteins as well as the induction of a number of genes (Knebel *et al.*, 1996; Fornace, 1992). Some of these pathways are activated in response to DNA damage, but signalling is not exclusively initiated by such lesions since it has been demonstrated that membrane receptors can mediate the transmission of such signals (Karin and Hunter, 1995). Signals initiated in these pathways can elicit DNA repair, lead to cell cycle arrest or direct cells to undergo apoptosis. The decision to induce cell cycle arrest or apoptosis is dependent upon the expression of other gene products that control cell growth and proliferation (Clarke *et al.*, 1993; Linke *et al.*, 1997). In this chapter radiation signal transduction and its role in apoptosis will be discussed.

RADIATION-INDUCED DNA DAMAGE

DNA appears to be the major target for radiation-induced cell killing (Ward 1985). The lesions induced by ionizing radiation include single and double strand breaks in the phosphodiester backbone of DNA, base and sugar modification as well as cross-links between DNA strands and between DNA and proteins (Van der Schans *et al.*, 1982; Ward, 1985; Téoule, 1987). The strand breaks in DNA arise either directly by interaction with atoms in the phosphodiester backbone causing the ejection of electrons from these atoms or indirectly through water radiolysis which generates free radical species capable of abstracting hydrogen atoms from DNA, resulting in strand breakage (Blok and Lohman, 1973). Single strand interruptions in the DNA backbone are rapidly and efficiently repaired in mammalian cells (Téoule, 1987). The probability of random ionization events being close to one another on opposite DNA strands increases with increasing radiation dose, resulting in double strand breaks in DNA. While a variety of lesions occur in DNA post-irradiation it seems likely that the double strand break is the most significant lesion for cell killing (Radford, 1986a, 1986b; Bryant, 1985). The description of a number of human and rodent cell lines characterized by hypersensitivity to radiation and reduced ability to repair double strand breaks in DNA, post-irradiation, provides further support for the importance of this lesion (Badie *et al.*, 1995; Lees-Miller *et al.*, 1995). A cell line, 180 BR, derived from a patient with acute lymphoblastic leukaemia and an adverse reaction to radiotherapy, represents the first known example of human fibroblasts that are both hypersensitive to ionizing radiation and deficient in double strand break repair (Badie *et al.*, 1995). While it seems unlikely that the molecular defect in the human genetic disorder ataxia-telangiectasia (A-T) is directly concerned with repair of breaks in DNA there is evidence that approximately 10% of double strand breaks persist up to 72 h post-irradiation (Cornforth and Bedford, 1985; Foray *et al.*, 1995, 1997). It is suggested that an intrinsic abnormality in chromatin structure in A-T cells causes a more proficient translation of DNA damage into chromosomal damage accounting for the higher level of chromosome aberrations in these cells (Pandita and Hittleman, 1992, 1994).

RADIATION INDUCED CELL KILLING

Hypersensitivity to radiation is associated with 11 complementation groups in rodent cell mutants and 4 of these are complemented by human genes involved in double strand break repair (Thompson and Jeggo, 1995; Zdzienicka, 1995). Mutations in other groups include those in the DNA-dependent protein kinase (DNA-PK) multi-protein complex; the Ku subunits (Ku 70 and Ku 86) that bind to the free ends of DNA and the catalytic subunit of DNA-PK, DNA-PKcs, which is recruited to DNA breaks and sites of damage by the Ku heterodimer (Gottlieb and Jackson, 1994; Kirchgessner *et al.*, 1995; Taccioli *et al.*, 1993). The *scid* mutation which occurs in DNA-PKcs (Blunt *et al.*, 1995; Boubnov and Weaver, 1995; Kirchgessner *et al.*, 1995) is characterized not only by a defect in double strand break repair, and as a consequence hypersensitivity to ionizing radiation (Biedermann *et al.*, 1991), but also by defective V(D)J recombinant and immunodeficiency (Bosma *et al.*, 1983; Hendrickson *et al.*, 1988; Lieber *et al.*, 1988). Another gene predisposing to radiation hypersensitivity is the gene (ATM) mutated in the human genetic disorder ataxia-telangiectasia (A-T) (Savitsky *et al.*, 1995; Lavin and Shiloh, 1997). The defect in A-T appears to be largely due to a failure to respond to or recognize DNA damage and as a consequence a fraction of double-strand breaks remain unrepaired (Cornforth and Bedford, 1985; Foray *et al.*, 1997). Meyn (1995) proposed a damage surveillance hypothesis which bestows on ATM the ability to activate an elaborate system responsible for cell cycle arrest, DNA repair and protection against apoptosis induced by DNA damage. In A-T cells, it is suggested that the threshold for p53-induced apoptosis is lowered and apoptosis accounts for the majority of cell death post-irradiation.

IMPORTANCE OF PRE-EXISTING PROTEINS

Evidence accumulated from cells in culture, cell-free extracts and enucleated cells demonstrate that the factors required for apoptosis are already present in the cell and no new RNA or protein synthesis is required (Duke *et al.*, 1983; Martin and Cotter, 1991; Nakajima *et al.*, 1995; Raff *et al.*, 1993; Sellins and Cohen, 1991). A greater knowledge of this process has helped to resolve the apparent differences that existed between murine thymocytes on the one hand and a variety of human cell lines on the other hand. Several reports have provided evidence that murine thymocytes are dependent on *de novo* protein synthesis for apoptosis to occur (Wyllie *et al.*, 1984; Sellins and Cohen, 1991; Cohen *et al.*, 1985; McConkey *et al.*, 1988). In these experiments evidence for a reliance on protein synthesis was provided by the use of inhibitors such as cycloheximide (Wyllie *et al.*, 1984; Cohen *et al.*, 1985; McConkey *et al.*, 1988). On the other hand not only did this compound fail to prevent apoptosis in human cells (Chang *et al.*, 1989; Kelley *et al.*, 1992) but it was shown in some cases to exacerbate the process (Baxter and Lavin, 1992; Baxter *et al.*, 1989; Collins *et al.*, 1991; Martin *et al.*, 1990). Inhibitors have been useful to delineate a variety of metabolic steps but have the disadvantage that they are differentially cytotoxic in different cell types. The evidence that emerges in several systems undergoing apoptosis is that pre-existing proteins are of key importance (Chang *et al.*, 1989; Kaufmann, 1989; Nazareth *et al.*, 1991; Song *et al.*, 1996). In this respect comparison can be made to signal transduction pathways involving receptor-ligand interaction which leads to a cascade of events that are largely controlled by phosphorylation and activation of proteins such as the caspase cascade (Grupp *et al.*, 1995; Kapeller and Cantley, 1994). We have shown previously that apoptosis induced by either heat treatment or ionizing radiation exposure of human cell lines is accompanied by dephosphorylation of a limited number of specific proteins (Baxter and Lavin, 1992). In addition inhibitors of phosphatases 1 and 2A, okadaic acid and calyculin A, delayed the process of apoptosis in response to a variety of damaging agents (Baxter and Lavin, 1992; Song *et al.*, 1993). At longer times after treatment, okadaic acid, because of its toxicity, induced apoptosis. Ohoka *et al.* (1993) have also demonstrated, using okadaic acid, that protein dephosphorylation is an essential step for glucocorticoid-induced apoptosis in murine T-cell hybridomas. Paradoxically okadaic acid is also capable of inducing apoptosis in myeloid cells (Boe *et al.*, 1995; Gjertsen *et al.*, 1994; Kiguchi *et al.*, 1994; Lerga *et al.*, 1995). Ishida *et al.* (1992) demonstrated the late onset of apoptosis in myeloid leukaemia cells exposed to okadaic acid. It is evident that the window of inhibition for okadaic acid depends very much on the cell type, presumably reflecting the responses of the cell to the specific apoptotic stimulus and the stage at which phosphatase activity is critical. Further evidence for the involvement of pre-existing factors is provided from *in vitro* experiments employing cell-free extracts incubated with nuclei (Martin *et al.*, 1995). These data show that the apoptotic program is predominantly extranuclear with cytoplasmic extracts from cells undergoing apoptosis being capable of causing condensation of chromatin in isolated nuclei. The use of specific inhibitors suggest that these changes are mediated by caspases. While the activation of these caspases by various stimuli is complex it is now evident that mitochondria play a key role in the activation of these enzymes (Orrenius *et al.*, 1997). The role of mitochondria in apoptosis is discussed at length elsewhere in this book (D Green chapter 10). In the present context the opening of mitochondrial megachannels by permeability transitions release an apoptosis-inducing factor which is apparently responsible for the activation of caspases.

PLASMA MEMBRANE INVOLVEMENT IN RADIATION-INDUCED SIGNALLING

As outlined above, DNA is a major target for cell killing in response to radiation damage. This is achieved by direct interaction with the DNA or by the creation of free radicals that indirectly damage DNA (Ward, 1985). However, as is obvious, H_2O, a target for ionization is not confined to the nucleus. While the biological significance remains uncertain, it has been long shown that radiation and the free radicals arising can alter membrane proteins and lipids (Wallach, 1972). Since the plasma membrane contains in excess of 60% protein, these molecules are clearly potential targets. The oxidation state of disulfide links is altered by radiation exposure and multiple amino acid residues have been shown to be damaged when proteins are irradiated in the solid state or in solution (Wallach, 1972). Membrane lipids can also be altered by the generation of lipid peroxides. However, radiation doses that alter the properties of membranes such as permeability are considerably higher (100-500 Gy) than those that exert significant damage in DNA (Kankura *et al.*, 1969). More recently with the advance of more sensitive methodology it has been possible to detect other aspects of the response of cells to radiation damage at the level of the membrane and cytoplasm (Devary *et al.*, 1993; Karin and Hunter, 1995; Uckun *et al.*, 1992; Haimovitz-Friedman *et al.*, 1994). Uckun *et al.*, (1992) showed that ionizing radiation activated a signal transduction pathway, presumably at the level of a membrane receptor, as evidenced by the enhanced tyrosine phosphorylation of multiple substrates in B-lymphocyte precursors. The protein tyrosine kinase inhibitors genistein and herbimycin prevented radiation-induced tyrosine phosphorylation, DNA fragmentation and cell death suggesting that protein phosphorylation was an important mediator of radiation-induced apoptosis. This pathway was further delineated when it was demonstrated that radiation caused stimulation of phosphatidylinositol turnover; activation of serine/ threonine kinases, including protein kinase C as well as activation of the stress responsive transcription factor NF-*k*B (Uckun *et al.*, 1993). Uckun *et al.* (1996) have more recently identified a protein tyrosine kinase, Bruton's tyrosine kinase (BTK), involved in the process of apoptosis. DT-40 lymphoma B cells, made BTK deficient by targeted disruption of the *btk* gene, failed to undergo apoptosis after radiation exposure (4 Gy or 8 Gy). On the other hand cells in which the Src protein tyrosine kinase genes *lyn* or *syk* were disrupted continued to die by apoptosis post-irradiation. Increased numbers of immature B cells in BTK-deficient mice points to an inability to remove unwanted clones of B cells by apoptosis (Kerner *et al.*, 1996). The mechanism by which ionizing radiation triggers activation of BTK remains unresolved but it is most likely initiated at the level of the membrane. Stevenson *et al.* (1994) have shown that ionizing radiation and H_2O_2 are capable of stimulating MAP kinase activity which may account for the activation of early and late-response genes by oxidative stress. It is well established that numerous transcription factors including c-jun, c-fos and NF-*k*B are activated by UV light and different oxidants (Stein *et al.*, 1989; Devary *et al.*, 1992; Sachsenmaier *et al.*, 1994). In turn a variety of DNA-damage-inducible genes are induced (Fornace, 1992). While there is considerable overlap in the spectrum of genes induced by different DNA damaging agents some genes such as heme oxygenase, tumour necrosis factor α, interleukin I (IL-I), RP2 and RP8 are primarily induced by oxidants such as H_2O_2 and ionizing radiation (Fornace, 1992). Boothman *et al.* (1993) have described 12 X-ray-inducible transcripts (Xips), differentially expressed 8-230 fold in irradiated radioresistant melanoma cells. They subsequently observed increased DNA binding activity to oligonucleotide consensus sequences for CREB, NF-*k*B and Sp1 with extracts from irradiated cells which could account for the alterations in gene expression (Sahijdak *et al.*, 1994). The spectrum of genes induced by ionizing radiation is broad and complex and is responsible for regulating additional transcription events (late response); cell cycle control; proliferation and apoptosis. In some cases such as for NF-*k*B it has been shown to be involved in protecting against apoptosis (Antwerp *et al.*, 1996; Wang *et al.*, 1996; Wu *et al.*, 1996) while in others, as for p53, it mediates the process of apoptosis in some cell types (Clarke *et al.*, 1993; Lotem and Sachs, 1993; Lowe *et al.*, 1993).

RESPONSE THROUGH CERAMIDE

As described above radiation-induced signalling is initiated from the plasma membrane via receptors such as EGFR (Knebel *et al.*, 1996). Ionizing radiation mediates its action not only through these membrane receptors but also by activating membrane sphinogomyelinase to cleave sphingomyelin to ceramide and phosphorylcholine (Haimovitz-Friedman *et al.*, 1994). Several cytokines including TNFα signal through the sphingomyelin pathway by activating membrane neutral sphingomyelinase to produce ceramide which in turn activates a specific serine/threonine kinase and other intermediates during the process of apoptosis (Obeid *et al.*, 1993; Jarvis *et al.*, 1994). Exposure of bovine endothelial cells to 10 Gy of ionizing radiation caused a rapid degradation of sphingomyelin and a corresponding increase in ceramide (Haimovitz-Friedman *et al.*, 1994). Phorbol ester (TPA) activation of protein kinase C (PKC) has been shown to protect against apoptosis under conditions where ceramide is produced (Obeid *et al.*, 1993; Jarvis *et al.*, 1994). As predicted this compound protected endothelial cells from radiation-induced apoptosis and inhibited production of ceramide. Direct involvement of ceramide in the process was shown by adding C-2 ceramide in the presence of TPA to restore the apoptotic response (Haimovitz-Friedman *et al.*, 1994).

Ceramide activates a proline-directed protein kinase which in turn causes phosphorylation of EGFR (Jarvis *et al.*, 1994). Both TNFα and ceramide enhance the phosphorylation of EGFR at Thr 669 (Joseph *et al.*, 1993). The localization of the protein kinase to the membrane and its activation in a cell-free system suggest that it may be a direct target for ceramide (Hannun and Obeid, 1995). Ceramide also activates mitogen-activated protein kinase (MAPK) which suggests that it is an intermediate in the process (Raines *et al.*, 1993). *In vitro* studies have also demonstrated that ceramide activates ceramide-activated protein phosphatase (CAPP), a member of the PP2A family of ser/thr protein phosphatases (Dobrowsky and Hannun, 1992). The observation that okadaic acid inhibits this enzyme at low concentrations (Dobrowsky and Hannun, 1992) supports results that show this compound interferes with apoptosis induced by ionizing radiation and other agents (Song *et al.*, 1993). Ceramide also has other effects on cell cycle progress inhibiting DNA synthesis and blocking cells in G0/G1 phase, apparently by leading to the dephosphorylation of Rb (Hannun, 1996; Jayader *et al.*, 1995). A direct role in apoptosis stems from the observation that ceramide activates the ICE-like proteases in apoptosis (Smyth *et al.*, 1996; Martin *et al.*, 1995).

Treatment of the T-cell leukaemia cell line, CEM, with the cell permeable synthetic ceramide, C-2 ceramide, caused apoptosis and cleavage of fodrin (non-erythroid spectrin) to a 120 kDa fragment (Martin *et al.*, 1995). Recent results demonstrate that fodrin is cleaved both by caspase-3 and calpain during apoptosis (Waterhouse *et al.*, 1997) so it seems likely that ceramide activates these proteases, presumably indirectly to cleave fodrin and other critical substrates. The observation that overexpression of Bcl-2 in leukaemia cells blocked vincristine-induced apoptosis, but not the accumulation of ceramide induced by vincristine, indicates that Bcl-2 functions downstream of ceramide (Zhang *et al.*, 1996).

Sensitivity to ionizing radiation has been shown to correlate with the early accumulation of ceramide in a number of different tumour cell lines (Michael *et al.*, 1997). In one radiosensitive Burkitt's lymphoma line, ceramide increased 4-fold by 10 min post-irradiation (10 Gy) and in the moderately sensitive HL-60 leukaemia cells, ceramide accumulated 2.5-fold above basal levels. In contrast, in all radioresistant tumour cells examined, including several Burkitt's lymphoma lines and a glioma cell line, ceramide did not accumulate post-irradiation. The ability to abrogate ceramide production, by pretreatment with the tumour promoter, 12-O-tetradecanoylphorbol 13-acetate (TPA), conferred resistance to radiation-induced apoptosis in the sensitive cells. An isogenic subline of the sensitive line was resistant to both C8-ceramide (20 μM) and ionizing radiation-induced apoptosis. Since bypassing the block in radiation-induced ceramide production, by the addition of exogenous ceramide was not sufficient to induce apoptosis, this suggests the existence of a second ceramide-associated signalling defect in these radioresistant cells which confers resistance to ceramide-induced apoptosis. While many of the effector molecules have been identified, the pathway by which ceramide induces apoptosis remains to be elucidated.

CELL CYCLE CHECKPOINT ACTIVATION

Exposure of mammalian cells to ionizing radiation causes a delay in progression of cells from G1 into S phase, inhibition of DNA synthesis and a delay in progression from G2 phase into mitosis (Leeper *et al.*, 1972; Konig and Baisch, 1980). In most organisms, DNA damage leads to the rapid and dose-dependent inhibition of DNA synthesis. This dose-dependent inhibition of DNA synthesis appears to have at least two components. One is the blockage of replication by the damage, which has been demonstrated *in vivo* and *in vitro*, and appears to be due to a delay in fork progression (Painter, 1985). The other is the blockage of replicon initiation for a period of time. The relative contribution of the two components depends on the type of damage.

Yeast mutants defective in checkpoint control either at the G1/S phase or G2/M transitions lose chromosomes spontaneously and are hypersensitive to ionizing radiation and/or fail to maintain the dependence on completion of prior cell cycle events (Weinert and Hartwell, 1990; Li and Murray, 1991; Jimenez *et al.*, 1992; Al-Khodairy and Carr, 1992). A recent report has shown that Chk1, a putative protein kinase, is a cell cycle transition-specific effector in the Rad3-dependent DNA damage pathway (Walworth and Bernards, 1996). Rad3 is a member of the phosphatidylinositol 3-kinase family and thus related to the ATM gene mutated in ataxia-telangiectasia and like this gene a non-essential gene. Another gene, MEC1 from Saccharomyces is required for both the G1/S and G2/M checkpoints and meiotic recombination (Weinert *et al.*, 1994). Cells expressing a partial loss-of-function allele for this gene fail to inhibit DNA replication after exposure of cells to radiation, reminiscent of A-T cells (Paulovich and Hartwell, 1995). Overexpression of the RAD53/SAD1 checkpoint kinase suppresses *mec*1^{-}lethality in *S.cerevisiae* (Sanchez *et al.*, 1996). RAD53 encodes a protein kinase and is required for the G1/S and G2/M damage checkpoints, control of DNA replication and transcriptional response to DNA damage (Weinhert and Hartwell, 1988).

Exposure of eukaryotic cells to ionizing radiation leads to the activation of several checkpoints that control the progression of cells between different phases of the cell cycle (Kastan *et al.*, 1991). Progression of cells from G1 phase to S phase is normally controlled by cyclin-dependent kinases (cyclin E - cdk 2 and cyclin Dl-cdk4), the phosphorylation status of these kinases, the substrates of these kinases such as the retinoblastoma protein and cyclin-kinase inhibitors (Hartwell and Kastan, 1994; Pardee, 1989). Signals arising due to DNA damage activate the product of the tumour suppressor gene p53 to prevent the progression of cells from G1 to S phase (Kastan *et al.*, 1991; Pardee *et al.*, 1989). This is achieved by p53-activated induction of the cdk inhibitor p21/WAF1 which binds to and inhibits cyclin-kinase complexes preventing the phosphorylation of specific substrates and in turn the passage of cells from G1 to S phase (Harper *et al.*, 1993; El-Deiry *et al.*, 1993; Xiong *et al.*, 1993). Wild type p53 blocks progression of cells from G1 to S phase in response to DNA damage (Kastan *et al.*, 1991; Kastan, 1993) or when it is overexpressed in cells (Zambetti and Levine, 1993). After exposure of cells to DNA damaging agents, p53 is translocated from the cytoplasm to the nucleus where it can bind to specific DNA sequences to positively or negatively regulate transcription (El-Deiry *et al.*, 1993). After treatment with radiation and other agents the half-life of p53 increases from 10–30min to several hours (Reich and Levine, 1984). Since p53 has multiple phosphorylation sites it has been suggested that phosphorylation is the means of stabilization (Ullrich *et al.*, 1992a,b). When rat embryo fibroblasts transfected with activated *ras* are exposed to phorbol ester co-operation with wild type p53, to bring about growth arrest, is observed (Delphin and Baudier, 1994). Under these conditions a specific enhancement of wild type p53 phosphorylation occurs and there is an enhancement of p53-DNA binding. The PKC mode of phosphorylation *in vitro* also stimulates p53 DNA-binding activity. Delphin and Baudier (1994) propose that PKC and p53 participate in a negative feedback control of phosphoinositide signals common to mitogenic stimulation. Wild type p53 leads to either G1 arrest or apoptosis in different cell types, but no G1 arrest is observed with mutant p53 and apoptosis can be prevented (Kastan *et al.*, 1991; Lotem and Sachs, 1993).

ATAXIA-TELANGIECTASIA

The enhanced chromosomal aberrations and the reduced survival of A-T cells after exposure to ionizing radiation suggests that the defect in A-T cells is due to a deficiency in the removal of radiation damage in DNA (Chen *et al.*, 1978; Taylor *et al.*, 1975). Paterson *et al.*, (1976) reported that some A-T cell lines had a reduced capacity to remove thymine glycol damage from DNA at high radiation doses. Several other studies failed to reveal a defect in the repair of single strand breaks or other forms of damage in DNA (Fornace and Little, 1980; Lavin and Davidson, 1981; Lehmann *et al.*, 1982; Shiloh *et al.*, 1983; Taylor *et al.*, 1975).

Considerable effort has been expended to determine whether the sensitivity to radiation in A-T is associated with defective double-strand break repair. A variety of earlier reports utilizing neutral sucrose gradients, neutral elution and pulse-field gel electrophoresis failed to reveal any significant defect in double-strand break rejoining at short times after irradiation (Blocher *et al.*, 1991; Lehmann and Stevens, 1979; Taylor *et al.*, 1975; Van der Schans, 1982). However, Coquerelle and Weibezahn (1981) and Coquerelle *et al.* (1987) reported that rejoining of γ-radiation-induced DNA double-strand breaks was slower in A-T fibroblasts than in controls. More support for a defect in double-strand break rejoining was provided by Cornforth and Bedford (1985) when they demonstrated the existence of residual breaks in A-T cells utilizing the premature chromatin condensation assay.

Such hypersensitivity would be explained by the continued existence of 5–10% of the total breaks induced. Now that we know the identity of the A-T gene, ATM (Savitsky *et al.*, 1995), its possible mode of action in cell cycle control and in detection of DNA damage (Lavin and Shiloh, 1996; Lavin and Shiloh, 1997) another explanation is possible. If ATM is a "sensor" of DNA damage then it is likely that inability to recognize the damage is responsible for the residual DNA breaks and their translation into chromosome breaks rather than a repair defect *per se*. Abnormalities in chromatin structure might be indirect due to absence of ATM or the presence of a mutated form of ATM. Either way the residual chromosome breaks may be at least partially responsible for the radiosensitivity in A-T and may also contribute to the genome instability and cancer predisposition.

A-T cells are characterized by a failure to activate either the Gl/S or G2/M checkpoints at short times after radiation exposure and they exhibit radioresistant DNA synthesis (Beamish and Lavin, 1994; Houldsworth and Lavin, 1980; Nagasawa and Little, 1983; Painter and Young, 1980; Scott and Zampetti-Bosseler, 1982). The inability of A-T cells to activate these checkpoints efficiently ultimately leads to the accumulation of irradiated cells in G2/M phase where they die. The molecular nature of the defect in cell cycle control in A-T cells has been extensively studied and it is evident that a pathway mediated by the ATM gene functioning through p53 is defective in A-T cells (Artuso *et al.*, 1995; Kastan *et al.*, 1992; Keegan *et al.*, 1996; Khanna and Lavin, 1993; Lu and Lane, 1993; Mirzayans *et al.*, 1995). The defect is evident not only at the level of radiation-induced stabilization of p53 but also at the level of its downstream effectors WAF1, gadd 45 and MDM2 (Canman *et al.*, 1994; Khanna *et al.*, 1995). Increased WAF1, as a consequence of exposure of control cells to radiation leads to an inhibition of cyclin-dependent kinase activity (cyclinE cdk2 at the Gl/S transition point) (Dulic *et al.*, 1994). This inhibition is caused by a change in the stoichiometry of binding of WAF1 to a complex that includes cyclinE, cdk2, PCNA and WAF1 (Xiong *et al.*, 1993). Inhibition of cyclinE-cdk2 activity prevents phosphorylation of substrates required for entry into S phase causing G1 arrest. These substrates include the retinoblastoma protein, which when phosphorylated disassociates from the transcription factor E2F and allows for the transcription of key enzymes such as thymidine kinase and ribonucleotide reductase necessary for S phase. In A-T cells WAF1 does not increase in response to radiation or is delayed in its induction, consequently, inhibition of cyclin E - cdk2 is not observed and cells progress from Gl into S phase without delay (Khanna *et al.*, 1995). Lack of inhibition of cyclin E - cdk2 in irradiated A-T cells appears to be due to little or no change in WAF1 association with the kinase (Beamish *et al.*, 1996). However, since the S phase and G2/M checkpoints are also defective in A-T one might expect some overlap given the likely involvement of ATM in control of the other checkpoints. Resistance of several different cyclin-dependent kinase activities, that control the different checkpoints, to radiation-induced inhibition in A-T cells points to a wider role for the p53 pathway in radiation signal transduction (Beamish *et al.*, 1996). The failure of A-T cells to mount a p53 response effective in delaying the progression of cells from Gl to S phase is unlikely to provide an explanation for the enhanced radiosensitivity in these cells. Others have not observed a correlation between absence of p53 or mutated (non-functional) p53 and propensity to radiation sensitivity (Clarke *et al.*, 1993; Lee and Bernstein, 1993; Lowe *et al.*, 1993; Slichenmyer *et al.*, 1993). It is more likely that the defective p53 response in G1 phase cells leads to increased chromosomal instability as a consequence of these cells entering S phase prior to repairing damage in DNA. Indeed chromosomal instability and predisposition to develop leukaemias, lymphomas and to a lesser extent solid tumours are very characteristic of A-T (Boder and Sedgwick, 1963; Hecht and Hecht, 1990; Spector *et al.*, 1982). Some evidence has been provided that p53 may also play a role in other checkpoints in S phase and G2/M (Agarwal *et al.*, 1995; Aloni-Grinstein *et al.*, 1994; Beamish *et al.*, 1996). Flow cytometric analysis has shown that when A-T cells are irradiated in Gl or S phase and allowed to proceed through the cycle they accumulate essentiallv, irreversibly in the following G2/M where they die (Beamish and Lavin, 1994). These results suggest that by ignoring the cell cycle checkpoints, A-T cells ultimately carry chromosome damage into G2 phase where they are incapable of normal chromosomal segregation and cell division. Treatment of irradiated A-T cells with caffeine allows them to bypass G2 delay but leads to massive chromosomal fragmentation and cell death, supporting the hypothesis that A-T chromosomes accumulate damage and as a consequence are incapable of proceeding to mitosis (Bates *et al.*, 1985). Further elucidation of the role of the ATM protein will provide a greater understanding of the basis for radiosensitivity in A-T.

REFERENCES

Agarwal, M.L., Agarwal, A., Taylor, W.R. and Stark, G.R. (1995) p53 controls both the G2/M and the G1 cell cycle checkpoints and mediates reversible growth arrest in human fobroblasts. *Proc. Natl. Acad. Sci. USA*, **92**, 8493–8497.

Al-Khodairy, F. and Carr, A.M. (1992) DNA repair mutants defining G2 checkpoint pathways in Schizosaccharomyces pombe. *EMBO J.*, **11**, 1343–1350.

Alnemri, E.S., Livingston, D.J., Nicholson, D.W., Salvesen, G., Thornberry, N.A., Wong, W.W. and Yuan, J. (1996) Human ICE/CED-3 protease nomenclature. *Cell*, **87**, 171.

Aloni-Grinstein, R., Schwartz, D. and Rotter, V. (1994) Accomodation of wild-type p53 protein upon γ-irradiation induces a G2 arrest-dependent immunoglobulin **k** light chain gene expression. *EMBO J.*, **14**, 1392–1401.

Artuso, M., Esteve, A., Bresil, H., Vuillaume, M. and Hall, J. (1995) The role of the ataxia-telangiectasia gene in the p53, WAF1/ClPl(p21) GADD45-mediated response to DNA damage produced by ionizing radiation. *Oncogene*, **8**, 1427–1435.

Badie, C.G., Iiiakis, N., Foray, G., Alsbeih, B., Pantelias, R., Okayasu, N., Cheong, N.S., Russel, A.C., Begg, C.F., Arlett, C.F. and Malaise, E.P. (1995) Defective repair of DNA double-strand breaks and chromosome damage in fibroblasts from a radiosensitive leukemia patient. *Cancer Res.*, **55**, 1232–1234.

Bates, P.R., Imray, F.P. and Lavin, M.F. (1985) Effect of caffeine on γ-ray induced G2 delay in ataxia telangiectasia. *Int. J. Radiat. Biol.*, **47**, 713–722.

Baxter, G.D., Collins, R.J., Harmon, B.V., Kumar, S., Prentice, R.L., Smith, P.J. and Lavin, M.F. (1989) Cell death by apoptosis in acute leukaemia. *J. Pathol.*, **158**, 123–129.

Baxter, G.D. and Lavin, M.F. (1992) Specific protein dephosphorylation in apoptosis induced by ionizing radiation and heat-shock in human lymphoid cells. *J. Immunol.*, **148**, 1949–1954.

Beamish, H. and Lavin. M.F. (1994) Radiosensitivity in ataxia-telangiectasia: anomalies in radiation-induced cell cycle delay. *Int. J. Radiat. Biol.*, **65**, 175–184.

Beamish, H., Williams, R., Chen, P. and Lavin, M.F. (1996) Defect in multiple cell cycle checkpoints in ataxia-telangiectasia post-irradiation. *J. Biol. Chem.*, **271**, 20486–20493.

Biedermann, K.A., Sun, J-R., Giaccia, A.J., Tosto, L.M. and Brown, J.M. (1991) SCID mutation in mice confers hypersensitivity to ionizing radiation and a deficiency in DNA double-strand break repair. *Proc. Natl. Acad. Sci. USA*, **88**, 1394–1397.

Blocher, D., Sigut, D. and Hannan, M.A. (1991) Fibroblasts from ataxia-telangiectasia (A-T) and A-T heterozygotes show an enhanced level of residual DNA double-strand breaks after low dose-rate γ-radiation as assayed by pulsed field gel electrophoresis. *Int. J. Radiat. Biol.*, **60**, 791–802.

Blok, J. and Lohman, H. (1973) The effects of γ-radiation in DNA. *Current Topics in Radiation Research*, **9**, 165–245.

Blunt, T., Finnie, N.J., Taccioli, G., Smith, G.C.M., Demengeot, J., Gottlieb, T., Mizuta, R., Varghese, A.J., Alt, F., Jeggo, P.A. and Jackson, S.P. (1995) Defective DNA-dependent protein kinase activity is linked to V(D)J recombination and DNA repair defects associated with the murine *scid* mutation. *Cell*, **80**, 813–823.

Boder, E. and Sedgwick, R.P. (1963) Ataxia-telangiectasia. A review of 101 cases. In: G. Walsh, Ed.: Little Club Clinics in Develop. Med. No. 8. London, Heinemann Medical Books pp. 110–118.

Boe, R., Gjertsen, B.T., Doskeland, S.O. and Vintermyr, O.K. (1995) 8-chloro-cAMP induces apoptotic cell death in a human mammary carcinoma cell (MCF-7) line. *Brit. J. Cancer*, **72**, 1151–1159.

Boldin, M.P., Goncharov, T.M., Goltsev, Y.V. and Wallach, D. (1996) Involvement of MACH, a novel MORT1/FADD-interacting protease, in Fas/APO-1-and TNF receptor-induced cell death. *Cell*, **85**, 803–815.

Boothman, D.A., Bouvard, I. and Hughes, E.N. (1993) Identification and characterization of X-ray-induced proteins in human cells. *Cancer Res.*, **49**, 2871–2878.

Bosma, C.C., Custer, R.R. and Bosma, M.J. (1983) A severe combine immunodeficiency mutation in the mouse. *Nature*, **301**, 527–530.

Boubnov, N.V. and Weaver, D.T. (1995) *Scid* cells are deficient in Ku and replication protein A phosphorylation by the DNA-dependent protein kinase. *Mol. Cell. Biol.*, **15**, 5700–5706.

Bryant, P.E. (1985) Enzymatic restriction of mammalian cell DNA: evidence for double-strand breaks as potentially lethal lesions. *Int. J. Radiat. Biol.*, **48**, 55–60.

Canman, C.E., Wolff, A.C., Chen, C.Y., Fornace Jr, A.J. and Kastan, M.B. (1994) The p53-dependent G1 cell cycle checkpoint pathway and ataxia-telangiectasia. *Cancer Res.*, **54**, 5054–5058.

Casciola-Rosen, L.A., Miller, D.K., Anhalt, G.J. and Rosen, A. (1994) Specific cleavage of the 70 kDa protein component of the U1 small nuclear ribonucleoprotein is a characteristic biochemical feature of apoptotic cell death. *J. Biol. Chem.*, **269**, 39757–39760.

Chang, M., Bramhall, J., Graves, S., Bonavida, B. and Wisnieski, B.J. (1989) Internucleosomal DNA cleavage precedes diptheria toxin-induced cytolysis. *J. Biol. Chem.*, **264**, 15261–15267.

Chen, P.C., Lavin, M.F., Kidson, C. and Moss D. (1978) Identification of ataxia telangiectasia heterozygotes, a cancer prone population. *Nature*, **274**, 484–486.

Chinnaiyan, A.M., O'Rourke, K., Tewari, M. and Dixit, V.M. (1995) FADD, a novel death domain-containing protein, interacts with the death domain of Fas and initiates apoptosis. *Cell*, **81**, 505–512.

Clarke, A.R., Purdie, C.A., Harrison, D.J., Morris, G., Bird, C.C., Hooper, M.L. and Wyllie, A.H. (1993) Thymocyte apoptosis induced by p53-dependent and independent pathways. *Nature*, **362**, 849–852.

Cohen, J.J., Duke, R.C., Chervenak, R., Sellins, K.S. and Olson, L.K. (1985) DNA fragmentation in targets of CTL: an example of programmed cell death in the immune system. *Adv. Exp. Med. Biol.*, **184**, 493–508.

Collins, R.J., Harmon, B.V., Souvlis, T., Pope, J.H. and Kerr, J.F.R. (1991) Effects of cycloheximide on B-chronic lymphocytic leukaemia and normal lymphocytes *in vitro*: induction of apoptosis. *Br. J. Cancer*, **64**, 518–522.

Coquerelle, T.M. and Weibezahn, K.F. (1981) Rejoining of DNA double-strand breaks in human fibroblasts and its impairement in one Ataxia-Telangiectasia and two Fanconi strains. *J. Supro. Structure Cell. Biolchem.*, **17**, 369–376.

Coquerelle, T,M, Weibezahn, K.F. and Lucke-Huhle. (1987) Rejoining of double strand breaks in normal human and ataxia-telangiectasia fibroblasts after exposure to ^{60}Co γ-rays, ^{241}Am α-particles or bleomycin. *Int. J. Radiat. Biol.*, **51**, 209–21.

Cornforth, M.W. and Bedford, J.S. (1985) On the nature of a defect in cells from individuals with ataxia-telangiectasia. *Science*, **227**, 1589–1591.

Delphin, C. and Baudier, J. (1994) The protein kinase C activator, phorbol ester, cooperates with the wild-type p53 species of Ras-transformed embryo fibroblasts. *J. Biol. Chem.*, **47**, 29579–29587.

Devary, Y., Gottlieb, R.A., Smeal, T. and Karin, M. (1992) The mammalian ultraviolet response is triggered by activation of Src tyrosine kinases. *Cell*, **71**, 1081–1091.

Devary, Y., Rosette, C., Didonato, J.A. and Karin, M. (1993) NF-**k**B activation by ultraviolet light not dependent on a nuclear signal. *Science*, **261**, 1442–1445.

Dobrowsky, R.T. and Hannun, Y.A. (1992) Ceramide stimulates a cytosolic protein phosphatase. *J. Biol. Chem.*, **267**, 5048–5051.

Duke, R.C., Chervenak, R. and Cohen, J.J. (1983) Endogenous endonuclease induced DNA fragmentation: an early event in cell mediated cytolysis. *Proc. Natl. Acad. Sci. USA*, **80**, 6361–6365.

Dulic, V., Kaufmann, W.K., Wilson, S.J., Tisty, T.D., Lees, E., Wade Harper, J., Elledge, S.J. and Reed, S.I. (1994) p53-dependent inhibition of cyclin-dependent kinase activities in human fibroblasts during radiation-induced G1 arrest. *Cell*, **76**, 1013–1023.

El-Deiry, W.S., Tokino, T., Velculescu, V.E., Levy, D.B., Parsons, R., Trent, J.M., Lin, D., Mercer, W.E., Kinzler, K.W. and Vogelstein, B. (1993) WAF1 a potential mediator of p53 tumor suppression. *Cell*, **75**, 817–825.

Emoto, Y., Manome, Y., Meinhardt, G., Kisaki, H., Kharbanda, S., Robertson, M., Ghayur, T., Wong, W.W., Kamen, R., Weichselbaum, R. and Kufe, D. (1995) Proteolytic activation of protein kinase C δ by an ICE-like protease in apoptotic cells. *EMBO J.*, **14**, 6148–6156.

Fernandez-Alnemri, T. *et al.* (1995b) Mch3, a novel human apoptotis cysteine protease highly related to CPP32. *Cancer Res.*, **55**, 6045–6052.

Foray, N., Arlett, C.F. and Malaise, E.P. (1995) Dose-rate effect on induction and repair rate of radiation-induced DNA double-strand breaks in a normal and an ataxia-telangiectasia human fibroblast cell line. *Biochimie.*, **77**, 900–905.

Foray, N., Priestley, A., Alsbeih, G., Badie, C., Capulas, E.P., Arlett, C.F. and Malaise. E.P. (1997) Hypersensitivity of ataxia-telangeictasia fibroblasts to ionizing radiation is associated with a repair deficiency of DNA double-strand breaks. *Int. J. Radiat. Biol.*, (in press).

Fornace, A.J., Jr. and Little, J.B. (1980) Normal repair of DNA single-strand breaks in patients with ataxia-telangiectasia. *Biochem. Biophys. Acta.*, **607**, 432–437.

Fornace, A.J. (1992) Mammalian genes induced by radiation; activation of genes associated with growth control. *Annu. Rev. Genet.*, **26**, 507–526.

Gjertsen, B.T., Vressey, L.I., Ruchaud, S., Houge, G., Lanotte, M. and Doskeland, S.O. (1994) Multiple apoptotic death types triggered through activation of separate pathways by cAMP and inhibitors of protein phosphatases in one (IPC leukemia) cell line. *J. Cell. Sci.*, **107**, 3363–3377.

Gottlieb, T.M. and Jackson, S.P. (1994) Protein kinases and DNA damage. *Trends Biochem. Sci.*, **19**, 500–503.

Grupp, S.A., Mitchell, R.N., Schreiber, K.L., McKean, D.J. and Abbas. A.K. (1995) Molecular mechanisms that control expression of the B lymphocyte antigen receptor complex. *J. Exp. Med.*, **181**, 161–168.

Haimovitz-Friedman, A., Kan, C.C., Ehleiter, D., Persaud, R.S., McLouglin, M., Fuks, Z., Koesnick, R.N. (1994) Ionizing radiation acts on cellular membranes to generate ceramide and initiate apoptosis. *J. Exp. Med.*, **180**, 525–535.

Hallahan, D.E., Sukhatme, V.P., Sherman, M.L., Virudachalam, S., Kufe, D. and Weicheselbaum, R.R. (1991) Protein kinase C mediates x-ray inducibility of nuclear signal transducers EGR1 and JUN. *Proc. Natl. Acad. Sci. USA*, **88**, 2156–2160.

Hannun, Y.A. and Obeid, L.M. (1995) Ceramide: an intracellular signal for apoptosis. *TIBS*, **20**, 73–77.

Hannun, Y.A. (1996) Function of ceramide in coordinating cellular responses to stress. *Science*, **274**, 1855–1859.

Harper, J.W., Adami, G.R., Wei, N., Keyomarsi, K. and Elledge, S.J. (1993) The p21 cdk-interacting protein cip1 is a potent inhibitor of G1 cyclin-dependent kinases. *Cell*, **75**, 805–816.

Hartwell, L.H. and Kastan, M.B. (1994) Cell cycle control and cancer. *Science*, **266**, 1821–1828.

Hecht, F. and Hecht, B.K. (1990) Cancer in ataxia-telangiectasia patients. *Cancer Genet. Cytogenet.*, **46**, 9–19.

Hendrickson, E.A., Schatz, D.G. and Weaver, D.T. (1988) The *scid* gene encodes a trans-acting factor that mediates the rejoining event of Ig gene rearrangement. *Genes Dev.*, **2**, 817–829.

Houldsworth, J. and Lavin, M.F. (1980) Effect of ionizing radiation on DNA synthesis in ataxia telangiectasia cells. *Nucleic Acids Res.*, **8**, 3709–3720.

Hsu, H., Xiong, J. and Goeddel, D.V. (1995) The TNF receotpr 1-associated protein TRADD signals cell death and NF-**kB** activation. *Cell*, **81**, 495–504.

Ishida, Y., Furukawa, Y., Decaprio, J.A., Saito, M. and Griffin, J.D. (1992) Treatment of myeloid leukemic cells with the phosphatase inhibitor okadaic acid induces cell cycle arrest at either G1/S or G2/M depending on dose. *J. Cell Physiol.*, **150**, 484–492.

Jarvis, W.D., Kolesnick, R.N., Fornari, F.A., Traylor, R.S., Gewirtz, D.A. and Grant, S. (1994) Induction of apoptotic DNA damage and cell death by activation of the sphinogomyelin pathway. *Proc. Natl. Acad. Sci. USA*, **91**, 73–77.

Jayader, S., Liu, B., Bielawaska, A.E., Lee, J.Y., Nazaire, F., Pushkareva, M.Y., Obeid, L.M. and Hannun, Y.A. (1995) Role for ceramide in cell cycle arrest. *J. Biol. Chem.*, **270**, 2047–2052.

Jimenez, G., Yucel, J., Rowley, R. and Subramani, S. (1992) The rad3$^+$ gene of Schizosaccharomyces pombe is involved in multiple checkpoint functions and in DNA repair. *Proc. Natl. Acad. Sci. USA*, **89**, 4952–4956.

Joseph, C.K., Byun, H.S., Bittman, R. and Kolesnick, R.N. (1993) Substrate recognition by ceramide-activated protein kinase. Evidence that kinase activity is proline-directed. *J. Biol. Chem.*, **268**, 20002–20006.

Kankura, T., Nakamura, W., Ero, H. and Nakao, M. (1969) Effect of ionizing radiation on passive transport of sodium ion into human erythrocytes. *Int. J. Radiat. Biol.*, **15**, 125.

Kapeller, R. and Cantley, L.C. (1994) Phosphatidylinositol 3-kinase. *Bio. Essays*, **16**, 565–576.

Karin, M. nd Hunter, T. (1995) Transcriptional control by protein phosphorylation: signal transmission from cell surface to the nucleus. *Curr. Biol.*, **5**, 747–757.

Kastan, M.B., Oneykwere, O., Sidransky, D., Vogelstein, B. and Craig, R.W. (1991) Participation of p53 protein in the cellular response to DNA damage. *Cancer Res.*, **51**, 6304–6311.

Kastan, M.B., Zhan, O., EL-Deiry, W.S., Carrier, F., Jacks, T., Walsh, W.V., Plunkett, B.S., Vogelstein, B. and Fornace, A.J. (1992) A mammalian cell cycle checkpoint pathway utilizing p53 and GADD45 is defective in ataxia-telangiectasia. *Cell*, **71**, 587–597.

Kastan, M.B. (1993) p53: a determinant of the cell cycle response to DNA damage. *Adv. Exp. Med. Biol.*, **339**, 291–293.

Kaufmann, S.W. (1989) Induction of endonucleolytic DNA cleavage in human acute myelogenous leukemia cells by etoposide, camptothecin, an other cytotoxic anticancer drugs: A cautionary note. *Cancer Res.*, **49**, 5870–5878.

Keegan, K.S., Holtzman, D.A., Plug, A.W., Christenson, E.R., Brainerd, E.E., Flaggs, G., Bentley, N.J., Taylor, E.M., Meyn, M.S., Moss, S.B., Carr, A.M., Ashley, T. and Hoekstra, M.F. (1996) The Atr and Atm protein kinases associate with different sites along meiotically pairing chromosomes. *Genes & Development*, **10**, 2423–2437.

Kelley, L.L., Koury, M.J. and Bondurant, M.C. (1992) Regulation of programmed death in erythroid progenitor cells by erythropoietin: effects of calcium and of protein and RNA synthesis. *J. Cell Physiol.*, **151**, 487–496.

Kerner, J.D., Appleby, M.W., Mohr, R.N., Chien, S., Rawlings, D.J., Maliszewski, C.R., Witte, O.N. and Perlmutter, R.M. (1996) Impaired expansion of mouse B cell progenitors lacking Btk. *Immunity*, **3**, 301–312.

Khanna, K.K. and Lavin, M.F. (1993) Ionizing radiation and UV induction of p53 protein by different pathways in ataxia-telangiectasia cells. *Oncogene*, **8**, 3307–3312.

Khanna, K.K., Beamish, H., Yan, J., Hobson, K., Williams, R., Dunn, I. and Lavin, M.F. (1995) Nature of GS/1 cell cycle checkpoint defect in ataxia-telangiectasia. *Oncogene*, **11**, 609–618.

Kiguchi, K., Glesne, D., Chubb, C.H., Fujiki, H. and Huberman, E. (1994) Differential induction of apoptosis in human breast tumour cells by okadaic acid and related inhibitors of protein phosphatases 1 and 2A. *Cell Growth Differentiation*, **5**, 995–1004.

Kirschkel, F.C., Hellbardt, S., Behrmann, I., Germer, M., Panolita, M., Krammer, P.H. and Peter, M.E. (1995) Cytoxicity-dependent APO-1 (Fas/CD95)-associated proteins form a death-inducing signalling complex (DISC) with the receptor. *EMBO J.*, **14**, 5579–5588.

Kirchgessner, C.U., Patil, C.K., Evans, J.W., Cuomo, C.A., Fried, L.M., Carter, T., Oettinger, M.A. and Brown, J.M. (1995) DNA-dependent kinase (p350) as a candidate gene for the murine SCID defect. *Science*, **267**, 1178–1182.

Knebel, A., Rahmsdorf, H.J., Ullrich, A. and Herrlich, P. (1996) Dephosphorylation of receptor tyrosine kinases as targets of regulation by radiation, oxidants or alkylating agents. *EMBO J.*, **15**, 5314–5325.

Konig, K. and Baisch, H. (1980) DNA synthesis and cell cycle progression of synchronized L-cells after irradiation in various phases of the cell cycle. *Radiat. Environ. Biophys.*, **18**, 257–265.

Lavin, M.F. and Davidson, M. (1981) Repair of strand breaks in superhelical DNA of ataxia telangiectasia lymphoblastoid cells. *J. Cell Sci.*, **48**, 383–391.

Lavin, M.F. and Shiloh Y. (1996) Ataxia-telangiectasia: A multifacet genetic disorder associated with defective signal transduction. *Curr. Opin. Immunol.*, **8**, 459–464.

Lavin, M.F. and Shiloh, Y. (1997) The genetic defect in ataxia-telangiectasia *Ann. Rev. Immunol.*, **15**, 177–202.

Lazebnik, Y.A., Kaufmann, S.H., Desmoyners, S., Poirier, G.G. and Earnshaw, W.C. (1994) Cleavage of poly (ADP-ribose) polymerase by a proteinase with properties like ICE. *Nature*, **371**, 346–347.

Lee, J.H. and Bernstein, A. (1993) p53 mutation increases resistance to ionizing radiation. *Proc. Natl. Acad. Sci. USA*, **90**, 5742–5746.

Leeper, D.B., Schneiderman, M.H. and Dewey, D.C. (1972) Radiation-induced division delay in synchronized Chinese hamster ovary cells in monolayer culture. *Radiat. Res.*, **50**, 401–417.

Lees-Miller, S.P., Godbout, R., Chan, D.W., Weinfeld, M., Day, R.S. III., Barron, G.M., Turner, J.A. (1995) Absence of p350 subunit of DNA-activated protein kinase from a radiosensitive human cell line. *Science*, **267**, 1183–1185.

Lehmann, A.R. and Stevens, S. (1979) The response of ataxia-telangiectasia cells to bleomycin. *Nucleic Acids Res.*, **6**, 1953–1960.

Lehmann, A.R., James, M.R. and Stevens, S. (1982) In: *A Cellular and Molecular Link with Cancer*. Bridges, B.A. and Harnden, D.G. (eds.). John Wiley and Sons Ltd, New York, 347–353.

Lerga, A., Belandia, B., Delgado, M.D., Cuadrado, M.A., Richard, C., Ortiz, J.M., Martin-Perez, J. and Leon, J. (1995) Down-regulation of *c-myc* and *max* genes is associated with inhibition of protein phosphatase 2A in K562 human leukemia cells. *Biochem. Biophys. Res. Commun.*, **215**, 889–895.

Li, R. and Murray, A.W. (1991) Feedback control of mitosis in budding yeast. *Cell*, **66**, 519–531.

Lieber, M.R., Hessem, J.T., Lewis, S., Bosma, G.C., Rosenberg, N., Mizuuchi, K., Bosma, M.J. and Gellert, M. (1988) The defect in murine severe combined immune deficiency; joining of signal sequences but not coding segments in V(D)J recombination. *Cell*, **55**, 7–16.

Linke, S.P., Clarkin, K.C. and Wahl, G.M. (1997) p53 mediates permanent arrest over multiple cell cycles in response to γ-irradiation. *Cancer Res.*, **57**, 1171–1179.

Lotem, J. and Sachs, L. (1993) Hematopoietic cells from mice deficient in wild-type p53 are more resistant to induction of apoptosis by some agents. *Blood*, **82**, 1092–1096.

Lowe, S., Schmitt, E., Smith, S., Osborne, B. and Jacks, (1993) p53 is required for radiation-induced apoptosis in mouse thymocytes. *Nature*, **362**, 847–849.

Lu, X. and Lane. D.P. (1993) Differential induction of transcriptionally active p53 following UV or ionizing radiation: Defects in chromosome instability syndromes? *Cell*, **75**, 765–778.

Martin, S.J., Lennon, S.V., Bonham, A.M. and Cotter, T.G. (1990) Induction of apoptosis (programmed cell death) in human leukemic HL-60 cells by inhibition of RNA or protein synthesis. *J. Immunol.*, **145**, 1859–1867.

Martin, S.J. and Cotter, T.G. (1991) Ultraviolet-B-irradiation of human leukemia HL-60 cells in vitro induces apoptosis. *Int. J. Radiat. Biol.*, **59**, 1001–1016.

Martin, S.J., O'Brien, G.A., Nishioka, W.K., McGahon, A.J., Mahboubi, A., Saido, T.C. and Green and D.R. (1995) Proteolysis of fodrin (Non-erythroid spectrin) during apoptosis. *J. Biol. Chem.*, **270**, 6425–6428.

Martin, S.J. and Green, D.R. (1996) Protease activation during apoptosis: death by a thousand cuts. *Cell*, **82**, 349–352.

McConkey, D.J., Hartzell, P., Duddy, S.K., Hakansson, H. and Orrenius, S. (1988) 2,3,7,8-Tetracholordibenzo-*p-dioxin* kills immature thymocytes by Ca^{2+}-mediated endonuclease activation. *Science*, **242**, 256–259.

Meyn, M.S. (1995) Ataxia-telangiectasia and cellular responses to DNA damage. *Cancer Res.*, **55**, 5591–6001.

Michael, J.M., Lavin, M.F. and Watters, D.J. (1997) Resistance to radiation-induced apoptosis in Burkitt's lymphoma cells is associated with defective ceramide signalling. *Cancer Res.*, **57**, 3600–3605.

Mirzayans, R., Famulski, K.S., Enns, L., Fraser, M. and Paterson, M.C. (1995) Characterization of the signal transduction pathway mediating γ-ray-induced inhibition of DNA synthesis in human cells: indirect evidence for involvement of calmodulin but not protein kinase C nor p53. *Oncogene*, **8**, 1597–1605.

Nagasawa, H. and Little, J.B. (1983) Comparison of kinetics of X-ray-induced cell killing in normal, ataxia-telangiectasia and hereditary retinoblastoma fibroblasts. *Mutat. Res.*, **109**, 297–308.

Nakajima, H., Golstein, P. and Henkart, P.A. (1995) The target cell nucleus in not required for cell-mediated granzyme -or Fas-based cytotoxicity. *J. Exp. Med.*, **181**, 1905–1909.

Nazareth, L.V., Harbour, D.V. and Thompson, E.B. (1991) Mapping the human glucocorticoid receptor for leukemic cell death. *J. Biol. Chem.*, **266**, 12976–12980.

Nicholson and Thornberry (1997) Caspases: killer proteases. *Trends Biochem. Sci.*, **22**, 299–306.

Obeid, L.M., Linardi, C.M., Karolak, L.A. and Hannun, Y.A. (1993) Programmed cell death induced by ceramide. *Science*, **259**, 1769–1771.

Ohoka, Y., Nakai, Y., Mukai, M. and Iwata, M. (1993) Okadaic acid inhibits glucocorticoid-induced apoptosis in T cell hybridomas at its late stage. *Biochem. Biophys. Res. Commun.*, **197**, 916–921.

Orrenius, S., Burgess, D.H., Hampton, M.B. and Zhivotovsky, B. (1997) Mitochondria as the focus of apoptosis research. *Cell Death and Diff.*, **4**, 427–428.

Painter, R.B. and Young, B.R. (1980) Radiosensitivity in ataxia-telangiectasia: A new explanation. *Proc. Natl. Acad. Sci. USA*, **77**, 7315–7317.

Painter, R.B. (1985) Radiation sensitivity and cancer in ataxia-telangiectasia. *Ann. N.Y. Acad. Sci.* **459**, 382–386.

Pandita, T.K. and Hittleman, W.N. (1992) Initial chromosome damage but not DNA damage is greater in ataxia telangiectasia cells. *Radiation Res.*, **130**, 94–103.

Pandita, T.K. and Hittleman, W.N. (1994) Increased initial levels of chromosome damage and heterogeneous chromosome repair in ataxia-telangiectasia heterozygote cells. *Mutation Res.*, **310**, 1–13.

Pardee, A.B. (1989) G1 events and regulation of cell proliferation. *Science*, **246**, 603–608.

Paterson, M.C., Smith, B.P., Lohman, P.H. , Andrews, A.K. and Fishman, I. (1976) Defective excision repair of gamma ray damaged DNA in human (ataxia-telangiectasia) fibroblasts. *Nature*, **260**, 444–447.

Paulovich, A.G. and Hartwell, L.H. (1995) A checkpoint regulates the rate of progression through S phase in *S. cerevisiae* in response to DNA damage. *Cell*, **82**, 841–847.

Radford, I.R. (1986a) Evidence for a general relationship between the induced level of DNA double-strand breakage and cell killing after X-irradiation of mammalian cells. *Int. J. Radiat. Biol.*, **49**, 611–620.

Radford, I.R. (1986b) Effect of radiomodifying agents on the ratios of X-ray induced lesions in cellular DNA: use in lethal lesion determination. *Int. J. Radiat. Biol.*, **49**, 621–637.

Raff, M.C., Barres, B.A., Burne, J.F., Coles, H.S., Ishizake, Y. and Jacobson, M.D. (1993) Programmed cell death and the control of cell survival: lessons from the nervous system. *Science*, **262**, 695–700.

Raines, M.A., Kolesnick, R.N. and Golde, D.W. (1993) Sphingomyelinase and ceramide activate mitogen-activated protein kinase in myeloid HL60 cells. *J. Biol. Chem.*, **268**, 14572–14575.

Reich, N.C. and Levine, A.J. (1984) Growth regulation of a cellular tumour antigen, p53, in nontransformed cells. *Nature*, **308**, 199–201.

Sachsenmaier, C., Radler-Pohl, A., Zinck, R., Nordheim, A., Herrlich, P., Rahmsdorf, H.J. (1994) Involvement of growth factor receptors in the mammalian UVC response. *Cell*, **78**, 963–972.

Sahijdak, W.M., Yang, C.R., Zuckerman, J.S., Meyers, M. and Boothman, S. (1994) Alterations in transcription factor binding in radioresistant human melanoma cells after ionizing radiation. *Radiation Res.*, **138**, 47–51.

Sanchez, Y., Desany, B.A., Jones, W.J., Liu, Q., Wang, B. and Elledge, S. (1996) Regulation of *RAD53* by the ATM-like kinases *MEC1* and *TEL1* in yeast cell cycle checkpoint pathways. *Science*, **271**, 357–359.

Savitsky, K., Bar-Shira, A., Gilad, S., Rotman, G., Ziv, Y., Vanagaite, L., Tagle, D.A., Smith, S., Uziel, T., Sfez, S., Ashkenazi, M., Pecker, I., Frydman, M.,Harnik, R., Patanjali, S.R., Simmons, A., Clines, G.A., Sartiel, A., Gatti, R.A., Chessa, L., Sanal, O., Lavin, M.F., Jaspers, N.G.J., Taylor, A.M.R., Arlett, C.F., Miki, T., Weissman, S.M., Lovett, M., Collins, F.S. and Shiloh, Y. (1995) A single ataxia-telangiectasia gene with a product similar to PI-3 kinase. *Science*, **268**, 1749–1753.

Scott, D. and Zampetti-Bosseler, F. (1982) Cell cycle dependence of mitotic delay in X-irradiated normal and ataxia-telangiectasia fibroblasts. *Int. J. Radiat. Biol.*, **42**, 679–683.

Sellins, K.S. and Cohen, J.J. (1991) Hyperthermia induces apoptosis in thymocytes. *Radiat. Res.*, **126**, 88–95.

Shiloh, Y., Tabor, E. and Becker, Y. (1983) Abnormal response of ataxia-telangiectasia cells to agents that break the deoxyribose moiety of DNA via a targeted free radical mechanism. *Carcinogenesis*, **4**, 1317–1322.

Slichenmyer, W.J., Nelson, W.G., Slebos, R.J. and Kastan, M.B. (1993) Loss of a p53 associated G1 checkpoint does not decrease cell survival following DNA damage. *Cancer Res.*, **53**, 4164–4168.

Song, Q. and Lavin, M.F. (1993) Calyculin A, a potent inhibitor of phosphatases-1 and -2A, prevents apoptosis. *Biochem. Biophys. Res. Commun.*, **190**, 47–55.

Song, Q., Lees-Miller, S.P., Kumar, S., Zhang, N., Chan, D.W., Smith, G.C.M., Jackson, S.P., Alnemri, E.S., Litwack, G. and Lavin, M.F. (1996) DNA-dependent protein kinase catalytic subunit: A target for ICE-like protease in apoptosis. *EMBO J.*, **15**, 3238–3246.

Smyth, M.J., Perry, D.K., Zhang, J., Poirier, G.G., Hannun, Y.A. and Obeid, L.M. (1996) prICE: a downstream target for ceramide-induced apoptosis and for the inhibitory action of Bcl-2. *Biochem. J.*, **316**, 25–28.

Spector, B.D., Filipovich, A.H., Perry, G.S. lll. and Kersey, J.H. (1982) Epidemiology of cancer in ataxia-telangiectasia. In B.A. Bridges and D.G. Harnden (Eds) Ataxia-telangiectasia. New York: J. Wiley and Sons Ltd.

Stanger, B.Z., Leder, P., Lee, T-H., Kim, E. and Seed, B. (1995) RIP: A novel protein containing a death domain that interacts with Fas/APO-1 (CD95) in yeast and causes cell death. *Cell*, **81**, 513–523.

Stein, B., Kramer, M., Rahmsdorf, H.J., Ponta, H. and Herrlich, P. (1989) UV-induced transcription from the human immunodeficiency virus type 1 (HIV-1) long terminal repeat and UV-induced secretion of an extracellular factor that induces HIV-1 transcription in nonirradiated cell. *J. Virol.*, **63**, 4540–4544.

Stevenson, M.A., Pollock, S.S., Coleman, C.N. and Calderwood, S.K. (1994) X-irradiation, phorbol esters, and H_2O_2 stimulate mitogen-activated protein kinase activity in NIH-3T3 cells through the formation of reactive oxygen intermediates. *Cancer Res.*, **54**, 12–15.

Strasser, A., Harris, A.W., Jacks, T. and Cory, S. (1994) DNA damage can induce apoptosis in proliferating lymphoid cells via p53-independent mechanisms inhibitable by BCl-2. *Cell*, **79**, 329–339.

Taccioli, G.E., Rathbun, G., Oltz, G., Stamato, T., Jeggo, P. and Alt, F.W. (1993) Impairment of V(D)J recombination in double-strand break repair mutants. *Science*, **260**, 207–210.

Taylor, A.M., Harnden, D.G., Arlett, C.F., Harcourt, S.A., Lehmann, A.R., Stevens, S. and Bridges. B.A. (1975) Ataxia-telangiectasia: a human mutation with abnormal radiation sensitivity. *Nature*, **4**, 427–429.

Téoule, R. (1987) Radiation-induced DNA damage and its repair. *Int. J. Radiat. Biol.*, **51**, 573–589.

Thompson, L.H. and Jeggo, P.A. (1995) Nomenclature of human genes involved in ionizing radiation sensitivity. *Mutation Res.*, **337**, 131–134.

Uckun, F.M., Tuel-Ahlgren, L., Song, C.W., Waddick, K., Myers, D.E., Kirihara, J., Ledbetter, J.A. and Schieven, G.L. (1992) Ionizing radiation stimulates unidentified tyrosine-specific protein kinases in human B-lymphocytes precursors, triggering apoptosis and clonogenic cell death. *Proc. Natl. Acad. Sci. USA*, **89**, 9005–9009.

Uckun, F.M., Schieven, G.L., Tuel-Ahlgren, L.M., Dibirdik, I., Myers, D.E., Ledbetter, J.A. and Song, C.W. (1993) Tyrosine phosphorylation is a mandatory proximal step in radiation-induced activation of the protein kinase C signalling pathway in human B-lymphocyte precursors. *Proc. Natl. Acad. Sci. USA*, **90**, 252–256.

Uckun, F.M., Wadick, K.G., Mahajan, S., Jun, X., Takata, M., Bolen, J. and Kurosaki, T. (1996) BTK as a mediatior of radiation-induced apoptosis in DT-40 lymphoma B cells. *Science*, **273**, 1096–1100.

Ullrich, S.J., Anderson, C.W., Mercer, W.E. and Appella, E. (1992a) The p53 tumour suppressor protein, a modulator of cell proliferation. *J. Biol. Chem.*, **267**, 15259–15262.

Ullrich, S.J., Mercer, W.E. and Appella, E. (1992b) Human wild-type p53 adopts a unique conformational and phosphorylation state *in vivo* during growth arrest of glioblastoma cells. *Oncogene*, **7**, 1635–1643.

Van Antwerp, D.J., Martin, S.J., Kafri, T., Green, D.R. and Verma, I.M. (1996) Suppression of TNF-alpha-induced apoptosis by NF-kappaB. *Science*, **274**, 787–789.

Van der Schans, G.P., Centen, H.B. and Lohman, P.H.M. (1982) The induction and repair of double-strand DNA breaks in normal and ataxia-telangeictasia cells exposed to ^{60}Co-γ-radiation, 4-nitroquinoline-1-oxide or bleomycin. In *Ataxia-Telangiectasia* – A cellular and Molecular link between cancer, Neuropathology and immune deficiency. Edited by: BA Bridges and DG Harnden. (Chichester, Wiley) pp.291–303.

Wallach, D.F.H. (1972) The plasma membrane: Dynamic perspectives, genetics and pathology. Springer-Verlag, New York, Inc., p145–179.

Walworth, N., Bernards, R. (1996) *rad*-dependent response of the *chk*1-encoded protein kinase at the DNA damage checkpoint. *Science*, **271**, 353–356.

Wang, C.Y., Mayo, M.W. and Baldwin, A.S. Jr. (1996) TNF-and cancer therapy-induced apoptosis: potentiation by inhibition of NF-**k**B. *Science*, **274**, 784–787.

Ward, J.F. (1985) Biochemistry of DNA lesions. *Radiation Res.*, **104**, 103–111.

Waterhouse, N., Kumar, S., Strike, P., Sparrow, L., Song, Q., Dreyfuss, G., Alnemri, E., Litwack, G., Lavin, M.F. and Watters, D. (1996) Heteronuclear ribonuleoproteins C1 and C2, components of the spliceosome, are specific targets of ICE-like proteases in apoptosis. *J. Biol. Chem.*, **271**, 29335–29341.

Waterhouse, N., Finucane, D., Green, D.R., Elce, J.S., Kumar, K.K., Khanna, K.K., Alnemri, E., Litwack, G., Lavin, M.F. and Watters, D. (1997) Calpain is activated before the caspases in radiation-induced apoptosis. *Cell Death and Diff.* (in press).

Weinhert, T.A. and Hartwell, L.H. (1988) The *RAD9* gene controls the cell cycle response to DNA damage in *Saccharomyces cerevisiae*. *Science*, **241**, 317–322.

Weinert, T.A., Hartwell, L.H. (1990) Characterization of *RAD9* of *Saccharomyces cerevisiae* and evidence that its function acts post translationally in cell cycle arrest after DNA damage. *Mol. Cell. Biol.*, **10**, 6554–6564.

Weinhert, T.A., Kiser, G.L. and Hartwell, L.H. (1994) Mitotic checkpoint genes in budding yeast and the dependence of mitosis on DNA replication and repair. *Genes Dev.*, **8**, 652–665.

Wu, M., Lee, H., Bellas, R.E., Schauer, S.L., Arsura, M., Katz, D., Fitzgerald, M.J., Rothstein, T.L., Sherr, D.H. and Sonenshein, G.E. (1996) Inhibition of NF-**k**B/Rel induces apoptosis of murine B cells. *EMBO J.*, **15**, 4682–4690.

Wyllie, A.H., Morris, R.G., Smith, A.L. and Dunlop, D. (1984) Chromatin cleavage in apoptosis: association with condensed morphology and dependence on macromolecular synthesis. *J. Pathol.*, **142**, 67–77.

Xiong, Y., Zhang, H. and Beach, D. (1993) Subunit rearrangement of the cyclin-dependent kinases is associated with cellular transformation. *Genes Dev.*, **7**, 1572–1583.

Zambetti, G.P. and Levine, A.J. (1993) A comparison of biological activaties of wild-type and mutant p53. *FASEB J.*, **7**, 855–865.

Zdzienicka, M.Z. (1995) Mammalian mutants defective in the response to ionizing radiation-induced DNA damage. *Mutation Res.*, **336**, 203–213.

Zhang, J., Alter, N., Reed, J.C., Bomer, C., Obeid, L.M. and Hannun, Y.A. (1996) Bcl-2 interrupts the ceramide-mediated pathway of cell death. *Proc. Natl. Acad. Sci. USA*, **93**, 5325.

Part 2
REGULATION OF APOPTOSIS

4. KINASE CASCADES AND APOPTOSIS

J.F. HANCOCK[*][†]

[*]*Dept. of Pathology, University of Queensland Medical School, Herston Road, Brisbane, QLD 4006*

KEY WORDS: stress activated protein kinase, mitogen activated protein kinase, phosphoinositide 3-OH kinase, Akt, protein kinase B.

INTRODUCTION

There are a number of extracellular signals or insults that induce apoptosis and a variety of growth factors and cytokines that can promote cell survival in the presence of an apoptotic signal. The cell execution machinery and the immediate upstream regulators of this machinery are rapidly being unravelled. An important question then, is what signal transduction pathways link membrane events to the apoptotic effector machinery? There is good evidence that stress activated protein kinases (SAPKs) are activated by signals that induce apoptosis and that constitutive activation of the SAPKs can induce apoptosis. It is also clear that activation of phosphoinositide 3-OH kinase (PI3K) and the Akt serine/threonine kinase can promote cell survival in the presence of stimuli that normally induce apoptosis. A similar protective role for the mitogen activated protein kinase (MAPK) cascade has been proposed. While it is clear that these well defined signalling cascades play a role in regulating apoptosis, the critical targets of these cascades are largely unknown, in other words, the wiring diagrams connecting to the effectors of apoptosis are incomplete. The aim of this chapter is to briefly review the organization and mechanisms of activation of the PI3K/Akt, MAPK and SAPK pathways and discuss some of the data that implicate these pathways in signalling for cell survival or apoptosis. Other pathways that signal for apoptosis are reviewed elsewhere in this book.

PHOSPHOINOSITIDE 3-OH KINASE AND AKT

Mechanism Of Activation Of PI3K And Akt

Recruitment and activation of phosphoinositide 3-OH kinase (PI3K) occurs when ligands including EGF, PDGF, bFGF, IGF-1 and insulin bind to their cognate receptor tyrosine kinases (Figure 4.1). PI3K is a heterodimer comprising a non-catalytic p85 subunit and a catalytic p110 subunit, both of which have several isoforms (Carpenter and Cantley, 1996). The SH2 domains of the p85 subunit recruit

---

† *Corresponding Author*: Tel.: +61-7-3365-5340, Fax: +61-7-3365-5511, e-mail: j.hancok @Mailbox. uq. edu.qu

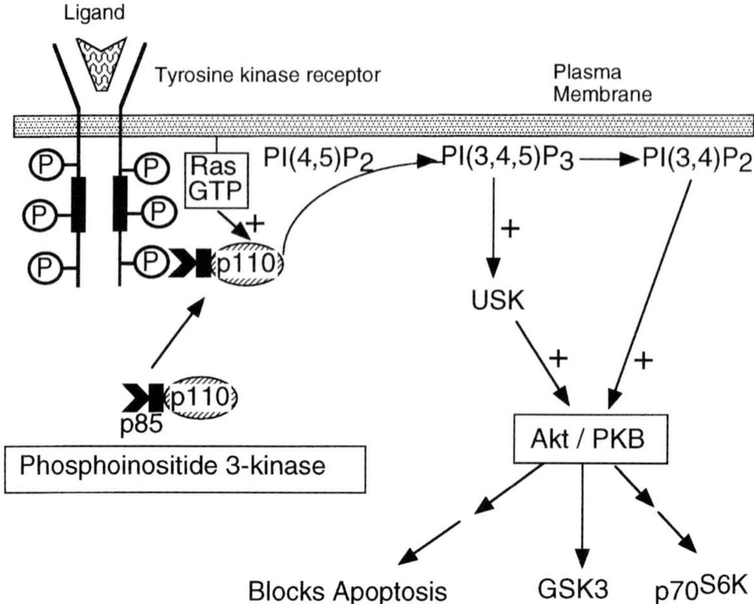

Figure 4.1 Activation of phosphoinositide 3-OH kinase and Akt/PKB. Cartoon of the critical events leading to the activation of the serine/threonine kinase Akt/PKB by upstream kinase (USK) and D3-phosphorylated phosphatidylinositides generated within the inner leaflet of the plasma membrane by PI3K.

PI3K to the plasma membrane by binding to specific phosphotyrosine docking sites, either on the C-terminal tail of the receptor tyrosine kinase or on the IRS-1 adapter protein. The docking of p85 onto phosphotyrosine residues partially activates the associated p110 catalytic subunit, but additional activating inputs are provided by Src and Ras (Rodriguez Viciana *et al.*, 1994.; Carpenter and Cantley, 1996; Rodriguez Viciana *et al.*, 1996). Activated PI3K phosphorylates the D3 position of inositol phospholipids. A major in vivo substrate of PI3K is phosphatidylinositol 4-5 bisphosphate (PtdIns 4,5 P_2), which is converted to PtdIns 3,4,5 P_3. A phospholipid 5'-phosphatase then generates PtdIns 3,4 P_2. PtdIns 3,4,5 P_3 and PtdIns 3,4 P_2 bind pleckstrin homology (PH) domains with high affinity and so the generation of these PtdIns in the inner leaflet of the plasma membrane provides novel docking sites for signalling molecules that contain PH domains (Hemmings, 1997b). Two such molecules are the Akt serine/threonine kinase (also called Protein Kinase B, PKB) (Bos, 1995; Burgering and Coffer, 1995; Datta *et al.*, 1996)and its activating kinase, PKBK (PKB kinase; also called Upstream Kinase, USK). Akt binds to PtdIns 3,4,5 P_3 or PtdIns 3,4 P_2 in the inner leaflet the plasma membrane via its PH domain (Franke *et al.*, 1995; James *et al.*, 1996; Franke *et al.*, 1997; Hemmings, 1997a; Klippel *et al.*, 1997). The interaction between the Akt PH domain and PtdIns 3,4,5 P_3 or PtdIns 3,4 P_2 induces a conformational change in the protein that allows PKBK to phosphorylate T308 and partially activate Akt (Alessi *et al.*, 1997; Stokoe *et al.*, 1997). The binding of PtdIns 3,4 P_2 also induces Akt dimerization, which may contribute to activation (Franke *et al.*, 1997). Full activation of Akt requires phosphorylation on S473 of the partially activated molecule (Alessi *et al.*, 1997; Hemmings, 1997c; Stokoe *et al.*, 1997). Although the in vivo S473 protein kinase has not yet been identified, it is interesting to note that MAPKAP-K2, a SAPK2 target, can phosphorylate S473 and partially activate Akt in vitro (Alessi *et al.*, 1996). Activated Akt plays a critical role in signalling for cell survival through as yet unidentified signalling pathways; these data will be examined in more detail in the next section. In addition Akt can directly phosphorylate and inactivate GSK3, thus stimulating glycogen synthesis (Cross *et al.*, 1995; Cross *et al.*, 1997), and indirectly activate $p70^{S6K}$, thus upregulating protein synthesis (Burgering and Coffer, 1995).

Other roles of PI3K include: regulation of the actin cytoskeleton and activation of Rac, most likely through the plasma membrane recruitment of a Rac exchange factor containing a PH domain (Kotani *et al.*, 1994; Wennstrom *et al.*, 1994; Hawkins *et al.*, 1995; Nobes and Hall, 1995; Rodriguez Viciana *et al.*, 1997). In addition, certain calcium independent, atypical isoforms of protein kinase C are activated by PtdIns 3,4 P_2 and PtdIns 3,4,5 P_3 (Carpenter and Cantley, 1996).

Role Of PI3K And Akt/PKB In Anti-Apoptotic Signalling

The PI3K/Akt pathway is now firmly established as an important cell survival signalling pathway. The story has evolved rapidly over the past few years, having started with observations that insulin-like growth factor 1 (IGF-1) and PDGF promote cell survival in the presence of diverse apoptotic stimuli and progressed most recently to identification of Akt/PKB as the critical effector.

Identification of IGF-1 and PDGF as survival factors

Hemopoietic growth factors promote survival of progenitor cells by preventing apoptosis, however, IGF-1 can partially replace the requirement for such factors. For example, human erythroid precursors growing in vitro rapidly undergo apoptosis when switched to serum free medium. If erythropoietin is included as the sole growth factor, >75% of the cells survive, while inclusion of IGF-1 as sole growth factor rescues >50% of the cells (Muta and Krantz, 1993). Similarly, IGF-1 promotes the proliferation and survival of IL3-dependent FDCP-1/Mac-1 murine myeloid progenitors in the absence of exogenous IL3 (Minshall *et al.*, 1996). Thus IGF-1 partially protects against apoptosis induced by the withdrawal of growth factors from haemopoietic cell lines. Similar observations have been made for IGF-1 and apoptosis induced by chemotherapeutic drugs. For example, mouse BALB/c 3T3 cells overexpressing IGF-1-R arrest in S phase and undergo apoptosis when treated with the topoisomerase I inhibitor etoposide. However, when grown in the presence of IGF-1, these cells still arrest in S phase when exposed to etoposide but are resistant to apoptosis. The survival effect of IGF-1 requires a functional IGF-1-R because IGF-1 provides minimal protection to parental BALB/c 3T3 cells expressing wild type levels of IGF-1-R and has no protective effect on fibroblasts with a targeted disruption of the IGF-1-R locus (Sell *et al.*, 1995).

Deregulated expression of the c-myc oncogene induces apoptosis in serum starved fibroblasts (Evan *et al.*, 1992). However, cells overexpressing c-myc are protected from apoptosis when grown in low serum containing IGF-1 or PDGF (Harrington *et al.*, 1994). IGF-1 also functions as an in vitro survival factor for primary cultures of cerebellar neurons (Dudek *et al.*, 1997) and fibroblasts exposed to UV irradiation (Kulik *et al.*, 1997). The survival effect of IGF-1 is independent of its proliferative action since it protects cells both before and after the G1/S checkpoint (Harrington *et al.*, 1994; Sell *et al.*, 1995).

PI3K mediates cell survival

Many studies have now shown that the survival signalling pathway downstream from the IGF-1 receptor requires PI3K. For example, primary cultures of cerebellar neurons maintained in IGF-1 have elevated PI3K activity and undergo apoptosis when treated with the PI3K inhibitors wortmannin or LY294002 (Dudek *et al.*, 1997). Rat-1 cells growing in IGF-1 are protected against apoptosis induced by UV-B: the protection is lost when the cells are treated with wortmannin but is unaffected when the cells are treated with the MEK inhibitor PD98059 (Kulik *et al.*, 1997). In addition, Rat-1 cells that express constitutively active PI3K are protected against UV-B to the same extent as those grown in IGF-1, but this protection remains wortmannin sensitive (Kulik *et al.*, 1997).

Kauffmann Zeh *et al.*, (1997) found that expression of activated Ras sensitizes fibroblasts to myc-induced apoptosis and used Ras effector site mutants to identify the effector pathway responsible. A Ras effector mutant that activates only the Raf/MEK/MAPK cascade sensitizes, and a Ras effector site mutant that activates only PI3K protects against myc-induced apoptosis. A Ras effector mutant that activates only RalGDS does not protect or sensitize. Consistent with these results, expression of constitutively activated PI3K also protects fibroblasts from apoptosis induced by myc-expression (Kauffmann Zeh *et al.*, 1997).

Other data show that PI3K is also the critical mediator of NGF-induced cell survival. Nerve growth factor (NGF) induces both differentiation and survival of neuronal cells. These two biological effects of NGF are mediated by different signalling pathways: NGF induced differentiation of the PC12 pheochromocytoma cell line requires the Ras/Raf/MEK/MAPK pathway (Cowley *et al.*, 1994; Marshall, 1995; Yao and Cooper, 1995), while Ras is not required for the NGF-mediated survival of PC12 cells in serum-free medium (Yao and Cooper, 1995). The survival effect of NGF is completely abrogated by the PI3K inhibitors wortmannin or LY294002. Heterologous expression of wild type PDGFR in PC12 cells renders them sensitive to the survival effect of PDGF. However, PC12 cells expressing a mutant PDGFR, which is unable to bind and activate PI3K, are not protected from undergoing apoptosis by PDGF. Taken together these data strongly implicate PI3K as a critical signalling protein in NGF and PDGF induced cell survival (Yao and Cooper, 1995).

Before examining events downstream of PI3K, it is worth noting that survival signalling is not all mediated through PI3K. For example, IGF-1 and IL3 both efficiently promote survival of IL3-dependent murine myeloid precursors and both activate PI3K (Minshall *et al.*, 1996). Cells maintained in IGF-1, but not those in IL3 undergo apoptosis when treated with wortmannin at concentrations that completely block PI3K activity. Thus in addition to the PI3K mediated survival pathway, IL3 must activate a second PI3K independent survival pathway in hemopoietic cells (Minshall *et al.*, 1996). The second IL3 pathway remains to be elucidated, but as for the PI3K mediated pathway (Harrington *et al.*, 1994; Yao and Cooper, 1995; Kulik *et al.*, 1997) it does not require new protein synthesis (Minshall *et al.*, 1996).

Akt/PKB is the critical PI3K target

As discussed above there are three kinases known to be downstream of PI3K. Akt/PKB and USK are direct targets of PI3K, while GSK3 and $p70^{S6K}$ are, respectively, direct and indirect targets of Akt/PKB. PI3K undoubtedly has multiple other targets, i.e. any protein with a PH domain. Nevertheless, attention has been sharply focused on whether Akt/PKB and $p70^{S6K}$ are the PI3K targets involved in signalling for cell survival.

PC12 cells growing in NGF, or Rat-1 cells growing in serum, both undergo apoptosis when treated with the PI3K inhibitors wortmannin and LY294002, but are unaffected by treatment with rapamycin at concentrations that demonstrably inhibit $p70^{S6K}$ activation (Yao and Cooper, 1996). The survival of primary cultures of cerebellar neurons, which is promoted by IGF-1 and blocked by wortmannin and LY294002, is not compromised by rapamycin (Dudek *et al.*, 1997). Similarly the IGF-1 mediated survival of Rat-1 cells exposed to UV-B, and the PI3K mediated survival of c-myc expressing fibroblasts are unaffected by rapamycin (Kauffmann Zeh *et al.*, 1997; Kulik *et al.*, 1997). It is clear from these studies that $p70^{S6K}$ is not involved in signalling for cell survival.

In contrast, Akt constitutively targeted to the plasma membrane by myristoylation protects COS 7 cells against apoptosis induced by UV-B (Kulik *et al.*, 1997). Similarly, a gag-Akt fusion protein, constructed to mimic v-Akt, the constitutively active retroviral protein, protects fibroblasts against myc-induced apoptosis, while a kinase inactive gag-Akt fusion renders no protection. The protection afforded by the gag-Akt fusion protein is insensitive to LY294002, demonstrating that Akt is downstream from PI3K (Kauffmann Zeh *et al.*, 1997). Akt is also activated by IGF-1 in cerebellar neurons and overexpression of wild type Akt in cerebellar neurons is sufficient to promote their survival in the absence of IGF-1. Furthermore, expression of dominant negative forms of Akt in cerebellar neurons abrogates the anti-apoptotic effect of IGF-1 (Dudek *et al.*, 1997). Essentially identical observations have recently been made in epithelial cells induced to undergo apoptosis by detachment from the extracellular matrix (so called anoikis). Activated Ras, activated PI3K and activated Akt all protect epithelial cells against anoikis. The protection afforded by Ras and PI3K, but not that afforded by Akt is wortmannin sensitive (Khwaji *et al.*, 1997) .

Taken together these data clearly implicate PI3K mediated activation of Akt as an important anti-apoptotic signalling pathway. What then are the critical targets of Akt that regulate the apoptotic effector machinery? Since $p70^{S6K}$ clearly plays no role and GSK3 has never been implicated in apoptotic control, the answer must await the identification of novel Akt targets. However, somewhat intriguingly, a recent study has shown that IGF-1 can suppress apoptosis caused by overexpression of interleukin-1β-converting enzyme (Jung *et al.*, 1996). The study did not address the signalling pathways involved, other than demonstrating that IGF-1 did not alter the expression of Bcl-2, Bcl-x, or Bax. In the light of the data discussed here, it is tempting to implicate Akt as the critical effector for this IGF-1 survival effect, which opens up the possibility that Akt may, inter alia, directly regulate a caspase cascade.

MAP KINASE CASCADES

Organization Of The MAPK Cascade

This conserved signalling pathway, delineated by genetic experiments in invertebrates and biochemistry in mammalian cells, comprises a basic module of Raf->MAPKKinase (MKK)->MAPK with tyrosine kinases and Ras as upstream activators (Figure 4.2). A large number of growth factors and cytokines, e.g. EGF, PDGF, FGF, Insulin, GM-CSF, M-CSF, IL3, activate this module in conjunction with other signalling pathways. In addition to being required and sufficient for DNA synthesis, activation of the MAPK cascade may promote cell survival and depending on the cellular context, stimulate cell differentiation. MAPK activation occurs rapidly after the engagement of receptor by ligand but the duration of activation varies for different ligands.

The critical initiating event in MAPK activation is recruitment of the Ras exchange factor hSos1 to the plasma membrane (Aronheim *et al.*, 1994) where Ras is localized; hSos1 then activates Ras by catalysing the exchange of GDP for GTP (Chardin *et al.*, 1993). The exchange factor hSos1 is recruited to the plasma membrane in a complex with adapter proteins Grb2 or Grb2/Shc (Buday and Downward, 1993; Egan *et al.*, 1993; Li *et al.*, 1993). The adapter proteins bind to specific phosphotyrosine residues on the cytoplasmic tail of activated tyrosine kinases. The kinases may be the growth factor receptor per se, or in the case of cytokine receptors, a recruited cytoplasmic non-receptor tyrosine kinase. Recently, a membrane protein, FRS2, which is tyrosine phosphorylated following FGF stimulation, has been shown to be an alternative plasma membrane docking site for Grb2/Sos complexes (Kouhara *et al.*, 1997). Once RasGTP is generated at the plasma membrane multiple effector proteins are activated, including Raf (Moodie *et al.*, 1993), PI3K (Rodriguez-Viciana *et al.*, 1994), RalGDS (Kikuchi *et al.*, 1994) and AF6 (Kuriyama *et al.*, 1996).

There are three Raf serine threonine kinases, c-Raf-1, A-Raf and B-Raf (Daum *et al.*, 1994). c-Raf-1 is widely expressed while A-Raf and B-Raf have more restricted expression patterns. The mechanism of activation of the Raf kinases is not fully resolved, but is best worked out for c-Raf-1. RasGTP recruits Raf from the cytoplasm to the plasma membrane by binding the N-terminal CR1 domain of Raf (Traverse *et al.*, 1993; Van Aelst *et al.*, 1993; Vojtek *et al.*, 1993; Zhang *et al.*, 1993). This recruitment of Raf to the plasma membrane is sufficient for partial activation (Leevers *et al.*, 1994; Stokoe *et al.*, 1994; Wartmann and Davis, 1994). Once at the plasma membrane Raf is available for phosphorylation by tyrosine kinases on residues Y340 and Y341 which leads to further activation (Fabian *et al.*, 1993; Marais

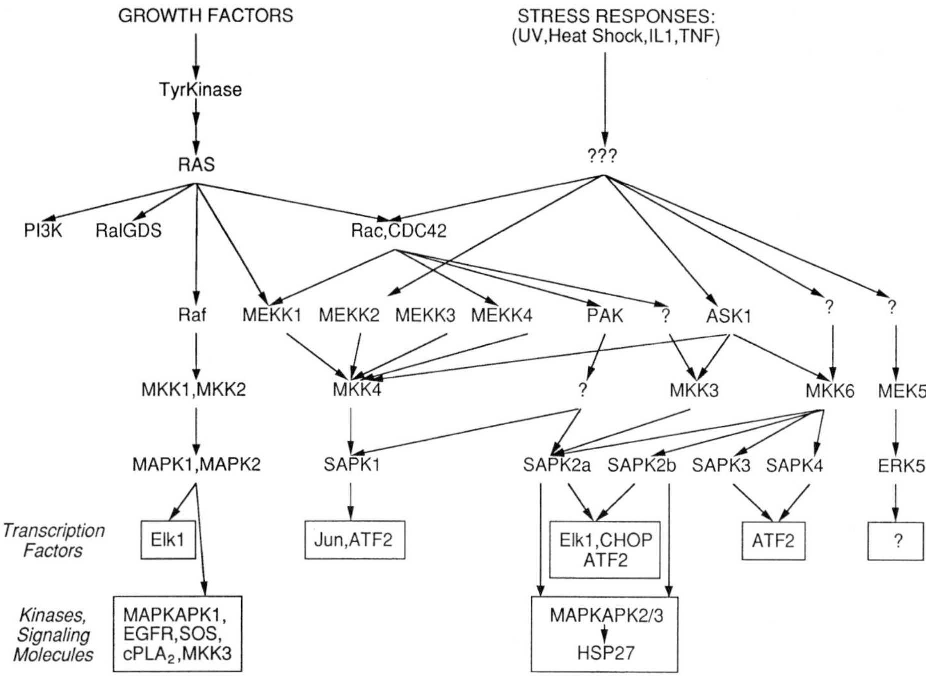

Figure 4.2 Activation of MAPK and SAPK cascades. Overview of the organization of the MAPK and SAPK cascades. Question marks indicate "missing kinases", that have not yet been identified, or activation mechanisms that have not been elucidated. Alternative names of the various kinases are given in the text.

et al., 1995). Ras also has an additional role in potentiating the activity of tyrosine phosphorylated Raf through an interaction with the Raf zinc finger (Jelinek *et al.*, 1996; Luo *et al.*, 1997; Roy *et al.*, 1997). Other activating inputs into Raf may include: phosphatidylserine binding to the Raf zinc finger (Ghosh *et al.*, 1996), displacement of 14-3-3 from the Raf N-terminus (Rommel *et al.*, 1996), phosphorylation of T269 by the ceramide activated protein kinase, Ksr1 (Yao *et al.*, 1995; Zhang *et al.*, 1997), phosphorylation of S499 by PKC (Kolch *et al.*, 1993) and potentially Raf dimerization (Farrar *et al.*, 1996; Luo *et al.*, 1996). The relative contribution of each of these mechanisms to A-Raf and B-Raf activation is probably different from c-Raf-1 since both of these Raf isoforms have negatively charged residues in place of one or both regulatory tyrosines. The presence of these negatively charged residues significantly increases the basal kinase activity of A-Raf and B-Raf but A-Raf and B-Raf still exhibit Ras dependent activation following membrane recruitment (Marais *et al.*, 1997). Recent experiments have also defined an alternative mechanism for the activation of B-Raf in PC12 cells, that is stimulated by cAMP and mediated through the Ras subfamily GTPase, Rap1 (Vossler *et al.*, 1997). These in vivo data are consistent with earlier observations that showed activation of a B-Raf/14-3-3 complex by GTP loaded Rap1 in a cell free system (Ohtsuka *et al.*, 1996).

All Raf kinases activate the dual specificity kinase MEK1 (MAP Kinase Kinase 1, MKK1) by phosphorylating residues S218 and S222 (Dent *et al.*, 1992; Kyriakis *et al.*,

1992; Alessi *et al.*, 1994; Zheng and Guan, 1994). c-Raf-1 but not A-Raf also activates the closely related dual specificity kinase MEK2 (MKK2) (Wu *et al.*, 1996). MKK1 and MKK2 in turn activate p44 MAPK1 (ERK1) and p42 MAPK2 (ERK2) by phosphorylating the threonine and tyrosine residues within a regulatory TEY motif in the kinase domain (Crews *et al.*, 1992; Howe *et al.*, 1992; Macdonald *et al.*, 1993; Marshall, 1994). Activated MAPK1 and MAPK2 have a plethora of substrates which include transcription factors, e.g. Elk1 (Janknecht *et al.*, 1993), other kinases $p90^{rsk}$ (MAPKAP-K1) (Blenis, 1993), signal transduction molecules, e.g. Sos1, MEK1, $cPLA_2$ (Lin *et al.*, 1993; Cadwallader *et al.*, 1994; Buday *et al.*, 1995), and regulators of translation. The MAPK phosphorylations either activate or downregulate the activity of each substrate. The net effect of MAPK activation depends on both the cell type and the duration of activation (Marshall, 1995). For example, in NIH3T3 fibroblasts activation of the Raf/MEK/MAPK cascade is sufficient to stimulate DNA synthesis and cell division (Cowley *et al.*, 1994; Mansour *et al.*, 1994). In neuronal PC12 cells transient MAPK activation as stimulated by EGF results in cell division, while sustained activation as stimulated by NGF is accompanied by nuclear translocation and results in differentiation (Cowley *et al.*, 1994; Marshall, 1995).

The downregulation of the Raf/MEK/MAP kinase pathway is less well understood. MAPK phosphorylation of the C-terminus of hSos1 may cause dissociation of the Grb2/Shc/EGFR complex (Buday *et al.*, 1995; Porfiri and McCormick, 1996), so terminating the activation of Ras. Ras GAPs compete with Ras effectors for binding to RasGTP and when bound, catalyse rapid GTP hydrolysis, returning Ras to the inactive ground state (McCormick *et al.*, 1988; McCormick, 1989). It is unclear, however, how activated Raf, that is no longer complexed with Ras at the plasma membrane, is deactivated and returned to the cytoplasm. There is clearly a role for phosphatases (Dent *et al.*, 1995; Dent *et al.*, 1996) and possibly also for 14-3-3 proteins.

Organization Of The SAPK Cascades

There are four Stress Activated Protein Kinases (SAPKs) that are activated by cellular stress: heat and osmotic shock, protein synthesis inhibition, DNA damaging agents, UV radiation and proinflammatory cytokines IL1 and TNF (Kyriakis and Avruch, 1996). The basic module is SAPK Kinase Kinase (SAPKKK)->SAPK Kinase (SKK, also frequently referred to as MKK)->SAPK, but the details of how specific stimuli activate the various SAPK cascades is not nearly as well worked out as the MAPK cascade (Figure 4.2). The nomenclature of the SAPK cascades suffers from multiple designations for all of the kinases; in this discussion alternate names will usually be given only once, when first mentioned.

SAPK1 (also called Jun N-terminal Kinases, JNKs) and SAPK2a (also called: RK, p38, CSBP kinase, Mxi2) activation is under the control of the Rho family GTPases Rac and Cdc42 (Coso *et al.*, 1995; Minden *et al.*, 1995; Olson *et al.*, 1995). The details of how Rho family GTPases are activated are not yet worked out. Extrapolating from the regulation of Ras by hSos1, it is postulated that colocalizing Rac and Cdc42 with their cognate exchange factors causes activation (van Leeuwen *et al.*, 1995; Michiels *et al.*, 1997). Several exchange factors have been cloned, including Dbl (Hart *et al.*, 1991) and other proteins containing Dbl-like homology domains: Vav (Katzav *et al.*, 1989; Adams *et al.*, 1992), Tiam2 (Habets *et al.*, 1994), Isc (Whitehead *et al.*, 1996) and possibly hSos1 (Chardin *et al.*, 1993). How these exchange factors are recruited to the plasma membrane to activate Rac or Cdc42 is unclear. However, the presence of pleckstrin homology domains in many Dbl-like exchange factors suggests that PtdIns 3,4 P_2 or PtdIns 3,4,5 P_3 might constitute the membrane docking site (Zheng *et al.*, 1996; Michiels *et al.*, 1997; Wang *et al.*, 1997). Consistent with this model, it has recently been shown that Rac activation is downstream of PI3K (Hawkins *et al.*, 1996), but other mechanisms must also operate, because tyrosine phosphorylation is required to activate the Vav exchange factor (Han *et al.*, 1997).

The functional homologue of Raf in the MAPK cascade, is MEKK1, and as with Raf the regulation of MEKK1 is not fully resolved. In vitro, MEKK1 can bind both Ras and the Rho subfamily GTPases, Rac and Cdc42 (Lange Carter and Johnson, 1994; Russell *et al.*, 1995), however, in vivo, activated Rac and Cdc42 robustly activate the SAPK1 cascade while Ras does so weakly (Coso *et al.*, 1995; Minden *et al.*, 1995; Olson *et al.*, 1995). To add further complexity, there are at least three other isoforms of MEKK (MEKK2-4) with near identical kinase domains (96% conserved), which all activate SAPK1 when conditionally expressed (Ellinger Ziegelbauer *et al.*, 1997; Gerwins *et al.*, 1997). The structure of their regulatory domains varies and only MEKK1 and MEKK4 can bind Rac and Cdc42, thus each MEKK may be subject to different mechanisms of regulation and transduce distinct upstream signals (Fanger *et al.*, 1997).

MEKK1-4 all activate the dual specificity kinase MKK4 (MAPK Kinase 4, SKK1 or SEK1, SAPK Kinase 1) and when overexpressed can activate MKK1(Lange Carter *et al.*, 1993; Minden *et al.*, 1994; Yan *et al.*, 1994). In addition, MEKK1 is an IκB kinase capable of triggering IκB degradation and NF-κB activation (Liu *et al.*, 1996; Lee *et al.*, 1997). Thus signals which activate MEKK1 will coordinately activate c-Jun and NF-κB dependent gene transcription. Other SAPKKKs with uncharacterized upstream activators include, mixed lineage kinase (MLK-3) which activates MKK4 and possibly MKK3/MKK6 and Tpl2 which also activates MKK4 (Salmeron *et al.*, 1996; Tibbles *et al.*, 1996). Germinal center kinase is a B cell specific Ste20 homologue that activates MKK4/SAPK1 via an as yet unidentified SAPKKK (Pombo *et al.*, 1995).

MKK4 activates SAPK1 by phosphorylating tyrosine and threonine residues in the TGY regulatory sequence, which is a signature motif of all SAPKs (Sanchez *et al.*, 1994; Yan *et al.*, 1994). There are three closely related isoforms of SAPK1 (JNKs) which phosphorylate c-Jun on residues S63 and S73 and upregulate its transcriptional activity (Derijard *et al.*, 1994; Kyriakis *et al.*, 1994). They also phosphorylate ATF2 (Gupta *et al.*, 1995; van Dam *et al.*, 1995; Beyaert *et al.*, 1996; Hazzalin *et al.*, 1996).

The dual specificity kinase MKK3 (SKK2) is a specific activator of SAPK2a while the dual specificity kinase MKK6 (SKK3, MEK6) activates both SAPK2a and SAPK2b (Cuenda *et al.*, 1996; Han *et al.*, 1996; Cuenda *et al.*, 1997). SAPK2a (RK, p38, CSBP kinase, Mxi2) (Galcheva Gargova *et al.*, 1994; Han *et al.*, 1994; Lee *et al.*,

1994; Rouse *et al.*, 1994) and SAPK2b (p38β) (Jiang *et al.*, 1996) phosphorylate and activate the transcription factors Elk-1 and CHOP (Price *et al.*, 1996; Wang and Ron, 1996). In addition these SAPKs phosphorylate and activate the protein kinases MAP-KAP-K2 (MAPK activated protein kinase2) and MAP-KAP-K3 (Rouse *et al.*, 1994; Clifton *et al.*, 1996; McLaughlin *et al.*, 1996), which in turn phosphorylate Hsp27 and the transcription factor CREB (Huot *et al.*, 1995; Tan *et al.*, 1996).

The MKK3/SAPK2a pathway is activated in vivo by activated Cdc42 and Rac (Zhang *et al.*, 1995) but not via MEKK1-4. Candidate regulators of this pathway are the p21 activated kinases. This group of kinases, PAK1 (hPAK65) and related isoforms (PAK2 and PAK3) are activated by binding RacGTP or Cdc42-GTP, and can stimulate SAPK1 and SAPK2 activation in vivo (Manser *et al.*, 1994; Manser *et al.*, 1995; Martin *et al.*, 1995; Pombo *et al.*, 1995; Zhang *et al.*, 1995). Yeast genetic studies place Ste20, the homologue of hPAK1 immediately upstream of Ste11, the homologue of MEKK1 (Herskowitz 1995), however, how PAK1 activates SAPK1 and SAPK2 is presently unclear. It does not directly activate MEKK1 and although its most likely targets are MKK4 and MKK3 or MKK6, PAK1 has not yet been shown to phosphorylate any of them (Fanger *et al.*, 1997). Another candidate for upstream regulator of the SAPK2 (and SAPK1) pathway is ASK1, a kinase distantly related to MEKK1, that activates MKK4, MKK3 and MKK6 (Ichijo *et al.*, 1997). Whether ASK1 is regulated by Rac and Cdc42 has not yet been determined, although preliminary data suggest it may be activated by TNFα (Ichijo *et al.*, 1997). Also cAbl is required for SAPK1 and SAPK2 activation in response to some DNA damaging agents but not UV light (Pandey *et al.*, 1996), but it is far from clear at what level this tyrosine kinase feeds into the SAPK activation module.

Less well characterised SAPKs are SAPK3 (Erk6) and the recently cloned SAPK4, which are both activated by MKK6 and which both phosphorylate ATF2; no other substrates have yet been identified (Mertens *et al.*, 1996; Cuenda *et al.*, 1997; Goedert *et al.*, 1997). Finally, Erk5 (Bmk1) is specifically activated by MEK5, but has no known physiological substrates (Abe *et al.*, 1996).

Clearly there are many SAPK substrates other than the few transcription factors and kinases known to date that need to be identified before the full biochemical consequences of activating SAPK cascades will be known.

Role Of The MAPK And SAPK Cascades In Apoptotic Signalling

Does the balance of MAPK vs SAPK activation determine cell fate?

There is significant overlap in the stress stimuli that activate the SAPK cascades and those that can induce apoptosis. One of the first studies that implicated activation of the SAPKs and MAPKs in determining cell fate was that of Xia *et al.* (1995). They showed that 6h after the withdrawal of NGF from PC12 cells and prior to any of the characteristic signs of apoptosis, there is sustained activation of SAPK1 and SAPK2 and inhibition of MAPK. Expression of constitutively activated MEKK1, or coexpression of activated MKK3 and SAPK2 induces apoptosis of PC12 cells in the presence of NGF, while expression of dominant negative c-jun, dominant negative MKK4 or dominant negative MKK3 blocks apoptosis induced by NGF withdrawal or activated MEKK1. Conversely, activated MKK1 prevents apoptosis induced by NGF withdrawal (Xia *et al.*, 1995). These results suggested that SAPK activation is pro-apoptotic while MAPK activation is anti-apoptotic and that the balance of activities in these two pathways determines cell fate. Two other studies lend support to this yin and yang hypothesis. First, in HL60 cells, ceramide, a second messenger induced by Fas ligand and TNFα induces activation of SAPK1 and apoptosis. This outcome is modified by sphingosine-1-phosphate which reduces SAPK activation and stimulates MAPK activation (Cuvillier *et al.*, 1996). Second, TNFα robustly activates SAPK1, weakly activates MAPK and induces apoptosis in L929 cells. In contrast, FGF-2 activates MAPK and protects L929 cells against the apoptotic effect of TNFα without affecting TNFα-induced SAPK1 activation. The protective effect of FGF-2 is abrogated if MAPK activation is blocked with a MEK1 inhibitor (Gardner and Johnson, 1996). Interestingly, in this study it was also observed that oncogenic mutant Ras, which does not activate MAPK in L929 cells, also partially protects against TNFα-induced apoptosis; in hindsight this is probably explained by the activation of PI3K and Akt by Ras.

Does MAPK activation promote cell survival?

Other data also argue for an anti-apoptotic effect of activated MAPK. Thus, dominant negative MEK inhibits IL3-induced MAPK activation in BAF3 bone marrow derived cells and increases the level of IL3 required to prevent apoptosis. However, the expression of constitutively activated MEK does not prevent apoptosis when IL3 is withdrawn (Perkins *et al.*, 1996). Leverrier *et al.* (1997) examined this system in more detail and concluded that MAPK activation is required for the induction of Bcl-X gene expression that correlates with the longterm survival of BAF3 cells returned to IL3 after a period of deprivation (Leverrier *et al.*, 1997). In another model system, GM-CSF prevents apoptosis of haemopoietic cells that express wild type GM-CSF receptor but not cells that express a truncated GM-CSF receptor that does not activate Ras; this deficiency is complemented by coexpression of oncogenic mutant Ras, which restores MAPK activation (Kinoshita *et al.*, 1995). Again these results might also be explained by the restoration of Ras dependent PI3K activation. Cardiotrophin-1, a potent survival factor for cardiac myocytes, provides another example; expression of dominant negative MEK1 or treatment of myocytes with the MEK inhibitor PD098059 both block the survival effect of cardiotrophin-1 (Sheng *et al.*, 1997).

Other groups have concluded that MAPK has no role in signalling for cell survival. The survival effect of IGF-1 on Rat-1 cells is not replicated by EGF or thrombin, which both robustly activate MAPK. And although overexpression of the EGFR in Rat-1 cells does render EGF anti-apoptotic against UV-B, this effect is still blocked by wortmannin (Kulik *et al.*, 1997). These results argue against a significant role for MAP kinase in protecting against UV-B induced apoptosis. Raf targeted to the plasma membrane using the Ras localization sequences (RafCAAX), robustly activates MAPK (Leevers *et al.*, 1994; Stokoe *et al.*, 1994) but does not protect 32.D3 cells from apoptosis following withdrawal of IL3 (Wang *et al.*, 1996). Conversely, Raf targeted to the mitochondrial membrane using the Bcl-2 localization sequences does not activate MAPK, but does protect 32.D3 cells from apoptosis following IL3 withdrawal, probably by stimulating the phosphorylation of BAD (Wang *et al.*, 1996). This study reveals one of the few known links between signalling molecules and the apoptotic effector machinery, but it also demonstrates that MAPK activation cannot promote cell survival. Dudek *et al.* (1997) reached similar conclusions when they observed that growth factors such as BDNF, which strongly activate MAPK, but which do not activate PI3K, could not promote the in vitro survival of cerebellar neurons.

Finally, there are studies showing that, persistent MAPK activation can actually enhance apoptosis. For example, the expression of RafCAAX or Ras mutants that selectively activate Raf enhances myc-induced apoptosis in fibroblasts (Kauffmann Zeh *et al.*, 1997). Similarly bufalin induces MAPK activation and apoptosis in U937 cells but the effects of bufalin are abrogated if MAPK activation is blocked using an antisense cDNA against MEK1 (Watabe *et al.*, 1996).

Taken together these data are not generally supportive of a significant role for MAPK in signalling cell survival. The few examples where a positive effect of MAPK is seen may be peculiar to the cell system and apoptotic stimulus. In these cases it is noteworthy that sustained MAPK activation and nuclear translocation correlated with survival, suggesting that MAPK may have some role in regulating the commitment to apoptosis when new gene expression is required.

Is SAPK activation required for apoptosis?

From the evidence available there is clearly no consensus that MAPK activation has a consistent role in promoting cell survival. If we then exclude the concept that the balance of activity in the MAPK and SAPK pathways determines cell fate, what of the simpler hypothesis that SAPK activation is pro-apoptotic?

The hypothesis is supported by two lines of evidence. First, activation of SAPK1 and SAPK2 precedes apoptosis in response to stimuli that include TNFα, UV and γ-irradiation, Fas ligand, chemotherapeutic drugs and other cellular stresses (Xia *et al.*, 1995; Chen *et al.*, 1996; Cuvillier *et al.*, 1996; Heidenreich and Kummer, 1996; Kyriakis and Avruch, 1996; Verheij *et al.*, 1996; Wilson *et al.*, 1996; Chauhan *et al.*, 1997). Secondly, expression of constitutively activated MEKKs or MKKs (SKKs) can induce apoptosis while dominant interfering MEKKs or MKKs (SKKs) can inhibit apoptosis, for example the data of Xia *et al.* (1995) in PC12 cells discussed earlier. Likewise, dominant negative SEK1(MKK4) or dominant negative c-jun protect U937 cells from apoptosis induced by ceramide, UV-C, heat shock, ionizing radiation and hydrogen peroxide(Verheij *et al.*, 1996) and dominant negative SEK1(MKK4) can block apoptosis induced by heat shock and cis-platinum (Zanke *et al.*, 1996). Moreover, ASK1, a novel SAPKKK that activates both SAPK1 and SAPK2, induces apoptosis when conditionally overexpressed in mink lung epithelial cells (Mv1Lu). Interestingly, endogenous ASK1 is activated when Mv1Lu cells are treated with TNFα and Mv1Lu cells are partially protected from TNFα-induced apoptosis by dominant negative ASK1 (Ichijo *et al.*, 1997). These data are consistent with SAPK activation being required and sufficient for apoptosis. In contrast, constitutively activated MEKK1 induces apoptosis when microinjected into Swiss 3T3 and REF52 fibroblasts and enhances the apoptotic response of these cells to UV-C when conditionally expressed. But although MEKK1 activates SAPK1, dominant negative SAPK1 does not suppress the apoptotic effect of activated MEKK1 in Swiss 3T3 and REF52 fibroblasts. Thus while MEKK1 activation is apoptotic, this effect appears not to be mediated via SAPK activation (Johnson *et al.*, 1996), indicating that MEKK1 must have substrates other than SEK1 that are relevant for apoptotic signalling.

The contribution of SAPK activation to TNFα and Fas-induced apoptosis is also debated. The issue is not simple because, as discussed elsewhere in this book, Fas and the TNFR, in addition to activating SAPKs, also activate the FADD/FLICE/caspase death pathway; Fas directly interacts with FADD while TNFR does so via the adapter protein, TRADD. It has recently been shown that Fas activates the SAPK pathway via a novel protein called Daxx, although how Daxx actually regulates SAPKKKs is unclear (Yang *et al.*, 1997). Accumulating evidence suggests that the FADD pathway is the more important one for signalling Fas or TNFα-induced apoptosis. For example, Liu *et al.* (1996) show that while FADD overexpression is sufficient to induce apoptosis it does not activate either SAPK or NFκB, leading them to conclude that SAPK activation is not required for TNFα-induced apoptosis. Consistent with this conclusion, dominant negative FADD blocks TNFα-induced apoptosis in Hela cells but has no effect on TNFα-induced SAPK activation (Natoli *et al.*, 1997). Similarly, crosslinking Fas induces SAPK activation and apoptosis in Jurkat cells, but whereas dominant negative SEK1(MKK4) effectively abrogates SAPK activation, it does not block apoptosis (Lenczowski *et al.*, 1997). The observation that Daax overexpression robustly activates SAPK but does not induce apoptosis, is also consistent with there being no role for SAPK activation in Fas-induced apoptosis. However, dominant negative Daax abrogates Fas-induced apoptosis of HeLa and 293 cells, clearly invoking a contribution of SAPK activation to Fas-induced apoptosis (Yang *et al.*, 1997).

Caspases and SAPKs, who regulates whom?

Are the caspases downstream targets of the SAPKs? In favour of this hypothesis, ICE inhibitors have been shown to block apoptosis induced by NGF withdrawal from PC12 cells, or etoposide treatment of U937 cells, but to have no effect on the coincident SAPK activation (Park *et al.*, 1996; Seimiya *et al.*, 1997). In direct contrast, other data place SAPK activation downstream of the caspases. For example, ICE inhibitors and crmA can block both Fas-induced apoptosis and Fas-induced SAPK activation in Jurkat cells as well as a variety of T and B cell lines (Cahill *et al.*, 1996; Lenczowski *et al.*, 1997). Juo *et al.* (1997) have extended these observations. They observe that Fas activation of SAPK2 correlates with the onset of apoptosis in Jurkat cells, and that the expression of constitutively activated MKK3 potentiates Fas-induced apoptosis. Pharmacological ICE inhibitors completely block Fas-induced activation of SAPK2 and Fas induced apoptosis, but interestingly do not block SAPK2 activation induced by osmotic shock. These results demonstrate that SAPK2 is activated by at least two mechanisms: a caspase dependent mechanism activated by Fas, and a caspase independent mechanism activated by osmotic shock

(Juo *et al.*, 1997). As with Fas-induced apoptosis, anoikis in epithelial cells is accompanied by activation of both ICE/LAP3 caspase and SAPK1. Expression of Bcl-2 or crmA protects against anoikis and also abrogates both caspase and SAPK activation (Frisch *et al.*, 1996), which also places SAPK1 downstream of caspase activation. A mechanism for the control of SAPK activation by the caspases has recently been uncovered (Cardone *et al.*, 1997). MEKK1 is activated when cells undergo anoikis, but MEKK1 activation induced by disruption of cell matrix results from a specific proteolytic event that generates an activated 78kDa C-terminal MEKK1 cleavage product. The cleavage event and MEKK1 activation are both blocked by caspase inhibitors that specifically target caspases which recognize DEVD motifs. Cleavage resistant MEKK1, constructed by mutating the caspase recognition sequence, inhibits caspase 7 activation and partially protects MDCK cells against anoikis. Taken together these data suggest that MEKK1/MKK3/SAPK and caspase 7 comprise a positive feed back loop, where activated MEKK1 activates caspase 7 via the SAPK pathway, and activated caspase 7 in turn activates more MEKK1 (Cardone *et al.*, 1997). As discussed earlier, MEKK1 also has additional substrates, notably IκB: thus MEKK1 activation will result in coincident SAPK, caspase and NFκB activation. The precise outcome of these events will likely be influenced by cell context.

Effect of cell context on the outcome of SAPK activation

A striking example of the effect of cell context on the outcome of activation of the MEKK/MKK4/SAPK1 cascade is that of lymphocytes. In B lymphocytes CD40 ligation activates SAPK1, but not MAPK, while cross-linking IgM activates MAPK. CD40 ligation rescues B cells from apoptosis induced by anti-IgM without affecting CD40 stimulated SAPK activation (Sakata *et al.*, 1995). Similarly, B cell receptor activation in WEHI-231 cells strongly activates MAPK and weakly activates SAPK1, while CD40 crosslinking strongly activates SAPK1 and SAPK2. B cell receptor activation induces apoptosis of these cells which is blocked by activating CD40; thus apoptosis correlates with MAPK activation while cell survival correlates with coactivation of SAPK1,2 and MAPK (Sutherland *et al.*, 1996).

Nishina *et al.*, (1997) deleted SEK1(MKK4) in ES cells by homologous recombination. In SEK1–/–ES cells, SAPK1 is no longer activated by heat shock or the protein synthesis inhibitor anisomycin but is activated normally by UV irradiation and osmotic stress, while SAPK2 is activated normally in response to anisomycin and UV irradiation. Interestingly, T cells derived from SEK1–/–mice are no less sensitive to cell death induced by γ-irradiation, UV irradiation, heat shock, anisomycin or cisplatinum than SEK1+/+ mice, but are more sensitive to apoptosis induced by TCR activation. Thus a SEK1 regulated pathway protects T cells from TCR-induced apoptosis.

Summary and speculation

Some of the apparent discrepancies between the studies discussed here may be explained by the different end points used to score apoptosis, ranging from relative early events, such as morphological changes and DNA fragmentation assays to longer term cell survival assays. Given these differences it is not unreasonable to make the general conclusion that SAPK activation is involved in signalling for apoptosis.

A specific model can be advanced, however, that embraces most of the data on SAPK activation discussed in this chapter. The model envisages activation of MEKK1 as the key event in triggering an apoptotic cascade. Since MEKK1 has IκB kinase activity and can phosphorylate and activate MKK4, it is tempting to speculate that it has other, as yet unidentified substrates that directly activate the apoptotic effector machinery and are essential for its apoptotic effect. The main role of SAPK activation then becomes one of participating in a positive feed back loop for activating caspases. Activated caspases further activate MEKK1 and so accentuate both SAPK activation and the activation of the unknown MEKK1 substrates. In this model SAPK activation is not essential for MEKK1 triggered apoptosis but facilitates the process by accelerating caspase activation. Inhibiting SAPK activation would, therefore, normally delay rather than prevent apoptosis. However, it is possible to envisage cell fate being altered by inhibiting SAPK activation if signalling in cell survival pathways is sufficiently great to balance the pro-apoptotic effects of MEKK1, weakened by loss of the SAPK/caspase feed back loop. Relevant survival signals would include Akt activity, and the anti-apoptotic effect of MEKK1-activated NFκB. Conversely, overexpression of constitutively activated MKK4/SAPK1 stimulates inappropriate caspase activation which cleaves MEKK1 and sets in train MEKK1 directed apoptosis. SAPK activation in one sense is a surrogate marker for MEKK1 activation whose substrate(s) other than SAPK may be the critical ones for inducing apoptosis. Caveats invoking cell and stimulus contexts must still be applied to this model, not least to account for the different role that SAPK activation plays in promoting survival of some lymphoid cells.

CONCLUDING REMARKS

Cell fate is determined by the prevailing balance of survival and pro-apoptotic signals. An overview of the studies discussed here leads to the following conclusions about the roles of PI3K/Akt, SAPKs and MAPKs in transducing these signals.

A major signal transduction pathway that promotes cell survival comprises PI3K mediated activation of Akt. The upstream activation mechanisms of this pathway are well understood but nothing yet is known about the Akt targets that negatively regulate the apoptotic effector machinery.

There is no consensus that MAPK activation has any significant role in promoting cell survival. It is probable that the few examples where MAPK does appear to be anti-apoptotic are peculiar to the cell system and apoptotic stimulus.

Apoptosis is almost universally accompanied by SAPK activation. A strong case can be made for a major pro-apoptotic signalling pathway that involves MEKK1 activating as yet unknown substrates. Nevertheless, SAPKs also play a role in ensuring rapid and efficient cell suicide probably by participating in a positive feedback loop that activates caspases which in turn feed back onto MEKK1.

The present challenge in this exciting area of biology is to extend these signal transduction pathways onto the direct regulators of apoptosis. The key lies in as yet unidentified subtrates of Akt/PKB and possibly also of MEKK1.

ACKNOWLEDGEMENTS

I would like to thank: Rob McPherson for compiling Figure 4.2, Kum Kum Khanna, Neal Walker, Glenda Gobe and Rob McPherson for critically evaluating the manuscript, and the Royal Children's Hospital Foundation for continuing support.

REFERENCES

Abe, J.I., Kusuhara, M., Ulevitch, R.J., Berk, B.C. and Lee, J.D. (1996) Big mitogen activated protein kinase (BMK1) is a redox sensitive kinase. *J. Biol. Chem.*, **271**, 16586–16590.

Adams, J.M., Houston, H., Allen, J., Lints, T. and Harvey, R. (1992) The hematopoietically expressed vav proto-oncogene shares homology with the dbl GDP-GTP exchange factor, the bcr gene and a yeast gene (CDC24) involved in cytoskeletal organization. *Oncogene*, **7**, 611–8.

Alessi, D.R., Andjelkovic, M., Caudwell, B., Cron, P., Morrice, N., Cohen, P. and Hemmings, B.A. (1996) Mechanism of activation of protein kinase B by insulin and IGF-1. *EMBO J.*, **15**, 6541–51.

Alessi, D.R., James, S.R., Downes, C.P., Holmes, A.B., Gaffney, P.R., Reese, C.B. and Cohen, P. (1997) Characterization of a 3-phosphoinositide-dependent protein kinase which phosphorylates and activates protein kinase Balpha. *Curr. Biol.*, **7**, 261–9.

Alessi, D.R., Saito, Y., Campbell, D.G., Cohen, P., Sithanandam, G., Rapp, U., Ashworth, A., Marshall, C.J. and Cowley, S. (1994) Identification of the sites in MAP kinase kinase-1 phosphorylated by p74raf-1. *EMBO J.*, **13**, 1610–9.

Aronheim, A., Engelberg, D., Li, N., al Alawi, N., Schlessinger, J. and Karin, M. (1994) Membrane targeting of the nucleotide exchange factor Sos is sufficient for activating the Ras signalling pathway. *Cell*, **78**, 949–61.

Beyaert, R., Cuenda, A., Vanden Berghe, W., Plaisance, S., Lee, J.C., Haegeman, G., Cohen, P. and Fiers, W. (1996) The p38/RK mitogen-activated protein kinase pathway regulates interleukin-6 synthesis response to tumor necrosis factor. *EMBO J.*, **15**, 1914–23.

Blenis (1993) Proceed at your own risk. *Proc. Natl. Acad. Sci. USA*, **90**, 5889–5892.

Bos, J.L. (1995) A target for phosphoinositide 3-kinase, Akt/PKB. *Trends Biochem. Sci.*, **20**, 441–2.

Buday, L. and Downward, J. (1993) Epidermal growth factor regulates p21ras through the formation of a complex of receptor, Grb2 adaptor protein, and Sos nucleotide exchange factor. *Cell*, **73**, 611–620.

Buday, L., Warne, P.H. and Downward, J. (1995) Downregulation of the Ras activation pathway by MAP kinase phosphorylation of Sos. *Oncogene*, **11**, 1327–31.

Burgering, B.M. and Coffer, P.J. (1995) Protein kinase B (c-Akt) in phosphatidylinositol-3-OH kinase signal transduction. *Nature*, **376**, 599–602.

Cadwallader, K., Paterson, H., Macdonald, S.G. and Hancock, J.F. (1994) N-terminally myristoylated Ras proteins require palmitoylation or a polybasic domain for plasma membrane localization. *Mol. Cell. Biol.*, **14**, 4722–4730.

Cahill, M.A., Peter, M.E., Kischkel, F.C., Chinnaiyan, A.M., Dixit, V.M., Krammer, P.H. and Nordheim, A. (1996) CD95 (APO-1/Fas) induces activation of SAP kinases downstream of ICE-like proteases. *Oncogene*, **13**, 2087–96.

Cardone, M.H., Salvesen, G.S., Widmann, C., Johnson, G. and Frisch, S.M. (1997) The regulation of Anoikis, MEKK1 activation requires cleavage by caspases. *Cell*, **90**, 315–323.

Carpenter, C.L. and Cantley, L.C. (1996) Phosphoinositide 3-kinase. *Curr. Opin. Cell Biol.*, **8**, 153–158.

Chardin, P., J.H., C., Gale, N.W., Van Aelst, L., Schlessinger, J., Wigler, M.H. and Bar-Sagi, D. (1993) Human Sos 1, a guanine nucleotide exchange factor for Ras that binds to GRB2. *Science*, **260**, 1338–1343.

Chauhan, D., Kharbanda, S., Ogata, A., Urashima, M., Teoh, G., Robertson, M., Kufe, D.W. and Anderson, K.C. (1997) Interleukin-6 inhibits Fas-induced apoptosis and stress-activated protein kinase activation in multiple myeloma cells. *Blood*, **89**, 227–34.

Chen, Y.R., Meyer, C.F. and Tan, T.H. (1996) Persistent activation of c-Jun N-terminal kinase 1 (JNK1) in gamma radiation-induced apoptosis. *J. Biol. Chem.*, **271**, 631–4.

Clifton, A.D., Young, P.R. and Cohen, P. (1996) A comparison of the substrate specificity of MAPKAP kinase-2 and MAPKAP kinase-3 and their activation by cytokines and cellular stress. *FEBS Lett.*, **392**, 209–14.

Coso, O., Chiariello, M., Yu, J.-C., Teramoto, H., Crespo, P., Xu, N., Miki, T. and Gutkind, J.S. (1995) The small GTP-binding proteins Rac1 and Cdc42 regulate the activity of the JNK/SAPK signalling pathway. *Cell*, **81**, 1137–1146.

Cowley, S., Paterson, H., Kemp, P. and Marshall, C.J. (1994) Activation of MAP kinase kinase is necessary and sufficient for PC12 differentiation and for transformation of NIH 3T3 cells. *Cell*, **77**, 841–52.

Crews, C.M., Alessandrini, A. and Erickson, R.L. (1992) The primary structure of MEK, a protein kinase that phosphorylates the Erk gene product. *Science*, **258**, 478–480.

Cross, D.A., Alessi, D.R., Cohen, P., Andjelkovich, M. and Hemmings, B.A. (1995) Inhibition of glycogen synthase kinase-3 by insulin mediated by protein kinase B. *Nature*, **378**, 785–9.

Cross, D.A., Watt, P.W., Shaw, M., van der Kaay, J., Downes, C.P., Holder, J.C. and Cohen, P. (1997) Insulin activates protein kinase B, inhibits glycogen synthase kinase-3 and activates glycogen synthase by rapamycin-insensitive pathways in skeletal muscle and adipose tissue. *FEBS Lett.*, **406**, 211–5.

Cuenda, A., Alonso, G., Morrice, N., Jones, M., Meier, R., Cohen, P. and Nebreda, A.R. (1996) Purification and cDNA cloning of SAPKK3, the major activator of RK/p38 in stress-and cytokine-stimulated monocytes and epithelial cells. *EMBO J.*, **15**, 4156–64.

Cuenda, A., Cohen, P., Buee Scherrer, V. and Goedert, M. (1997) Activation of stress-activated protein kinase-3 (SAPK3) by cytokines and cellular stresses is mediated via SAPKK3 (MKK6); comparison of the specificities of SAPK3 and SAPK2 (RK/p38) *EMBO J.*, **16**, 295–305.

Cuvillier, O., Pirianov, G., Kleuser, B., Vanek, P.G., Coso, O.A., Gutkind, S. and Spiegel, S. (1996) Suppression of ceramide-mediated programmed cell death by sphingosine-1-phosphate. *Nature*, **381**, 800–3.

Datta, K., Bellacosa, A., Chan, T.O. and Tsichlis, P.N. (1996) Akt is a direct target of the phosphatidylinositol 3–kinase. Activation by growth factors, v-src and v-Ha-ras, in Sf9 and mammalian cells. *J. Biol Chem.*, **271**, 30835–9.

Daum, G., Eisenmann Tappe, I., Fries, H.W., Troppmair, J. and Rapp, U.R. (1994) The ins and outs of Raf kinases. *Trends Biochem. Sci.*, **19**, 474–80.

Dent, P., Haser, W., Haystead, T.A., Vincent, L.A., Roberts, T.M. and Sturgill, T.W. (1992) Activation of mitogen-activated protein kinase kinase by v-Raf in NIH 3T3 cells and in vitro [see comments]. *Science*, **257**, 1404–7.

Dent, P., Jelinek, T., Morrison, D.K., Weber, M.J. and Sturgill, T.W. (1995) Reversal of Raf-1 activation by purified and membrane-associated protein phosphatases. *Science*, **268**, 1902–6.

Dent, P., Reardon, D.B., Wood, S.L., Lindorfer, M.A., Graber, S.G., Garrison, J.C., Brautigan, D.L. and Sturgill, T.W. (1996) Inactivation of raf-1 by a protein-tyrosine phosphatase stimulated by GTP and reconstituted by Galphai/o subunits. *J. Biol. Chem.*, **271**, 3119–23.

Derijard, B., Hibi, M., Wu, I.H., Barrett, T., Deng, T., Karin, S.M. and Davis, R.J. (1994) JNK1 a protein kinase stimulated by UV light and Ha-ras that binds and phosphorylates the c-jun activation domain. *Cell*, **76**, 1025–1037.

Dudek, H., Datta, S.R., Franke, T.F., Birnbaum, M.J., Yao, R., Cooper, G.M., Segal, R.A., Kaplan, D.R. and Greenberg, M.E. (1997) Regulation of neuronal survival by the serine-threonine protein kinase Akt. *Science*, **275**, 661–5.

Egan, S.E., Giddings, B.W., Brooks, M.W., Buday, L., Sizeland, A.M. and Weinberg, R.A. (1993) Association of Sos Ras exchange protein with Grb2 is implicated in tyrosine kinase signal transduction and transformation. *Nature*, **363**, 45–51.

Ellinger Ziegelbauer, H., Brown, K., Kelly, K. and Siebenlist, U. (1997) Direct activation of the stress-activated protein kinase (SAPK) and extracellular signal-regulated protein kinase (ERK) pathways by an inducible mitogen-activated protein Kinase/ERK kinase kinase 3 (MEKK) derivative. *J. Biol. Chem.*, **272**, 2668–74.

Evan, G.I., Wylilie, A.H., Gilbert, C.S., Littlewood, T.D., Land, H., Brooks, M., Waters, C.M., Penn, L.Z. and Hancock, D.C. (1992) Induction of apoptosis in fibroblasts by c-myc protein. *Cell*, **69**, 119–128.

Fabian, J.R., Daar, I.O. and Morrison, D.K. (1993) Critical tyrosine residues regulate the enzymatic and biological activity of Raf-1 kinase. *Mol. Cell. Biol.*, **13**, 7170–9.

Fanger, G.R., Gerwins, P., Widmann, C., Jarpe, M.B. and Johnson, G.L. (1997) MEKKs, GCKs, MLKs, PAKs, TAKs and Tpls, upstream regulators of the c-Jun amino-terminal kinases. *Curr. Opin. Genet. Dev.*, **7**, 67–74.

Farrar, M.A., Alberol, I. and Perlmutter, R.M. (1996) Activation of the Raf-1 kinase cascade by coumermycin-induced dimerization. *Nature*, **383**, 178–81.

Franke, T.F., Kaplan, D.R., Cantley, L.C. and Toker, A. (1997) Direct regulation of the Akt proto-oncogene product by phosphatidylinositol-3, 4-bisphosphate. *Science*, **275**, 665–8.

Franke, T.F., Yang, S.I., Chan, T.O., Datta, K., Kazlauskas, A., Morrison, D.K., Kaplan, D.R. and Tsichlis, P.N. (1995) The protein kinase encoded by the Akt proto-oncogene is a target of the PDGF-activated phosphatidylinositol 3-kinase. *Cell*, **81**, 727–36.

Frisch, S.M., Vuori, K., Kelaita, D. and Sicks, S. (1996) A role for Jun-N-terminal kinase in anoikis; suppression by bcl-2 and crmA. *J. Cell. Biol.*, **135**, 1377–82.

Galcheva Gargova, Z., Derijard, B., Wu, I.H. and Davis, R.J. (1994) An osmosensing signal transduction pathway in mammalian cells. *Science*, **265**, 806–8.

Gardner, A.M. and Johnson, G.L. (1996) Fibroblast growth factor-2 suppression of tumor necrosis factor alpha-mediated apoptosis requires Ras and the activation of mitogen-activated protein kinase. *J. Biol. Chem.*, **271**, 14560–6.

Gerwins, P., Blank, J.L. and Johnson, G.L. (1997) Cloning of a novel mitogen-activated protein kinase kinase kinase, MEKK4, that selectively regulates the c-Jun amino terminal kinase pathway. *J. Biol. Chem.*, **272**, 8288–95.

Ghosh, S., Strum, J.C., Sciorra, V. A., Daniel, L. and Bell, R.M. (1996) Raf-1 kinase possesses distinct binding domains for phosphatidylserine and phosphatidic acid. Phosphatidic acid regulates the translocation of Raf-1 in 12-O-tetradecanoylphorbol-13-acetate-stimulated Madin-Darby canine kidney cells. *J. Biol. Chem.*, **271**, 8472–80.

Goedert, M., Cuenda, A., Craxton, M., Jakes, R. and Cohen, P. (1997) Activation of the novel stress-activated protein kinase SAPK4 by cytokines and cellular stresses is mediated by SKK3 (MKK6); comparison of its substrate specificity with that of other SAP kinases. *EMBO J.*, **16**, 3563–3571.

Gupta, S., Campbell, D., Derijard, B. and Davis, R.J. (1995) Transcription factor ATF-2 regulation by the JNK signal transduction pathway. *Science*, **267**, 389–393.

Habets, G.G., Scholtes, E.H., Zuydgeest, D., van der Kammen, R.A., Stam, J.C., Berns, A. and Collard, J.G. (1994) Identification of an invasion-inducing gene, Tiam-1, that encodes a protein with homology to GDP-GTP exchangers for Rho-like proteins. *Cell*, **77**, 537–49.

Han, J., Das, B., Wei, W., Van Aelst, L., Mosteller, R.D., Khosravi Far, R., Westwick, J.K., Der, C.J. and Broek, D. (1997) Lck regulates Vav activation of members of the Rho family of GTPases. *Mol. Cell. Biol.*, **17**, 1346–53.

Han, J., Lee, J.D., Bibbs, L. and Ulevitch, R.J. (1994) A MAP kinase targeted by endotoxin and hyperosmolarity in mammalian cells. *Science*, **265**, 808–11.

Han, J., Lee, J.D., Jiang, Y., Li, Z., Feng, L. and Ulevitch, R.J. (1996) Characterization of the structure and function of a novel MAP kinase kinase (MKK6) *J. Biol. Chem.*, **271**, 2886–91.

Harrington, E.A., Bennett, M.R., Fannidi, A. and Evan, G.I. (1994) c-Myc induced apoptosis in fibroblasts is inhibited by specific cytokines. *EMBO J.*, **13**, 3286–3295.

Hart, M.J., Eva, A., Evans, T., Aaronson, S.A. and Cerione, R.A. (1991) Catalysis of guanine nucleotide exchange on the CDC42Hs protein by the *dbl* oncogene product. *Nature*, **354**, 311–314.

Hawkins, P.T. and al, e. (1995) PDGF stimulates an increase in GTP-Rac via activation of phosphoinositide 3-kinase. *Curr. Biol.*, **5**, 393–403.

Hazzalin, C.A., Cano, E., Cuenda, A., Barratt, M.J., Cohen, P. and Mahadevan, L.C. (1996) p38/RK is essential for stress-induced nuclear responses, JNK/SAPKs and c-Jun/ATF-2 phosphorylation are insufficient. *Curr. Biol.*, **6**, 1028–31.

Heidenreich, K.A. and Kummer, J.L. (1996) Inhibition of p38 mitogen-activated protein kinase by insulin in cultured fetal neurons. *J. Biol. Chem.*, **271**, 9891–4.

Hemmings, B.A. (1997a) Akt signalling, linking membrane events to life and death decisions. *Science*, **275**, 628–30.

Hemmings, B.A. (1997b) PH domainsa universal membrane anchor. *Science*, **275**, 1899.

Hemmings, B.A. (1997c) PtdIns(3,4,5) gets its message across. *Science*, **277**, 534.

Herskowitz, I. (1995) MAP kinase pathways in yeast, for mating and more. *Cell*, **80**, 187–197

Howe, L.R., Leevers, S.J., Gomez, N., Nakielny, S., Cohen, P. and Marshall, C.J. (1992) Activation of the MAP kinase pathway by the protein kinase raf. *Cell*, **71**, 335–42.

Huot, J., Lambert, H., Lavoie, J.N., Guimond, A., Houle, F. and Landry, J. (1995) Characterization of 45-kDa/54-kDa HSP27 kinase, a stress-sensitive kinase which may activate the phosphorylation-dependent protective function of mammalian 27-kDa heat-shock protein HSP27. *Eur. J. Biochem.*, **227**, 416–27.

Ichijo, H., Nishida, E., Irie, K., ten Dijke, P., Saitoh, M., Moriguchi, T., Takagi, M., Matsumoto, K., Miyazono, K. and Gotoh, Y. (1997) Induction of apoptosis by ASK1, a mammalian MAPKKK that activates SAPK/JNK and p38 signalling pathways. *Science*, **275**, 90–4.

James, S.R., Downes, C.P., Gigg, R., Grove, S.J., Holmes, A.B. and Alessi, D.R. (1996) Specific binding of the Akt-1 protein kinase to phosphatidylinositol 3,4,5-trisphosphate without subsequent activation. *Biochem. J.*, **315**, 709–13.

Janknecht, R., Ernst, W.H., Pingoud, V. and Nordheim, A. (1993) Activation of ternary complex factor Elk-1 by MAP kinases. *EMBO J.*, **12**, 5097–104.

Jelinek, T., Dent, P., Sturgill, T.W. and Weber, M.J. (1996) Ras-induced activation of Raf-1 is dependent on tyrosine phosphorylation. *Mol. Cell. Biol.*, **16**, 1027–34.

Jiang, Y., Chen, C., Li, Z., Guo, W., Gegner, J.A., Lin, S. and Han, J. (1996) Characterization of the structure and function of a new mitogen-activated protein kinase (p38β) *J. Biol. Chem.*, **271**, 17920–6.

Johnson, N.L., Gardner, A.M., Diener, K.M., Lange Carter, C A., Gleavy, J., Jarpe, M.B., Minden, A., Karin, M., Zon, L.I. and Johnson, G.L. (1996) Signal transduction pathways regulated by mitogen-activated/extracellular response kinase kinase kinase induce cell death. *J. Biol. Chem.*, **271**, 3229–37.

Jung, Y.K., Miura, M. and Yuan, J. (1996) Suppression of interleukin-1b-converting enzyme mediated cell death by insulin-like growth factor. *J. Biol. Chem.*, **271**, 5112–5117.

Juo, P., Kuo, C.J., Reynolds, S.E., Konz, R.F., Raingeaud, J., Davis, R.J., Biemann, H.P. and Blenis, J. (1997) Fas activation of the p38 mitogen-activated protein kinase signalling pathway requires ICE/CED-3 family proteases. *Mol. Cell. Biol.*, **17**, 24–35.

Katzav, S., Martin-Zanca, D. and Barbacid, M. (1989) vav, a novel human oncogene derived from a locus ubuquitously expressed in hematopoietic cells. *EMBO J.*, **8**, 2283–2290.

Kauffmann Zeh, A., Rodriguez Viciana, P., Ulrich, E., Gilbert, C., Coffer, P., Downward, J. and Evan, G. (1997) Suppression of c-Myc-induced apoptosis by Ras signalling through PI(3)K and PKB. *Nature*, **385**, 544–8.

Khwaji, A., Rodriguez-Viciana, P., Wennstrom, S., Warne, P.H. and Downward, J. (1997) Matrix adhesion and Ras transformation both activate a phosphoinisotide 3-OH kinase and protein kinase B/Akt cellular survival pathway. *EMBO J.*, **16**, 2783–2793.

Kikuchi, A., Demo, S.D., Ye, Z.-H., Chen, T.-W. and Wiliams, L.T. (1994) RalGDS family members interact with the effector loop of ras p21. *Mol. Cell. Biol.*, **14**, 7483–7491.

Kinoshita, T., Yokota, T., Arai, K. and Miyajima, A. (1995) Suppression of apoptotic death in hematopoietic cells by signalling through the IL-3/GM-CSF receptors. *EMBO J.*, **14**, 266–75.

Klippel, A., Kavanaugh, W.M., Pot, D. and Williams, L.T. (1997) A specific product of phosphatidylinositol 3-kinase directly activates the protein kinase Akt through its pleckstrin homology domain. *Mol. Cell. Biol.*, **17**, 338–44.

Kolch, W., Heidecker, G., Kochs, G., Hummel, R., Vahidi, H., Mischak, H., Finkenzeller, G., Marme, D. and Rapp, U.R. (1993) Protein kinase C alpha activates RAF-1 by direct phosphorylation. *Nature*, **364**, 249–52.

Kotani, K., Yonezawa, K., Hara, K., Ueda, H., Kitamura, Y., Sakaue, H. *et al.*, (1994) Involvement of phosphinositide 3-kinase in insulin or IGF-1 induced membrane ruffling. *EMBO J.*, **13**, 2313–2321.

Kouhara, H., Hadari, Y.R., Spivak-Kroizman, Schilling, J., Bar-Sagi, D., Lax, I. and Schlessinger, J. (1997) A lipid anchored Grb2 binding protein that links FGF-receptor activation to the Ras/MAPK signalling pathway. *Cell*, **89**, 693–702.

Kulik, G., Klippel, A. and Weber, M.J. (1997) Anti-apoptotic signalling by the insulin-like growth factor I receptor, phosphatidylinositol 3-kinase, and Akt. *Mol. Cell. Biol.*, **17**, 1595–606.

Kuriyama, M., Harada, N., Kuroda, S., Yamamoto, T., Nakafuku, M., Iwamatsu, A., Yamamoto, D., Prasad, R., Croce, C., Canaani, E. and Kaibuchi, K. (1996) Identification of AF-6 and canoe as putative targets for Ras. *J. Biol. Chem.*, **271**, 607–10.

Kyriakis, J.M., App, H., Zhang, X.-F., Banerjee, P., Brautigan, D.L., Rapp, U.R. and Avruch, J. (1992) Raf1 activates MAP kinase kinase. *Nature*, **358**, 417–421.

Kyriakis, J.M. and Avruch, J. (1996) Protein kinase cascades activated by stress and inflammatory cytokines. *Bioessays*, **18**, 567–77.

Kyriakis, J.M., Banerjee, P., Nikolakaki, E., Dai, T., Rubie, E.A., Ahmad, M.F., Avruch, J. and Woodgett, J.R. (1994) The stress-activated protein kinase subfamily of c-Jun kinases. *Nature*, **369**, 156–60.

Lange Carter, C.A. and Johnson, G.L. (1994) Ras-dependent growth factor regulation of MEK kinase in PC12 cells. *Science*, **265**, 1458–61.

Lange Carter, C.A., Pleiman, C.M., Gardner, A.M., Blumer, K.J. and Johnson, G.L. (1993) A divergence in the MAP kinase regulatory network defined by MEK kinase and Raf. *Science*, **260**, 315–9.

Lee, F.S., Hagler, J., Chen, Z.J. and Maniatis, T. (1997) Activation of the IκBα Kinase complex by MEKK1, a kinase of the JNK pathway. *Cell*, **88**, 213–222.

Lee, J.C., Laydon, J.T., McDonnell, P.C., Gallagher, T.F., Kumar, S., Green, D., McNulty, D., Blumenthal, M.J., Heys, J.R., Landvatter, S.W. and *et al.* (1994) A protein kinase involved in the regulation of inflammatory cytokine biosynthesis. *Nature*, **372**, 739–46.

Leevers, S.J., Paterson, H.F. and Marshall, C.J. (1994) Requirement for Ras in Raf activation is overcome by targeting Raf to the plasma membrane. *Nature*, **369**, 411–4.

Lenczowski, J.M., Dominguez, L., Eder, A.M., King, L.B., Zacharchuk, C.M. and Ashwell, J.D. (1997) Lack of a role for Jun kinase and AP-1 in Fas-induced apoptosis. *Mol. Cell. Biol.*, **17**, 170–81.

Leverrier, Y., Thomas, J., Perkins, G.R., Mangeney, M., Collins, M.K. and Marvel, J. (1997) In bone marrow derived Baf-3 cells, inhibition of apoptosis by IL-3 is mediated by two independent pathways. *Oncogene*, **14**, 425–30.

Li, N., Batzer, A., Daly, R., Yajnik, V., Skolnik, E., Chardin, P., Bar Sagi, D., Margolis, B. and Schlessinger, J. (1993) Guanine-nucleotide-releasing factor hSos1 binds to Grb2 and links receptor tyrosine kinases to Ras signalling. *Nature*, **363**, 85–8.

Lin, L., Wartmann, M., Lin, A., Knopf, J.L., Seth, A. and Davis, R.J. (1993) cPLA2. *Cell*, **72**, 269–278.

Liu, Z., Hsu, H., Goeddel, D.V. and Karin, M. (1996) Dissection of TNF receptor 1 effector functions, JNK activation is not linked to apoptosis while NFκB activation prevents cell death. *Cell*, **87**, 565–576.

Luo, Z., Diaz, B., Marshall, M. and Avruch, J. (1997) An intact zinc finger is required for optimal binding to processed Ras and for Ras dependent Raf activation in situ. *Mol. Cell. Biol.*, **17**, 46–53.

Luo, Z., Tzivion, G., Belshaw, P.J., Vavvas, D., Marshall, M. and Avruch, J. (1996) Oligomerization activates c-Raf-1 through a Ras-dependent mechanism. *Nature*, **383**, 181–5.

Nishina, H., Fischer, K.D., Radvanyi, L., Shahinian, A., Hakem, R., Rubie, E.A., Bernstein, A., Mak, T.W., Woodgett, J.R., Penninger, J.M., (1997) Stress-signalling kinase Sek1 protects thymocytes from apoptosis mediated by CD95 and CD3. *Nature*, **385**, 350–3

Macdonald, S.G., Crews, C.M., Wu, L., Driller, J., Clark, R., Erikson, R.L. and McCormick, F. (1993) Reconstitution of the Raf-1-MEK-ERK signal transduction pathway in vitro. *Mol. Cell. Biol.*, **13**, 6615–6620.

Manser, E., Chong, C., Zhao, Z.S., Leung, T., Michael, G., Hall, C. and Lim, L. (1995) Molecular cloning of a new member of the p21-Cdc42/Rac-activated kinase (PAK) family. *J. Biol. Chem.*, **270**, 25070–8.

Manser, E., Leung, T., Salihuddin, H., Zhao, Z.S. and Lim, L. (1994) A brain serine/threonine protein kinase activated by Cdc42 and Rac1. *Nature*, **367**, 40–46.

Mansour, S.J., Matten, W.T., Hermann, A.S., Candia, J.M., Rong, S., Fukasawa, K., Vande Woude, G.F. and Ahn, N.G. (1994) Transformation of mammalian cells by constitutively active MAP kinase kinase. *Science*, **265**, 966–70.

Marais, R., Light, Y., Paterson, H.F. and Marshall, C.J. (1995) Ras recruits Raf-1 to the plasma membrane for activation by tyrosine phosphorylation. *EMBO J.*, **14**, 3136–45.

Marais, R., Light, Y., Paterson, H.F., Mason, C.S. and Marshall, C.J. (1997) Differential regulation of Raf-1, A-Raf, and B-Raf by oncogenic ras and tyrosine kinases. *J. Biol. Chem.*, **272**, 4378–83.

Marshall, C.J. (1994) Signal transduction. Hot lips and phosphorylation of protein kinases. *Nature*, **367**, 686.

Marshall, C.J. (1995) Specificity of receptor tyrosine kinase signalling, transient versus sustained extracellular signal-regulated kinase activation. *Cell*, **80**, 179–85.

Martin, G.A., Bollag, G., McCormick, F. and Abo, A. (1995) A novel serine threonine kinase activated by rac1/CDC42Hs-dependent autophosphorylation is related to PAK65 and STE20. *EMBO J.*, **14**, 1970–1978.

McCormick, F. (1989) ras GTPase activating protein, signal transmitter and signal terminator. *Cell*, **56**, 5–8.

McCormick, F., Adari, H., Trahey, M., Halenbeck, R., Koths, K., Martin, G.A., Crosier, W.J., Watt, K., Rubinfeld, B. and Wong, G. (1988) Interaction of ras p21 proteins with GTPase activating protein. *Cold Spring Harb. Symp. Quant. Biol*., **2**, 849–854.

McLaughlin, M.M., Kumar, S., McDonnell, P.C., Van Horn, S., Lee, J.C., Livi, G.P. and Young, P.R. (1996) Identification of mitogen-activated protein (MAP) kinase-activated protein kinase-3, a novel substrate of CSBP p38 MAP kinase. *J. Biol. Chem.*, **271**, 8488–92.

Mertens, S., Craxton, M. and Goedert, M. (1996) SAP kinase-3, a new member of the family of mammalian stress-activated protein kinases. *FEBS Lett.*, **383**, 273–6.

Michiels, F., Stam, J.C., Hordijk, P.L., van der Kammen, R.A., Ruuls Van Stalle, L., Feltkamp, C.A. and Collard, J.G. (1997) Regulated membrane localization of Tiam1, mediated by the NH2-terminal pleckstrin homology domain, is required for Rac-dependent membrane ruffling and C-Jun NH2-terminal kinase activation. *J. Cell. Biol.*, **137**, 387–98.

Minden, A., Lin, A., Claret, F.-X., Abo, A. and Karin, M. (1995) Selective activation of the JNK signalling cascade and c-jun transcriptional activity by the small GTPases Rac and Cdc42Hs. *Cell*, **81**, 1147–1157.

Minden, A., Lin, A., McMahon, M., Lange Carter, C., Derijard, B., Davis, R.J., Johnson, G.L. and Karin, M. (1994) Differential activation of ERK and JNK mitogen-activated protein kinases by Raf-1 and MEKK. *Science*, **266**, 1719–23.

Minshall, C., Arkins, S., Freund, G.G. and Kelley, K.W. (1996) Requirement for phosphatidylinositol 3'-kinase to protect hemopoietic progenitors against apoptosis depends upon the extracellular survival factor. *J. Immunol.*, **156**, 939–47.

Moodie, S.A., Willumsen, B.M., Weber, M.J. and Wolfman, A. (1993) Complexes of Ras.GTP with Raf-1 and mitogen-activated protein kinase kinase. *Science*, **260**, 1658–61.

Muta, K. and Krantz, S.B. (1993) Apoptosis of human erythroid colony-forming cells is decreased by stem cell factor and insulin-like growth factor I as well as erythropoietin. *J. Cell. Physiol.*, **156**, 264–71.

Natoli, G., Costanzo, A., Ianni, A., Templeton, D.J., Woodgett, J.R., Balsano, C. and Levrero, M. (1997) Activation of SAPK/JNK by TNF receptor 1 through a noncytotoxic TRAF2-dependent pathway. *Science*, **275**, 200–3.

Nobes, C.D. and Hall, A. (1995) Rho, rac, and cdc42 GTPases regulate the assembly of multimolecular focal complexes associated with actin stress fibers, lamellipodia, and filopodia. *Cell*, **81**, 53–62.

Ohtsuka, T., Shimizu, K., Yamamori, B., Kuroda, S. and Takai, Y. (1996) Activation of brain B-Raf protein kinase by Rap1B small GTP-binding protein. *J. Biol. Chem.*, **271**, 1258–61.

Olson, M.F., Ashworth, A. and Hall, A. (1995) An essential role for Rho, Rac and Cdc42 GTPases in cell cycle progression through G1. *Science*, **269**, 1270–1272.

Pandey, P., Raingeaud, J., Kaneki, M., Weichselbaum, R., Davis, R.J., Kufe, D. and Kharbanda, S. (1996) Activation of p38 mitogen-activated protein kinase by c-Abl-dependent and -independent mechanisms. *J. Biol. Chem.*, **271**, 23775–9.

Park, D.S., Stefanis, L., Yan, C.Y.I., Farinelli, S.E. and Greene, L.A. (1996) Ordering the cell death pathway. Differential effects of BCL2, an interleukin-1-converting enzyme family protease inhibitor, and other survival agents on JNK activation in serum/nerve growth factor-deprived PC12 cells. *J. Biol. Chem.*, **271**, 21898–905.

Perkins, G.R., Marshall, C.J. and Collins, M.K. (1996) The role of MAP kinase kinase in interleukin-3 stimulation of proliferation. *Blood*, **87**, 3669–75.

Pombo, C.M., Kehrl, J.H., Sanchez, I., Katz, P., Avruch, J., Zon, L.I., Woodgett, J.R., Force, T. and Kyriakis, J.M. (1995) Activation of the SAPK pathway by the human STE20 homologue germinal centre kinase. *Nature*, **377**, 750–754.

Porfiri, E. and McCormick, F. (1996) Regulation of epidermal growth factor receptor signalling by phosphorylation of the ras exchange factor hSOS1. *J. Biol. Chem.*, **271**, 5871–7.

Price, M.A., Cruzalegui, F.H. and Treisman, R. (1996) The p38 and ERK MAP kinase pathways cooperate to activate Ternary Complex Factors and c-fos transcription in response to UV light. *EMBO J.*, **15**, 6552–63.

Rodriguez Viciana, P., Warne, P.H., Dhand, R., Vanhaesebroeck, B., Gout, I., Fry, M.J., Waterfield, M.D. and Downward, J. (1994) Phosphatidylinositol-3-OH kinase as a direct target of Ras. *Nature*, **370**, 527–32.

Rodriguez-Viciana, P., Warne, P.H., Dhand, R., Vanhaesebroeck, B., Gout, I., Fry, M.J., Waterfield, M.D. and Downward, J. (1994) Phosphatidylinositol-3-OH kinase as a direct target of Ras. *Nature*, **370**, 527–532.

Rodriguez Viciana, P., Warne, P.H., Vanhaesebroeck, B., Waterfield, M.D. and Downward, J. (1996) Activation of phosphoinositide 3-kinase by interaction with Ras and by point mutation. *EMBO J.*, **15**, 2442–51.

Rodriguez Viciana, P., Warne, P.H., Khwaja, A., Marte, B.M., Pappin, D., Das, P., Waterfield, M.D., Ridley, A. and Downward, J. (1997) Role of phosphoinositide 3-OH kinase in cell transformation and control of the actin cytoskeleton by Ras. *Cell*, **89**, 457–67.

Rommel, C., Radziwill, G., Lovric, J., Noeldeke, J., Heinicke, T., Jones, D., Aitken, A. and Moelling, K. (1996) Activated Ras displaces 14-3-3 protein from the amino terminus of c-Raf-1. *Oncogene*, **12**, 609–19.

Rouse, J., Cohen, P., Trigon, S., Morange, M., Alonso Llamazares, A., Zamanillo, D., Hunt, T. and Nebreda, A.R. (1994) A novel kinase cascade triggered by stress and heat shock that stimulates MAPKAP kinase-2 and phosphorylation of the small heat shock proteins. *Cell*, **78**, 1027–37.

Roy, S., Lane, A., Yan, J., McPherson, R. and Hancock, J.F. (1997) Activity of plasma membrane recruited Raf-1 is regulated by Ras via the Raf zinc finger. *J. Biol. Chem.*, **272**, 20139–20145.

Russell, M., Lange Carter, C.A. and Johnson, G.L. (1995) Direct interaction between Ras and the kinase domain of mitogen-activated protein kinase kinase kinase (MEKK1) *J. Biol. Chem.*, **270**, 11757–60.

Sakata, N., Patel, H.R., Terada, N., Aruffo, A., Johnson, G.L. and Gelfand, E.W. (1995) Selective activation of c-Jun kinase mitogen-activated protein kinase by CD40 on human B cells. *J. Biol. Chem.*, **270**, 30823–8.

Salmeron, A., Ahmad, T.B., Carlile, G.W., Pappin, D., Narsimhan, R.P. and Ley, S.C. (1996) Activation of MEK-1 and SEK-1 by Tpl-2 proto-oncoprotein, a novel MAP kinase kinase kinase. *EMBO J.*, **15**, 817–26.

Sanchez, I., Hughes, R.T., Mayer, B.J., Yee, K., Woodgett, J.R., Avruch, J., Kyriakis, J.M. and Zon, L.I. (1994) Role of SAPK/ERK kinase-1 in the stress-activated pathway regulating transcription factor c-Jun. *Nature*, **372**, 794–8.

Seimiya, H., Mashima, T., Toho, M. and Tsuruo, T. (1997) c-Jun NH2-terminal kinase-mediated activation of interleukin-1beta converting enzyme/CED-3-like protease during anticancer drug-induced apoptosis. *J. Biol. Chem.*, **272**, 4631–6.

Sell, C., Baserga, R. and Rubin, R. (1995) Insulin like growth factor (IGF-1) and the IGF-1 receptor prevent etoposide-induced apoptosis. *Cancer Res.*, **55**, 303–306.

Sheng, Z., Knowlton, K., Chen, J., Hoshijima, M., Brown, J.H. and Chien, K.R. (1997) Cardiotrophin 1 (CT-1) inhibition of cardiac myocyte apoptosis via a mitogen-activated protein kinase-dependent pathway. Divergence from downstream CT-1 signals for myocardial cell hypertrophy. *J. Biol. Chem.*, **272**, 5783–91.

Stokoe, D., Macdonald, S.G., Cadwallader, K., Symons, M. and Hancock, J.F. (1994) Activation of Raf as a result of recruitment to the plasma membrane. *Science*, **264**, 1463–1467.

Stokoe, D., Stephens, L.R., Copeland, T., Gaffney, P.R.J., Reese, C.B., Painter, G.F., Holmes, A.B., McCormick, F. and Hawkins, P.T. (1997) Dual Role of phaphatidylinositol-3,4,5-trisphosphate in the activation of protein kinase B. *Science*, **277**, 567–570.

Sutherland, C.L., Heath, A.W., Pelech, S.L., Young, P.R. and Gold, M.R. (1996) Differential activation of the ERK, JNK, and p38 mitogen-activated protein kinases by CD40 and the B cell antigen receptor. *J. Immunol.*, **157**, 3381–90.

Tan, Y., Rouse, J., Zhang, A., Cariati, S., Cohen, P. and Comb, M.J. (1996) FGF and stress regulate CREB and ATF-1 via a pathway involving p38 MAP kinase and MAPKAP kinase-2. *EMBO J.*, **15**, 4629–42.

Tibbles, L.A., Ing, Y.L., Kiefer, F., Chan, J., Iscove, N., Woodgett, J.R. and Lassam, N.J. (1996) MLK-3 activates the SAPK/JNK and p38/RK pathways via SEK1 and MKK3/6. *EMBO J.*, **15**, 7026–35.

Traverse, S., Cohen, P., Paterson, H., Marshall, C., Rapp, U. and Grand, R.J. (1993) Specific association of activated MAP kinase kinase kinase (Raf) with the plasma membranes of ras-transformed retinal cells. *Oncogene*, **8**, 3175–81.

Van Aelst, L., Barr, M., Marcus, S., Polverino, A. and Wigler, M. (1993) Complex formation between RAS and RAF and other protein kinases. *Proc. Natl. Acad. Sci. USA.*, **90**, 6213–7.

van Dam, H., Wilhelm, D., Herr, I., Steffen, A., Herrlich, P. and Angel, P. (1995) ATF-2 is preferentially activated by stress-activated protein kinases to mediate c-jun induction in response to genotoxic agents. *EMBO J.*, **14**, 1798–811.

van Leeuwen, F.N., van der Kammen, R.A., Habets, G.G. and Collard, J.G. (1995) Oncogenic activity of Tiam1 and Rac1 in NIH3T3 cells. *Oncogene*, **11**, 2215–21.

Verheij, M., Bose, R., Lin, X.H., Yao, B., Jarvis, W.D., Grant, S., Birrer, M.J., Szabo, E., Zon, L.I., Kyriakis, J.M., Haimovitz Friedman, A., Fuks, Z. and Kolesnick, R.N. (1996) Requirement for ceramide-initiated SAPK/JNK signalling in stress-induced apoptosis. *Nature*, **380**, 75–9.

Vojtek, A.B., Hollenberg, S.M. and Cooper, J.A. (1993) Mammalian Ras interacts directly with the Serine/Threonine kinase Raf. *Cell*, **74**, 205–214.

Vossler, M.R., Yao, H., York, R.D., Pan, M.G., Rim, C.S. and Stork, P.J. (1997) cAMP activates MAP kinase and Elk-1 through a B-Raf-and Rap1-dependent pathway. *Cell*, **89**, 73–82.

Wang, D.S., Deng, T. and Shaw, G. (1997) Membrane binding and enzymatic activation of a Dbl homology domain require the neighboring pleckstrin homology domain. *Biochem. Biophys. Res. Commun.*, **234**, 183–9.

Wang, H.G., Rapp, U.R. and Reed, J.C. (1996) Bcl-2 targets the protein kinase Raf-1 to mitochondria. *Cell*, **87**, 629–38.

Wang, X.Z. and Ron, D. (1996) Stress-induced phosphorylation and activation of the transcription factor CHOP (GADD153) by p38 MAP Kinase. *Science*, **272**, 1347–9.

Wartmann, M. and Davis, R.J. (1994) The native structure of the activated Raf protein kinase is a membrane-bound multi-subunit complex. *J. Biol. Chem.*, **269**, 6695–701.

Watabe, M., Masuda, Y., Nakajo, S., Yoshida, T., Kuroiwa, Y. and Nakaya, K. (1996) The cooperative interaction of two different signalling pathways in response to bufalin induces apoptosis in human leukemia U937 cells. *J. Biol. Chem.*, **271**, 14067–72.

Wennstrom, S., Hawkins, P., Cooke, F., Hara, K., Yonezawa, K., Kasuga, M., Jackson, T., Claesson-Welsh, L. and Stephens, L. (1994) Activation of phoshoinositide 3-kinase is required for PDGF stimulated membrane ruffling. *Curr. Biol.*, **4**, 385–393.

Whitehead, I.P., Khosravi Far, R., Kirk, H., Trigo Gonzalez, G., Der, C.J. and Kay, R. (1996) Expression cloning of lsc, a novel oncogene with structural similarities to the Dbl family of guanine nucleotide exchange factors. *J. Biol. Chem.*, **271**, 18643–50.

Wilson, D.J., Fortner, K.A., Lynch, D.H., Mattingly, R.R., Macara, I.G., Posada, J.A. and Budd, R.C. (1996) JNK, but not MAPK, activation is associated with Fas-mediated apoptosis in human T cells. *Eur. J. Immunol.*, **26**, 989–94.

Wu, X., Noh, S.J., Zhou, G., Dixon, J.E. and Guan, K.-L. (1996) Selective activation of MEK1 but not MEK2 by A-Raf from epidermal growth factor stimulated Hela ells. *J. Biol. Chem.*, **271**, 3265–3271.

Xia, Z., Dickens, M., Raingeaud, J., Davis, R.J. and Greenberg, M.E. (1995) Opposing effects of ERK and JNK-p38 MAP kinases on apoptosis. *Science*, **270**, 1326–31.

Yan, M., Dai, T., Deak, J.C., Kyriakis, J.M., Zon, L.I., Woodgett, J.R. and Templeton, D.J. (1994) Activation of stress-activated protein kinase by MEKK1 phosphorylation of its activator SEK1. *Nature*, **372**, 798–800.

Yang, X., Khosravi-Far, R., Chang, H.Y., and Baltimore, D. (1997) Daxx, a novel Fas-binding that activates JNK and apoptosis. Cell **89**, 1067–1076.

Yao, B., Zhang, Y., Delikat, S., Mathias, S., Basu, S. and Kolesnick, R. (1995) Phosphorylation of Raf by ceramide-activated protein kinase. *Nature*, **378**, 307–10.

Yao, R. and Cooper, G.M. (1995) Requirement for phosphatidylinositol-3 kinase in the prevention of apoptosis by nerve growth factor. *Science*, **267**, 2003–6.

Yao, R. and Cooper, G.M. (1996) Growth factor-dependent survival of rodent fibroblasts requires phosphatidylinositol 3-kinase but is independent of pp70S6K activity. *Oncogene*, **13**, 343–51.

Zanke, B.W., Boudreau, K., Rubie, E., Winnett, E., Tibbles, L.A., Zon, L., Kyriakis, J., Liu, F.F. and Woodgett, J.R. (1996) The stress-activated protein kinase pathway mediates cell death following injury induced by cis-platinum, UV irradiation or heat. *Curr. Biol.*, **6**, 606–13.

Zhang, S., Han, J., Sells, M.A., Chernoff, J., Knaus, U.G., Ulevitch, R.J. and Bokoch, G.M. (1995) Rho family GTPases regulate p38 mitogen-activated protein kinase through the downstream mediator Pak1. *J. Biol. Chem.*, **270**, 23934–6.

Zhang, X., Settleman, J., Kyriakis, J.M., Takeuchi-Suzuki, E., Elledge, S.J., Marshall, M.S., Bruder, J.T., Rapp, U.R. and Avruch, J. (1993) Normal and oncogenic p21ras proteins bind to the amino-terminal regulatory domain of c-Raf-1. *Nature*, **364**, 308–365.

Zhang, Y., Yao, B., Delikat, S., Bayoumy, S., Lin, X.H., Basu, S., McGinley, M., Chan Hui, P.Y., Lichenstein, H. and Kolesnick, R. (1997) Kinase suppressor of Ras is ceramide-activated protein kinase. *Cell*, **89**, 63–72.

Zheng, C.F. and Guan, K.L. (1994) Activation of MEK family kinases requires phosphorylation of two conserved Ser/Thr residues. *EMBO J.*, **13**, 1123–31.

Zheng, Y., Zangrilli, D., Cerione, R.A. and Eva, A. (1996) The pleckstrin homology domain mediates transformation by oncogenic dbl through specific intracellular targeting. *J. Biol. Chem.*, **271**, 19017–20.

5. PROTEIN KINASE C ISOENZYMES: EVIDENCE FOR SELECTIVITY IN THE REGULATION OF APOPTOSIS

JANET M. LORD†, ELIZABETH M. DEACON, LORNA McMILLAN, GARETH GRIFFITHS, HEMA CHAHAL AND JUDIT PONGRACZ

Department of Immunology, Birmingham University Medical School, Birmingham B15 2TT, UK

Several disease model systems are now under active consideration for intervention at the level of the apoptotic programme. For example, the induction of apoptosis offers an alternative to cytostatic or differentiation therapies in the treatment of cancer. Many of the studies to identify those factors regulating apoptosis, that may be suitable for therapeutic intervention, have centred upon characterisation of the signal transduction pathways involved in the regulation of cell proliferation and apoptosis. One signalling element implicated in the regulation of cell proliferation and apoptosis, and already a recognised target for therapeutic modulation, is protein kinase C (PKC). PKC is a multigene family consisting of 11 isoenzymes which are regulated independently and are proposed to play specific roles in the regulation of cell functions, including apoptosis. We review here the experimental evidence that has been published concerning the involvement of PKC isoenzymes in the regulation of apoptosis. Whilst there is clearly still much contradictory data in this field of research, we propose that the weight of evidence suggests that the activation of PKC-alpha and beta constitutes a survival signal in many cell types. Whether the two splice forms of PKC-beta may have differential roles in the control of proliferation and apoptosis is also discussed. In contrast we have concluded that one PKC isoenzyme, PKC-delta, is activated during apoptosis and represents a pro-apoptotic PKC. The role of PKC and apoptosis in disease pathogenesis is also considered, with emphasis placed upon colorectal cancer.

PROTEIN KINASE C

The realisation that apoptosis is a fundamental cellular process in development and morphogenesis (Kerr *et al.*, 1974; Kerr *et al.*, 1987; Kerr *et al.*, 1972; Savill *et al.*, 1993; Smith *et al.*, 1989; Wyllie, 1980) and moreover, that its improper regulation impacts upon the initiation and progression of disease states(Savill *et al.*, 1989; Williams, 1991), has led to the broadening of disease model systems under consideration for intervention at the level of the apoptotic programme. For example, it has become clear that several anti-cancer agents already in use are potent inducers of apoptosis (Dive and Hickman, 1991; Fisher, 1994). Many of the studies to identify those factors regulating apoptosis, have focussed upon identification of the signal transduction pathways involved in the regulation of cell proliferation, differentiation and apoptosis. More specifically, efforts have been concentrated on determining exactly how these pathways interact and how their disregulation leads to pathogenesis (Martin, 1982; Whitehouse *et al.*, 1982; Grigg *et al.*, 1991; Bursch *et al.*, 1992; Groux *et al.*, 1992; Kerr *et al.*, 1994). One signalling element implicated in the regulation of cell proliferation, differentiation and apoptosis, and already a

† *Corresponding Author*: Tel.: 44–121–414 4399. Fax: 44–121–414 3599. e-mail: Lord JM @novell2. bham.ac.uk.

target for the design of novel therapeutic modalities, is protein kinase C (Gescher, 1992).

Protein kinase C (PKC) is a serine/threonine kinase, which functions as a key element in signalling pathways regulating a variety of cell functions (Hug and Sarre, 1993; Nishizuka, 1992). PKC was first discovered in rat brain extracts by Nishizuka and co-workers (Inoue *et al.*, 1977) who described it as a histone protein kinase that displayed a phospholipid and calcium dependency and in addition required diacylglycerol (DAG) for full enzyme activity. Interest in PKC as a signalling molecule increased very dramatically, following the discovery that this kinase was the cellular receptor for the tumour-promoting phorbol esters (Castagna *et al.*, 1982). The range of cell functions known to be regulated through the PKC signalling pathway is extensive and includes, cell proliferation and differentiation (Clemens *et al.*, 1992), secretion (Lord and Ashcroft, 1984), cytoskeleton function (Jaken *et al.*, 1989; Owen *et al.*, 1996), cell-cell contacts (Barry and Critchley, 1994), gene expression (Baudier *et al.*, 1992; Li *et al.*, 1992)and cell survival (Lotem *et al.*,1991; Pongracz *et al.*, 1994; Lucas and Sanchez-Margalet, 1995).

The pleiotropic involvement of PKC in the control of cellular activity raises the question of how specificity of biological action can be maintained, if a wide range of receptors are linked to the PKC signalling pathway. This paradox began to be resolved when the screening of rat cDNA libraries revealed the existence of several different, but closely related, isoenzymes of PKC (Parker *et al.*, 1989). As most cells express several PKC isoenzymes and their amino acid sequences show a high degree of conservation across mammalian species (Dekker and Parker, 1994), it is likely that the individual PKC isoenzymes have specific rather than overlapping cell functions.

PKC Isoenzymes

The PKC isoenzyme family consists of multiple genes, defined by sequence homology and distinct sequence motifs (Figure 5.1). The 11 members identified to date, display differential properties with regard to tissue distribution. PKC-α, -δ and -ζ are the most widely distributed (Nishizuka, 1984; Wada *et al.*, 1989; Wetsel *et al.*, 1992) of the PKC isoenzymes so far assessed. In contrast, PKC-γ is expressed exclusively in the central nervous system (Nishizuka, 1988), PKC-θ predominantly in skeletal muscle and haemopoietic cells (Osada *et al.*, 1992) and PKC-η expression is greatest in skin and lung tissue, with only low levels detected in the brain and spleen (Osada *et al.*, 1990). In addition, the isoenzymes of PKC differ in their susceptibility to regulation by co-factors, subcellular location, substrate specificity and down-regulation by proteolytic cleavage (reviewed by (Hug and Sarre, 1993)). Thus, although all of the isoenzymes transduce intracellular signals which involve the generation of lipid species, they appear to be regulated independently to effect an isoenzyme specific response (Nishizuka, 1992; Dekker and Parker, 1994). Detailed descriptions of the biochemistry and regulation of PKC have been covered in several recent reviews (Nishizuka, 1992; Hug and Sarre, 1993; Dekker and Parker, 1994; Goodnight *et al.*, 1994; Newton, 1995), and will consequently require only a brief consideration here.

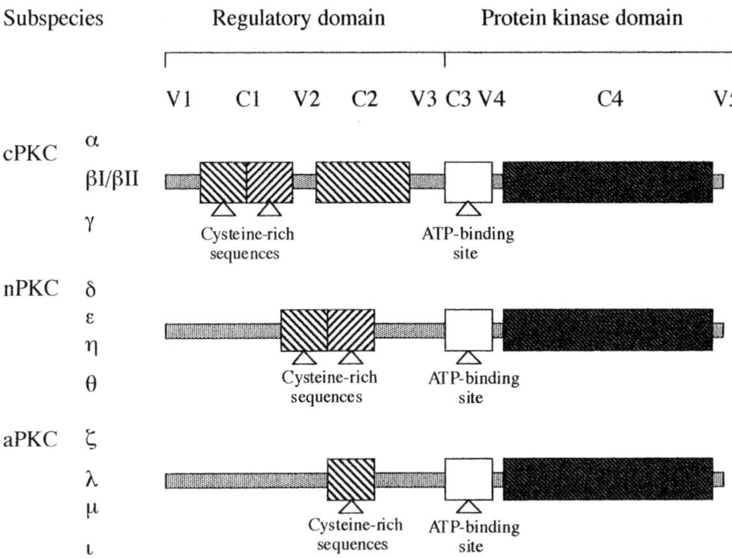

Figure 5.1 Structure of the 3 main groups of the PKC isoenzyme family. (Key: c, classical; n, novel; a, atypical; V, variable region; C, constant region)

Activation of PKC isoenzymes

Whilst all PKC isoenzymes contain an amino terminal phosphatidylserine (PS)-binding domain, the PKC isoenzyme family can be divided into three main groups, based on their additional activation requirements and ability to respond to phorbol esters (Figure 5.1). The latter bind to the regulatory domain of PKC, mimicking the acute physiological activation by DAG:

- Classical, c-PKCs: -α , -β1(II), -β2(I)* and -γ are calcium dependent and activated by phorbol esters;
- Novel, n-PKCs: -δ, -ε, -η and -θ are calcium independent, but can be activated by phorbol esters;
- Atypical, a-PKCs: -ζ, -λ/i and -μ are calcium independent and do not bind phorbol esters. Suggested physiological activators of these isoenzymes include ceramide (Lozano *et al.*, 1994; Muller *et al.*, 1995) and phosphatidylinositol 3,4,5 trisphosphate* (Nakanishi *et al.*, 1993).

* PKC-β isoenzymes are designated throughout in Arabic numerals according to the human cDNA nomenclature, with the Roman numeral designation for rat cDNA in parentheses, even if the original papers employed only one of the designations.

DAG which is generated as a result of phospholipid hydrolysis will be able potentially to regulate both the c-PKCs and the n-PKCs. However, agonist stimulated phosphoinositide hydrolysis produces DAG and a calcium signal, via the generation of inositol 1,4,5-trisphosphate (IP3). Both of these second messengers would be required for the full activation of the c-PKC isoenzymes. In contrast, the n-PKCs do not require the calcium signal and are able to respond to DAG produced in the absence of calcium. In support of this assertion, Ha and Exton have shown that signalling through thrombin and PDGF receptors results in the selective activation of n-PKCs, via DAG produced by the action of phospholipase D upon phosphatidylcholine (Ha and Exton, 1993).

The PKC isoenzymes show another level of complexity of regulation in that they appear to be differentially responsive to lipid species. For example, activation of c-PKCs is further enhanced by cis-unsaturated fatty acids and lysophosphatidylcholine; free fatty acids activate PKC-ε and -ζ, but inhibit PKC-δ (Kikkawa *et al.*, 1988; Lee and Bell, 1991; Hug and Sarre, 1993) and PKC-η is most effectively activated by cholesterol sulphate (Gschwendt *et al.*, 1994). Ceramide is reported to activate PKC-ζ (Muller *et al.*, 1995), but inhibits the activation of PKC-α (Lee *et al.*, 1996). Thus the activation of specific PKC isoenzymes will vary dependent upon the lipid species generated as second messengers following receptor ligation.

In addition, analysis of total cellular DAG in Swiss 3T3 fibroblasts, has revealed a complex mixture of 27 different molecular species, with different fatty acid compositions. Stimulation of these cells with Epidermal Growth Factor induced an increase in only a few DAG species (Pettit *et al.*, 1994). Although differential activation of PKC isoenzymes by DAG species has not been reported as yet, this could clearly provide another route to selectivity in signalling through PKC. Our own preliminary studies have shown that stearoyl-arachidonyl glycerol production is associated with activation of PKC-β during the G2/M stage of cell cycle in U937 cells (E.M. Deacon, unpublished observations). Furthermore, the phospholipase D-derived DAG species generated in PAE cells, following treatment with LPA, did not activate PKC (Pettit *et al.*, 1997). Further analyses are now required to establish the isoenzyme selectivity of lipid species generated in vivo and to confirm in vitro data, as the mode of presentation of lipid co-factors to PKC can affect their response (Palmer *et al.*, 1995).

Although a variety of lipids are able to modulate the activity of PKC isoenzymes selectively, it is unlikely that specificity of biological response is generated solely through differences in second messenger generation. It is clear that subcellular targeting following enzyme activation (Kiley *et al.*, 1995) and substrate availability (Jaken, 1996), are key factors generating the functional specificity of the PKC isoenzymes.

PKC translocation and subcellular targetting

All PKC isoenzymes possess catalytic and regulatory domains separated by a hinge region, V3, (Figure 5.1) which is subject to proteolysis (Kishimoto *et al.*, 1983). Dependent upon the class of PKC isoenzyme, the regulatory domain can include a calcium binding and a DAG/phorbol ester binding site. Activation of PKC isoenzymes requires their interaction with membrane phospholipid, anionic species such as PS being most effective (Kikkawa *et al.*, 1988). In addition, inactive PKCs are located predominantly, though not exclusively, in the cytosol, though many of the known substrates for PKC are cytoskeletal or membrane proteins (Jaken, 1996). Thus activation of PKC is associated with its redistribution and association with intracellular membranes, to allow interaction with PS and substrates.

In vitro studies have shown that recombinant PKC isoenzymes show very little substrate preference (Fujise *et al.*, 1994) and a degree of pre-positioning of PKC isoenzymes may consequently be required to achieve the selective phosphorylation of substrates reported for in vivo studies. Differential subcellular localisation of PKC isoenzymes has been reported in a variety of cell types. PKC-α is localised to focal contacts in normal REF52 cells (Liao *et al.*, 1994b), PKC-δ is associated with vimentin intermediate filaments in differentiated HL60 cells (Owen *et al.*, 1996) and PKC-β2(l) is localised to the microtubule cytoskeleton in proliferating U937 cells (Kiley and Parker, 1995). Clearly the location of substrates such as talin, vimentin and microtubule associated proteins will contribute to the targeting of PKC isoenzymes following their activation. This proposal is supported by data from a variety of studies reporting that the direction of translocation of PKC isoenzymes is highly variable. For example, in Swiss 3T3 cells treated with bombesin, PKC-α translocates to the cell plasma membrane. In the same cells treated with IGF-1, PKC-α is associated with the nuclear membrane (Divecha *et al.*, 1991). PKC-β translocates to the nucleus when K562 cells are treated with bryostatin-1 and to the plasma membrane following treatment with phorbol dibutyrate (Hocevar and Fields, 1991). Clearly such differences will influence the actions of PKC isoenzymes, most notably regarding substrate availability (Kiley *et al.*, 1995) and will dictate the role of PKC isoenzymes in cell function.

Exactly how targeting of PKCs before and after activation are related and effected at the molecular level is not clear. Several laboratories have used approaches such as interaction cloning and yeast-two hybrid screening to identify proteins that may localise PKCs to particular regions of the cell. These studies have identified a variety of proteins, including known PKC substrates and novel PKC-interacting proteins (Chapline *et al.*, 1993; Staudinger *et al.*, 1995), that shed some light on this problem. For example, several PKC binding proteins are not substrate proteins and appear to act to localise PKC following its activation. Such proteins include the receptors for activated C kinase (RACKs) first described by Mochly-Rosen and co-workers (Ron *et al.*, 1994) and the pleckstrin homology domain of Brutons tyrosine kinase, which binds PKC-μ (Sidorenko *et al.*, 1996). Furthermore, Dong *et al.*, showed that one of the PKC binding proteins, 35H, was phosphorylated following agonist-stimulated second messenger generation and that the phosphorylated protein had a greatly reduced affinity for PKC (Dong *et al.*, 1995). These observations have led Jaken to propose that the relocation of PKC isoenzymes following activation (Kiley and Parker, 1995), may be a consequence of release of inactive PKCs from binding proteins and their diacylglycerol-induced association with substrates in the membrane/cytoskeleton (Jaken, 1996).

If more of these PKC binding proteins are identified and shown to be located at specific sites within the cell, they may provide the selective localisation that would allow discrete activation of PKCs and interactions with specific substrates. Recently, proteins have been identified which effect the compartmentalisation of cAMP-dependent protein kinase (PKA). The A-kinase anchoring proteins (AKAP) are a diverse group of proteins which bind the regulatory subunit of type II PKA isoforms and are located at distinct sites within cells, including the plasma membrane (Gray *et al.*, 1997), sarcoplasmic reticulum (McCartney *et al.*, 1995), nuclear matrix (Zhang *et al.*, 1996) and actin cytoskeleton (Dransfield *et al.*, 1997). Moreover, the AKAP proteins have been shown to bind protein phosphatase and PKC at a site distinct from PKA (Nauert *et al.*, 1997) and may therefore also act as scaffolding proteins for PKCs and other signalling elements. Whether the different AKAP proteins show selectivity in their binding of PKC isoenzymes has not yet been reported.

The regulatory domain of PKC appears to play a crucial role in targeting of inactive PKC and substrate interactions (Pears *et al.*, 1991). For example, removal of the pseudosubstrate domain decreases its affinity for PKC binding proteins (Liao *et al.*, 1994a). The regulatory domain of PKC contains several regions that are able to bind PS and diacylglycerol contributes to the translocation of PKC by increasing its affinity for PS (Newton, 1995). PKC substrates also contain PS binding domains (Chapline *et al.*, 1993; Hyatt *et al.*, 1994) and substrate proteins, such as the MARCKS protein, which are able to localise PS into high density regions. Such domains would allow PKC binding to PS and to other proteins, in effect PS may act as a bridge between PKC and its substrates in a ternary complex located at cell membranes (Jaken, 1996). Thus, substrate localisation is not the only factor affecting PKC translocation, the PS binding domains are important in enzyme activation (see above) and protein interactions. For example, the regulatory domain of PKC-α masks nuclear localisation sequences in inactive PKC (Liao *et al.*, 1994a) and the cysteine rich motif zinc finger domain of PKC-ε mediates its localisation to the Golgi (Lehel *et al.*, 1995).

In summary, a model may be proposed in which differential generation of second messengers dictates which PKCs are activated in response to receptor ligation and the subcellular targeting of PKC isoenzymes pre- and post-activation, mediated by the PS binding domains on PKCs and PKC-binding proteins, will determine isoenzyme function in vivo. Interestingly, this model may already require slight modification to resolve the most recent data in the literature. Sawai and co-workers (Sawai *et al.*, 1997) have reported that PKC-δ and ε were translocated from the membrane to the cytosol in response to the generation of ceramide, following treatment of cells with TNF-α and prior to the subsequent induction of apoptosis . However, rather than concluding that these isoenzymes were inhibited by ceramide, the authors suggest that the translocation to the cytosol leads to their proteolytic activation by cytosol based ICE proteases (see below). In this case, a second messenger has induced translocation of PKC and has possibly effected enzyme activation, but with translocation directed away from cell membranes. Whether the proteolytically-activated PKC isoenzymes in these studies ultimately translocated towards the membrane/cytoskeleton was not determined.

Proteolysis and Downregulation of PKC

The regulatory domain of PKC also includes a motif resembling the consensus site for PKC substrates, xRxxS/TxRx, in which a serine/threonine residue is changed to alanine. This motif represents a pseudosubstrate site which blocks the catalytic site in inactive PKCs. Binding of an activator such as DAG, induces a conformational change removing any inhibition by the pseudosubstrate site and also renders the hinge region open to proteolytic cleavage (Newton, 1995). Proteolysis generates two distinct fragments, the regulatory domain and a catalytically active protein kinase domain known as PKM (Huang and Huang, 1986; Schaap *et al.*, 1990). In the case of the c-PKCs, the hinge region contains cleavage sites for the calcium-dependent neutral proteases calpains I and II (Kishimoto *et al.*, 1989). Schaap *et al.* (Schaap *et al.*, 1990) have also reported a cleavage site for trypsin in PKC-ε. Persistent activation of PKC, resulting from treatment with phorbol esters or certain growth factors (Olivier and Parker, 1994), leads to the downregulation of PKC protein. In most cases it is not known whether the sustained activation of PKC, or its subsequent downregulation, is important for the elicitation of cellular responses. Interestingly the PKC isoenzymes vary in their susceptibility to such cleavage. PKC-α has been shown to be relatively resistant to proteolysis mediated downregulation, whereas PKC-β and PKC-γ are more easily inactivated (Kochs *et al.*, 1993). As will be discussed later, it has been shown recently that a n-PKC, PKC-δ, can be cleaved by the ICE-like protease CPP32 to generate a catalytically active 40kDa PKC (Emoto *et al.*, 1996). This property was reported to be unique to PKC-δ, though a recent publication has claimed that PKC-ε is also susceptible (Sawai *et al.*, 1997). Whether PKC isoenzymes are targets for other proteases remains to be established, but clearly this mode of activation could involve PKC in an even wider range of signalling pathways.

The preceding information illustrates the complexity of PKC regulation and the factors that must be taken in to account when considering its role in apoptosis and disease. PKC involvement will be affected by the isoenzyme content of the tissue being studied, the second messengers generated following receptor ligation, degree of downregulation, direction of translocation and substrate availability. To date the majority of studies that have considered the role of PKC in apoptosis have not taken these factors into account. However, it is becoming clear that PKC does regulate apoptosis and that specific isoenzymes are involved in signals that promote or delay apoptosis. The following is an attempt to review the current literature in this area, with emphasis placed upon studies that have considered the role of individual PKC isoenzymes.

PKC ISOENZYMES AND APOPTOSIS

Apoptosis is a multi-step process, which can be broken down to four distinct stages:

- signal generation and interpretation. Signals can be either intrinsic or extrinsic to the cell and will dictate whether a particular cell proliferates, differentiates or dies;
- the commitment phase. This stage is variable in length and commences once the pro-apoptotic signal is received;
- execution of the apoptotic programme. This stage includes all the characteristic morphological changes of apoptosis and is complete within 15–60 minutes (reviewed in (Earnshaw, 1995));

- recognition and removal of apoptotic cells. Changes that occur at the cell membrane during apoptosis allow recognition of dying cells by professional phagocytes or neighbouring cells (Savill *et al.*, 1993).

It is therefore possible that individual PKC isoenzymes may play different roles in regulating cell death and survival. Evidence supporting a role for PKC in the regulation of apoptosis is substantial, though contradictory (Table 5.1). Data concerning specific PKC isoenzymes are less abundant, but may provide the greatest insight into the involvement of PKC in cell survival processes. Studies to determine the role of PKC isoenzymes in cell survival have involved three broad approaches:

(a) the use of pharmacological modulators of PKC activity with varying degrees of isoenzyme selectivity;
(b) correlative approaches, assessing PKC isoenzyme expression and/or activation status with cell survival;
(c) assessment of phenotypic changes in cells overexpressing wild type PKC isoenzymes, constitutively active or kinase defective PKC mutants.

Rather than attempt to review the findings with regard to these distinct methodological approaches or with respect to specific cell types, this article will consider the data in relation to the role of the three functional classes of PKC isoenzymes. This approach should facilitate conclusions regarding individual isoenzymes and highlight those PKCs which appear to play a consistent role in apoptosis, rather than a cell specific involvement in cell survival. In addition, we will consider briefly the known substrates for the relevant PKC isoenzymes, in an attempt to identify potential down-stream targets that may mediate the effects of PKC upon cell survival. The chapter will then close with a discussion of the role of apoptosis in disease pathogenesis, concentrating upon examples in which alterations to PKC isoenzymes have also been documented.

Table 5.1 Effects of TPA on Apoptosis.

Induced by TPA	*Prevented by TPA*
Hydrocortisone-treated thymoctyes (Moore *et al.*, 1992)	IL-2-deprived T-lymphocytes (Rodriguez-Tarducy and Lopez-Rivas, 1989)
T cell hybridomas treated with anti-CD3 (Iseki *et al.*, 1991)	IL-3-deprived promyeloid cells (Lotem *et al.*, 1991)
Human prostate cancer cells (Powell *et al.*, 1996)	Germinal center B cells (Knox *et al.*, 1993)
Nitric oxide effects on HL60 cells (Jun *et al.*, 1997)	Radiation-induced T cell apoptosis (Radford, 1994)

c-PKC Isoenzymes

This class of PKC isoenzymes are responsive to phorbol esters, calcium and a range of lipid co-factors, including diacylglycerols. The c-PKCs might therefore be expected to mediate the effects of ligands that induce apoptosis via generation of intracellular calcium and/or lipid hydrolysis products, or to be the target for phorbol ester mediated changes in cell survival. There is now an increasing body of data suggesting that PKC-α may be classed as an anti-apoptotic PKC, the picture concerning PKC-β is less clear and there is little data concerning the role of PKC-γ. As the latter is predominantly expressed in neuronal tissue, its function may be very restricted and relate to brain function.

PKC-alpha

PKC-α appears to be expressed universally and has been implicated in mitogenesis by several authors. For example, PKC-α was shown to translocate to the nucleus following treatment of llC9 fibroblasts with α-thrombin (Leach *et al.*, 1992) and more recently, PKC-α was shown to interact with Raf-1, leading to activation of the MAP kinase cascade and proliferation of NIH 3T3 cells (Li *et al.*, 1995). Whilst the regulation of proliferation and apoptosis appear to be closely integrated (Evan *et al.*, 1992), they are ultimately mutually exclusive events. By inference PKC-α activation would be expected to constitute a survival signal.

Grant and co-workers showed that pre-treatment of HL60 cells with bryostatin 1 for 24 hours produced a significant down-regulation of c-PKC isoenzymes and potentiated Ara-C-induced apoptosis. Mezerein, another potent PKC activator, did not produce the same degree of down-regulation and induced differentiation of HL60 cells, with no potentiation of Ara-C-induced apoptosis (Grant *et al.*, 1996). These studies did not consider individual PKC isoenzymes and can only infer a protective role for c-PKCs as they did not assess changes in the n-PKC isoenzymes.

However, studies in endothelial cells have suggested that PKC-α is involved specifically in mediating the protective effects of b-FGF against radiation-induced apoptosis (Haimovitz-Friedman *et al.*, 1994). The protective effects of bFGF were mimicked by short term treatment of cells with TPA and bFGF was shown to translocate PKC-α within 30 seconds of treatment. Moreover, depletion of PKC-α abrogated the effects of bFGF.

More recently, modulation of PKC-α has been shown to be involved in the induction of apoptosis by ceramide. Ceramide is a lipid second messenger produced from the hydrolysis of sphingomyelin, following the activation of sphingomyelinase by extracellular ligands such as TNF-α and IFN-γ (Dressler *et al.*, 1992). Ceramide mediates the effects of these extracellular ligands on cell differentiation, growth arrest and apoptosis. Two groups have shown that ceramide inhibits PKC-α, though they disagree on the molecular basis of this inhibition. Jones and Murray, working with mouse epidermal (HEL37) and human skin fibroblast (SF3155) cells, showed that ceramide induced apoptosis and inhibited the translocation of PKC-α (Jones and Murray, 1995). The effect appeared to be specific to PKC-α , with no effect on PKC-ε. In contrast, Lee and co-workers showed that ceramide did not inhibit translocation of PKC-α in MOLT-4 cells and could not inhibit PKC directly in vitro (Lee *et al.*, 1996). Inhibition of PKC enzyme activity by ceramide was in fact achieved by dephosphorylation of PKC-α (Lee *et al.*, 1996).

Phosphorylation is an important mode of regulation of PKC, which is first synthesised as a dephosphorylated, inactive precursor (Borner *et al.*, 1989). Several groups have shown that phosphorylation of PKC within the activation loop leads to activation of the enzyme and dephosphorylation within this domain inhibits enzyme activity (Cazoubon *et al.*, 1994; Orr and Newton, 1994). One of the cellular targets of ceramide is a ceramide-activated protein phosphatase, CAPP (Dobrowsky and Hannun, 1992). CAPP shares several properties with protein phosphatase 2A (PP2A). Interestingly, PP2A was able to dephosphorylate PKC (Dutil *et al.*, 1994) and okadaic acid was able to reverse the actions of ceramide on PKC-α (Lee *et al.*, 1996). These data suggest that PKC-α is a down-stream target in the ceramide survival regulatory pathway and that its inhibition allows for growth inhibition and the induction of apoptosis.

As with most of the literature concerning PKC and apoptosis, the data concerning PKC-α does not allow for a uniform interpretation to be made. For example, induction of apoptosis by TPA in androgen-sensitive human prostate cells (LnCap) was accompanied by a 12 fold increase in PKC-α mRNA (Powell *et al.*, 1996). TPA resistant LnCap cells had reduced levels of PKC-α and raised levels of PKC-μ (Powell *et al.*, 1996). In addition, Devente *et al.*, showed that MCF-7 cells over-expressing PKC-α were induced to apoptose when treated with TPA. This is in contrast to wild type MCF-7 cells which differentiated in response to TPA (Devente *et al.*, 1995).

However, the majority of the current literature suggests that PKC-α is involved in mitogenic signalling pathways in many cells. Indeed, PKC-α expression and/or activation status is raised in several cancers (Lord and Pongracz, 1995). Reduction in PKC-α expression, achieved by the use of anti-sense oligonucleotides, results in decreased proliferation in rat coronary vascular smooth muscle cells (Leszczynski *et al.*, 1996). Encouragingly, antisense to PKC-α has been used very recently as a therapeutic approach to the treatment of glioblastoma. Reduced levels of PKC-α led to a significant reduction in tumour mass. Whether the reduced tumour growth was due to alterations in proliferation or apoptosis was not determined in these studies (Yazaki *et al.*, 1996).

PKC-beta

PKC-β, as stated earlier, exists as two alternate splice forms β1(II) and β2(I). The two splice forms differ by only approximately 50 amino acids in the C terminal regulatory domain, but this would appear to be sufficient to influence their distribution within the cell. For example, studies in U937 cells have shown that PKC-β1(II) was present in the cytoplasm of proliferating cells and was specifically localised to vesicles containing β2-integrin molecules. In contrast, PKC-β2(I) was associated with the microtubule network, via binding to microtubule-associated proteins (Kiley and Parker, 1995). As targeting of PKC isoenzymes to specific locations within the cell may be a factor in determining isoenzyme specific function in vivo (Kiley *et al.*, 1995), PKC-β1 and β2 may be differentially involved in the processes of cell differentiation, proliferation and apoptosis. Unfortunately, only very few studies have assessed the role of the individual PKC-β splice forms simultaneously and it is therefore too early to be able to state whether or not they have differential roles in the regulation of cell survival.

What is clear from several sources is that PKC-β plays a major role in the co-ordination of cell proliferation, differentiation and apoptosis. Studies of PKC isoenzyme expression in non-apoptotic and apoptotic cells showed that PKC-β expression was increased in apoptotic U937 cells (Pongracz *et al.*, 1995b). PKC isoenzyme expression was also assessed with relation to the expression of Bcl-2 in human tonsil, these data revealed a negative correlation between expression of PKC-β and Bcl-2 (Knox *et al.*, 1993). In addition, HL60 cells which were unable to differentiate or apoptose in response to phorbol esters, were shown to lack expression of PKC-β. Induction of PKC-β expression following treatment of cells with 1,25-dihydroxyvitamin D3, restored responsiveness to phorbol esters (Macfarlane and Manzel, 1994). HL60 cells stop proliferating in response to phorbol esters such as TPA. Growth arrest is rapid (complete within 12h) and precedes acquisition of the differentiated monocyte phenotype and apoptosis. Precisely which stage of the differentiation/apoptosis process required PKC-β was not determined in these studies. It will be important to consider both PKC-β1 and PKC-β2 in studies such as these, if we are to determine their precise role in cell proliferation and survival.

A series of detailed studies from Fields and co-workers, primarily using HL60 cells, have established a role for PKC-β1(ll) in cell proliferation (see also PKC substrates section). They have shown that PKC-β1(ll) is a mitotic lamin kinase in haemopoietic cells (Goss *et al.*, 1994). Analysis of enzyme translocation through the cell cycle revealed that this isoenzyme was activated and translocated to the nucleus at G2/M. The nuclear lamin proteins form a polymeric network which is regulated by multisite phosphorylation. Whilst $p34^{cdc2}$ kinase is able to phosphorylate nuclear lamin proteins in vitro to bring about dissolution of the lamina complex (Peter *et al.*, 1990), in vivo PKC-β1(ll) also phosphorylates lamin B in G2/M at a site distinct from $p34^{cdc2}$ (Goss *et al.*, 1994). The most recent report from this group has shown that PKC-β1(ll) was necessary for progression of cells through G2/M. Furthermore, lamin phosphorylation sites were analysed and site-directed mutagenesis employed to show that the sites phosphorylated by PKC were required to maintain lamin B protein in the cytoplasm during mitosis (Thompson and Fields, 1996).

Thus, PKC-β1(ll) would appear to be primarily involved in the regulation of proliferation, at least in haemopoietic cells. This proposal is supported by data from studies investigating the molecular basis of the suppression of apoptosis by the oncogene v-abl. Transfection of a temperature sensitive v-abl construct in to an IL3-dependent haemopoietic cell line, abrogated apoptosis induced following growth factor withdrawal when cells were kept at the permissive temperature and v-Abl tyrosine kinase was expressed and activated (Evans *et al.*, 1993). Prevention of apoptosis was later shown to be associated with the activation and translocation of PKC-β1(ll) towards the nucleus (Evans *et al.*, 1995). Pre-treatment of cells with calphostin C prevented nuclear translocation of PKC-β1(ll) and inhibited the suppression of apoptosis by v-Abl. Thus, like PKC-β, PKC-β1(ll) would appear to predominantly regulate cell proliferation and promote cell survival. Whether the inappropriate activation of PKC-β1(ll) in proliferating cells could in certain instances lead to growth arrest and apoptosis is one possibility that could be considered. As this isoenzyme is normally activated at G2/M, activation at any other stage of the cell cycle could induce a form of "mitotic catastrophe", postulated by several authors to result in apoptosis in cycling cells (Shi *et al.*, 1994).

With regard to PKC-β2(l) and apoptosis, the data are equivocal. Treatment of U937 cells with the the phorbol ester, TPA, induced growth arrest and differentiation towards monocytes. Treatment of cells with the deoxyphorbol ester, Doppa, which is selective for the activation of PKC-β in vitro (Ryves *et al.*, 1991), did not induce differentiation and increased apoptosis (Pongracz *et al.*, 1996). PKC-β2(l), the dominant splice form in these cells, was selectively translocated at concentrations of Doppa at or below 50nM. However, this effect was transient and after 1 hour other isoenzymes were also translocated. Furthermore, other authors have reported that Doppa is not PKC-β selective in vivo (Roivainen and Messing, 1993) and these data must therefore be interpreted with caution. We have found more recently that upregulation of PKC-β occurred in apoptotic human neutrophils (Pongracz *et al.*, unpublished observations). These cells are fully differentiated and therefore do not proliferate and die spontaneously by apoptosis within 24–48 hours of leaving the bone marrow. Interestingly, this change in expression was restricted to the β2(l) isoform, with levels of PKC-β1(ll) remaining unchanged. Recent studies in small lung carcinoma cells (NC1 H209) also suggest a role for PKC-β and suggest that the effects of c-myc upon cell proliferation may be mediated via this isoenzyme (Barr *et al.*, 1997). However, the data from these studies suggest a role for PKC-β2(l), rather than PKC-β1(ll), in the regulation of proliferation and apoptosis. PKC-β1(ll) and PKC-β2(l) were transfected into cells individually, either before or after transfection with c-myc. Overexpression of c-myc alone produced a significant increase in cell proliferation which was not affected by co-expression of PKC-β1(ll). In contrast, transfection of PKC-β2(l) alone extended cell doubling time and increased the fraction of cells in G0/G1. Co-expression of c-myc with PKC-β2(l) improved survival of cells in low serum, with cells in G0/G1 being particularly affected.

In conclusion, PKC-β1 and β2 have both been implicated in the regulation of apoptosis, but the role of each splice form may vary with cell type and the compliment of isoenzymes expressed . At this stage, the few individual studies that have considered both beta isoforms, do appear to suggest their differential involvement in the regulation of apoptosis within a particular cell type. At this stage no further generalisations can be made as to their individual roles in the regulation of apoptosis.

n-PKC Isoenzymes

It is unlikely that apoptosis is regulated only by the classical PKC isoenzymes. Apoptosis is not associated with changes in intracellular calcium in all cells (Beaver and Waring, 1994) and although PKC-α is expressed in a very wide range of cells, many cells do not express PKC-β or PKC-γ. We have used a cell free approach to identify PKC isoenzymes that are required for apoptosis in human neutrophils. Nuclei isolated from healthy, non-apoptotic neutrophils were combined with the cytosol of neutrophils at an early stage of apoptosis. DNA fragmentation occurred within 30 minutes and was inhibited if PKC-δ was removed from the cytosol by immunoprecipitation (Pongracz *et al.*, unpublished observations). Therefore, PKC-δ is implicated in the apoptotic process in neutrophils and studies from other laboratories are now also suggesting a principal role for this isoenzyme in the promotion of apoptosis.

PKC-delta

PKC-δ, like alpha, appears to be expressed in almost every cell tested (Hug and Sarre, 1993). Several lines of evidence suggest that activation of PKC-δ may inhibit cell growth or cell cycle progression and induce apoptosis. Overexpression of PKC-δ in Chinese hamster ovary cells and NIH 3T3 cells induced growth arrest in G2/M (Watanabe *et al.*, 1992). The marine compound Bistratene A also induced growth arrest in G2/M in HL60 cells (Griffiths *et al.*, 1996), which was followed by apoptosis (Lord *et al.*, 1995). Bistratene A also induced apoptosis in a Burkitts lymphoma cell line BM13674 (Song *et al.*, 1992). We have shown recently that Bistratene A activated PKC-δ specifically, using both in vitro and whole cell PKC assay methods (Griffiths *et al.*, 1996). These data suggest that activation of PKC-δ will cause cells to accumulate in G2/M and will eventually lead to apoptosis in haemopoietic cells.

The argument in favour of a role for PKC-δ in the induction of apoptosis has been supported most clearly by a series of experiments reported by Kufe and co-workers. Emoto *et al.* (Emoto *et al.*, 1995) originally purified and characterised a 40kDa protein kinase activity that was induced by a variety of agents which caused apoptosis in U937 cells, including Ara-C, TNF-α, Fas-ligation and UV-irradiation (Datta *et al.*, 1996). Sequencing of the purified kinase revealed it to be a cleaved form of PKC-δ, containing the catalytic domain. Further analysis showed that full length PKC-δ was cleaved at $DMQD_{330}N$ during apoptosis, to produce a catalytically active PKC. The finding that PKC-δ was cleaved at a site adjacent to aspartic acid suggested the potential involvement of aspartate-specific cysteine proteases (Caspases), which are known to be activated during apoptosis. CPP32 is one of the caspases known to be involved in the execution of the apoptotic programme, its substrates include the DNA repair enzyme polyADP-ribose polymerase. Ghayur *et al.* (Ghayur *et al.*, 1996) showed that a CPP32-like protease activity was responsible for the proteolytic activation of PKC-δ. Furthermore, transfection of U937 cells with the protease-generated fragment of PKC-δ was sufficient to induce apoptosis. Full length PKC-δ, or a kinase defective PKC-δ fragment, did not induce apoptosis. Finally, inhibition of apoptosis by overexpression of Bcl-2 or Bcl-XL was also associated with abrogation of PKC-δ cleavage (Emoto *et al.*, 1995). Data from another group has suggested recently that the effects of Fas-ligation and TNF-α on apoptosis and PKC-δ cleavage may be mediated by ceramide (Sawai *et al.*, 1997). Treatment of U937 and HPB-ALL cells with anti-Fas antibody, TNF-α or cell permeable ceramide analogues, raised cellular levels of ceramide and induced apoptosis. These agents also induced translocation of PKC-δ and ε from the membrane to the cytosol. As caspases are located within the cytosol, this translocation may well effect the proteolytic activation of PKC-δ and ε. Although these authors did not present the relevant data, in the discussion section of this paper they report that translocation of PKC-δ and PKC-ε was accompanied by their proteolytic cleavage (Sawai *et al.*, 1997). This latter observation is however in disagreement with published data, which have shown that the proteolytic activation of PKC in cells undergoing apoptosis was restricted to PKC-δ and did not include PKC-α, PKC-ε or PKC-ζ (Emoto *et al.*, 1996; Emoto *et al.*, 1995).

Whilst the majority of available data suggest a pro-apoptotic role for PKC-δ, there are some exceptions in the literature. Leszczynski has used a PKC-δ antisense oligonucleotide to reduce expression of this isoenzyme selectively in rat vascular smooth muscle cells (Leszczynski *et al.*, 1995). Loss of PKC-δ in these studies resulted in increased apoptosis, leading to the suggestion that down-regulation, rather than activation, of PKC-δ was a signal for apoptosis. These studies need to be repeated in a variety of cell types before any firm conclusions can be made, but at the present time this proposal would appear to be at odds with the rest of the literature, which suggests a positive role for PKC-δ in the promotion of apoptosis.

PKC-epsilon

PKC-ε is the only PKC isoenzyme that has been shown to have full oncogenic potential, its overexpression in NIH 3T3 cells produces a transformed phenotype (Mischak *et al.*, 1993). Despite this observation, the literature on the role of PKC-ε in the regulation of apoptosis is relatively sparse. Recent preliminary data have suggested that PKC-ε may function as an "early response" protein involved in entry of quiescent cells into cell cycle and DNA replication (Watson *et al.*, 1996). This proposal is supported by the data of Li *et al.* (Li *et al.*, 1996), which shows that PKC-ε is required for induction of c-myc and DNA replication in erythroleukaemia cells by erythropoietin. Mihalik *et al.* (Mihalik *et al.*, 1996) have also shown that low concentrations of TPA (0.5ng/ml) produced growth arrest of HT58 human lymphoblastic cells in G1, which was correlated with the downregulation of PKC-ε. In addition, PKC-ε is expressed at high levels in tumorigenic rat colon epithelial cells (Perletti *et al.*, 1996), neoplastic rat prostate cancer cells (Hrzenjak and Shain, 1995) and MCF-7 breast cancer cells (Manni *et al.*, 1996).

From the above data it is tempting to conclude that PKC-ε is a mitogenic isoenzyme, with a resulting inhibitory effect on apoptosis. However, an equal number of studies have produced evidence to the contrary and suggest that PKC-ε is pro-apoptotic in function. Kiss and Anderson (Kiss and Anderson, 1994) showed that carcinogens reduced the expression of PKC-ε in mouse embryo fibroblasts. The same group reported that the inhibition of apoptosis in prostatic carcinoma cells could be achieved by the activation and downregulation of PKC-ε (Rusnak and Lazo, 1996). Our own studies have shown that PKC-ε was significantly decreased in human colon cancer tissue (Pongracz *et al.*, 1995a). Finally, glucocorticoid-induced apoptosis in immature thymocytes is dependent upon PKC. Iwata *et al.*, have shown that PKC-ε was translocated from the cytosol to the particulate fraction in immature thymocytes in response to glucocorticoids, PKC-α and β were unaffected (Iwata *et al.*, 1994). If PKC-ε is indeed an early response protein, then the outcome of activation of this isoenzyme may well depend upon the genes that are subsequently transcribed.

a-PKC Isoenzymes

The atypical PKCs have only recently been implicated in the regulation of proliferation and apoptosis. It is likely that our incomplete understanding of the factors regulating these isoenzymes has restricted the study of their role in cell regulation. However, one signalling intermediate that was identified as an activator of PKC-ζ in vitro was ceramide (Muller *et al.*, 1995). Bearing in mind the weight of evidence suggesting a major role for ceramide in mediating signals for cell death inducing ligands, further investigation of the role of PKC-ζ and other a-PKCs is warranted.

PKC-zeta

Although PKC-ζ was shown to be activated by ceramide in in vitro studies, evidence from several sources suggest that this isoenzyme is in fact involved in mitogenesis and is generally anti-apoptotic. Berra *et al.* have demonstrated that PKC-ζ interacts with key elements of the mitogenic signalling cascade downstream of Ras. Transfection of Cos cells with a constitutively active mutant of PKC-ζ led to the activation of MAP kinase and MEK and a kinase-defective dominant negative mutant of PKC-ζ impaired the activation of these kinases by serum and TNF-α (Berra *et al.*, 1995). Moreover, the stimulation of stress-activated protein kinase (SAPK) was shown to be independent of PKC-ζ in these studies, giving further support to the proposal that this isoenzyme is not involved in signalling pathways leading to apoptosis.

Whilst PKC-ζ does not appear to be involved in the pro-apoptotic pathway, it is apparent that to initiate apoptosis, it may be necessary to inhibit signalling through PKC-ζ. The expression of PKC-ζ is markedly reduced in apoptotic U937 cells (Pongracz *et al.*, 1995b) and loss of this isoenzyme in rat vascular smooth muscle cells, following treatment with an anti-sense oligonucleotide, resulted in apoptosis (Leszczynski *et al.*, 1995). More recently, Diazmeco *et al.*, have shown that the product of the PAR-4 gene, which is induced during apoptosis in a variety of cells, interacts specifically with a-PKCs (Diazmeco *et al.*, 1996). Interaction of PAR-4 with the regulatory domains of a-PKCs, including PKC-ζ, dramatically inhibited their enzymatic activity. Therefore, in contrast to the data regarding the c-PKCs and n-PKCs, the studies that have considered the a-PKCs appear to unanimously suggest a proliferative and survival role for this group of isoenzymes.

PKC Substrates

In determining a role for PKC in the regulation of a specific cell function, such as apoptosis, it is clearly important to establish whether a particular biological effect is correlated with activation or downregulation of a specific isoenzyme. In addition, significant progress will only be made if we can define precisely how PKC isoenzymes mediate their effects on cell regulation. In this respect it is the identification of PKC substrates in whole cell studies that will increase our understanding of PKC in the future. PKC is known to phosphorylate a number of different substrates in vitro, which include other proteins involved in signal transduction (Berra *et al.*, 1995), proteins regulating DNA synthesis (Baudier *et al.*, 1992; Li *et al.*, 1992), DNA modifying enzymes (Sahyoun *et al.*, 1986; Bauer *et al.*, 1992) and proteins involved in cell cycle control (Goss *et al.*, 1994). The modification of any of these substrates could be expected to be involved in the execution of the apoptotic programme. For example, cell cycle arrest at the G1/S boundary induced by TNF-α, has been shown to involve dephosphorylation of Rb and to require the inactivation of PKC-α, possibly mediated by ceramide activated protein phosphatase (Lee *et al.*, 1996). In addition, it is clear that several of the PKC substrates that have been identified in whole cell assays are cytoskeletal elements (Jaken *et al.*, 1989; Murti *et al.*, 1992; Owen *et al.*, 1996). As the disassembly of the cells cytoskeletal structure is a feature of apoptosis, PKC may play a central role in this aspect of cell death. For example, PKC has been suggested to play a role in cell cycle control and studies by Fields and co-workers have identified PKC-β1(II) as a mitotic lamin kinase (Goss *et al.*, 1994). Moreover, phosphorylation of lamin B by PKC was shown to contribute to the disassembly of the nuclear lamina at mitosis. Whether phosphorylation of lamin proteins occurs during apoptosis has not been established, though the ultimate fate of these proteins is proteolytic degradation (Earnshaw, 1995). If a PKC isoenzyme is involved in lamina disassembly during apoptosis, it is unlikely to be PKC-β, as the majority of data support an anti-apoptotic role for this isoenzyme (see earlier). Our own preliminary data have shown an association of PKC-δ with nuclear lamins prior to apoptosis (Lord *et al.*, 1995). Bearing in mind the evidence in favour of a role of this isoenzyme in the induction of apoptosis, these observations now require further consideration.

Two tumour suppressor proteins that have been identified as PKC substrates are p53 and bcl-2. p53 is a tumour suppressor protein which is activated in response to DNA-damage, inducing a G1/S stage growth arrest (Donehawer and Bradley, 1993). p53 is phosphorylated at several sites in vivo and by various protein kinases in vitro, including PKC-α and β (Meek, 1994), leading to activation of the DNA-binding function of p53. However, studies in SV3T3 cells suggest that p53 and PKC do not actually interact in vivo to effect DNA-binding of p53 (Milne *et al.*, 1996). Whether PKC may affect other functions of p53 has yet to be determined. Bcl-2-alpha is serine phosphorylated during suppression of apoptosis in growth factor dependent-haemopoietic cells by interleukin-3, or the PKC activator bryostatin-1 (Mays *et al.*, 1994). PKC inhibitors prevent the hyperphosphorylation of bcl-2-alpha and the suppression of apoptosis (Mays *et al.*, 1994). Interestingly, these authors also reported that purified PKC was able to phosphorylate bcl-2-alpha in vitro in a calcium-dependent manner, suggesting that the regulation of bcl-2-alpha was mediated by c-PKCs . These examples illustrate the relevance of studies aiming to identify not only the PKC isoenzymes involved in regulation, but also their physiologically relevant substrates.

PKC, APOPTOSIS AND DISEASE

As stated at the beginning of this article, PKC has become of major interest in recent years as a target for therapeutic intervention in a range of different diseases.

Several of these studies have centred on the role of PKC in the regulation of apoptosis. The removal of a diseased cell by inducing apoptosis has many advantages, not least that apoptotic cells are efficiently removed by phagocytosis, preventing the release of potentially toxic cell contents. Whilst the emphasis in the literature is given to apoptosis and cancer, other disease states clearly include disregulation of apoptosis as a significant pathogenic factor (reviewed in Deacon *et al.*, 1997). These diseases include: disorders of the immune system such as chronic granulomatous disease, in which neutrophil (Coxon *et al.*, 1996) apoptosis is reduced; neurodegenerative disorders, including Alzheimer's disease, which involves increased apoptosis of neuronal cells (Jenner and Olanow, 1996); vascular disease (Hamet *et al.*, 1996); AIDS (Gougeon, 1996) and Ataxia Telangiectasia (Meyn, 1995) which are both associated with accelerated lymphocyte apoptosis.

Alterations in PKC expression or function have been established in several of these diseases, for example, in the case of Alzheimers disease, decreased expression of PKC-β (Chachin *et al.*, 1996), PKC-ε (Matsushima *et al.*, 1996) and -α and -ζ (Bergamaschi *et al.*, 1995; Govoni *et al.*, 1996; Greenwood *et al.*, 1996) have been reported. However, with the exception of cancer and Ataxia Telangiectasia, the connecting factor between changes in PKC and disease pathogenesis does not appear to involve apoptosis. For example, defective secretion of amyloid-β protein has now been identified as the function directly affected by the alteration in PKC-α in Alzheimers disease (Bergamaschi *et al.*, 1995). However, it is still likely that future studies that consider alterations in specific PKC isoenzymes in relation to disease states and apoptosis may reveal more positive connections. For example, chronic granulomatous disease (CGD) is associated with defective neutrophil apoptosis and activation, with the latter associated with reduced PKC activity (Curnutte *et al.*, 1994).

In the case of neoplasia, the role for PKC in disease progression is better defined, though relatively few studies have considered individual PKC isoenzymes. For this reason, and the vast nature of the general literature concerning PKC and cancer, we have chosen to highlight only one cancer in which the role for PKC isoenzymes has been well documented. Colorectal cancer (CRC) is the second most common malignancy occurring in the Western world. Incidence rates for this disease vary 20-fold world-wide, from 25–35 per 100 000 in North America and Western Europe, to 1–3 per 100 000 in India (Potter, 1995). This variation is just one indication that a combination of genetic (Fearon and Vogelstein, 1990) and environmental (Nagengast *et al.*, 1995) factors play an important role in the progression of CRC. Epidemiological data indicate that diets which are high in fat give an increased risk of CRC (Reddy and Wynder, 1977). Bile acids, which are elevated by such diets, are raised in patients with CRC (Reddy and Wynder, 1977). Bile acids have been shown to activate PKC in vitro, with the secondary bile acids showing selectivity in their activation of individual isoenzymes (Pongracz *et al.*, 1995a). PKC-β and ε were activated most effectively by the secondary bile acids deoxycholic acid, ursodeoxycholic acid and lithocholic acid. In contrast, PKC-α was only activated by primary bile acids. Importantly, it is the secondary bile acids which show the greatest increase in patients with CRC (Imray *et al.*, 1992).

Many studies have compared total PKC levels in normal and neoplastic colonic mucosa and have found consistently that levels are lower in cancerous tissue (Levy *et al.*, 1993; Doi *et al.*, 1994; Pongracz *et al.*, 1995a). Analyses of PKC isoenzymes at the level of mRNA (Doi *et al.*, 1994) and protein (Pongracz *et al.*, 1995a) have shown a decrease in PKC-β, leading to the suggestion that raised levels of secondary bile acids within the colon lead to persistent activation and ultimate downregulation of specific PKC isoenzymes, notably PKC-β (Pongracz *et al.*, 1995a). As this isoenzyme appears to play a central role in the co-ordinated regulation of proliferation and apoptosis (see previous sections), its reduction in the colonic epithelium may be a key factor leading to the disruption of the normal balance between cell proliferation and loss that has been noted for CRC (Sinicrope *et al.*, 1996).

CONCLUSIONS

It has not been possible in this review to cover all the literature concerning PKC and its involvement in apoptosis, though certain conclusions can be drawn from the published data at this time. Thus, we propose that PKC isoenzymes are differentially involved in the regulation of apoptosis and whilst there are likely to be variations between cell types, the literature at present suggests that PKC-δ is involved in the execution of the apoptotic programme. In contrast, PKC-α and -ζ are frequently associated with cell survival and suppression of apoptosis. The role of PKC-βI and -βII remains to be clarified, though both are intimately associated with signalling pathways that modulate cell proliferation and apoptosis.

Clearly the role of PKC in apoptosis and disease will only be fully understood when individual family members are considered. Access to specific inhibitors and activators of PKC isoenzymes will also improve our understanding of PKC isoenzyme function and may additionally uncover therapeutically useful compounds (Ishii *et al.*, 1996). In addition, the increased use of specific antisense oligonucleotides and constitutively active PKC isoenzyme mutants, mentioned briefly in this review, has already begun to increase understanding of PKC isoenzyme function. Studies of PKC function such as these, will allow informed decisions to be made regarding the suitability of PKC as a target for drug therapy.

ACKNOWLEDGEMENTS

EMD, JP and LM are supported by grants from the Leukaemia Research Fund, BBSRC and MAFF, respectively. HC is supported by the United Birmingham Hospitals Endowment Fund. GG holds an MRC PhD scholarship and JML is a Royal Society University Research Fellow.

REFERENCES

Barr, L.F., Campbell, S.E. and Baylin, S.B. (1997) Anti-protein kinase C β2 inhibits cycling and decreases c-myc-induced apoptosis in small lung carcinoma cells. *Cell Growth Differ.*, **8**, 381–392.

Barry, S.T. and Critchley, D.R. (1994) The Rho A-dependent assembly of focal adhesions in Swiss 3T3 cells is associated with tyrosine phosphorylation and the recruitment of both PP125 FAK and protein kinase C-delta to focal adhesions. *J. Cell Sci.*, **107**, 2033–2045.

Baudier, J., Delphin, C., Grunwald, D., Knochbin, S. and Lawrence, J.J. (1992) Character isolation of the tumour suppressor protein p53 as a protein kinase C substrate and a s100b-binding protein. *Proc. Natl. Acad. Sci. USA*, **89,** 11627–11631.

Bauer, P.I., Farkas, G., Buday, L., Mikala, G., Meszaros, G., Kun, E. and Farago, A. (1992) Inhibition of DNA binding by the phosphorylation of poly ADP-ribose polymerase protein catalysed by protein kinase C. *Biochem. Biophys. Res. Comm.*, **187**, 730–736.

Beaver, J.P. and Waring, P. (1994) Lack of correlation between early intracellular calcium ion rises and the onset of apoptosis in thymocytes. *Immunol. Cell. Biol.*, **72**, 489–499.

Bergamaschi, S., Binetti, G., Govoni, S., Wetsel, W.C., Battaini, F., Trabucchi, M., Bianchetti, A. and Racchi, M. (1995) Defective phorbol ester-stimulated secretion of beta-amyloid precusor protein from Alzheimers-disease fibroblasts. *Neuroscience Lett.*, **201**, 1–4.

Berra, E., Diaz-Meco, M.T., Lozano, J., Frutos, S., Municio, M.M., Sanchez, P., Sanz, L. and Moscat, J. (1995) Evidence for a role of MEK and MAPK during signal transduction by protein kinase C-zeta. *EMBO J.*, **14,** 6157–6163.

Borner, C., Filipuzzi, I., Wartmann, M., Eppenberger, U. and Fabbro, D. (1989) Biosynthesis and post-translational modifications of protein kinase C in human breast cancer cells. *J. Biol. Chem.*, **264**, 13902–13909.

Bursch, W., Oberhammer, F. and Schulte-Hermann, R. (1992) Cell death by apoptosis and its protective role against disease. *Trends Pharmacol. Sci.*, **13**, 245–251.

Castagna, M., Takai, Y., Kaibuchi, K., Sano, K., Kikkawa, U. and Nishizuka, Y. (1982) Direct activation of calcium activated, phospholipid dependent protein kinase by tumour promoting phorbol esters. *J. Biol. Chem.*, **257**, 7847–7851.

Cazoubon, S., Borrancin, F. and Parker, P.J. (1994) Threonine 497 is a critical site for permissive activation of protein kinase C-alpha. *Biochem. J.*, **301**, 443–448.

Chachin, M., Shimohama, S., Kunugi, Y.U. and Taniguchi, T. (1996) Assessment of protein kinase C messenger-RNA levels in Alzheimers disease brains. *Jpn. J. Pharmacol.*, 71, 175–177.

Chapline, C., Ramsay, K., Klauck, T. and Jaken, S. (1993) Interaction cloning of protein kinase C substrates. *J. Biol. Chem.*, **268**, 6858–6861.

Clemens, M.J., Trayner, I. and Menaya, J. (1992) The role of protein kinase C isoenzymes in the regulation of cell proliferation and differentiation. *J. Cell. Sci.*, **103**, 881–887.

Coxon, A., Barkalow, F.J., Askari, S., Rieu, P., Sharpe, A.H., Vonandrian, U., Arnaout, M.A. and Mayadas, T.N. (1996) A novel role for the beta(2) integrin CD11B/CD18 in neutrophil apotosis. *J. Leukocyte Biol.*, SS, 307.

Curnutte, J.T., Erickson, R.W., Ding, J.B. and Badwey, J.A. (1994) Reciprocal interactions between protein kinase C and components of the NADPH oxidase complex may regulate superoxide production by neutrophils stimulated with phorbol ester. *J. Biol. Chem.*, **269**, 10813–10819.

Datta, R., Banach, D., Kojuma, H., Talanian, R.V., Alnemri, E.S., Wong, W.W. and Kufe, D.W. (1996) Activation of the CPP32 protease in apoptosis induced by 1-β-D-arabinofuranosylcytosine and other DNA-damaging agents. *Blood*, **88**, 1936–1943.

Deacon, E.M., Pongracz, J., Griffiths, G. and Lord, J.M. (1997) Protein kinase C isoenzymes: differential involvement in apoptosis and pathogensis. *J. Clin. Pathol: Mol. Pathol.*, In Press.

Dekker, L.V. and Parker, P.J. (1994) Protein kinase C – a question of specificity. *Trends Biochem. Sci.*, **19**, 73–77.

Devente, J.E., Kukoly, C.A., Bryant, W.O., Posekany, K.Y., Chen, J.M., Fletcher, D.J., Parker, P.J., Pettit, G.J., Lozano, G., Cook, P.P. and Ways, D.K. (1995) Phorbol esters induce death in MCF-7 breast cancer cells with altered expression of protein kinase C isoforms – role for p53-independent induction of GADD45 in initiating death. *J. Clin. Invest.*, **96**, 1874–1886.

Diazmeco, M.T., Municio, M.M., Frutos, S., Sanchez, P., Lozano, J., Sanz, L. and Moscat, J. (1996) The product of PAR-4, a gene induced during apoptosis, interacts selectively with the atypical isoforms of protein-kinase C. *Cell*, **86**, 777–786.

Divecha, N., Banfic, H. and Irvine, R.F. (1991) The polyphosphoinositide cycle exists in the nuclei of Swiss 3T3 cells under the control of a receptor (for IGF-1) in the plasma-membrane, and stimulation of the cycle increases nuclear diacylglycerol and apparently induces translocation of protein kinase-C to the nucleus. *EMBO J.*, **10**, 3207–3214.

Dobrowsky, R.T. and Hannun, Y.A. (1992) Ceramide stimulates a cytosolic protein phosphatase-activity. *J. Biol. Chem.*, **267**, 5048–5051.

Doi, S., Goldstein, D., Hug, H. and Weinstein, I.B. (1994) Expression of multiple isoforms of protein kinase C in normal human colon mucosa and colon tumours and decreased levels of protein kinase C β1 and ηmRNAs in the tumours. *Mol. Carcinogenesis*, **11**, 197–203.

Donehawer, L.A. and Bradley, A. (1993) The tumour suppressor p53. *Biochem. Biophys. Acta.*, **1155**, 181–205.

Dong, L., Chapline, C., Mousseau, B., Fowler, L., Ramsay, K., Stevens, J.L. and Jaken, S. (1995) 35H, a sequence isolated as a protein kinase C binding protein, is a novel member of the adducin family. *J. Biol. Chem.*, **270**, 25534–25540.

Dransfield, D.T., Bradford, A.J., Smith, J., Martin, M., Roy, C., Mangeat, P.M. and Goldenring, J.R. (1997) Ezrin is a cyclic AMP dependent protein kinase anchoring protein. *EMBO J.*, **16**, 35–43.

Dressler, K.A., Mathias, S. and Kolesnick, R.N. (1992) Tumour necrosis factor-alpha activates the sphingomyelin signal transduction pathway in a cell free system. *Science*, **255**, 1715–1718.

Dutil, E.M., Keranen, L.M., Paoli-Roach, A.A.D. and Newton, A.C. (1994) In vivo regulation of protein kinase C by trans-phosphorylation followed by autophosphorylation. *J. Biol. Chem.*, **269**, 29359–29362.

Earnshaw, W.C. (1995) Nuclear changes in apoptosis. *Current Opinion Cell Biol.*, 7, 337–343.

Emoto, Y., Kisaki, H., Manome, Y., Khardanda, S. and Kufe, D. (1996) Activation of protein kinase C-delta in human myeloid leukemia cells treated with 1 beta-D arabinofuranosylcytosine. *Blood*, **87**, 1990–1996.

Emoto, Y., Manome, Y., Meinhardt, G., Kisaki, H., Kharbanda, S., Robertson, M., Ghayur, T., Wong, W.W., Kamen, R., Weichselbaum, R. and Kufe, D. (1995) Proteolytic activation of protein kinase C δ by an ICE-like protease in apoptotic cells. *EMBO J.*, **14**, 6148–6156.

Evan, G.I., Wyllie, A.H., Gilbert, C.S., Littlewood, T.D., Land, H., Brooks, M., Waters, C.M., Penn, L.Z. and Hancock, D.C. (1992) Induction of apoptosis in fibroblasts by c-myc protein. *Cell*, **89**, 119–128.

Evans, C.A., Lord, J.M., Owen-Lynch, P.J., Johnson, G., Dive, C. and Whetton, A.D. (1995) Suppression of apotosis by v-ABL protein-tyrosine kinase is associated with nuclear translocation and activation of protein-kinase-C in an interleukin-3-dependent hematopoietic cell line. *J. Cell Sci.*, **108**, 2591–2598.

Evans, C.A., Owen-Lynch, P.J., Whetton, A.D. and Dive, C. (1993) Activation of the abelson tyrosine kinase activity is associated with suppression of apoptosis in haemopoietic cells. *Cancer Res.*, **53**, 1735–1738.

Fearon, E.R. and Vogelstein, B. (1990) A genetic model for colorectal tumorigenesis. *Cell*, **61**, 759–767.

Fujise, A., Mizuno, K., Ueda, Y., Osada, S., Hirai, S., Takayanui, A., Shimuzu, N., Owada, M.K., Nakajima, H. and Ohno, S. (1994) Specificity of the high affinity interaction of protein kinase C with a physiological substrate, myristoylated alanine-rich protein kinase C substrate. *J. Biol. Chem.*, **269**, 31642–31648.

Gescher, A. (1992) Towards selective pharmacological modulation of protein-kinase C – opportunities for the development of novel antineoplastic agents. *British J. Cancer*, **66**, 10–19.

Ghayur, T., Hugunin, M., Talanian, R.V., Ratnofsky, S., Quinlav, C., Emoto, Y., Pandey, P., Datta, R., Huang, Y.Y., Kharbanda, S., Allen, H., Kamen, R., Wong, W. and Kufe, D. (1996) Proteolytic activation of protein kinase C delta by an ICE/CED 3-like protease induces characteristics of apoptosis. *J. Exp. Med.*, **184**, 2399–2404.

Goodnight, J., Mishak, H. and Mushinski, J.F. (1994) Selective involvement of protein kinase C in differentiation and neoplastic transformation. *Adv. Cancer Res.*, **64**, 159–209.

Goss, V.L., Hocevar, B.A., Thompson, L.J., Stratton, C.A., Burns, D.J. and Fields, A.P. (1994) Identification of nuclear bII protein kinase C as a mitotic lamin kinase. *J. Biol. Chem.*, **269**, 190719080.

Gougeon, M.L. (1996) Apoptosis in AIDS – genetic control and relevance for AIDS pathogenesis. *Biochem. Soc. Trans.*, **24**, 1055–1058.

Govoni, S., Racchi, M., Bergamaschi, S., Trabucchi, M., Battaini, F., Bianchetti, A. and Binetti, G. (1996) Defective protein kinase C-alpha leads to impaired secretion of soluble beta-amyloid precursor protein from Alzheimers-disease fibroblasts. *Ann. N. Y. Acad. Sci.*, **777**, 332–337.

Grant, S., Turner, A.J., Freemerman, A.J., Wang, Z.L., Kramer, L. and Jarvis, W.D. (1996) Modulation of protein kinase C activity and calcium-sensitive isoform expression in human myeloid leukemia cells by bryostatin-1 - relationship to differentiation and Ara-C induced apoptosis. *Exp. cell. res.*, **228**, 65–75.

Gray, P.C., Tibbs, V.C., Cotterall, W.A. and Murphy, B.J. (1997) Identification of a 15kDa cAMP-dependent protein kinase-anchoring protein associated with skeletal muscle L-type calcium channels. *J. Biol. Chem.*, **272**, 6297–6302.

Greenwood, A.F., Powers, R.E. and Jope, R.S. (1996) Phosphoinositide hydrolysis, G-alpha-Q, phospholipase C, and protein kinase C in post-mortem interval, subject age, and Alzheimers disease. *Neuroscience*, **69**, 125–138.

Griffiths, G., Garrone, B., Deacon, E., Owen, P., Pongracz, J., Mead, G., Bradwell, A., Watters, D. and Lord, J. (1996) The polyether bistratene A activates protein kinase C-delta and induces growth arrest in HL60 cells. *Biochem. Biophys. Res. Comm.*, **222**, 802–808.

Grigg, J.M., Savill, J.S., Sarraf, C., Haslett, C. and Silverman, M. (1991) Neutrophil apoptosis and clearance from neonatal lungs. *Lancet*, **338**, 720–722.

Groux, H., Torpier, G., Monte, D., Mouton, Y., Capron, A. and Ameisen, J.C. (1992) Activation-induced death by apoptosis in CD4+ T cells from human-immunodeficiency-virus infected asymptomatic individuals. *J. Exp. Med.*, **175**, 331–340.

Gschwendt, M., Kielbassa, K., Kittstein, W. and Marks, F. (1994) PKC substrates: differential phosphorylation by PKC isoenzymes. *J. Cell. Biochem.*, 18, 83D.

Ha, K.S. and Exton, J.H. (1993) Differential translocation of protein kinase C isoenzymes by thrombin and platelet-derived growth factor. A possible function for phosphatidylcholine-derived diacylglycerol. *J. Biol. Chem.*, **268**, 10534–10539.

Haimovitz-Friedman, A., Balaban, N., McLoughlin, M., Ehleiter, D., Michaeli, J., Vlodavsky, I. and Fuks, Z. (1994) Protein kinase C mediates basic fibroblast growth factor protection of endothelial cells against radiation-induced apoptosis. *Cancer Res.*, **54**, 2591–2597.

Hocevar, B.A. and Fields, A.P. (1991) Selective translocation of bII PKC to the nucleus of human promyelocytic (HL60) leukemia cells. *J. Biol. Chem.*, **266**, 28–33.

Hrzenjak, M. and Shain, S.A. (1995) Protein-kinase-C-dependent and protein-kinase-C-independent pathways of signal-transduction in prostate cancer cells – fibroblast growth-factor utilization of a protein kinase C-independent pathway. *Cell Growth Diff.*, **6**, 1129–1142.

Huang, K.P. and Huang, F.L. (1986) Conversion of protein kinase-C from a Ca2+-dependent to an independent form of phorbol ester binding protein by digestion with trypsin. *Biochem. Biophys. Res. Comm.*, **139**, 320–326.

Hug, H. and Sarre, T.F. (1993) Protein kinase C isoenzymes: divergence in signal transduction? *Biochem. J.*, **291**, 329–343.

Hyatt, S.L., Liao, L., Aderem, A., Naim, A. and Jaken, S. (1994) Correlation between protein kinase C binding proteins and substrates in REF52 cells. *Cell Growth Differ*, **5**, 495–502.

Imray, C.H.E., Radley, S., Davis, A., Barker, G., Hendrickse, C.W., Donovan, I.A., Lawson, A.M., Baker, P.R. and Neoptolemos, J.P. (1992) Fecal unconjugated bile acids in patients with colorectal cancer or polyps. *Gut*, 33, 1239–1245.

Iseki, R., Mukai, M. and Iwata, M. (1991) Regulation of lymphocyte -T apoptosis – signals for the antagonism between activation-induced and glucocorticoid-induced death. *J. Immunol.*, **147**, 4286–4292.

Ishii, H., Jurousek, M.R., Koya, D., Takagi, C., Xia, P., Clermont, A., Bursell, S.E., Kern, T.S., Ballas, L.M., Heath, W.F., Stramm, L.E., Feener, E.P. and King, G.L. (1996) Amelioration of vascular dysfunction in diabetic rats by an oral PKC-beta inhibitor. *Science*, **272**, 728–731.

Iwata, M., Iseki, R., Sato, K., Tozawa, Y. and Ohoka, Y. (1994) Involvement of protein kinase C-εpsilon in glucocorticoid-induced apoptosis in thymocytes. *Int. Immunol*, **6**, 431–438.

Jaken, S. (1996) Protein kinase C isozymes and substrates. *Current Opinion Cell Biol.*, **8**, 168–173.

Jaken, S., Leach, K. and Klauck, T. (1989) Association of type 3 protein kinase C with focal contacts in rat embryo/fibroblasts. *J. Cell Biol.*, **109**, 697–704.

Jenner, P. and Olanow, C.W. (1996) Oxidative stress and the pathogenesis of Parkinsons-disease. *Neurology*, 47, S161–S170.

Jones, M.J. and Murray, A.W. (1995) Evidence that ceramide selectively inhibits protein kinase C alpha translocation and modulates bradykinin activation of phospholipase D. *J. Biol. Chem.*, **270**, 5007–5013.

Jun, C.D., Park, S.J., Choi, B.M., Kwak, H.J., Park, Y.C., Kim, M.S., Park, R.K. and Ching, H.T. (1997) Potentiation of the activity of nitric oxide by the protein kinase C activator phorbol ester in human myeloid leukemia HL60 cells: association with enhanced fragmentation of mature genomic DNA. *Cell Immunol.*, **176**, 41–49.

Kerr, J.F.R., Harmon, B. and Searle, J. (1974) An electron-microscope study of cell deletion in the anuran tadpole tail during spontaneous meta-morphosis with special reference to apoptosis of striated muscle fibres. *J. Cell. Sci.*, **14**, 571–585.

Kerr, J.F.R., Searle, J., Harmon, B.V. and Bishop, C.J. (1987) Apoptosis. In *Perspectives on mammalian cell death*, edited by C.S.Potten, pp. 93–128. Oxford: Oxford Science Publications.

Kerr, J.F.R., Winterford, C.M. and Harmon, B.V. (1994) Apoptosis – its significance in cancer and cancer therapy. *Cancer*, **73**, 2013-2026.

Kerr, J.F.R., Wyllie, A.H. and Currie, A.R. (1972) Apoptosis: a basic biological phenomenon with wide ranging implications in tissue kinetics. *Br. J. Cancer*, **26**, 239-257.

Kikkawa, U., Ogita, K., Shearman, M.S., Ase, K., Sekiguchi, K., Naor, Z., Kishimoto, A., Nishizuka, Y., Saito, N., Tanaka, C., Ono, Y., Fujii, T. and Igarashi, K. (1988) The family of protein kinase-C - its molecular heterogeneity and differential expression. *Cold Spring Harbor Symp. Quant. Biol.*, **53**, 97-102.

Kiley, S., Jaken, S., Whelan, R. and Parker, P.J. (1995) Intracellular targetting of protein kinase C isoenzymes – functional implications. *Biochem. Soc. Trans.*, **23**, 601–605.

Kiley, S.C. and Parker, P.J. (1995) Differential localisation of protein kinase C isozymes in U937 cells: evidence for distinct isozyme functions during monocyte differentiation. *J. Cell. Sci.*, **108**, 1003–1016.

Kishimoto, A., Kajikawa, N., Shiota, M. and Nishizuka, Y. (1983) Proteolytic activation of calcium activated, phospholipid-dependent neutral protease. *J. Biol. Chem.*, **258**, 1156–1164.

Kishimoto, A., Mikawa, I.L., Hashimoto, K., Yasuda, I. and Tanaka, S.I. (1989) Limited proteolysis of protein kinase C subspecies by calcium dependent neutral protease (calpain). *J. Biol. Chem.*, **204**, 4088–4092.

Kiss, Z. and Anderson, W.H. (1994) Selective down-regulation of protein kinase C-epsilon by carcinogens does not prevent stimulation of phospholipase D by phorbol ester and platelet-derived growth-factor. *Biochem. J.*, **300**, 751–756.

Knox, K.A., Johnson, G.D. and Gordon, J. (1993) A study of protein kinase C isoenzyme ditribution in relation to Bcl-2 expression during apoptosis of epithelial cells in vivo. *Exp. Cell Res.*, **207**, 68–73.

Kochs, G., Hummel, R., Fiebich, B., Sarre, T.F., Marme, D. and Hug, H. (1993) Activation of purified human PKC alpha and beta-1 isoenzymes in vitro by Ca2+, phosphatidyl inositol and phosphatidyl inositol 4,5-biphosphate. *Biochem. J.*, **291**, 627–633.

Leach, K.L., Ruff, V.A., Jarpe, M.B., Adams, L.D., Fabbro, D. and Raben, D.M. (1992) Alpha-thrombin stimulates nuclear diacylglyceride levels and differential nuclear localization of protein kinase C isozymes in llC9 cells. *J. Biol. Chem.*, **267**, 21816–21822.

Lee, J.Y., Hannun, Y.A. and Obeid, L.M. (1996) Ceramide inactivates cellular protein kinase C-alpha. *J. Biol. Chem.*, 271, 13169–13174.

Lee, M.H. and Bell, R.M. (1991) Mechanism of protein-kinase-C activation by phosphatidylinositol 4, 5-biphosphate. *Biochemistry*, **30**, 1041–1049.

Lehel, C., Olah, Z., Jakob, G., Szallasi, Z., Petrovics, G., Harta, G., Blumberg, P.M. and Anderson, W.B. (1995) Protein kinase C-ε subcellular localisation domains and proteolytic degradation sites. A model for protein kinase C conformational changes. *J. Biol. Chem.*, **270**, 19651–19658.

Leszczynski, D., Joen-Vaara, S. and Foegh, M.L. (1996) Protein kinase C-alpha regulates proliferation but not apoptosis in rat coronary vascular smooth muscle cells. *Life Sci.*, **58**, 599–606.

Leszczynski, D., Joen-Vaara, S. and Foegh, M.L. (1995) Apoptosis of rat vascular smooth muscle cells is regulated by δ and ζ but not by α and ε isozymes of protein kinase C. *Mol. Biol. Cell*, 6 (suppl), 246a.

Levy, M.F., Pocsidio, J., Guillem, J.G., Forde, K., LoGerfo, P. and Weinstein, I.B. (1993) Decreased levels of protein kinase C enzyme activity and protein kinase C mRNA in primary colon tumours. *Dis. Colon Rectum*, **36**, 913–921.

Li, L., Zhou, J., James, G., Heller-Harrison, R., Czech, M.P. and Olson, E.N. (1992) FGF inactivates helix-loop-helix proteins through phosphorylation of a conserved protein kinase C site in their DNA binding domains. *Cell*, **71**, 1181–1194.

Li, S.F., Janosch, P., Tanji, M., Rosenfeld, G.C., Waymire, J.C., Mischak, H., Kolch, W. and Sedivy, J.M. (1995) Regulation of Raf-1 kinase activity by the 14-3-3 family of proteins. *EMBO J.*, **14**, 685–696.

Li, Y.K., Davis, K.L. and Sytkowski, A.J. (1996) Protein-kinase C-epsilon is necessary for erythropoietins up-regulation of c-myc and for factor dependent DNA synthesis evidence for discrete signals for growth and differentiation. *J. Biol. Chem.*, **271**, 27025–27030.

Liao, L., Hyatt, S.L., Chapline, C. and Jaken, S. (1994a) Protein kinase C domains involved in interaction with other proteins. *Biochemistry*, **33**, 1229–1233.

Liao, L., Ramsay, K. and Jaken, S. (1994b) Protein kinase C isozymes in progressively transformed rat embryo fibroblasts. *Cell Growth Differ.*, **5**, 1185–1194.

Lord, J.M. and Ashcroft, S.J.H. (1984) Identification and characterization of Ca2+-phospholipid-dependent protein kinase in rat islets and hamster β-cells. *Biochem. J.*, **219**, 547–551.

Lord, J.M., Garrone, B., Griffiths, G. and Watters, D. (1995) Apoptosis induced by Bistratene A in HL60 cells involves the selective activation of PKC-delta and its association of nuclear lamins and DNA replication sites. *J. Cell Biochem.*, **S19B**, 275.

Lord, J.M. and Pongracz, J. (1995) Protein kinase C: a family of isoenzymes with distinct roles in pathogenesis. *J. Clin. Pathol. Mol. Pathol.*, **48**, M57–M64.

Lotem, J., Cragoe, E.J. and Sachs, L. (1991) Rescue from programmed death in leukaemic and normal myeloid cells. *Blood*, **78**, 953–960.

Lozano, J., Berra, E., Municio, M.M., Diaz-Meco, M.F., Dominguez, I., Sanz, L. and Muscat, J. (1994) Protein kinase C-ζ is critical for NF-κB-dependent promotor activation by sphingomyelinase. *J. Biol. Chem.*, **269**, 19200–19202.

Lucas, M. and Sanchez-Margalet, V. (1995) Protein kinase C involvement in apoptosis. *Gen. Pharmacol.*, **26**, 881–887.

Macfarlane, D.E. and Manzel, L. (1994) Activation of the beta-isozyme of protein kinase C is necessary and sufficient for phorbol ester induced differentiation of HL60 promyelocytes. *J. Biol. Chem.*, **269**, 4327.

Manni, A., Buckwalter, E., Etindi, R., Kunselman, S., Rossini, A., Mauger, D., Dabbs, D. and Demers, L. (1996) Induction of a less aggressive breast-cancer phenotype by protein kinase C-alpha and C beta overexpression. *Cell Growth Diff.*, **7**, 1187–1198.

Martin, J.B. (1982) Huntingdons disease: genetically programmed cell death in the human central nervous system. *Nature*, **299**, 205–208.

Matsushima, H., Shimohama, S., Chachin, M., Taniguchi, T. and Kimura, J. (1996) Ca2+-dependent and Ca2+-independent protein kinase C changes in the brains of patients with Alzheimers-disease. *J. Neurochem.*, **67**, 317–323.

Mays, W.S., Tyler, P.G., Ito, T., Armstrong, D.K., Qatsha, K.A. and Davidson, N.E. (1994) Interleukin-3 and bryostatin 1 mediate hyperphosphorylation of bcl-2 alpha in association with suppression of apoptosis. *J. Biol. Chem.*, **269**, 26865–26870.

McCartney, S., Little, B.M., Landeberg, L.K. and Scott, J.D. (1995) Cloning and characterisation of A-kinase anchor protein-100 (AKAP-100) – a protein that targets A-kinase to the sarcoplasmic reticulum. *J. Biol. Chem.*, **270**, 9327–9333.

Meek, D.W. (1994) Post-translational modification of p53. *Seminars Cancer Biol*, 5, 203–210.

Meyn, M.S. (1995) Ataxia-Telangiectasia and cellular response to DNA damage. *Cancer Res.*, **55**, 5991–6001.

Mihalik, R., Farkas, G., Kopper, L., Benczur, M. and Farago, A. (1996) Possible involvement of protein kinase C-epsilon in phorbol ester-induced growth inhibition of human lymphoblastic cells. *Int. J. Biochem. Cell Biol.*, **28**, 925–933.

Milne, D.M., McKendrick, L., Jardine, L.J., Deacon, E., Lord, J.M. and Meek, D.W. (1996) Murine p53 is phosphorylated within the Pab421 epitope by protein kinase C in vitro, but not in vitro, even after stimulation with the phorbol ester 12-O-tetradecanoylphorbol 13 acetate. *Oncogene*, **13**, 205–211.

Mischak, H., Goodnight, J., Kolch, W., Martinybaron, G., Schaechtle, C., Kazanietz, M.G., Blumberg, P.M., Pierce, J.H. and Mushinski, J.F. (1993) Overexpression of protein kinase-C-delta and kinase-epsilon in NIH 3T3 cells induces opposite effects on growth, morphology, anchorage dependence and tumorigenicity. *J. Biol. Chem.*, **268**, 6090–6096.

Moore, N.C., Jenkinson, E.J. and Owen, J.J. (1992) Effects of the thymic microenvironment on the response of thymocytes to stimulation. *Eur. J. Immunol.*, **22**, 2533–2537.

Muller, G., Ayoub, M., Storz, P., Rennecke, J., Fabbro, D. and Pfizenmaier, K. (1995) PKC ζ is a molecular switch in signal transduction of TNF-α, bifunctionally regulated by ceramide and arachidonic acid. *EMBO J.*, **14**, 1961–1969.

Murti, K.G., Kaur, K. and Goorha, R.M. (1992) Protein kinase C associates with intermediate filaments and stress fibres. *Exp. Cell Res.*, 202, 36–44.

Nagengast, F.M., Grubben, M.J.A.L. and Munster, I.P.v. (1995) Role of bile acids in colorectal carcinogenesis. *Eur. J. Cancer*, **31A**, 1067–1070.

Nakanishi, H., Brewer, K.A. and Exton, J.H. (1993). Activation of the ζ-isozyme of protein kinase C by phospatidylinositol 3, 4, 5-triphosphate. *J. Biol. Chem.*, **268**, 13–16.

Nauert, J.B., Klauck, T.M., Langeburg, L.K. and Scott, J.D. (1997) Gravin, an autoantigen recognized by serum from myasthenia gravis patients, is a kinase scaffold protein. *Current Biol.*, **7**, 52–62.

Newton, A.C. (1995) Protein kinase C: Structure, function, and regulation. *J. Biol. Chem.*, **270**, 28495–28498.

Nishizuka, Y. (1992) Intracellular signalling by hydrolysis of phospholipids and activation of protein kinase *C. Science*, **258**, 607–614.

Olivier, A.R. and Parker, P.J. (1994) Bombesin, platelet-derived growth factor and diacylglycerol induce selective membrane association and downregulation of protein kinase C isotypes in Swiss 3T3 cells. *J. Biol. Chem.*, **269**, 2758–2763.

Orr, J.W. and Newton, A.C. (1994) Requirement for negative charge on activation loop of protein kinase C. *J. Biol. Chem.*, **269**, 27715–27718.

Osada, S., Mizuno, K., Saido, T.C., Akita, Y., Suzuki, K., Kuroki, T. and Ohno, S. (1990) A phorbol ester receptor protein kinase, nPKC-eta, a new member of the protein kinase C family predominantly expressed in lung and skin. *J. Biol. Chem.*, **265**, 22434–22440.

Osada, S., Muzuno, K., Saido, T.C., Suzuki, K., Kuroki, T. and Ohno, S. (1992) A new member of the protein kinase C family, nPKCθ, predominately expressed in skeletal muscle. *Mol. Cell Biol.*, **12**, 3930–3938.

Owen, P.J., Johnson, G.D. and Lord, J.M. (1996) Protein kinase C-delta associates with vimentin intermediate filaments in differentiated HL60 cells. *Exp. Cell Res.*, **225**, 366–373.

Palmer, R.H., Dekker, L.V., Woschalski, R., Good, J.A.L., Gigg, R. and Parker, P.J. (1995) Activation of PRK1 by phosphatidylinositol 3,4,5-trisphosphate – a comparison with protein kinase C isotypes. *J. Biol. Chem.*, **270**, 22412–22416.

Parker, P.J., Kour, G., Marais, R.M., Mitchell, F., Pears, C., Schaap, D. *et al.* (1989) Protein kinase C - a family affair. *Mol. Cell Endocrinol.*, 65, 1–11.

Pears, C., Schaap, D. and Parker, P.J. (1991) The regulatory domain of protein kinase C-ε restricts the catalytic domain specificity. *Biochem. J.*, **276**, 257–260.

Perletti, G.P., Foloni, M., Lin, H.C., Mischak, H., Piccinini, F. and Tashjian, A.H. (1996) Overexpression of protein kinase C-epsilon is oncogenic in rat colonic epithelial cells. *Oncogene*, **12**, 847–854.

Peter, M., Nakagawa, J., Dance, M., Labbe, J.C. and Nigg, E.A. (1990) In vitro disassembly of the nuclear lamina and M phase specific phosphorylation of lamins by cdc 2 kinase. *Cell*, **61**, 591–602.

Pettit, T.R., Martin, A., Horton, T., Liossis, C., Lord, J.M. and Wakelam, M.J.O. (1997) Diacylglycerol and phosphatidate generated by phospholipases C and D respectively have distinct fatty acid compositions: phospholipase D-derived diacylglycerol does not activate protein kinase C in PAE cells. *J. Biol. Chem.*, In Press.

Pettit, T.R., Zaqqa, M. and Wakelam, M.J.O. (1994) Epidermal growth factor stimulates distinct diacylglcerol species generation in Swiss 3T3 fibroblasts – evidence for a potential phosphatidylcholine-specific phospholipase C catalyzed pathway. *Biochem. J.*, **298**, 655–660.

Pongracz, J., Clark, P., Neoptolemous, J.P. and Lord, J.M. (1995a) Expression of protein kinase C isozymes in colorectal cancer tissue and their differential activation by different bile acids. *Int. J. Cancer*, **61**, 35–39.

Pongracz, J., Deacon, E.M., Johnson, G.D., Burnett, D. and Lord, J.M. (1996) Doppa induces cell death but not differentiation of U937 cells: Evidence for the involvement of PKC-beta-1 in the regulation of apoptosis. *Leukemia Res.*, **20**, 319–326.

Pongracz, J., Johnson, G.D., Crocker, J., Burnett, D. and Lord, J.M. (1994) The role of protein kinase C in myeloid cell apoptosis. *Biochem. Soc. Trans.*, **22**, 593–597.

Pongracz, J., Tuffley, W., Johnson, G.D., Deacon, E.M., Burnett, D., Stockley, R.A. and Lord, J.M. (1995b) Changes in protein kinase C isoenzyme expression associated with apoptosis in U937 myelomonocytic cells. *Exp. Cell Res.*, **218**, 430–438.

Potter, J.D. (1995) Risk factors for colon neoplasia – epidemiology and biology. *Eur. J. Cancer*, **31A**, 1033–1038.

Powell, C.T., Brittis, N.J., Stec, D., Hug, H., Heston, W.D.W. and Fair, W.R. (1996) Persistent membrane translocation of protein kinase C-alpha during 12-0-tetradecanoylphorbol-13-acetate-induced apoptosis of LNCAP human prostate cancer cells. *Cell Growth Diff.*, **7**, 419–428.

Radford, I.R. (1994) Phorbol esters can protect mouse pre-T cell lines from radiation induced rapid interphase apoptosis. *Int. J. Radiat. Biol.*, **65**, 345–355.

Reddy, B.S. and Wynder, E.L. (1977) Metabolic epidemiology of colon cancer. Faecal bile acids and neutral sterols in colon cancer patients and patients with adenomatous polyps. *Cancer*, **39**, 2533–2539.

Rodriguez-Tarducy, G. and Lopez-Rivas, A. (1989) Phorbol esters inhibits apoptosis in IL-2-dependent T lymphocytes. *Biochem. Biophys. Res. Comm.*, **164**, 1069–1075.

Roivainen, R. and Messing, R.O. (1993) The phorbol derivatives thymeleatoxin and 12 deoxyphorbol 13 phenylacetate 20 acetate cause translocation and downregulation of multiple protein kinase C isozymes. *Febs. Lett.*, **319**, 31–34.

Ron, D., Chen, C., Caldwell, J., Jamieson, L., Orr, E. and Mochly-Rosen, D. (1994) Cloning of an intracellular receptor for protein kinase C: a homolog of the b subunit of G proteins. *Proc. Natl. Acad. Sci. USA*, **91**, 839–843.

Rusnak, J.M. and Lazo, J.S. (1996) Downregulation of protein kinase C suppresses induction of apoptosis in human prostatic carcinoma cells. *Exp. Cell. Res.*, 224, 189–199.

Ryves, W.J., Evans, A.T., Olivier, A.R., Parker, P.J. and Evans, F.J. (1991) Activation of the PKC isotypes α, β1, γ, δ and ε by phorbol esters of different biological activities. *FEBS Lett.*, **288**, 5–9.

Sahyoun, N., Wolf, M., Besterman, J., Hsieh, T.S., Sander, M., Levine, H., Chang, K.J. and Cuatrecasas, P. (1986) Protein kinase C phosphorylates DNA topoisomerase II: Topoisomerase activation and its possible role in phorbol ester-induced differentiation of HL60 cells. *Proc. Natl. Acad. Sci. USA*, **83**, 1603–1607.

Savill, J., Fadok, V., Henson, P. and Haslett, C. (1993) Phagocyte recognition of cells undergoing apoptosis. *Immunology Today*, **14**, 131–136.

Savill, J.S., Wyllie, A.H., Henson, J.E., Walport, M.J.J., Henson, P.M. and Haslett, C. (1989) Macrophage phagocytosis of ageing neutrophils inflammation. Programmed cell death in the neutrophil leads to recognition by macrophages. *J. Clin. Invest.*, **83**, 865–875.

Sawai, H., Okazaki, T., Takeda, Y., Tashima, M., Sawada, H., Okuma, M., Kishi, S., Umehara, H. and Damae, N. (1997) Ceramide-induced translocation of PKC-δ and -ε to the cytosol: implication in apoptosis. *J. Biol. Chem.*, **272**, 2452–2458.

Schaap, D., Husan, J.M., Totty, N. and Parker, P.J. (1990) Proteolytic activation of PKC-epsilon. *Eur. J. Biochem.*, **191**, 431–435.

Shi, L., Nishioka, W.K., Thng, J., Bradbury, E.M., Litchfield, D.W. and Greenburgh, A.H. (1994) Premature P34 (cdc2) activation required for apoptosis. *Science*, **263**, 1143–1145.

Sidorenko, S.P., Law, C.L., Klaus, S.J., Chandron, K.A., Takata, M., Kurosaki, T. and Clark, E.A. (1996) Protein kinase C-mu (PKC-mu) associates with the B-cell antigen receptor complex and regulates lymphocyte signalling. *Immunity*, **5**, 353–363.

Sinicrope, F.A., Roddey, G., McDonnell, T.J., Shen, Y., Cleary, K.R. and Stephens, L.C. (1996) Increased apoptosis accompanies neoplastic deveopment in the human colorectum. *J. Clin. Cancer Res.*, **2**, 1999–2006.

Smith, C.A., Williams, G.T., Kingston, R.T., Jenkinson, R., Jenkinson., E.J. and Owen, J.J.T. (1989) Antibodies to CD3/T receptor complex induce death by apoptosis in immature thymic cultures. *Nature*, **337**, 181–184.

Song, Q.Z., Baxter, G.D., Kovacs, E.m., Findik, D. and Lavin, M.F. (1992) Inhibition of apoptosis in human tumour cells by okadaic acid. *J. Cell. Physiol.*, **153**, 550–556.

Staudinger, J., Zhou, J., Burgess, R., Elledge, S.J. and Olson, E.N. (1995) Pick 1: a perinuclear binding protein and substrate for protein kinase C isolated by the yeast two hybrid system. *J. Cell. Biol.*, **128**, 263–271.

Thompson, L.J. and Fields, A.P. (1996) Beta (II) protein kinase C is required for the G(2)/M phase-transition of cell cycle. *J. Biol. Chem.*, **271**, 15045–15053.

Wada, H., Ohno, S., Hubo, K., Taya, C., Txyi, S., Yonehara, S. and Suzuki, K. (1989). Cell type-specific expression of the genes for the protein kinase C family: Down-regulation of mRNAs for the PKC-α and PKC-ε upon *in vitro* differentiation of a mouse neuroblastoma cell line neuro 2a. *Biochem. Biophys. Res. Comm.*, **165**, 533–538.

Watanabe, T., Ono, Y., Taniyama, Y., Hazama, K., Igarashi, K., Ogita, K., Kikkawa, K. and Nishizuka, Y. (1992) Cell division arrest induced by phorbol ester in CHO cells overexpressing protein kinase C delta. *Proc. Natl. Acad. Sci. USA*, **84**, 10159–10163.

Watson, J., Littlebury, P. and Rumsby, M. (1996) Is synthesis of protein kinase C-epsilon under translational control in 3T3 and 3T6 fibroblasts. *Immunology*, **89**, G74.

Wetsel, W.C., Khan, W.A., Merchenthaler, I., Rivera, H., Halpern, A.E., Phung, H.M., Negrro-Vilar, A. and Hannun, Y.A. (1992). Tissue and cellular distribution of the extended family of protein kinase C isoenzymes. *J. Chem Biol.*, **117**, 121–133.

Whitehouse, P.J., Price, D.L., Struble, R.G., Clark, A.W., Coyle, J.T. and DeLong, M.R. (1982) Alzheimers disease and senile dementia: loss of neurons in the basal forebrain. *Science*, **215**, 1237–1246.

Williams, G.T. (1991) Programmed cell death: Apoptosis and carcinogenesis. *Cell*, **65**, 1097–1098.

Wyllie, A.H. (1980) Glucocorticoid induced thymocyte apoptosis is associated with endogenous endonuclease activation. *Nature*, **284**, 555–556.

Yazaki, T., Ahmad, S., Chahlavi, A., Zylberkatz, E., Dean, N.M., Rabkin, S.D., Martuza, R.L. and Glazer, R.I. (1996) Treatment of glioblastoma U-87 by systemic administration of an antisense protein-kinase C-alpha phosphorothioate oligodeoxynucleotide. *Mol. Pharmacol.*, **50**, 236–242.

Zhang, Q.H., Carr, D.W., Lerea, K.M., Scott, J.D. and Newman, S.A. (1996) Nuclear localisation of type-II cAMP dependent protein kinase during limb cartilage differentiation is associated with a novel developmentally regulated A-kinase anchoring protein. *Develop. Biol.*, **176**, 51–61.

6. APOPTOSIS IN *DROSOPHILA*

JOHN M. ABRAMS†, PO CHEN, WILLIAM NORDSTROM AND JOHNSON VARKEY

Department of Cell Biology and Neuroscience, The University of Texas Southwestern Medical Center, 5323 Harry Hinds Boulevard, Dallas, Texas 75235–9039

KEY WORDS: reaper, grim, head involution defective (hid), caspase, IAP, p35.

INTRODUCTION

The elimination of cells by apoptotic cell death seems to be a universal feature of development and aging in metazoans. How are otherwise healthy cells condemned to die and what is the mechanism underlying the execution of this self-destruct program? Approaches to these questions at the molecular level have begun to emerge from model genetic systems and several lines of evidence suggest that the physiology underlying cellular suicide is highly conserved (Jacobson *et al.*, 1997; Steller, 1995). These models could therefore facilitate treatments for human diseases that are caused, or exacerbated by, the misregulation of apoptosis.

Mutations isolated in the nematode *Caenorhabditis elegans* provided the first indisputable evidence that "naturally-occurring" cell death was indeed a gene-directed process. Loss of function mutations were found that prevented all cell deaths in this organism (Ellis and Horvitz, 1986) suggesting that a common mechanism of killing is utilized among all cells which die in this animal. The cloning of genes corresponding to several of these mutations has provided molecular entry points into at least three essential components of the apoptotic pathway. One gene which normally acts to suppress programmed cell death (PCD) in nematodes, *ced-9*, shows structural and functional homology to *bcl-2*, a proto-oncogene which suppresses apoptosis in mammals (Hengartner *et al.*, 1992; Hengartner and Horvitz, 1994). Another nematode gene, *ced-3*, is the founding member of a growing family of proteolytic enzymes (Yuan *et al.*, 1993) or Caspases, some of which contribute essential roles during apoptotic cell death throughout the animal kingdom. A third locus, *ced 4*, links activities of the upstream *ced9/bcl2* family to downstream death caspases (Chinnaiyan *et al.*, 1997; Spector *et al.*, 1997; Wu *et al.*, 1997) and biochemical evidence suggests that this function is conserved as well.

PCD is also a common feature in *Drosophila* development, yet in contrast to the nematode, many cell deaths in the fly are not strictly predetermined by lineage (Abrams, 1996; Abrams *et al.*, 1993; Wolff and Ready, 1991). Development in the fruit fly exhibits remarkable plasticity, including the ability to eliminate cells (by apoptosis) that are either damaged or unable to complete their differentiation

† *Corresponding Author*: Tel.: 214–648–9226. Fax: 214–648–8694. e-mail: Abrams @utsw. swmed.edu
Supported in part by a grant from the NIH: R01 12466

program (Abrams, 1996; Abrams *et al.*, 1993). *Drosophila* is thus uniquely suited for studying how cell interactions or environmental stresses can specify or trigger the cell death fate. This chapter offers a brief overview of studies documenting PCD in *Drosophila* and reviews more recent findings that focus upon the molecular basis of apoptosis during normal and abnormal development.

THE OCCURRENCE OF PCD DURING DEVELOPMENT

Cell death in *Drosophila* has been described in many tissues at virtually all developmental stages. Investigators have generally utilized histological stains (such as toluidine blue) on fixed preparations and/or vital stains on live preparations to selectively identify dying cells. The vital dye acridine orange (AO) can be used to visualize cell death in dissected imaginal discs (Spreij, 1971; Wolff and Ready, 1991; Wolff and Ready, 1993) and, with some adaptation, AO or nile blue can be used to visualize apoptosis in the embryo (Abrams *et al.*, 1993). In fixed preparations, PCD can be visualized by TUNEL labeling (Gavrieli *et al.*, 1992) which exploits the characteristic degradation of chromatin in apoptotic nuclei. The application of these methods to *Drosophila* tissue suggests that mechanisms responsible for DNA degradation during apoptosis are conserved between vertebrates and invertebrates.

Although many of descriptions of PCD in *Drosophila* occurred prior to common usage of the term "apoptosis", an examination of published ultrastructure often shows that the mode of death is clearly apoptotic (nuclear material condenses, cells become shrunken and cellular fragments are engulfed). For instance, an examination of PCD in ovarian chambers reported that the nuclei of some nurse cells were severely condensed and the resulting cellular debris was engulfed by neighboring follicle cells (Giorgi and Deri, 1976). Some of the earliest descriptions of cell death in *Drosophila* were reported by Fristrom (Fristrom, 1968; Fristrom, 1969) studying imaginal development of the eye and wing disc. These reports established that reproducible patterns of inappropriate cell death accounted for structural defects in the adults of a variety of mutant strains and, in essence, were among the first to establish that a mutant genotype could have predictable effects upon the incidence of cell death during development.

PCD is a very prominent feature of embryonic development (Campos-Ortega and Hartenstein, 1985) and, at the ultrastructural level, these cell deaths are strikingly similar to apoptotic deaths observed in vertebrate systems (Abrams *et al.*, 1993). Patterns of cell death in the embryo are dynamic and widespread among many different organs and tissues, yet a fairly stereotypical distribution and number are observed for each stage. After embryogenesis, conspicuous numbers of dying cells are not observed until metamorphic changes are initiated. In fact, the widespread histolysis of larval tissue during metamorphosis is often cited as a classic example of PCD (Bodenstein, 1950).

TREATMENTS THAT INDUCE APOPTOSIS

Ionizing radiation is the most commonly used agent to induce apoptosis in *Drosophila*. Irradiation has historically been used to generate mosaic animals during early larval development (Ashburner, 1989) and studies to characterize the biological consequences of these treatments cited clear effects upon the incidence of cell death. In wing discs, treatment with ionizing radiation showed evidence of both apoptosis and necrotic death (Abbot, 1983) whereas only apoptotic deaths were observed in embryos exposed to similar treatments (Abrams *et al.*, 1993; White *et al.*, 1994). Not surprisingly, the stages and cell types that are most vulnerable to radiation correlate with replicative potential (Fryxell and Kumar, 1993; Wurgler and Ulrich, 1976). Moreover, notable differences in the kinetics of apoptosis were associated with exposure doses (Abrams *et al.*, 1993). Induction of cell death by ionizing radiation may reflect a process of "altruistic suicide" or "cell-replacement repair" whereby the elimination of cells that may harbor damaging mutations stimulates the proliferation of healthy cells to replace them (Kondo, 1988).

SIGNALS THAT GOVERN CELL DEATH

Among several potential endocrine regulators of PCD in *Drosophila*, the ecdysteroid hormones are the most well understood. Regulation of cellular physiology and gene expression cascades by steroids are extremely well characterized at the molecular and genetic level (Ashburner, 1989; Baehrecke, 1996). Much of our knowledge regarding the regulation of PCD by ecdysteroids derives from analogy to studies of the moth *Manduca sexta* and the giant silkworm *Antheraea polyphemus* (Truman, 1984; Truman, 1992) which are particularly amenable to detailed physiological studies because of their large size. In these insects, metamorphic cell death is largely governed by falling titers of ecdysteroids. Withdrawal of this hormone can either directly trigger cell death (Schwartz and Truman, 1983) or, alternatively, lower ecdysone levels can "prime" cells for this fate (Schwartz and Truman, 1982). Because these deaths can be suppressed by cyclohexamide (Farbach and Truman, 1988) or actinomycin D (Truman, 1992) they probably require macromolecular synthesis. Other endocrine factors, including juvenile hormone (Schwartz, 1992) and eclosion hormone (Schwartz and Truman, 1982) have been implicated as cues that elicit PCD in larger insects which might serve similar functions in *Drosophila*.

Mutations in genes that encode receptors for ecdysone are providing exciting avenues for exploring hormonal mechanisms that regulate PCD. Probes that discriminate isoforms of the ecdysone receptor (Koelle *et al.*, 1991) have already provided some tantalizing clues. This receptor encodes three protein isoforms (Talbot *et al.*, 1993) that share DNA and hormone binding domains yet differ in their N-terminal regions. Expression patterns of each isoform show distinct tissue distributions that correlate with differential responses to ecdysone during metamorphosis (Talbot *et al.*, 1993). High expression levels of one isoform, EcR-A, specifically anticipates PCD in a heterogeneous group of cells in the central nervous system (Robinow *et al.*, 1993). Apparently, the predetermined fate of these doomed cells is reflected by preferential expression of this receptor isoform. The functional importance of this observation was substantiated by showing that treatment with ecdysone could block the death of these cells if provided 3 hrs. prior to degeneration. Withdrawal of edysone may trigger the onset of death in neurons expressing high levels of EcR-A by direct induction of death-related genes such as *reaper* (see below). Isoform specific differences of these receptors should therefore provide important clues to explain how cells chose to survive or die in response to a common hormonal signals.

MUTATIONS THAT CAUSE INAPPROPRIATE APOPTOSIS

Lesions at a variety of loci are associated with characteristic patterns of ectopic cell death in many regions of the developing fly. Estimates from one screen for cell death defective mutations suggest that nearly 20% of the genes in the fly genome can cause excessive PCD when mutated (White *et al.*, 1994). A survey of cloned genes associated with ectopic cell death phenotype suggests that most of these function in the specification of cell fate during development (reviewed in (Abrams, 1996)). This correlation suggests that the inability to differentiate properly will often elicit apoptotic responses (Abrams *et al.*, 1993). The fact that so many different mutations can influence the pattern of cell death argues that apoptosis can be the outcome of a default program triggered by conflicting developmental signals. This response may contribute to plasticity during cell fate determination and it is thus difficult to establish whether ectopic death phenotypes reflect secondary consequences of aberrant development or specific apoptosis functions. Mutations displaying the opposite phenotype (see below) are therefore more likely to reveal specific components of the apoptosis machinery.

MUTATIONS THAT REDUCE THE INCIDENCE OF APOPTOSIS

Some mutations in *Drosophila* cause a reduction of the incidence of PCD in specific tissues. These loci could affect components necessary to specify cell death, and may be similar to a class of genes in *C. elegans* that either determine or trigger PCD in the nematode (Ellis and Horvitz, 1986). Mutations at *apterous* were perhaps the first phenotypes associated with failures in PCD (Butterworth, 1972; Butterworth and King, 1965). In wild type adults, cells of the larval fat body disappear entirely within the first 4 days after emergence. Animals bearing lesions at *apterous* initially show delayed rates of adipose cell death and, eventually, these animals show a complete block to PCD in these cells. More than 20% of the larval cells that would otherwise have died were found to persist in *apterous* adults. Several experiments, including cell transplantations, suggested that the PCD phenotype was nonautonomous and possibly associated with endocrine, neuronal or cytolytic factors. Ironically, mutations at this same locus may also cause excessive PCD in other cell types (Sedlak and Manzo, 1984). Because *apterous* encodes a transcriptional regulator (Cohen *et al.*, 1992) it is possible that these phenotypes arise via the misregulation of secondary agents that trigger PCD.

At least three mutations affecting eye development are associated with reduced cell death phenotypes. One allele of *Notch* reduces the number of cell deaths in the developing retina (Cagan and Ready, 1989). This gene is required for a variety of local cell interactions that specify differentiation fates. PCD phenotypes associated with mutations at this locus could therefore result from misrouting of cell fate decisions rather than a direct failure to die. Two other mutations that disrupt retinal development, *roughest* and *echinus*, also appear to reduce the incidence of PCD. In the case of the former, mosaic analyses suggested that gene action is cell autonomous and may reflect failures in one of many signalling pathways that provoke apoptosis (Wolff and Ready, 1991). Sequence analysis shows that *roughest* is a large transmembrane protein containing several immunoglobulin-like motifs and that the allele which affects PCD disrupts the intracellular domain (Ramos *et al.*, 1993).

There are other genes that affect the rate of cell death. For example, after an adult fly emerges from the puparium, a group of muscles in the head degenerate within 12 hours. Two *mcd* (muscle cell death) mutations delay this process showing blocks to fragmentation and/or absorption of the muscle cell corpses. PCD in the head muscles of *mcd* mutants is clearly distinct from wild type, suggesting that these genes might function as part of the machinery which dismantles dying cells (Kimura and Tanimura, 1992).

THE *REAPER* REGION, A GENOMIC INTERVAL REQUIRED FOR APOPTOSIS

The *reaper (rpr)* region is required for all cell deaths which occur during embryonic development in *Drosophila* (Abrams *et al.*, 1993; White *et al.*, 1994). This locus was identified from a screen which sampled ~50% of the fly genome for cell death defective mutations (White *et al.*, 1994). No apoptotic cell death is observed in embryos deleted in the *rpr* interval. Although they develop fairly normal segmentation and cuticular structures these individuals ultimately show failures in head involution and nerve cord condensation and they do not hatch as larvae. Some of these defects probably reflect the anatomic consequences of failures in cell death commencing at earlier developmental stages.

Several lines of evidence argue that functions mapping to the *rpr* region are globally required PCD during *Drosophila* development. First, direct inspection using multiple histological methods (including electron microscopy and TUNEL labeling) shows a conspicuous absence of apoptotic cells. Second, the macrophages in *rpr* mutants are unusually small and devoid of internalized cell corpses with which they normally are engorged. Third, extra cells (that would otherwise die) persist in these mutants and they adopt differentiated fates typical of their surrounding tissue (Grether *et al.*, 1995; White *et al.*, 1994; Zhou *et al.*, in press). Together, these observations demonstrate a requirement for this genomic region during normal PCD.

Is the *reaper* interval also involved in cell deaths induced by exogenous agents or developmental defects? The answer to both is affirmative. When embryos deleted for the *rpr* region are exposed to X-irradiation, the numbers of induced cell deaths are far less than those observed for wild type siblings exposed to the same treatment (White *et al.*, 1994). Therefore, components in the *rpr* interval may not be part of the "apoptosis machinery" itself, but may instead trigger a set of commonly used effector molecules. This radiation resistant phenotype is reminiscent of cells that lack p53 (Clarke *et al.*, 1993; Lowe *et al.*, 1993) and demonstrates that: 1) loss of genes in the *rpr* region confer protection against the induction of ectopic death by at least one extrinsic agent and 2) some apoptotic deaths can occur in the absence of these functions. Apparently, ionizing radiation can induce apoptosis through at least two genetic pathways, one of which depends on functions that are utilized during normal embryonic development.

Defective development is often associated with ectopic or extra cell deaths (see above and (Abrams, 1996)) and deletions of the *rpr* interval will also prevent these (White *et al.*, 1994). For example, mutations at *crumbs* cause failures in the development of various tissues followed by massive apoptosis among epithelial cells that normally survive. When this mutation was introduced into a genetic background deleted for the *reaper* interval, the developmental failures still occurred but the associated ectopic deaths were prevented. Collectively, these observations argue that functions in the *reaper* interval, are not only necessary for PCD during normal development but are also required for abnormal deaths induced either by damaging agents or congenital defects.

The cell death phenotype was originally defined by a deletion interval that maps to genomic region 75C1, 2 on the third chromosome (see Figure 6.1). The H99 deficiency is the smallest deletion associated with a fully penetrant cell death defective phenotype and spans ~ 300kb of DNA. Other relevant deletions (X14 and X25) have breakpoints that map within the H99 interval and strains bearing these deletions exhibit cell death defects that are less severe than H99 homozygotes. Interestingly, the penetrance of each is markedly enhanced when placed in trans to the H99 chromosome. The complex genetic properties associated with these deletions led to suspicions that multiple cell death functions might reside in the obligate region, a prediction that has since been confirmed at the molecular level.

To identify the corresponding functions at the molecular level, a genomic walk through the obligate interval was used to establish a contig upon which deletion breakpoints and transcripts were mapped. *Reaper* was the first cell death gene identified in this interval. This gene is selectively expressed in cells that will later die and partially restores PCD to H99 embryos in transformation experiments (White *et al.*, 1994). Moreover, expression of *rpr* is sufficient to trigger apoptosis in both transgenic animals and cultured cells (Chen *et al.*, 1996a; Nordstrom *et al.*, 1996; Pronk *et al.*, 1996; White *et al.*, 1996). RPR is a 65 amino acid cytoplasmic protein bearing no obvious similarities to known proteins. Alignments between RPR and the death domain of tumor necrosis factor receptor 1 family members have led to suggestions that these proteins might share an ancestral heritage (Cleveland and Ihle, 1995; Golstein *et al.*, 1995a; Golstein *et al.*, 1995b). However, empirical tests have so far failed to obtain evidence supporting the notion that RPR and the TNFR1 death domain share common functions (Chen *et al.*, 1996a; Vucic *et al.*, 1997). Despite extensive efforts, no mutant fly strains bearing single gene lesions in *rpr* have, as yet, been isolated. The identification of mutations in this gene could be hampered by its rather small size and the occurrence of additional cell death genes in the region that probably serve redundant functions (see below).

Robinow and his colleagues have gathered some of the most convincing data in support of the idea that the expression of *rpr* RNA anticipates the onset of pro-

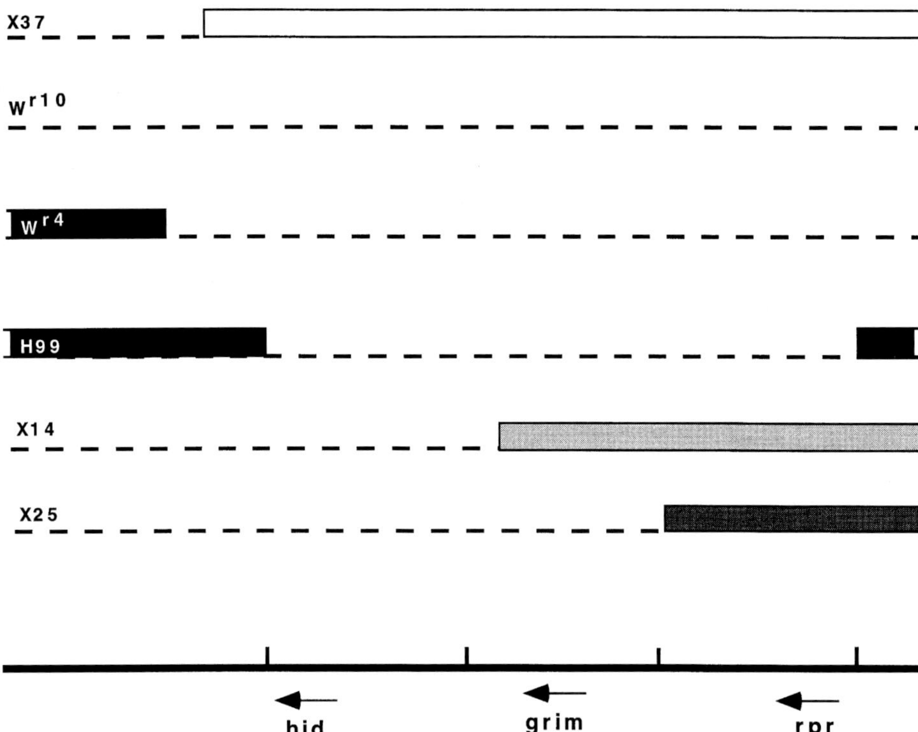

Figure 6.1 Physical map of the *Drosophila* cell death interval at 75C1,2. Deletion chromosomes are depicted above the genomic interval where each notch represents a physical distance of 100 kilobases. Relevant transcripts (not drawn to scale) are schematically represented below. Left-ward direction is toward the telomere. Dotted lines represent DNA that is missing in each deletion strain and thick bars represent portions that are intact. Strains with dark, solid bars display a fully penetrant cell death defective phenotype. Strain X37, represented by the open bar, shows no cell death phenotype and thus delimits a distal boundary for cell death functions. Strains that display partially penetrant cell death phenotypes, X14 and X25, are denoted with shaded bars. The intensity of shading indicates the severity of the phenotype (darker shading corresponds to increased penetrance; see text). The obligate interval defining the cell death function(s) thus resides inside the breakpoints of the H99 deletion. The transcript encoded by *hid* represents the only single-gene lethal function identified in this interval.

grammed cell death. These workers focus on a set of doomed neurons (n4 neurons) which are marked during pupation by high levels of the EcR-A ecdysone receptor isoform (see above) and die 4–8 hr. after eclosion (Robinow *et al.*, 1993). At eclosion, *rpr* RNA is undetectable in the n4 neurons yet high levels of *rpr* are observed in these same cells one hour later (Robinow *et al.*, 1993). Moreover, treatments that hormonally block these cell deaths also prevents the appearance of *rpr* expression. In these cells, the onset of *rpr* expression correlates with (and might define) a commitment to the death program.

Studies that monitor expression of *rpr* in contexts where ectopic cell death is provoked by exogenous damaging agents have also born out predictions of earlier

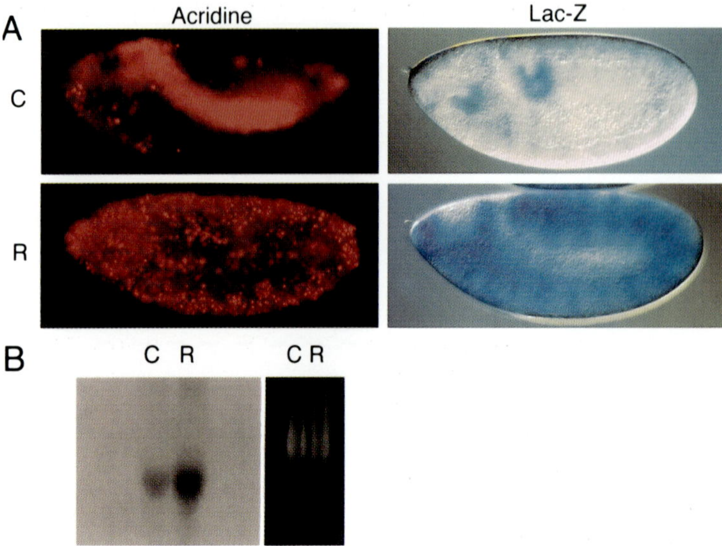

Figure 6.2 (A). Transactivation of *reaper* RNA anticipates radiation-induced cell killing. Left column extensive induction of apoptosis 1.5 hrs. after exposure to 4000 rads of γ-rays (R) compared to control (C) is detected by acridine orange staining. Note that longer exposure times for control embryo image (C) results in increased detection of yolk autoflorescence (denoted by arrow). Prominent induction at the *rpr* promoter is detected by staining for β-galactosidase in embryos bearing a *rpr*-promoter-lacZ reporter transgene (Rpr-11LacZ) 1.5 h after exposure to radiation (R) when compared to untreated samples (C). **(B). *rpr* RNA levels are induced by ionizing radiation**. Hybridization of a radio-labeled *rpr*-specific probe to RNA from irradiated embryos (lane R) and similarly staged control embryos (lane C) is shown in the left hand panel. Ethidium bromide staining of the gel (right panel) demonstrates that equivalent quantities of target RNA were assayed. (taken from (Nordstrom *et al.*, 1996))

genetic evidence. For instance, expression of *rpr* (shown in Figure 6.2) is acutely responsive to ionizing radiation such that a robust elevation of this RNA anticipates the onset of X-ray induced apoptosis (Nordstrom *et al.*, 1996). Similar cell specific induction is observed in models of defective development where inappropriate expression of *rpr* anticipates inappropriate cell deaths (Nordstrom *et al.*, 1996; Singer *et al.*, 1996).

The fact that *rpr* is sufficient to trigger apoptosis, together with observations showing that expression anticipates both programmed and induced cell death, imply that mechanisms regulating the occurrence of this transcript can profoundly influence the decision of a given cell to survive or die. In large measure, this regulation appears to be achieved at the transcriptional level. The behavior of *rpr* in models of inappropriate cell death is recapitulated by a lacZ reporter transgene (Nordstrom *et al.*, 1996) and, at the gross level this reporter also mirrors some (but not all) elements of the endogenous expression pattern (see Figure 6.2). These observations raise the possibility that the *rpr* promoter might define a molecular integration site for a number of distinct death signals. Alternatively, the integration of cues governing cell death might occur upstream of the *rpr* locus. Discriminating between these possibilities and determining the nature of the circuits that govern transactivation of the *rpr* locus will be exciting avenues of future exploration.

GRIM

As the physical map of the *reaper* region was compiled, evidence suggesting that the H99 interval might harbor more than a single cell death gene began to emerge. The most definitive observation in this regard was that internal deletion strains (DfX14 and DfX25) displayed markedly different phenotypes when placed in trans to H99 (Figure 6.1). The more severe phenotype associated with DfX25 could be explained by presuming that at least one additional cell death function maps between the proximal breakpoints of DfX14 and DfX25. Direct tests for such a function led to the identification of *grim* (Chen *et al.*, 1996b) a cell death gene which exhibits many features in common with *rpr*.

Several lines of evidence established that *grim* encodes an activator of PCD during development (Chen *et al.*, 1996b). First, germline transformation of genomic DNA spanning the *grim* locus resulted in restoration of cell death to H99 mutant embryos. Although only partial rescue of the cell death defect by the *grim* cosmid was observed, the extent of rescue was dose dependent and similar to levels of rescue observed for corresponding doses of genomic *rpr*. Second, induction of *grim* either in cultured cells or in ectopic tissues triggered extensive apoptosis as determined by methods that are selective for this form of cell death (AO and TUNEL). Third, these deaths were prevented by p35, a viral inhibitor of apoptosis which targets essential components of the death machinery (see below). Fourth, the distribution of *grim* transcript during embryogenesis is coincident with patterns of embryonic cell death. Moreover, *grim* RNA was occasionally detected inside phagocytes. This feature is an unusual staining characteristic that has also been observed with probes for *rpr* (White *et al.*, 1994; Zhou *et al.*, 1995) and *hid* (Grether *et al.*, 1995) and probably reflects hybridization to RNAs that persist within engulfed cell corpses. Finally, a burst of *grim* expression precedes the death of n4 neurons and, in this respect, also appears to mirror the behavior of *rpr* (Robinow *et al.*, 1993). These observations suggest that, like *rpr*, RNA expressed from the *grim* gene is a molecular prophet of death.

The deduced open reading frame (ORF) for *grim* is a protein of 138 amino acids with no predicted trans-membrane domains (Rost and Sander, 1994) and no extensive homologies to any sequence in the current database. However, the amino terminal end of GRIM shares very notable similarity to RPR (Chen *et al.*, 1996b). Ten of the first 14 residues of GRIM are identical to the corresponding position in RPR while three of the remainder are conserved substitutions. Interestingly, although a RPR mutant deleted for these residues is a less stable protein, it is still a fairly proficient activator of death in cell culture assays. This shared N terminal motif is therefore not essential for killing functions (at least with respect RPR) (Chen *et al.*, 1996a) and its functional significance remains somewhat obscure.

Studies on epitope-tagged versions of this protein suggest that GRIM is predominantly localized to the cytoplasmic compartment (Varkey and Abrams, unpublished observations).

HEAD INVOLUTION DEFECTIVE

At least two mutagenic assaults on the H99 cell death interval suggest that *head involution defective (hid)* is the only lethal complementation group accessible by chemically induced point mutations in this region (Abbott and Lengyel, 1991; Grether *et al.*, 1995; White *et al.*, 1994). The *hid* gene was originally described as a locus required for proper head and genitalia formation (Abbott and Lengyel, 1991). Embryos carrying single gene *hid* mutations fail to complete the final morphogenetic movements of the anterior head segments and the overwhelming majority of these embryos do not hatch as larvae. The rare *hid* individuals that do survive to adulthood (~ 5% of expected frequency) exhibit defects of the genitals and wing blades.

hid mutations do not exhibit gross cell death defects that phenocopy H99 but they do manifest partial failures in apoptosis (Grether *et al.*, 1995). The incidence of PCD (particularly in the developing head region) is noticeably decreased in these mutants and supernumerary cells are observed in larval optic nerve. Single-gene null alleles of *hid* cause phenotypes that are relatively mild when compared to the severity of defects associated with the H99 deletion interval. This difference is emphasized by the fact that null alleles of *hid* are only semi-lethal whereas H99 is an absolutely lethal deletion (all homozygous embryos die). These observations, and the complex complementation pattern of *hid* point mutations, establish that the *hid* gene function alone does not fully account for the severity apoptosis defects associated with 75C1,2 deletions. Nevertheless, it is also clear that expression of *hid* is sufficient to trigger extensive apoptosis in both transgenic animals (Grether *et al.*, 1995) and cell culture (Varkey and Abrams, unpublished observations). Also, like the previous models described for *rpr* and *grim*, these deaths are also prevented by co-expression of p35. The predicted *hid* gene product is a novel 410-amino-acid protein rich in serine and proline residues. Although *hid* mRNA occurs in many regions that are coincident with patterns of PCD, it does not behave like a "prophet of death" because expression of this gene occurs in cells that do not ultimately die and it is absent from at least one organ (the nerve cord) at a time when substantial PCD occurs there.

GENETICS OF THE H99 CELL DEATH INTERVAL

At least three cell death genes are uncovered by the H99 deletion. These share a common, short motif at the N terminus and are all transcribed in the same orientation (towards the telomere). Are there additional cell death genes in the H99 interval? Genetic approaches to this problem are hampered by the absence of single-gene mutations in *rpr* or *grim*. The fact that only alleles of *hid* were recovered from screens for lethal mutations in the H99 interval (Abbott and Lengyel, 1991; Grether *et al.*, 1995; White *et al.*, 1994) could suggest that null mutations in *rpr* or *grim* might be even more mild than *hid* and could, therefore, be inaccessible by screens for lethal or visible phenotypes. According to this scenario, *hid, rpr* and *grim* might encode partially redundant functions and the more severe (or noticeable) cell death phenotypes could require the disruption of at least two of the three known elements in the region. Conversely, if these three loci accounted for all apoptosis functions in the *reaper* region, simultaneous restoration of both *rpr* and *grim* functions might be expected to convert H99 towards a phenocopy of *hid* null mutations. Tests of this prediction, however, were negative. H99 embryos homozygous for *grim* and *rpr* cosmid-transgenes do not phenocopy *hid* with respect to levels of PCD (these individuals are indistinguishable from H99 embryos bearing four doses of either *grim* or *rpr*). However, since expression from cosmid-transgenes might be compromised, solid conclusions regarding the number of distinct cell death functions uncovered by H99 can not be drawn from these results.

In *Drosophila*, and in other species, tight linkage can occur among groups of genes that share closely related functions (Krumlauf, 1994; Lawrence, 1992). Typically, the individual members within conserved gene clusters share a common orientation of transcription and a considerable degree of sequence similarity. The H99 cell death interval therefore, shares some, but not all, of the classic features of a complex: *grim, hid* and *rpr* share commonality of orientation and function yet except for a short motif at the N terminus, these proteins share little or no sequence similarity. Since individual members within a gene cluster are typically thought to have arisen by evolutionary duplication and divergence (Kenyon, 1994), it is possible that organization of the H99 cell death cluster may have evolved by different mechanisms.

How do these genes function to elicit apoptosis? Investigating this question is certain to be a high priority for the future yet already several observations permit us to make some predictions in this regard. First, conditional expression of each is sufficient to elicit cell death in H99 homozygotes: *rpr*-mediated apoptosis does not require *grim* or *hid, grim*-mediated apoptosis does not require *rpr* or *grim* and *hid*-mediated apoptosis does not require *grim* or *rpr*. Thus *rpr, grim* and *hid can* each function independently to trigger apoptosis. Moreover, experiments that test for cross-regulation among these genes have obtained no evidence for feedback circuits at the level of gene expression (Chen *et al.*, 1996b; Grether *et al.*, 1995). However, at least one example occurs where *rpr* and *hid* cooperate to trigger apoptosis when neither alone was sufficient (Zhou *et al.*, in press). Therefore, although each can function independently to trigger apoptosis, there will certainly turn out to be combinatorial requirements and selective constraints that specify patterns of PCD in different contexts. A second important consideration is that death pathways triggered by either RPR, GRIM or HID are all similarly influenced by *Drosophila* homologs of the IAP genes, DIAP1 and DIAP2 (see below). Third, these activators of apoptosis proceed through a p35 inhibitable step, implicating a downstream requirement for Caspase function (see below). These gene products might thus be viewed as parallel switches that might ultimately activate a common circuit of apoptotic downstream effectors (see Figure 6.3). Alternatively, two or more parallel pathways, each of which is separately blocked by p35, could define the effector circuits downstream of these genes. In either case, *grim, rpr* and *hid* might represent alternate switches that are triggered by distinct, but overlapping, sets of apoptotic signals.

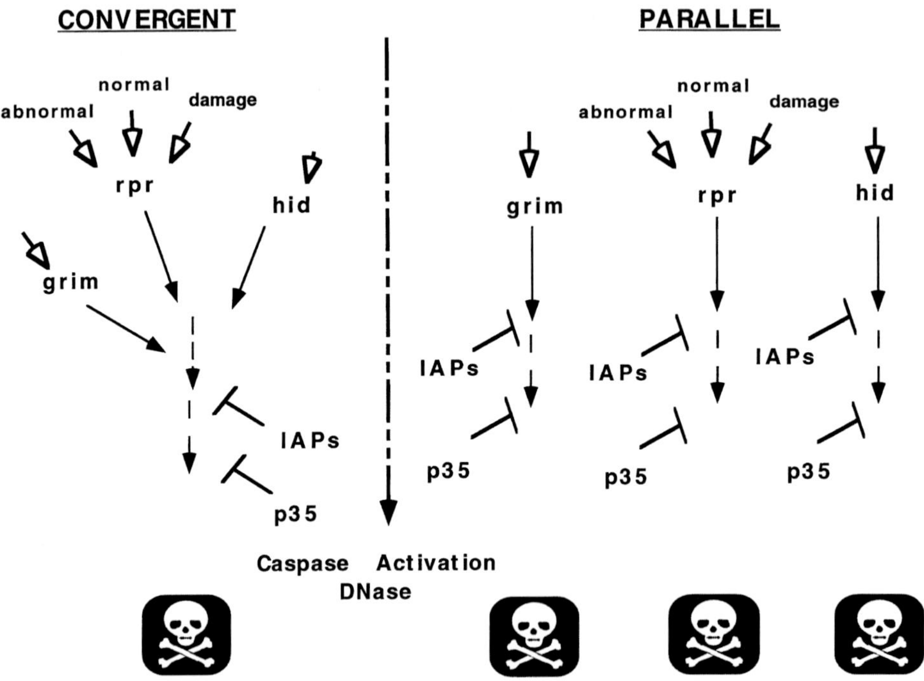

Figure 7.3 Alternative models of gene action for functions in the *reaper* region. Signals governing transcriptional activation (open arrows) are thought to regulate *rpr, grim* and *hid* expression (proven thus far only for *rpr*). These gene products might converge upon a singular hypothetical circuit that can be blocked by IAPs and p35 (left side) or might operate in parallel circuits, each of which can be separately blocked by IAPs and p35 (right side). Death induced by these activators ultimately triggers activation of Caspases and cleavage of DNA. Known functions are solid arrows, hypothetical ones are dotted. Some observations suggest that *grim* may engage the death circuitry further downstream than *rpr*.

These activators of death are, for the most, indistinguishable with respect to their activities in transgenic animals and cell culture assays, Yet, in at least one context, the functions of RPR and GRIM are not identical-ectopic *grim* induction triggers apoptosis at a time in early embryonic development when no such effects were observed with *rpr* (Chen *et al.*, 1996b). This distinction could reflect differences in cell death signalling that are uncovered during early development. For instance, a negative regulator of *rpr* that does not block *grim* might be present at this stage or, alternatively, an effector required for *rpr* but not for *grim* might be absent at this stage.

THE *DROSOPHILA* IAP GENES

The first members of the IAP (Inhibitors of Apoptosis) gene family were identified as baculoviral proteins that function to suppress the death of infected host cells (Clem and Miller, 1994). These viral IAPs were probably acquired from host genomes as an evolutionary strategy that maximizes productive replication. IAPs typically encode a RING finger motif at their carboxy terminus and one or more BIRs (Baculovirus IAP repeats) at the amino terminus (Duckett *et al.*, 1996; Hawkins *et al.*, 1996; Liston *et al.*, 1996; Uren *et al.*, 1996). The former may constitute a negative regulatory domain (Hay *et al.*, 1995), whereas the latter motif is essential for the anti-apoptotic activity associated with these proteins (Clem and Miller, 1994; Duckett *et al.*, 1996; Hawkins *et al.*, 1996; Hay *et al.*, 1995; Uren *et al.*, 1996). The mechanism by which these proteins prevent apoptosis is not yet known. Members of the IAP gene family occur in mice and humans and at least two have been identified in *Drosophila*.

In a screen for modifiers of *rpr*-mediated apoptosis in the eye, Hay and his colleagues identified mutations at DIAP1 (Hay *et al.*, 1995). Mutations at this locus are dominant enhancers of the death phenotype and were determined to be allelic to a previously identified gene referred to as thread (*th*). Sequence analyses prompted by the discovery of DIAP1 led to the identification of a related gene, referred to as DIAP2. Unlike DIAP1, deletions uncovering DIAP2 do not enhance *rpr*-mediated retinal cell death. Overexpression of DIAP1 or DIAP2 not only suppressed normal PCD in the eye, but also prevented excessive apoptosis due to overexpression of *rpr, hid* (Hay *et al.*, 1995) or *grim* (Chen and Abrams, unpublished observations).

What is the precise function of the IAP proteins during normal development? *In situ* hybridization experiments show that expression of DIAP1 is widespread throughout the embryo and the developing eye disc. However, embryos homozygous for loss-of-function mutations in either DIAP1 or DIAP2 showed no obvious cell death phenotypes, possibly owing to extensive maternal contributions of this product and/or functional redundancies. Results from clonal analyses in eye discs or ovaries suggest that DIAP1 could be required for cell survival, but the results could alternatively suggest a requirement for proliferation. Therefore, although the *Drosophila* IAP genes products exhibit potent activities in sensitized transgenic backgrounds, their function in normal development has yet to be determined.

PREVENTION OF APOPTOSIS BY P35

p35 is encoded by the *Autographa californica* nuclear polyhedrosis virus (AcMNPV) which normally infects lepidopteran insect cells. In a series of elegant studies (Clem and Miller, 1994; Clem *et al.*, 1991). L. Miller and her colleagues established that this protein functions to suppress apoptosis that would otherwise occur among infected host cells. p35 bears no obvious similarity to known proteins yet it is capable of suppressing cell death in a wide range of heterologous contexts.

In part, p35 prevents apoptotic death because it is an irreversible inhibitor of (and pseudo-substrate for) members of the ICE/CED-3 (Caspase) family of apoptotic proteases (Bertin *et al.*, 1996; Bump *et al.*, 1995; Xue *et al.*, 1996). Activation of Caspases may be a universal feature of apoptotic cell death and numerous studies argue that Caspases are essential effectors of the apoptotic pathway (reviewed in (Alnemri, 1997; Chinnaiyan and Dixit, 1996; Kumar, 1995; Nicholson, 1996)). The importance of these proteins as requisite components in apoptotic circuits is further bolstered by the fact that viral products (e.g. p35 and crmA) target these enzymes for inactivation (reviewed in (Teodoro and Branton, 1997)).

In both transgenic flies and cell culture models, p35 is a potent death suppressor in models of *rpr, grim* or *hid* – induced apoptosis (Chen *et al.*, 1996b; Grether *et al.*, 1995; Nordstrom *et al.*, 1996; White *et al.*, 1996). More striking, perhaps is the observation that expression of p35 in wild type embryos can phenocopy deletions of the entire *rpr* region (Hay *et al.*, 1994). These observations suggest that p35 intervenes downstream of functions in the *reaper* region and also imply that death triggered by *rpr, grim* and *hid* operates through *Drosophila* members of the Caspase gene family.

DROSOPHILA CASPASES

Cell killing induced by RPR, GRIM or HID triggers induction of proteolytic activity characteristic of members of the Caspase gene family. This activity is strikingly similar to mammalian counterparts because extracts from *Drosophila* cultures exhibit signature cleavage activity against the mammalian death substrate, poly-ADP-ribose-polymerase (PARP) (Chen and Abrams, unpublished observations).

At least three members of the Caspase family have been identified thus far in *Drosophila*. Two of these, DCP-1 (Song *et al.*, 1997) and drICE (Fraser and Evan, in press), are most closely related to CPP-32 and Mch2, whereas the third, DREDD (Chen *et al.*, in preparation), is most closely related to CED-3 and ICH-1/NEDD. A precise determination of the role of these enzymes during PCD will require detailed genetic and biochemical analyses. Precedents from genetic studies on DCP-1 (Song *et al.*, 1997), however, suggest that sophisticated analyses for characterizing multiply-mutated individuals will be required because PCD failures are not observed in embryos homozygous for DCP-1 loss-of-function mutations (these mutations cause larval lethality). Functional redundancy and/or maternal contributions of this gene product might account for the fact that zygotic DCP-1 is dispensable for PCD in the embryo. Larva homozygous for *dcp-1* show no direct failures in PCD defects. These individuals lack all imaginal discs and gonads and exhibit evidence of unusual self-reactive immune responses.

ENGULFMENT OF APOPTOTIC CELLS

Engulfment of dead cell corpses is the final stage of apoptotic death throughout the animal kingdom. The "professional phagocytes" in *Drosophila* are referred to as macrophages or hemocytes. Although these cells are most frequently observed to engulf late stage corpses, they display a remarkable capacity to recognize and discriminate dying cells, even at very early stages of apoptosis (Abrams *et al.*, 1993). Macrophages appear to be a homogeneous group of cells which derive from the mesoderm of the head prior to, and independently of, the first signs of PCD (Tepass *et al.*, 1994). Because abundant cell death can be observed in mutations that lack macrophages, these cells are apparently not essential for cell killing in the embryo (Tepass *et al.*, 1994). There are no *Drosophila* mutations yet described that show specific failures in the engulfment process but at least two genes have been suggested as candidate functions in this process.

The class C scavenger receptor, dSR-CI, is specifically expressed in *Drosophila* macrophages and might contribute to the recognition and processing of dying cells (Abrams *et al.*, 1992; Pearson *et al.*, 1995). Scavenger receptors endocytose a wide variety of polyanionic macromolecules (Krieger and Herz, 1994) and have been implicated in the recognition of apoptotic cells (Abrams *et al.*, 1992; Krieger and Herz, 1994; Savill *et al.*, 1993). A potential mechanism to account for the recognition of apoptotic cells involves phophatidylserine, a ligand for macrophage scavenger receptors (Nishikawa *et al.*, 1990) which is restricted to the inner monolayer in healthy cells. This asymmetric distribution is lost during apoptotic death, and may be recognized on the surface by one or more receptors (Fadok *et al.*, 1992), including the macrophage scavenger receptors. The recovery of dSR-CI mutant strains will permit a rigorous test of its function in the engulfment process.

Another gene that has been proposed to play a role in the recognition of apoptotic by macrophages is *peroxidasin*, a unique heme peroxidase that appears to be deposited in the extracellular matrix (Nelson *et al.*, 1994). This protein, originally referred to as "protein X" (Abrams *et al.*, 1993; Fessler and Fessler, 1989), is synthesized by macrophages and may have both intracellular and extracellular oxidation activity (Nelson *et al.*, 1994). These properties have led to the intriguing proposal that *peroxidasin* is anchored to basement membranes and could thus function to "mark" damaged or dying cells that might secrete H_2O_2 or other peroxidase substrates (Nelson *et al.*, 1994). Cells "marked" in this way could be recognized for disposal by scavenger receptors. This proposal is attractive because it relies upon multiple activities for the detection of apoptotic cells which could impart both high selectivity and fidelity to the recognition process.

FINAL REMARKS

During the past several years, apoptosis research has witnessed spectacular growth. A general lesson to be drawn from our knowledge thus far is that the parallels between apoptosis in *Drosophila* and mammals are striking. In both systems, the death of cells is regulated by hormones, occurs in reproducible patterns in development and is associated with a characteristic cytomorphology. Similarly, in both systems, ectopic apoptosis can be observed in association with defective development and induced by exposure to damaging agents. These parallels persist at the molecular level as well. In both systems, cell death is associated with Caspase activity,

DNA degradation and modulation by IAPs. Although our knowledge of apoptosis in *Drosophila* is still rudimentary, the prospects for rapid progress in this area are outstanding. Persuasive evidence argues for a high degree of evolutionary conservation in components required for PCD, and it is therefore likely that understanding apoptosis in *Drosophila* will have profound and direct relevance to the understanding and treatment of various human pathologies. Genes homologous to *rpr, grim* and *hid*, for instance, could provide excellent targets for the discovery of therapeutic drugs that are intended to block or induce apoptosis in humans.

REFERENCES

Abbot, L.A. (1983) Ultrastructure of cell death in Gamma-or X-irradiated imaginal wing discs of *Drosophila Radiat. Res.*, **96**, 611–627.

Abbott, M.K. and Lengyel, J.A. (1991) Embryonic head involution and rotation of male terminalia require the Drosophila locus head involution defective. *Genetics*, **129**, 783–9.

Abrams, J.M. (1996) Molecular and Genetic Control of Apoptosis in *Drosophila*. In "Apoptosis in Normal Development and Cancer" (M., Sluyser, Ed.), pp. 171–188. Taylor & Francis, London.

Abrams, J.M., Lux, A., Steller, H. and Krieger, M. (1992) Macrophages in *Drosophila* embryos and L2 cells exhibit scavenger receptor-mediated endocytosis. *Proc. Nat'l. Acad. Sci. USA*, **89**, 10375–10379.

Abrams, J.M., White, K., Fessler, L. and Steller, H. (1993) Programmed cell death during *Drosophila* embryogenesis. *Development*, **117**, 29–44.

Alnemri, E.S. (1997) Mammalian cell death proteases – a family of highly conserved aspartate specific cysteine proteases. *Journal of Cellular Biochemistry*, **64**, 33–42.

Ashburner, M. (1989) "Drosophila: A Laboratory Handbook." Cold Spring Harbor Laboratory Press, Cold Spring Harbor.

Baehrecke, E.H. (1996) Ecdysone signalling cascade and regulation of Drosophila metamorphosis. [Review] [69 refs]. *Archives of Insect Biochemistry & Physiology*, **33**, 231–44.

Bertin, J., Mendrysa, S.M., Lacount, D.J., Gaur, S., Krebs, J.F., Armstrong, R.C., Tomaselli, K.J. and Friesen, P.D. (1996) Apoptotic suppression by baculovirus p35 involves cleavage by and inhibition of a virus-induced ced-3/ice-like protease. *Journal of Virology*, **70**, 6251–6259.

Bodenstein, D. (1950) The Postembryonic Development of Drosophila. In "Biology of Drosophila" (Demerec, Ed.), pp. 275–364. John Wiley & Sons, Inc, New York.

Bump, N.J., Hackett, M., Hugunin, M., Seshagiri, S., Brady, K., Chen, P., Ferenz, C., Franklin, S., Ghayur, T., Li, P., Licari, P., Mankovich, J., Shi, L.F., Greenberg, A.H., Miller, L.K. and Wong, W.W. (1995) Inhibition of ICE family proteases by baculovirus antiapoptotic protein p35. *Science*, **269**, 1885–1888.

Butterworth, F.M. (1972) Adipose tissue of *Drosophila melanogaster*. V., Genetic and experimental studies of an extrinsic influence on the rate of cell death in the larval fat body. *Dev. Biol.*, **28**, 311–325.

Butterworth, F.M. and King, R.C. (1965) The developmental genetics of *apterous* mutants in *Drosophila melanogaster*. *Genetics*, **52**, 1153–1174.

Cagan, R.L. and Ready, D.F. (1989) *Notch* is required for successive cell decisions in the developing *Drosophila* retina. *Genes & Dev.*, **3**, 1099–1112.

Campos-Ortega, J.A. and Hartenstein, V. (1985) "The embryonic development of *Drosophila melanogaster*.", Springer-Verlag, Berlin.

Chen, P. and Abrams, J.M., (unpublished observations).

Chen, P., Rodriguez, A., Erskine, R., Thach, T. and Abrams, J.M., (in press).

Chen, P., Lee, P., Otto, L. and Abrams, J.M. (1996a) Apoptotic activity of REAPER is distinct from signalling by the tumor necrosis factor receptor 1 death domain. *J. Biol. Chem.*, **271**, 25735–25737.

Chen, P., Nordstrom, W., Gish, B. and Abrams, J.M. (1996b) Grim, a novel cell death gene in drosophila. *Genes & Development*, **10**, 1773–1782.

Chinnaiyan, A., O'Rourke, K., Lane, B.R. and Dixit, V.M. (1997) Interaction of ced-4 with ced-3 and ced-9: a molecular framework for cell death. *Science*, **275**, 1122–1126.

Chinnaiyan, A.M. and Dixit, V.M. (1996) The Cell-Death Machine [Review]. *Current Biology*, **6**, 555–562.

Clarke, A.R., Purdie, C.A., Harrison, D.J., Morris, R.J., Bird, C.C., Hooper, M.L. and Wylie, A.H. (1993) Thymocyte apoptosis induced by p53-dependent and independent pathways. *Nature*, **362**, 849–852.

Clem, R. and Miller, L.K. (1994) Control of Programmed Cell Death by the Baculovirus Genes *p35* and *iap*. *Molecular and Cellular Biology*, **14**, 5212–5222.

Clem, R.J., Fechheimer, M. and Miller, L.K. (1991) Prevention of apoptosis by a baculovirus gene during infection of insect cells. *Science*, **254**, 1388–1390.

Cleveland, J.L. and Ihle, J.N. (1995) Contenders in FasL/TNF death signalling. *Cell*, **81**, 479–82.

Cohen, B., McGuffin, M.E., Pfeifle, C., Segal, D. and Cohen, S.M. (1992) *apterous*, a gene required for imaginal disc development in *Drosophila* encodes a member of the LIM family of developmental regulatory proteins. *Genes & Dev.*, **6**, 715–729.

Duckett, C.S., Nava, V.E., Gedrich, R.W., Clem, R.J., Vandongen, J.L., Gilfillan, M.C., Shiels, H., Hardwick, J.M. and Thompson, C.B. (1996) A conserved family of cellular genes related to the baculovirus iap gene and encoding apoptosis inhibitors. *EMBO Journal*, **15**, 2685–2694.

Ellis, H.M. and Horvitz, H.R. (1986) Genetic control of programmed cell death in the nematode *C. elegans*. *Cell*, **44**, 817–829.

Fadok, V.A., Voelker, D.R., Campbell, P.A., Cohen, J.J., Bratton, D.L. and Henson, P.M. (1992) Exposure of phosphatidyl serine on the surface of lymphocytes triggers specific recognition and removal by macrophages. *J. Immunol.*, **148**, 2207–2216.

Farbach, S.E. and Truman, J.W. (1988) Cycloheximide inhibits ecdysteroid-regulated neuronal death in the moth *Manduca sexta*. *Society Neurosci. Abst.*, **14**, 368.

Fessler, J.H. and Fessler, L.I. (1989) *Drosophila* extracellular matrix. *Ann Rev. Cell. Biol.*, **5**, 309–339.

Fraser, A. and Evan, G.I., (in press). Identification of a *Drosophila melanogaster* ICE/CED-3-related protease.

Fristrom, D. (1968) Cellular degeneration in wing development of the mutant *vestigial* of *Drosophila melanogaster*. *J. Cell. Biol.*, **39**, 488–491.

Fristrom, D. (1969) Cellular degeneration in the production of some mutant phenotypes in *Drosophila melanogaster*. *Molec. Gen. Genetics*, **103**, 363–379.

Fryxell, K.J. and Kumar, J.P. (1993) Characterization of the radiation sensitive stage in the development of the compound eye of *Drosophila Mutat. Res.*, **285**, 181–189.

Gavrieli, Y., Sherman, Y. and Ben-Sasson, S.A. (1992) Identification of programmed cell death *in situ* via specific labeling of nuclear DNA fragmentation. *J. Cell. Biol.*, **119**, 493–501.

Giorgi, F. and Deri, P. (1976) Cell death in ovarian chambers of Drosophila melanogaster. *J. Embryol. exp. Morph.*, **35**, 521–533.

Golstein, P., Marguet, D. and Depraetere, V. (1995a) Fas bridging cell death and cytotoxicity: the reaper connection. *Immunol. Rev.*, **146**, 45–56.

Golstein, P., Marguet, D. and Depraetere, V. (1995b) Homology between Reaper and the cell death domains of Fas and TNFR1. *Cell*, **81**, 185–186.

Grether, M.E., Abrams, J.M., Agapite, J., White, K. and Steller, H. (1995) The *head involution defective* gene of *Drosophila melanogaster* functions in programmed cell death. *Genes & Development* **9**, 1694–708.

Hawkins, C.J., Uren, A.G., Hacker, G., Medcalf, R.L. and Vaux, D.L. (1996) Inhibition of interleukin 1-beta-converting enzyme-mediated apoptosis of mammalian cells by baculovirus iap. *Proceedings of the National Academy of Sciences of the United States of America*, **93**, 13786–13790.

Hay, B.A., Wassarman, D.A. and Rubin, G.M. (1995) Drosophila homologs of baculovirus inhibitor of apoptosis proteins function to block cell death. *Cell*, **83**, 1253–62.

Hay, B.A., Wolff, T. and Rubin, G.M. (1994) Expression of baculovirus P35 prevents cell death in Drosophila. *Development*, **120**, 2121–9.

Hengartner, M.O., Ellis, R.E. and Horvitz, H.R. (1992) *Caenorhabditis elegans* gene *ced-9* protects cells from programmed cell death. *Nature*, **356**, 494–499.

Hengartner, M.O. and Horvitz, H.R. (1994) *C. elegans* cell survival gene *ced-9* encodes a functional homolog of the mammalian proto-oncogene *bcl-2*. *Cell*, **76**, 665–676.

Jacobson, M.D., Weil, M. and Raff, M.C. (1997) Programmed Cell death in Animal Development. *Cell*, **88**, 347–354.

Kenyon, C. (1994) If birds can fly, why can't we? Homeotic genes and evolution. *Cell*, **78**, 175–180.

Kimura, K. and Tanimura, T. (1992) Mutants with delayed cell death of the ptilinal head muscles in *Drosophila. J. Neurogenet.*, **8**, 57–69.

Koelle, M.R., Talbot, W.S., Segraves, W.A., Bender, M.T., Cherbas, P. and Hogness, D.S. (1991) The *Drosophila* EcR gene encodes an ecdysone receptor, a new member of the steroid receptor superfamily. *Cell*, **67**, 59–77.

Kondo, S. (1988) Altruistic cell suicide in relation to radition hormesis. *Int. J. Radiat. Biol.*, **53**, 95–102.

Krieger, M. and Herz, J. (1994) Structures and functions of multiligand lipoprotein receptors: Macrophage scavenger receptors and LDL receptor-related protein (LRP). *Annu. Rev. Biochem.* **63**, 601–637.

Krumlauf, R. (1994) Hox genes in vertebrate development. [Review]. *Cell*, **78**, 191–201.

Kumar, S. (1995) ICE-like proteases in apoptosis. [Review]. *Trends in Biochemical Sciences*, **20**, 198–202.

Lawrence, P.A. (1992) "The making of a fly: the genetics of animal design." Blackwell Scientific Publications, London.

Liston, P., Roy, N., Tamai, K., Lefebvre, C., Baird, S., Cherton, H.G., Farahani, R., McLean, M., Ikeda, J.E., MacKenzie, A. and Korneluk, R.G. (1996) Suppression of apoptosis in mammalian cells by NAIP and a related family of IAP genes. *Nature*, **379**, 349–53.

Lowe, S.W., Schmitt, E.M., Smith, S.W., Osborne, B.A. and Jacks, T. (1993) p53 is required for radiation-induced apoptosis in mouse thymocytes. *Nature*, **362**, 847–849.

Nelson, R.E., Fessler, L.I., Takgi, Y., Blumberg, B., Keene, D.R., Olson, P.F., Parker, C.G. and Fessler, J.H. (1994) Peroxidasin: a novel enzyme-matrix protein of *Drosophila* development. *EMBO J.*, **13**, 3438–47.

Nicholson, D.W. (1996) ICE/CED3-Like Proteases as therapeutic targets for the control of inappropriate apoptosis [Review]. *Nature Biotechnology*, **14**, 297–301.

Nishikawa, K., Arai, H. and Inoue, K. (1990) Scavenger receptor-mediated uptake and metabolism of lipid vesicles containing acidic phospholipids by mouse peritoneal macrophages. *J. Biol. Chem.*, **265**, 5226–5231.

Nordstrom, W., Chen, P., Steller, H. and Abrams, J.M. (1996) Activation of the reaper gene during ectopic cell killing in drosophila. *Developmental Biology*, **180**, 213–226.

Pearson, A., Lux, A. and Krieger, M. (1995) Expression cloning of dSR-CI, a class C macrophage-specific scavenger receptor from *Drosophila melanogaster. Proceedings of the National Academy of Sciences of the United States of America*, **92**, 4056–60.

Pronk, G.J., Ramer, K., Amiri, P. and Williams, L.T. (1996) Requirement of an ice-like protease for induction of apoptosis and ceramide generation by reaper. *Science*, **271**, 808–810.

Ramos, R.G., Igloi, G.L., Lichte, B., Baumann, U., Maier, D., Schneider, T., Brandstatter, J.H., Frohlich, A. and Fischbach, K.F. (1993) The *irregular chiasm C-roughest* locus of *Drosophila*, which affects axonal projections and programmed cell death, encodes a novel immunoglobulin-like protein. *Genes & Dev.*, **7**, 2533–2547.

Robinow, S., Talbot, W.S., Hogness, D.S. and Truman, J.W. (1993) Programmed cell death in the Drosophila CNS is ecdysone-regulated and coupled with a specific ecdysone receptor isoform. *Development*, **119**, 1251–9.

Rost, B. and Sander, C. (1994) Combining evolutionary information and neural networks to predict protein secondary structure. *Proteins*, **19**, 55–72.

Savill, J., Fadok, V., Henson, P. and Haslett, C. (1993) Phagocyte recognition of cells undergoing apoptosis. *Immunolgy Today*, **14**, 131–136.

Schwartz, L.M. (1992) Insect muscle as a model for programmed cell death. *Journal of Neurobiology*, **23**, 1312–1326.

Schwartz, L.M. and Truman, J.W. (1982) Peptide and steroid regulation of muscle degeneration in an insect. *Science*, 1420–1421.

Schwartz, L.M. and Truman, J.W. (1983) Hormonal control of rates of metamorphic development in the tobacco hornworm, Manduca sexta. *Dev. Biol.*, **99**, 103–144.

Sedlak, B.J. and Manzo, R. (1984) Localized cell death in *Drosophila* imaginal wing disc epithelium caused by the mutation *apterous-blot. Dev. biol.*, **104**, 489–496.

Singer, J.B., Harbecke, R., Kusch, T., Reuter, R. and Lengyel, J.A. (1996) Drosophila brachyenteron regulates gene activity and morphogenesis in the gut. *Development*, **122**, 3707–3718.

Song, Z.W., Mccall, K. and Steller, H. (1997) Dcp-1, a drosophila cell death protease essential for development. *Science*, **275**, 536–540.

Spector, M.S., Desnoyers, S., Hoeppner, D.J. and Hengartner, M.O. (1997) Interaction between the c-elegans cell-death regulators ced-9 and ced-4. *Nature*, **385**, 653–656.

Spreij, T.E. (1971) Cell death during the development of the imaginal discs of Calliphora erythrocephala. *Netherlands J. Zool.*, **21**, 221–264.

Steller, H. (1995) Mechanisms and Genes of Cellular Suicide. *Science*, **267**, 1445–1449.

Talbot, W.S., Swyryd, E.A. and Hogness, D.S. (1993) *Drosophila* tissues with different metamorphic responses to ecdysone express different ecdysone receptor isoforms. *Cell*, **73**, 1323–1337.

Teodoro, J.G. and Branton, P.E. (1997) REGULATION OF APOPTOSIS BY VIRAL GENE PRODUCTS [Review]. *Journal of Virology*, **71**, 1739–1746.

Tepass, U., Fessler, L.I., Aziz, A. and Hartenstein, V. (1994) Embryonic origin of hemocytes and their relationship to cell death in *Drosophila. Development*, **120**, 1829–1837.

Truman, J. (1984) Cell death in invertebrate nervous systems. *Ann. Rev. Neurosci.*, **7**, 171–188.

Truman, J. (1992) Insect systems for the study of programmed neuronal death. *Experimental Gerontology* **27**, 17–28.

Uren, A.G., Pakusch, M., Hawkins, C.J., Puls, K.L. and Vaux, D.L. (1996) Cloning and expression of apoptosis inhibitory protein homologs that function to inhibit apoptosis and/or bind tumor necrosis factor receptor-associated factors. *Proceedings of the National Academy of Sciences of the United States of America*, **93**, 4974–4978.

Varkey, J. and Abrams, J.M., (unpublished observations).

Vucic, D., Seshagiri, S. and Miller, L.K. (1997) Characterization of reaper- and fadd-induced apoptosis in a lepidopteran cell line. *Molecular & Cellular Biology*, **17**, 667–676.

White, K., Grether, M., Abrams, J.M., Young, L., Farrell, K. and Steller, H. (1994) Genetic Control of Programmed Cell Death in *Drosophila. Science*, **264**, 677–683.

White, K., Tahaoglu, E. and Steller, H. (1996) Cell killing by the drosophila gene reaper. *Science*, **271**, 805–807.

Wolff, T. and Ready, D.F. (1991) Cell death in normal and rough eye mutants of *Drosophila. Dev. Biol.*, **113**, 825–839.

Wolff, T. and Ready, D.F. (1993) Pattern Formation in the *Drosophila* retina. In "The Development of *Drosophila melanogaster*" (M., Bate and A.M., Arias, Eds.), Vol. 2, pp. 1277–1326. Cold Spring Harbor Laboratory Press, Cold Spring Harbor.

Wu, D., Wallen, H.D. and Nunez, G. (1997) Interaction and regulation of subcellular localization of ced-4 by ced-9. *Science*, **275**, 1126–1129.

Wurgler, F.E. and Ulrich, H. (1976) Radiosensitivity of embryonic stages. In "The genetics and biology of *Drosophila*" (M., Ashburner and E., Novitski, Eds.), Vol. 1c, pp. 1269–1298. Academic Press, London and New York.

Xue, D., Shaham, S. and Horvitz, H.R. (1996) The caenorhabditis elegans cell-death protein ced-3 is a cysteine protease with substrate specificities similar to those of the human cpp32 protease. *Genes & Development*, **10**, 1073–1083.

Yuan, J., Shaham, S., Ledoux, S., Ellis, H.M. and Horvitz, H.R. (1993) The C., elegans cell death gene *ced-3* encodes a protein similar to mammalian Interleukin-1B-converting enzyme. *Cell* **75**, 641–652.

Zhou, L., Hashimi, H., Schwartz, L.M. and Nambu, J.R. (1995) Programmed cell death in the Drosophila central nervous system midline. *Current Biology*, **5**, 784–90.

Zhou, L., Schnitzler, A., Agapite, J., Schwartz, L.M., Steller, H. and Nambu, J.R. (1997). Cooperative functions of the *reaper* and *head involution defective* genes in programmed cell death of the *Drosophila* CNS midline cells. *Proc. Nat'l. Acad. Sci.*, **94**, 5131–5136.

7. BACULOVIRAL LESSONS IN APOPTOSIS

CHRISTINE J. HAWKINS[*†], ELIZABETH J. COULSON AND DAVID L. VAUX

[*]*Caltech Division of Biology 156–29, 1201 East California Blvd., Pasadena, CA 91125, USA*

Apoptosis is an evolutionarily conserved process of fundamental importance that is used in multicellular organisms for development, homeostasis and defence. Cells often respond to infection by intracellular pathogens by undergoing apoptosis, thus limiting replication of the pathogen and spread of the infection. Viruses have evolved that carry anti-apoptotic genes to disable this response. In baculoviruses, two families of such cell survival genes have been found that encode the p35 proteins and the inhibitor of apoptosis (IAP) proteins. p35 inhibits apoptosis by binding to activated caspases, the key apoptosis effector proteases. How IAPs work has not been determined, but their existence in viral genomes as well as those from insects and mammals indicates their function is highly conserved. One member of this family has been implicated in an inherited neurodegenerative disease, and it is possible that other similar diseases may result from lesions in other members of this family. By studying these viral anti-apoptotic genes and their cellular relatives, much can be learnt not only about the control and implementation of the apoptotic process, but hopefully also about the clinical manifestations of abnormal regulation of apoptosis.

INTRODUCTION

Functions Accomplished by Apoptosis

Apoptosis is a physiological process used to eliminate unwanted cells. It is used during development, to sculpt the body. In adult organisms, it fulfils a homeostatic role, to regulate numbers of cells in organs and tissues by balancing mitotic cell production. Apoptosis is also used as a defence mechanism. Cell suicide can provide protection both against dangerously altered cells of the organism itself, as well as against intracellular pathogens such as viruses.

The ability of a virally infected cell to detect the infection and respond by killing itself limits the ability of the virus to replicate and hence protects the organism as a whole from a more widespread infection.

As a counter strategy some viruses carry genes which interfere with the ability of infected cells to activate apoptotic pathways, thus allowing the virus to replicate and spread to other cells. This is evident from experiments in which such genes are mutated, resulting in reduced infectivity of the virus due to enhanced death of infected cells. Such a process of altruistic apoptosis is most readily understood evolutionarily in the context of multicellular organisms, although there are reports of similar phenomena occurring in single cellular organisms where genetically related cells nearby may be saved by an infected cell altruistically killing itself (Shub, 1994, Yarmolinsky, 1995, Cornillon *et al.*, 1994).

† *Corresponding Author*: Tel.: 16263956451. Fax: 1626449 0756. e-mail: hawkins @wehi.edu.au

VIRAL ANTI-APOPTOTIC GENES

Apoptosis is used as an anti-viral defence, and as one might expect, some viruses have evolved strategies which enable them to evade this host response. Various viruses have adopted different ways of achieving this. Some, such as Epstein Barr Virus, African Swine Fever Virus, and Kaposi sarcoma associated Human Herpes Virus 6 carry genes which resemble *bcl-2*, both in sequence and function (Henderson *et al.*, 1993; Neilan *et al.*, 1993; Cheng *et al.*, 1997; Sarid *et al.*, 1997).

A second molecular tactic used by viruses to foil their host's apoptotic response to infection is to carry genes encoding proteins which interfere with components of the cell death machinery, such as the ICE/ced-3 like proteases. These key effector proteases of apoptosis have been termed caspases because they are cysteine proteases which cleave after aspartate residues. The first members of this family to be described were the *C. elegans* gene *ced-3* (Yuan *et al.*, 1993) and mammalian interleukin-1β-converting enzyme (ICE) (Cerretti *et al.*, 1992; Thornberry *et al.*, 1992), but many other related genes have recently been cloned from mammals and insects (Alnemri, 1997; Song *et al.*, 1997; Ahmad *et al.*, 1997).

One method used by viruses to interfere with the host's apoptotic response to infection is to disable caspases. For example, the crmA, encoded by cowpox virus, binds with high affinity to ICE (caspase-1) and FLICE (caspase-8) (Ray *et al.*, 1992; Srinivasula *et al.*, 1996), acting as a pseudosubstrate to prevent their proteolytic activity (Ray *et al.*, 1992; Komiyama *et al.*, 1994). *In vitro*, crmA can also inhibit granzyme B, a serine protease expressed by cytotoxic T cells (Quan *et al.*, 1995).

The finding that lymphocytes from transgenic mice expressing crmA were resistant to apoptosis induced by ligation of CD95, but remained sensitive to apoptosis induced in other ways, such as by irradiation or treatment with dexamethasone, showed that apoptosis could occur by independent pathways, only some of which required crmA-inhibitable caspases (Smith *et al.*, 1996).

Recently a mammalian gene has been identified that closely resembles crmA. This protein, PI-9, can, like crmA, inhibit granzyme B *in vitro* (Sun *et al.*, 1996), but its activity with respect to the caspases has not been published to date.

p35 – A Baculovirus Apoptosis Inhibitory Protein

p35 was originally identified as one of a pair of non-overlapping genes from *Autographa californica* nuclear polyhedrosis virus (AcNPV) which were transcribed in opposite directions from the same region of DNA (Friesen and Miller, 1987). At the time it was cloned its function was unknown; its sequence did not reveal the presence of any recognisable motifs, and it showed no homology to any other gene.

The role of p35 was revealed from its recognition as the gene mutated in the "Annihilator" strain of the AcNPV baculovirus (Clem *et al.*, 1991). When susceptible insect cells are infected with Annihilator virus, in contrast to wild type baculovirus, many of the cells die by apoptosis (Hershberger *et al.*, 1992; Clem *et al.*, 1994). DNA encoding p35 was able to inhibit apoptosis caused by Annihilator virus. p35 was also able to prevent insect cell apoptosis triggered in other ways, such as treatment with actinomycin D (Crook *et al.*, 1993; Clem and Miller, 1994).

A p35 homologue has been found in one other baculovirus species to date, *Bombyx mori* nuclear polyhedrosis virus (BmNPV). The BmNPV p35 is very similar at to that from AcNPV, and can also prevent the apoptotic response of insect cells to infection. However, cellular homologues of p35 are yet to be identified.

Baculovirus p35 is able to prevent apoptosis in cells from organisms other than insects. Mammalian neuronal cells were protected from death induced by withdrawal of either serum or glucose, or treatment with a calcium ionophore (Rabizadeh *et al.*, 1993). The ability of p35 to prevent neuronal death was confirmed by Martinou *et al.* using sympathetic neurons dissociated from the superior cervical ganglia of newborn rats (Martinou *et al.*, 1995). In this system cells microinjected with p35 cDNA exhibited a survival advantage over control cells following nerve growth factor withdrawal. Bcl-2 also conferred this anti-apoptotic effect, but this could be blocked by coexpression of the Bcl-2 family pro-apoptotic member Bcl-X_S. The protection mediated by p35 was not blocked by Bcl-X_S, which indicated that p35 worked in a different way to Bcl-2.

In mammalian cells p35 was able to suppress apoptosis induced by ligation of Tumor Necrosis Factor Receptor (TNFR) family members such as TNFRI and CD95. This was demonstrated by Beidler *et al.* using cell lines sensitive to TNF and anti-CD95 antibodies (Beidler *et al.*, 1995). They showed that p35 not only inhibited the morphological changes associated with apoptosis, but also the downstream molecular events such as cleavage of PARP by caspases.

p35 can also prevent apoptosis in nematodes, as when expressed in *C. elegans*, it protected many cells from developmentally programmed apoptosis (Sugimoto *et al.*, 1994). It could also partially rescue the lethality of a *ced-9* mutation (Sugimoto *et al.*, 1994), suggesting that, as had been shown for Bcl-2 (Vaux *et al.*, 1992; Hengartner and Horvitz, 1994), p35 works at the same step or downstream of ced-9 in a well conserved apoptotic pathway.

In *Drosophila*, p35 has been shown to prevent developmental cell death (Hay *et al.*, 1994) and that induced by enforced expression of the fly cell death genes *rpr* (White *et al.*, 1996), *hid* (Grether *et al.*, 1995) and *grim* (Chen *et al.*, 1996).

The mechanism by which p35 protects cells from apoptosis was determined by Bump *et al.* (1995) and Xue and Horvitz (1995). p35, like crmA, binds to the caspases and inhibits them from cleaving other substrates. p35 is cleaved as a result of this binding and the two halves remain attached to the caspase molecule. For this reason, p35 must be present in at least an equimolar concentration with respect to the caspase(s) which it needs to inhibit (Xue and Horvitz, 1995; Bertin *et al.*, 1996). p35 can inhibit a wider range of caspases than crmA, as it has been shown to directly inhibit Caspases 1 (Xue and Horvitz, 1995; Bump *et al.*, 1995), 2 (Bump *et al.*, 1995), 3 (Bump *et al.*, 1995), 4 (Bump *et al.*, 1995) and *ced-3* (Xue and Horvitz, 1995) but not granzyme B (Bump *et al.*, 1995). Recently insect caspases from *Spodoptera frugiperda*; a baculoviral host species (Ahmad *et al.*, 1997) and *Drosophila* (Song *et al.*, 1997) have been identified, which may represent natural targets of p35.

THE IAP FAMILY OF APOPTOSIS INHIBITORS

Insect Virus IAP Proteins

In an attempt to isolate genes for proteins with similar activities to p35, a modified version of the complementation strategy which yielded p35 was repeated using DNA from a second baculovirus strain, *Cydia Pomenella* granulosis virus (CpGV), cloned into cosmids (Crook *et al.*, 1993). One cosmid was capable of preventing apoptosis induced by infection with the p35 mutant baculovirus. The gene identified was not related to p35, and was designated IAP, for Inhibitor of Apoptosis (Crook *et al.*, 1993).

Unlike p35, the sequence of IAP bore identifiable motifs. It contained a zinc finger of a previously described type referred to as a **Ring** Finger (Saurin *et al.*, 1996), similar to those found in PML (Kakizuka *et al.*, 1991), BMI-1 (Haupt *et al.*, 1991), and c-cbl (Blake *et al.*, 1991) among many others. It was initially speculated that, like some other zinc fingers, **Ring** Fingers may bind nucleic acid (Lovering *et al.*, 1993). This, however, has not been confirmed in physiological conditions.

Although the crystal structures of the RING finger containing molecules PML and XNF7 (Borden *et al.*, 1995a; Borden *et al.*, 1995b) have been solved, the function of the RING finger domain is still not known. It has been speculated that this domain mediates protein-protein interactions (Borden *et al.*, 1995a; Borden *et al.*, 1995b), but this has not been experimentally demonstrated. Another possibility is that the RING finger acts as a structural component that does not bind to other molecules, but acts to hold other folds of the polypeptide in place.

The RING finger motif found in IAP is slightly different from those found in other proteins, with the consensus sequence being $CX_2CX_{11}CXHX_3CX_2CX_{7\text{-}8}CPXCR$. In IAP, unlike most other RING finger-containing proteins, this domain is located in the carboxyl half of the protein.

The other discernible feature of the IAP sequence was a repeated motif in the amino terminal half of the molecule, which was termed BIR (for baculoviral IAP-like repeat). This motif, with the consensus $GX_2YX_4DX_3CX_2CX_6\ WX_9HX_{6\text{-}10}C$, was repeated twice in IAP (Crook *et al.*, 1993). Although the conservation of cysteine and histidine residues in this motif suggested metal binding properties (Birnbaum *et al.*, 1994), the function of the BIRs was unclear at the stage that the baculoviral IAPs were cloned.

The sequence of the IAP cloned from CpGV (CpIAP) resembled two stretches of DNA in the Genbank database, one from AcNPV and one from *Chilo iridescent* virus (CIV). These genes are designated AcIAP and CiIAP to reflect their sources. A fourth baculovirus IAP gene was isolated from *Orygia pseudosugata* NPV by complementation of the annihilator phenotype, and is called OpIAP. Of these IAPs, CpIAP and OpIAP can act like p35 to prevent apoptosis, while the less closely related AcIAP is inactive (Birnbaum *et al.*, 1994). It is possible that AcIAP has another function, unrelated to the control of apoptosis, which promotes its retention in the AcNPV virus, although an AcIAP deficient virus displayed a wildtype phenotype.

In addition to blocking apoptosis when expressed as part of the virus, CpIAP and OpIAP, like p35, also had anti-apoptotic properties when expressed from plasmids in the absence of the other viral genes (Clem and Miller, 1994). This showed that the IAPs were acting directly on components of the insect cells' apoptosis machinery, rather than in conjunction with other viral gene products. In experiments assaying the IAPs' ability to prevent actinomycin D induced cell death, OpIAP gave consistently better protection than CpIAP (Clem and Miller, 1994). As had been shown in the context of the whole virus, AcIAP was unable to prevent apoptosis when expressed from a plasmid, and further was unable to increase the protection conferred by p35 when both were expressed in the same cells (Clem and Miller, 1994). Interestingly, Bcl-2 and a viral homologue E1B-19K were incapable of suppressing apoptosis in this assay, suggesting that IAP and Bcl-2 act in separate pathways.

OpIAP, like p35, could also potently prevent apoptosis triggered in the insect cell line SF21 by overexpression of *rpr* (Vucic *et al.*, 1997), whereas the protection offered by Bcl-2 family members was negligible.

The creation of chimeric molecules between the functional CpIAP and the inactive AcIAP allowed the definition of regions which were essential for the protection offered by CpIAP (Clem and Miller, 1994). Regions at both end of CpIAP (both the BIRs and the RING finger) were required for its function, as replacing either region with the corresponding sequence from AcIAP abolished protection. Swapping the RING finger section with that from OpIAP however still produced an active molecule (Clem and Miller, 1994).

Protection by IAPs in Evolutionarily Divergent Species

To investigate whether the protection conferred by the baculoviral IAPs was restricted to insect cells, its ability to inhibit mammalian cell death was investigated. The most well studied component of the apoptotic machinery are the ICE-like cysteine proteases (caspases). Overexpression of caspases causes apoptosis in many cell types, as first demonstrated by Miura *et al.* (1993), who transfected cells with plasmids expressing caspases and the β-galactosidase enzyme, and assayed the viability of blue staining cells. Using this technique it was shown the OpIAP could protect against death induced by overexpression of Caspases 1 (Hawkins *et al.*, 1996, Duckett *et al.*, 1996) and 2 (Hawkins *et al.*, 1996), but not 7 (Hawkins *et al.*, 1996). OpIAP was also partially able to suppress death caused by overexpression of a component of the CD95 signalling machinery; FADD/MORT-1 (Boldin *et al.*, 1995, Chinnaiyan *et al.*, 1995, Hawkins *et al.*, 1996).

Using Sindbis virus, it was shown that OpIAP could inhibit apoptosis triggered by viral infection of mammalian cells (Duckett *et al.*, 1996). These results indicated that, like p35 and Bcl-2, IAP proteins could inhibit components of the apoptosis signal transduction machinery which are conserved between evolutionarily divergent species.

Cellular IAP Genes

As baculoviral IAPs can prevent death of mammalian cells, they must interact with components of the apoptotic machinery that are conserved between insects and

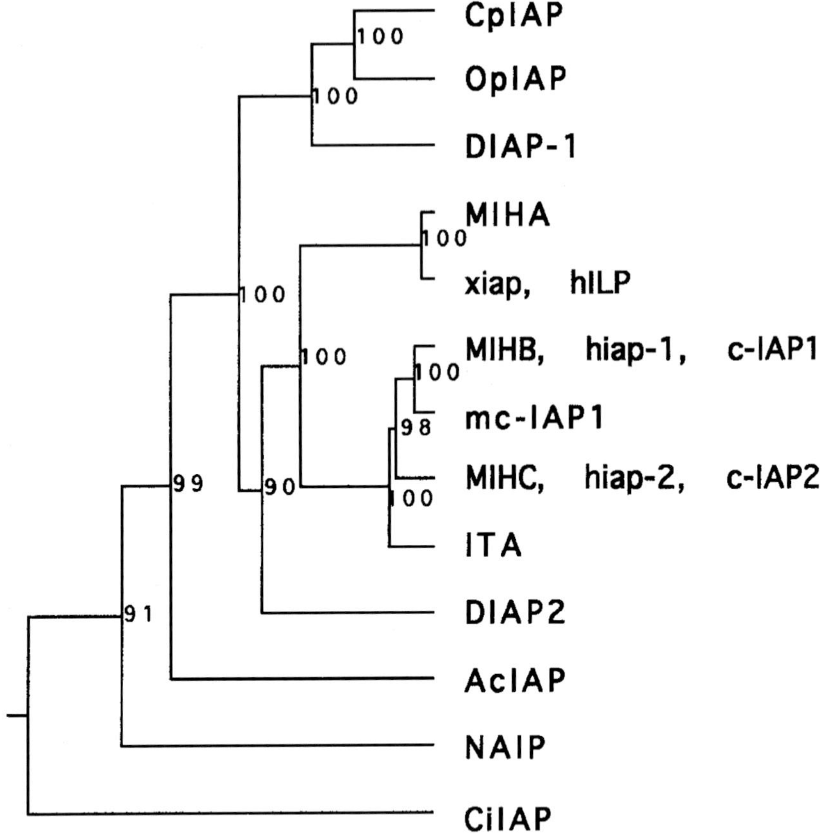

Figure 7.1 IAP Family Phylogenetic Tree. The full length amino acid sequences (except NAIP; for which only the BIR-containing portion was used) of the IAP family members identified to date were phylogenetically compared. Relationships were determined from a CLUSTAL5 alignment of the amino acid sequences using programs from the PHYLUP program package. The phylogeny shown was calculated by a rooted, distance matrix (PROTPARS, KITCH) on default setting with random input order. Bootstrap analysis using 100 replicas (SEQBOOT, CONCENSE) shows the degree of confidence for the position of each branch to the others. The Genbank assession numbers used for this analysis were L22564 (OpIAP), L05494 (CpIAP), M96361 (AcIAP), M81387 (CiIAP), L49440 (DIAP-1), L49441 (DIAP-2), U19251 (NAIP), U36842 (MIHA), U45880 (xiap), U37547 (MIHB), L49433 (mc-IAP-1) U37546 (MIHC), U27466 (ITA).

mammals. It seemed likely that the IAPs may have cellular homologues in insects (presumably these would be the ancestral source from which the baculoviruses obtained their IAP genes). It was also probable that there would be IAP homologues in mammals, since the insect virus IAPs inhibit mammalian cell death. Such cellular IAP homologues from mammals, insects and birds do in fact exist, and have been cloned independently by several groups. Figure 7.1 illustrates the similarities of the IAPs published to date.

Mammalian IAPs

The first cellular gene bearing homology to the viral IAPs was one of two candidate genes for Spinal Muscular Atrophy (SMA), a fatal autosomal recessive neuronal disease (Munsat, 1991; Roy *et al.*, 1995).

SMA is classified into three types based on severity and age of onset. The disease is characterised by neuronal degeneration, stemming from death of the anterior horn cells of the spinal cord. The gene(s) for SMA had been previously shown to lie in a region of chromosome 5q13. The Survival Motor Neuron (SMN) and Neuronal Apoptosis Inhibitory Protein (NAIP) genes were cloned by mapping the region and determining which genes were deleted or mutated in SMA sufferers but not in unaffected individuals. SMN mutations or deletions were found in almost all patients with SMA. Lesions in the NAIP gene however were less common, and tended to be found in patients with more severe SMA (Roy *et al.*, 1995; Wirth *et al.*, 1995; Burlet *et al.*, 1996; Hahnen *et al.*, 1995). Around 67% of patients with the more serious, early onset type I SMA, also known as Werdnig-Hoffmann disease, had defets in NAIP, compared with 42% for patients with types II or III SMA (Roy *et al.*, 1995).

Analysis of the sequence of NAIP revealed the presence of the BIR motif first described in IAP genes. As the SMA phenotype results from inappropriate death of specific neurons (Fidzianska *et al.*, 1990), the demonstration that one of the candidate SMA genes shared homology with IAPs suggested that the wild type NAIP protein may act by inhibiting cell death. Unlike the baculoviral IAPs, NAIP does not possess a RING finger. The remainder of the NAIP gene contains regions which resemble a GTP/ATP binding site and membrane spanning domains.

Three other mammalian IAP homologues were identified in EST databases by virtue of their similarity to baculoviral IAPs (Liston *et al.*, 1996; Uren *et al.*, 1996; Duckett *et al.*, 1996). These genes were given different names by the various laboratories. One, located on the X chromosome was named MIHA (Uren *et al.*, 1996), hILP (Duckett *et al.*, 1996) and xiap (Liston *et al.*, 1996). Other IAPs, closely related to each other were called MIHB/hiap-1 and MIHC/hiap-2 (Uren *et al.*, 1996; Liston *et al.*, 1996). For simplicity, the genes will be referred to as MIHA, MIHB and MIHC hereafter.

Another group cloned two of the mammalian IAPs while searching for proteins which formed a complex with the cytoplasmic domain of TNFRII. TNFRII (p75) is a member the TNFR superfamily which usually triggers cellular proliferation via activation of NF-κB, but can in certain circumstances signal death (Tartaglia *et al.*, 1991; Tartaglia and Goeddel, 1992; Grell *et al.*, 1993). Two TNF receptor associated factors (TRAF1 and TRAF2) had already been identified by co-immunoprecipitation, followed by protein sequencing (Rothe *et al.*, 1994). An additional protein was also associated with the TNFRII/TRAF complex, and when purified and sequenced, was found to resemble the baculoviral IAPs (Rothe *et al.*, 1995a). The gene encoding this protein, and a second highly related gene were cloned. They were called c-IAP1 and 2 (Rothe *et al.*, 1995a), and were identical to MIHB and C respectively.

Association of IAPs with TRAFs

To date the only IAPs that can bind to TRAF molecules are MIHB (c-IAP1) and MIHC (c-IAP2) which can both bind to TRAF1 and TRAF2. TRAF2 also binds to the intracellular region of TNFRII. TRAF1 binds to TNFRII indirectly, via its association with TRAF2. MIHB and C were shown to interact, via their BIR motifs, with the C-terminal half of TRAFs 1 and 2. By virtue of their association with the TRAFs, the IAP molecules are recruited to the TNF receptor upon ligand binding (Rothe *et al.*, 1995). In addition to binding IAPs and TNFRII, TRAF2 can also interact with TRADD, and via this association transduces signals for NF-κB and JNK activation from the TNFRI and CD40 (Rothe *et al.*, 1995b; Hsu *et al.*, 1996; Liu *et al.*, 1996). Despite its association with proteins which can transmit apoptotic signals (TNFRII and TRADD), TRAF2 does not appear to play a role in apoptosis signal transduction, as dominant negative truncated TRAF2 proteins that prevent activation of NF-κB by TNFR family members do not prevent induction of cell death by TNF (Rothe *et al.*, 1995b; Hsu *et al.*, 1996; Liu *et al.*, 1996).

Endogenous insect IAPs

Two *Drosophila* IAP genes have also been cloned. Hay *et al.* (1995) found DIAP1 in a screen for enhancers of cell death triggered by enforced ectopic expression of the cell death gene reaper in the fly eye. This group and others also cloned a related *Drosophila* gene (DIAP2/DIHA/dILP) through searches of gene databases (Hay *et al.*, 1995; Uren *et al.*, 1996; Liston *et al.*, 1996).

An Avian IAP gene

A T cell specific chicken IAP homologue has also been published (Digby *et al.*, 1996), but its anti-apoptotic potential has not been reported to date.

Protection Conferred by Cellular IAPs

Two of the mammalian IAPs (MIHA and B) can inhibit apoptosis mediated by overexpression of Caspase-1, but the protection conferred is not complete, and is not as great as that afforded by OpIAP (Uren *et al.*, 1996). Whe death is induced by overexpression of FADD, OpIAP rescues around half the FADD-sensitive transfected cells (Hawkins *et al.*, 1996) whereas the mammalian proteins gave negligible protection (Uren *et al.*, 1996). Confirming these results, Duckett *et al.* also showed that MIHA could suppress death induced by Caspase-1 overexpression, and additionally demonstrated protection against apoptosis triggered by Sindbis virus infection (Duckett *et al.*, 1996). Liston *et al.* described some level of protection by all three MIH genes against apoptosis induced by menadione treatment and serum withdrawal (Liston *et al.*, 1996). The Spinal Muscular Atrophy gene, NAIP, was also capable of inhibiting cell death in these assays (Liston *et al.*, 1996), with its strongest effect being against apoptosis induced by treatment with menadione.

Both DIAP1 and 2 could inhibit developmental (Hay *et al.*, 1995), *rpr* induced (Hay *et al.*, 1995; Vucic *et al.*, 1997) and *hid* induced *Drosophila* cell death (Hay *et al.*,

1995). Expression of a truncated form of MIHB in *Drosophila* eyes could partially alleviate *reaper* induced apoptosis, although the full length protein did not protect (Hay *et al.*, 1995).

Structure of Cellular IAPs

The overall structure of these cellular IAPs is similar to the viral proteins, except that all but DIAP1 possess three BIRs, rather than two, and NAIP lacks the RING finger motif. The homology between the IAPs is strongest in the BIR and RING finger domains, suggesting that these are functional motifs. There is controversy in the literature about the functional roles of these domains. The baculoviral IAPs require both the BIRs and RING finger to prevent apoptosis in both insect and mammalian systems (Clem and Miller, 1994; Hawkins *et al.*, 1996 and our unpublished results). However in the *Drosophila* eye, both DIAP1 and MIHB appeared more effective when their RING finger are removed (Hay *et al.*, 1995).

It was surprising that the mammalian IAPs were generally less efficient apoptosis inhibitors than the insect virus OpIAP (Uren *et al.*, 1996), although one group reported equivalent protection from OpIAP as MIHA (Duckett *et al.*, 1996). There are numerous possible explanations for a less potent anti-apoptotic effect of OpIAP than its cellular homologues. Possibly there are factors in the cell lines used which modulate the activity of MIH proteins, but are ineffective against IAPs encoded by insect viruses. It seems plausible that the cells may carry genes which regulate the activity of their own apoptosis inhibitors, and it also is likely that viruses would benefit from mutations in the death inhibitor genes they carry which render their products no longer susceptible to this control by the host.

It is curious that MIHB and MIHC are the most similar of the mammalian IAPs, but MIHB is much more potent that MIHC. The negligible protection offered by MIHC is reminiscent of AcIAP, which is also non-functional (with respect to the assays performed to date). It is possible that both of these proteins work to inhibit other apoptosis pathways than those tested. Alternatively they may have roles not related to the regulation of apoptosis, or they may have once been used to inhibit cell death, but with this role fulfilled by other genes (p35 in the case of AcNPV and perhaps MIHA or MIHB for mammals), there has been no selective pressure against mutations, leaving the products of the genes inactive.

The observation that some IAP genes could bind to TRAFs was intriguing, as it suggested a mechanism of action for the IAPs – to inhibit signalling by members of the TNFR family. However three lines of evidence make this theory less plausible. Neither the baculoviral IAPs nor MIHA can bind to the TRAFs assayed to date (Uren *et al.*, 1996), and our unpublished results indicate that the neither MIHA, B nor C can bind TRAFs 4, 5 and 6. Also, TRAF2 dominant negative constructs repress NF-κB and JNK activation, but do not affect apoptosis signalling (Rothe *et al.*, 1995b; Hsu *et al.*, 1996; Liu *et al.*, 1996). Thirdly, and less conclusively, proteins which interact with IAPs to inhibit cell death would be expected to be found in distantly related animals, as the IAPs are. A search of the Genbank databases revealed that the TRAF-C region of TRAF2 (which interacts with MIHB and MIHC) did not resemble any non-mammalian genes. This would be consistent with

TRAFs being an evolutionarily recent development, appearing first during mammalian evolution. Of course, mammalian genes may be over-represented in these databases, so there may be TRAFs from insects, for example, which have not yet been cloned.

During evolution IAPs were probably retained to control cell death. Presumably through a series of duplication events, mammals came to possess multiple copies. Possibly in mammals some of these like MIHA still act primarily to inhibit apoptosis, whereas other (like MIHB and C) have evolved other roles. With the evolution of the TRAFs, MIHB and C could have adapted to bind them and thereby influence signalling pathways controlled by TRAFs. Hopefully, gene deletion studies will reveal more of the roles of the cellular IAP homologs.

CONCLUSION

Valuable lessons about the regulation of cell death have been learnt from studies of baculoviral inhibitors of apoptosis. The elucidation of the mechanism of action of p35 has permitted an appreciation of the importance of the caspases in apoptosis, and will assist in research which could lead to the generation of therapeutic agents to modulate cell death. The search for the mechanism by which IAPs exert their anti-apoptotic effects is still underway; hopefully the answer to this problem may be as revealing and useful. The association of mutations in an IAP gene (NAIP) in Spinal Muscular Atrophy raises the possibility of the involvement of the IAP family in neurological and other degenerative diseases, possibly including Motor Neurone Disease (Jackson *et al.*, 1996).

ACKNOWLEDGMENTS

This work was supported by the NH&MRC of Australia. DLV is an Investigator of the Cancer Research Institute of New York and the Edward Dunlop Fellow of the ACCV.

REFERENCES

Ahmfad, M., Srinivasula, S.M., Wang, L.J., Litwack, G., Fernandes-Alnemri, T. and Alnemri, E.S. (1997) *Spodoptera frugiperda* caspase-1, a novel insect death protease that cleaves the nuclear immunophilin fkbp46, is the target of the baculovirus antiapoptotic protein p35. *J. Biol. Chem.*, **272**, 1421–1424.

Alnemri, E.S. (1997) Mammalian cell death proteases – a family of highly conserved asparate specific cysteine proteases. *J. Cell. Biochem.*, **64**, 33–42.

Beidler, D.R., Tewari, M., Friesen, P., Poirier, G. and Dixit, V.M. (1995) The baculovirus p35 protein inhibits Fas- and Tumor Necrosis Factor-induced apoptosis. *J. Biol. Chem.*, **270**, 16526–16528.

Bertin, J., Mendrysa, S.M., Lacount, D.J., Gaur, S., Krebs, J.F., Armstrong, R.C., Tomaselli, K.J. and Friesen, P.D. (1996) Apoptotic suppression by baculovirus p35 involves cleavage by and inhibition of a virus-induced ced-3/ICE-like protease. *J. Virol.*, **70**, 6251–6259.

Birnbaum, M.J., Clem, R.J. and Miller, L.K. (1994) An apoptosis-inhibiting gene from a nuclear polyhedrosis virus encoding a polypeptide with cys/his sequence motif. *J. Virol.*, **68**, 2521–2528.

Blake, T.J., Shapiro, M., Morse, H.C. and Langdon, W.Y. (1991) The sequences of the human and mouse *c-cbl* proto-oncogenes show *v-cbl* was generated by a large truncation encompassing a proline-rich domain and a leucine zipper-like motif. *Oncogene*, **6**, 653–657.

Boddy, M.N., Freemont, P.S. and Borden, K. (1994) The p53-associated protein mdm2 contains a newly characterized zinc binding domain called the RING finger. *Trends Biochem. Sci.*, **19**, 198–199.

Boldin, M.P., Goncharov, T.M., Goltsev, Y.V. and Wallach, D. (1996) Involvement of MACH, a navel MORT1/FADD-interacting protease, in Fas/APO-1-and TNF receptor-induced cell death. *Cell*, **85**, 803–815.

Boldin, M.P., Varfolomeev, E.E., Pancer, Z., Mett, I.L., Camonis, J.H. and Wallach, D. (1995) A novel protein that interacts with the death domain of Fas/APO-1 contains a sequence motif related to the death domain. *J. Biol. Chem.*, **270**, 7795–7798.

Borden, K., Lally, J.M., Martin, S.R., O'Reilly, N.J., Etkin, L.D. and Freemont, P.S. (1995a) Novel topology of a zinc-binding domain from a protein involved in regulating early *Xenopus* development. *EMBO J.*, **14**, 5947–5956.

Borden,. K.L., Boddy, M.N., Lally, J., O'Reilly, N.J., Martin, S., Howe, K., *et al.* (1995b) The solution structure of the RING finger domain from the acute promyelocytic leukaemia proto-oncoprotein PML. *EMBO J.*, **14**, 1532–41.

Bump, N.J., Hackett, M., Hugunin, M., Seshagiri, S., Brady, K., Chen, P., *et al.* (1995) Inhibition of ICE family proteases by baculovirus antiapoptotic protein p35. *Science*, **269**, 1885–1888.

Burlet, P., Burglen, L., Clermont, O., Lefebvre, S., Viollet, L., Munnich, A., *et al.* (1996) Large scale deletions of the 5q13 region are specific to Werdnig-Hoffmann disease. *J. Med. Gen.*, **33**, 281–3.

Cerretti, D.P., Kozlosky, C.J., Mosley, B., Nelson, N., Van Ness, K., Greenstreet, T.A., *et al.* (1992) Molecular cloning of the interleukin-1 beta converting enzyme. *Science*, **256**, 97–100.

Chen, P., Nordstrom, W., Gish, B. and Abrams, J.M. (1996) *Grim*, a novel cell death gene in *Drosophila. Genes & Dev.*, **10**, 1773–1782.

Cheng, E., Nicholas, J., Bellows, D.S., Hayward, G.S., Guo, H.G., Reitz, M.S. and Hardwick, J.M. (1997) A Bcl-2 homolog encoded by kaposi sarcoma-associated virus, human herpesvirus 8, inhibits apoptosis but does not heterodimerize with Bax or Bak. *Proc. Natl. Acad. Sci. USA*, **94**, 690–694.

Chinnaiyan, A.M., O'Rourke, K., Tewari, M. and Dixit, V.M. (1995) FADD, a novel death domain-containing protein, interacts with the death domain of Fas and initiates apoptosis. *Cell*, **81**, 505–512.

Clem, R.J., Fechheimer, M. and Miller, L.K. (1991) Prevention of apoptosis by a baculovirus gene during infection of insect cells. *Science*, **254**, 1388–90

Clem, R.J. and Miller, L.K. (1994) Control of programmed cell death by the baculovirus genes p35 and IAP. *Mol. Cell. Biol.*, **14**, 5212–22.

Clem, R.J., Robson, M. and Miller, L.K. (1994) Influence of infection route on the infectivity of baculovirus mutants lacking the apoptosis-inhibiting gene p35 and the adjacent gene p94. *J. Virol.*, **68**, 6759–6762.

Cornillon, S., Foa, C., Davoust, J., Buonavista, N., Gross, J.D. and Golstein, P. (1994) Programmed cell death in *Dictyostelium. J. Cell Sci.*, **107**, 2691–2704.

Cowan, W.M., Fawcett, J.W., O'Leary, D.D. and Stanfield, B.B. (1984) Regressive events in neurogenesis. *Science*, **225**, 1258–65.

Crook, N.E., Clem, R.J. and Miller, L.K. (1993) An apoptosis inhibiting baculovirus gene with a zinc finger like motif. *J. Virol.*, **67**, 2168–2174.

Digby, M.R., Kimpton, W.G., York, J.J., Connick, T.E. and Lowenthal, J.W. (1996) ITA, a vertebrate homologue of IAP that is expressed in T lymphocytes. *DNA&Cell Biol.*, **15**, 981–988.

Duckett, C.S., Nava, V.E., Gedrich, R.W., Clem, R.J., Van Dongen, J.L., Gilfillan, M.C., Shiels, H., *et al.* (1996) A conserved family of cellular genes related to the baculovirus IAP gene and encoding apoptosis inhibitors. *EMBO J.*, **15**, 2685–94.

Fidzianska, A., Goebel, H.H. and Warlo, I. (1990) Acute infantile spinal muscular atrophy. Muscle apoptosis as a proposed pathogenetic mechanism. *Brain*, **113**, 433–45.

Friesen, P.D. and Miller, L.K. (1987) Divergent transcription of early 35- and 94-kilodalton protein genes encoded by the HindIII K genome fragment of the baculovirus *Autographa californica* nuclear polyhedrosis virus. *J. Virol.*, **61**, 2264–2272.

Grell, M., Scheurich, P., Meager, A. and Pfizenmaier, K. (1993) TR60 and TR80 tumor necrosis factor (TNF)-receptors can independently mediate cytolysis. *Lymphokine & Cytokine Res.*, **12**, 143–148.

Grether, M.E., Abrams, J.M., Agapite, J., White, K. and Steller, H. (1995) The head involution defective gene of *Drosophila melanogaster* functions in programmed cell death. *Genes & Dev.*, **9**, 1694–1708.

Hahnen, E., Forkert, R., Marke, C., Rudnik-Schoneborn, S., Schonling, J., Zerres, K., *et al.* (1995) Molecular analysis of candidate genes on chromosome 5q13 in autosomal recessive spinal muscular atrophy: evidence of homozygous deletions of the SMN gene in unaffected individuals. *Human Mol. Gen.*, **4**, 1927–1933.

Hamburger, V. (1975) Cell death in the development of the lateral motor column of the chick embryo. *J. Comp. Neurol.*, **160**, 535–546.

Haupt, Y., Alexander, W.S., Barri, G., Klinken, S.P. and Adams, J.M. (1991) Novel zinc finger gene implicated as myc collaborator by retrovirally accelerated lymphomagenesis in E mμ-*myc* transgenic mice. *Cell*, **65**, 753–763.

Hawkins, C.J., Uren, A.G., Hacker, G., Medcalf, R.L. and Vaux, D.L. (1996) Inhibition of interleukin 1β-converting enzyme-mediated apoptosis of mammalian cells by baculovirus IAP. *Proc. Natl Acad. Sci. USA*, **93**, 13786–13790.

Hay, B.A., Wassarman, D.A. and Rubin, G.M. (1995) *Drosophila* homologs of baculovirus inhibitor of apoptosis proteins function to block cell death. *Cell*, **83**, 1253–1262.

Hay, B.A., Wolff, T. and Rubin, G.M. (1994) Expression of baculovirus p35 prevents cell death in *Drosophila*. *Development*, **120**, 2121–2129.

Hengartner, M.O. and Horvitz, H.R. (1994) *C. elegans* cell survival gene *ced-9* encodes a functional homolog of the mammalian proto-oncogene *bcl-2*. *Cell*, **76**, 665–676.

Henderson, S., Huen, D., Rowe, M., Dawson, C., Johnson, G. and Rickinson. A. (1993) Epstein barr virus coded BHRF1 protein, a viral homog of Bcl 2, protects human B cells from programmed cell death. *Proc. Natl. Acad. Sci. USA*, **90**, 9479–8483.

Hershberger, P.A., Dickson, J.A. and Friesen, P.D. (1992) Site-specific mutagenesis of the 35-kilodalton protein gene encoded by *Autographa californica* nuclear polyhedrosis virus: cell line-specific effects on virus replication. *J. Virol.*, **66**, 5525–5533.

Hsu, H., Shu, H.B., Pan, M.G. and Goeddel, D.V. (1996) TRADD-TRAF2 and TRADD-FADD interactions define two distinct TNF receptor 1 signal transduction pathways. *Cell*, **84**, 299–308.

Jackson, M., Morrison, K.E., Alchalabi, A., Bakker, M. and Leigh, P.N. (1996) Analysis of chromosome 5q13 genes in amyotrophic lateral sclerosis – homozygous NAIP deletion in a sporadic case. *Ann. Neurol.*, **39**, 796–800.

Kakizuka, A., Miller, W., Jr., Umesono, K., Warrell, R., Jr., Frankel, S.R., Murty, V.V., *et al.* (1991) Chromosomal translocation t(15;17) in human acute promyelocytic leukemia fuses RAR alpha with a novel putative transcription factor, PML. *Cell*, **66**, 663–674.

Kimura, S. and Shiota, K. (1996) Sequential changes of programmed cell death in developing fetal mouse limbs and its possible roles in limb morphogenesis. *J. Morphol.*, **229**, 337–346.

Komiyama, T., Ray, C.A., Pickup, D.J., Howard, A.D., Thornberry, N.A., Peterson, E.P. and Salvesen, G. (1994) Inhibition of interleukin-1β converting enzyme by the cowpox virus serpin CrmA. An example of cross-class inhibition. *J. Biol. Chem.*, **269**, 19331–19337.

Kuida, K., Lippke, J.A., Ku, G., Harding, M.W., Livingston, D.J., Su, M.S., *et al.* (1995) Altered cytokine export and apoptosis in mice deficient in interleukin-1 converting enzyme. *Science*, **267**, 2000–2003.

Kumar, S., Tomooka, Y. and Noda, M. (1992) Identification of a set of genes with developmentally down-regulated expression in the mouse brain. *Biochem. Biophys. Res. Commun.*, **185**, 1155–1161.

Liston, P., Roy, N., Tamai, K., Lefebvre, C., Baird, S., Cherton-Horvat, G., Farahani, R., McLean, M., Ikeda, J.E., MacKenzie, A. and Korneluk, R.G. (1996) Suppression of apoptosis in mammalian cells by NAIP and a related family of IAP genes. *Nature*, **379**, 349–353.

Liu, Z.G., Hsu, H.L., Goeddel, D.V. and Karin, M. (1996) Dissection of TNF receptor 1 effector functions – JNK activation is not linked to apoptosis while NF-κB activation prevents cell death. *Cell*, **87**, 565–576.

Lovering, R., Hanson, I.M., Borden, K.L., Martin, S., O'Reilly, N.J., Evan G.I., *et al.* (1993) Identification and preliminary characterization of a protein motif related to the zinc finger. *Proc. Natl. Acad. Sci. USA*, **90**, 2112–2216.

Martin, S.J., Amarantemendes, G.P., Shi, L.F., Chuang, T.H., Casiano, C.A., Obrien, G.A., *et al.* (1996) The cytotoxic cell protease granzyme B initiates apoptosis in a cell-free system by proteolytic processing and activation of the ICE/ced-3 family protease, CPP32, via a novel two-step mechanism. *EMBO J.*, **15**, 2407–2416.

Martinou, I., Fernandez, P.A., Missotten, M., White E., Allet, B., Sadoul, R., *et al.* (1995) Viral proteins E1B19K and p35 protects sympathetic neurons from cell death induced by NGF deprivation. *J. Cell Biol.*, **128**, 201–208.

Mittler, R. and Lam, E. (1995) Identification, characterization, and purification of a tobacco endonuclease activity induced upon hypersensitive response cell death. *Plant Cell*, **7**, 1951–1962.

Miura, M., Zhu, H., Rotello, R., Hartweig, E.A. and Yuan, J. (1993) Induction of apoptosis in fibroblasts by IL-1β-converting enzyme, a mammalian homolog of the *C. elegans* cell death gene *ced-3*. *Cell*, **75**, 653–660.

Munsat, T.L. (1991) Workshop report: International SMA collaboration. *Neuromuscular Disord.*, **1**, 81.

Neilan, J.G., Lu, Afonso, C.L., Kutish, G.F., Sussman, M.D. and Rock, D.L. (1993) An African swine fever virus gene with similarity to the proto-oncogene Bcl-2 and the Epstein-Barr virus gene BHRF1. *J. Virol*, **67**, 4391–4394.

Oppenheim, R.W. (1991) Cell death during development of the nervous system. *Annu. Rev. Neurosci.*, **14**, 453–501.

Quan, L.T., Caputo, A., Bleackley, R.C., Pickup, D.J. and Salvesen, G.S. (1995) Granzyme B is inhibited by the cowpox virus serpin cytokine response modifier A. *J. Biol. Chem.*, **270**, 10377–10379.

Rabizadeh, S., LaCount, D.J., Friesen, P.D. and Bredesen, D.E. (1993) Expression of the baculovirus p35 gene inhibits mammalian neural cell death. *J. Neurochem.*, **61**, 2318–2321.

Ray, C.A., Black, R.A., Kronheim, S.R., Greenstreet, T.A., Sleath, P.R., Salvesen, G.S., *et al.* (1992) Viral inhibition of inflammmation: cowpox virus encodes an inhibitor of the interleukin-1β converting enzyme. *Cell*, **69**, 597–604.

Rothe, M., Pan, M.G., Henzel, W.J., Ayres, T.M. and Goeddel, D.V. (1995a) The TNFR2-TRAF signalling complex contains two novel proteins related to baculoviral inhibitor of apoptosis proteins. *Cell*, **83**, 1243–1252.

Rothe, M., Sarma, V., Dixit, V.W. and Goeddel, D.V. (1995b) TRAF2-mediated activation of NF-κB by TNF receptor 2 and CD40. *Science*, **269**, 1424–1427.

Rothe, M., Wong, S.C., Henzel, W.J. and Goeddel, D.V. (1994) A novel family of putative signal transducers associated with the cytoplasmic domain of the 75 kDa tumor necrosis factor receptor. *Cell*, **78**, 681–692.

Roy, N., Mahadevan, M.S., McLean, M., Shutler, G., Yaraghi, Z., Farahani, R., *et al.* (1995) The gene for neuronal apoptosis inhibitory protein is partially deleted in individuals with spinal muscular atrophy. *Cell*, **80**, 167–178.

Sarid, R., Sato, T., Bohenzky, R.A., Russo, J.J. and Chang, Y. (1997) Kaposis sarcoma-associated herpesvirus encodes a functional bcl-2 homologue. *Nature Med.* **3**, 293–298.

Saurin, A.J., Borden, K., Boddy, M.N. and Freemont, P.S. (1996) Does this have a familiar ring. *Trends Biochem. Sci.*, 208–214.

Schatz, D.G., Oettinger, M.A. and Baltimore, D. (1989) The V(D)J recombination activating gene, RAG-1. *Cell*, **59**, 1035–1048.

Shub, D.A. (1994) Bacterial altruism? *Current Biology*, 4, 555–556.

Smith, K.G., Strasser, A. and Vaux, D.L. (1996) CrmA expression in T lymphocytes of transgenic mice inhibits CD95 (Fas/APO-1)-transduced apoptosis, but does not cause lymphadenopathy or autoimmune disease. *EMBO J.*, **15**, 5167–5176.

Song, Z.W., Mccall, K. and Steller, H. (1997) Dcp-1, a drosophila cell death protease essential for development. *Science*, **275**, 536–540.

Srinivasula, S.M., Ahmad, M., Fernandes-Alnemri, T., Litwack, G. and Alnemri, E.S. (1996) Molecular ordering of the FAS-apoptotic pathway – the FAS/APO-1 protease MCH5 is a CrmA-inhibitable protease that activates multiple ced-3/ICE-like cysteine proteases. *Proc. Natl. Acad. Sci. USA*, **93**, 14486–14491.

Sugimoto, A., Friesen, P.D. and Rothman, J.H. (1994) Baculovirus p35 prevents developmentally programmed cell death and rescues a *ced-9* mutant in the nematode *Caenorhabditis elegans. EMBO J.*, **13**, 2023–2028.

Sun, J.R., Bird, C.H., Sutton, V., McDonald, L., Coughlin, P.B., Dejong, T.A., *et al.* (1996) A cytosolic granzyme B inhibitor related to the viral apoptotic regulator cytokine response modifier A is present in cytotoxic lymphocytes. *J. Biol. Chem.*, **271**, 27802–27809.

Tagawa, M., Sakamoto, T., Shigemoto, K., Matsubara, H., Tamura, Y., Ito, T., *et al.* (1990) Expression of novel DNA-binding protein with zinc finger structure in various tumor cells. *J. Biol. Chem.*, **265**, 20021–20026.

Takahashi, M. and Cooper, G.M. (1987) Ret transforming gene encodes a fusion protein homologous to tyrosine kinases. *Mol. Cell. Biol.*, **7**, 1378–1385.

Tartaglia, L.A. and Goeddel, D.V. (1992) Two TNF receptors. *Immunology Today*, **13**, 151–153.

Tartaglia, L.A., Weber, R.F., Figari, I.S., Reynolds, C., Palladino, M., Jr., and Goeddel, D.V. (1991) The two different receptors for tumor necrosis factor mediate distinct cellular responses. *Proc. Natl. Acad. Sci. USA*, **88**, 9292–9296.

Thornberry, N.A., Bull H.G., Calaycay, J.R., Chapman, K.T., Howard, A.D., Kostura, M.J., *et al.* (1992) A novel heterodimeric cysteine protease is required for interleukin-1β processing in monocytes. *Nature*, **356**, 768–774.

Uren, A.G., Pakusch, M., Hawkins, C.J., Puls, K.L. and Vaux, D.L. (1996) Cloning and expression of apoptosis inhibitory protein homologs that function to inhibit apoptosis and/or bind tumor necrosis factor receptor-associated factors *Proc. Natl. Acad. Sci. USA*, **93**, 4974–4978.

Vaux, D.L., Weissman, I.L. and Kim, S.K. (1992) Prevention of programmed cell death in *Caenorhabditis elegans* by human Bcl-2. *Science*, **258**, 1955–1957.

Vucic, D., Seshagiri, S. and Miller, L.K. (1997) Characterization of reaper- and FADD-induced apoptosis in a lepidopteran cell line. *Mol. Cell. Biol.*, **17**, 667–676.

White, K., Tahaoglu, E. and Steller, H. (1996) Cell killing by the *Drosophila* gene reaper. *Science*, **271**, 805–807.

Wirth, B., Hahnen, E., Morgan, K., DiDonato, C.J., Dadze, A., Rudnik-Schoneborn, S., *et al.* (1995) Allelic association and deletions in autosomal recessive proximal spinal muscular atrophy: association of marker genotype with disease severity and candidate cDNAs. *Human Molecular Genetics*, **4**, 1273–1284.

Xue, D. and Horvitz, H.R. (1995) Inhibition of the *Caenorhabditis elegans* cell-death protease ced-3 by a ced-3 cleavage site in baculovirus p35 protein. *Nature*, **377**, 248–251.

Yarmolinsky, M.B. (1995) Programmed cell death in bacterial populations, *Science*, **267**, 836–837.

Yuan, J.Y., Shaham, S., Ledoux, S., Ellis, H.M. and Horvitz, H.R. (1993) The *C. elegans* cell death gene ced 3 encodes a protein similar to mammalian interleukin 1β converting enyzyme. *Cell*, **75**, 641–652.

8. THE MITOCHONDRION: DECISIVE FOR CELL DEATH CONTROL?

CATHERINE BRENNER* AND GUIDO KROEMER*†

*Centre National de la Recherche Scientifique, Unité Propre de Recherche 420, 19 rue Guy Môquet, F-94801 Villejuif, France

KEY WORDS: apoptosis, mitochondrial transmembrane potential, necrosis, permeability transition, programmed cell death.

INTRODUCTION

Apoptosis is a strictly regulated ("programmed") device which allows the removal of superfluous, aged, or damaged cells. Apoptosis constitutes a physiological mechanism, but its control can become deficient and lead to numerous pathologies. Thus, an abnormal resistance to apoptosis induction generates malformations, autoimmune disease or cancer due to the persistence of superfluous, self-specific, or mutated cells, respectively. In contrast, elevated apoptotic decay of cells participates in acute diseases (infection by toxin-producing microorganisms, ischemia-reperfusion damage, infarction, apoplexy), as well as in chronic pathologies (neurodegenerative and neuromuscular diseases, AIDS).

The process of apoptosis can be subdivided into a least three different phases: initiation, effector, and degradation. During the initiation phase, cells receive apoptosis-triggering stimuli. For example, in mammalian cells, such apoptosis-triggering stimuli include numerous toxins, suboptimal culture conditions, interventions on second messenger systems, and ligation of certain receptors (Fas/APO-1/CD95, TGF-R, TNF-R, etc.) or, in the case of obligate growth factor receptor, the absence of receptor occupancy (Barr and Tomei, 1994; Kroemer, 1995; Kroemer, 1997b; Kroemer *et al.*, 1997a; Kroemer *et al.*, 1995; Kroemer *et al.*, 1997b; Thompson, 1995; Wertz and Hanley, 1996). Non-specific or receptor-mediated death induction involves a stimulus-dependent ("private") biochemical pathway, and it is only after this initiation phase that common pathways come into action. In spite of the striking heterogeneity of apoptosis induction pathways, some characteristics of the apoptotic process are near-to-constant and do not depend on the induction protocol. It is generally assumed that the execution phase of apoptosis defines the "decision to die" at the "point-of-no-return" of the apoptotic cascade. It is at this level that the different private pathways converge into one (or few) common pathway(s) and that cellular processes (redox potentials, expression levels of oncogene products including Bcl-2 related proteins) have still a decisive regulatory function.

---

† *Corresponding Author*: 19, rue Guy Môquet, B.P. 8, F-94801 Villejuif, France. Tel.: 33–1–49 58 35 13. Fax: 33–1–49 58 35 09. e-mail: kroemer @ infobiogen.fr

During the execution phase, the "central executioner", "great integrator", or "apostat" would be activated, thus sealing the cell's fate. Subsequently, the cell has been irreversible committed to death and the different manifestations of degradation phase become apparent. This degradation phase is similar in all cell types. It is characterized by the action of catabolic enzymes, including specific proteases (caspases) and endonucleases, within the limits of a near-to-intact plasma membrane. Thus, the cell actively contributes to its removal in a "suicidal" fashion and undergoes stereotyped biochemical and ultrastructural alterations. These changes include certain nuclear features of apoptosis such as chromatin condensation and DNA fragmentation, plasma membrane alterations (exposure of phosphatidylserine residues on the outer leaflet), as well as cytoplasmic changes (cell shrinkage, hyperproduction of reactive oxygen species (ROS), activation and action of certain proteases or caspases) (Cohen, 1991; Kroemer, 1995; Kroemer *et al.*, 1995; Thompson, 1995).

The discovery that programmed cell death (PCD) may be induced in anucleate cells (cytoplasts) (Jacobson *et al.*, 1994; Nakajima *et al.*, 1995; Schulze-Osthoff *et al.*, 1994) has led to the postulation of a cytoplasmic (non-nuclear) effector or "central executioner" that would participate in life/death decision making and would be influenced by endogenous control mechanisms (Henkart, 1995; Jacobson *et al.*, 1994; Martin and Green, 1995; Oltvai and Korsmeyer, 1994). The exact nature of this cytoplasmic executioner is a matter of debate. In particular, the putative role of mitochondria in the apoptotic effector stage now emerges more clearly from studies based on cell-free systems and molecular functional analysis. The present review will summarize evidence indicating that mitochondria play a major role in the apoptotic effector phase.

NO NEED FOR MITOCHONDRIA IN APOPTOSIS?

Cells lacking mitochondria DNA can undergo apoptosis

The mitochondrial genome of human cells encodes RNA molecules necessary for mitochondrial protein synthesis (12 S and 16S ribosomal RNAs, 22 different transfer RNAs) and a few subunits of the multiprotein respiratory chain complex I (6 NADH dehydrogenase subunits), complex III (apocytochrome b), complex IV (cytochrome b oxidase subunits 1, 2 and 3), and the ATP synthase subunits 6 and 8 (Borst *et al.*, 1984; Gray, 1989). However, most mitochondrial proteins are synthesized in the cytosol as precursors and are selectively addressed to either of the two mitochondrial membranes, the intermembrane space or the matrix (Schatz and Dobberstein, 1996). Thus, mitochondria lacking mtDNA have a near-to-normal morphological aspect and can divide by fission to replicate indefinitely in the cytoplasm of the proliferating cell.

With respect to the status of their mtDNA, cells can be normal (ρ^+), carry deletions of part of the mitochondrial genome (ρ^- mutants) or may have deleted the entire mitochondrial genome ($\rho°$). The rationale for the isolation of human $\rho°$ cells is based on the use of inhibitors of mtDNA replication, such as the DNA intercalating dye ethidium bromide, which, at low concentrations (0.1 to 2 μg/ml), demonstrates a relatively selective incorporation into the mitochondrial *vs* nuclear DNA, resulting in either partial or complete inhibition of mtDNA replication without affecting nuclear DNA synthesis (King and Attardi, 1996; Marchetti *et al.*, 1996d). It may be important to note that $\rho°$ cells require relatively large amounts of glucose since they cannot generate ATP via oxidative phosphorylation and entirely rely on glycolytic ATP generation. Moreover, $\rho°$ cells are generally auxotrophic for uridine and pyruvate. Failure to provide such metabolites implies that ethidium bromide becomes toxic for cells. At least part of this toxicity involves induction of apoptosis (Baixeras *et al.*, 1994). Although the electron transport on the inner mitochondrial membrane is defective in cell lacking mitochondrial DNA, $\rho°$ cells possess a normal mitochondrial transmembrane potential ($\Delta\Psi_m$) (Marchetti *et al.*, 1996c; Skowronek *et al.*, 1992) and fulfill a number of metabolic functions including import of proteins encoded by nuclear genes. It is unclear how the $\Delta\Psi_m$ which is essentially a proton gradient is generated in $\rho°$ cells. Irrespective of the exact mechanism of $\Delta\Psi_m$ generation, it appears that inhibition of glycolysis by NaF causes an immediate disruption of the $\Delta\Psi_m$ (Skowronek *et al.*, 1992). Thus, maintenance of the $\Delta\Psi_m$ in $\rho°$ cells requires anaerobic glycolysis.

The finding that $\rho°$ cells can readily undergo apoptosis has been first reported by Martin Raff's group (Jacobson *et al.*, 1993) and then confirmed in numerous studies. Thus, human GM701 $\rho°$ fibroblasts die from apoptosis in response to anti-CD95 and staurosporine, a protein kinase inhibitor (Jacobson *et al.*, 1993). Similarly, U937 $\rho°$ cells rapidly undergo apoptosis upon stimulation with TNF or anti-CD95 (Gamen *et al.*, 1995; Marchetti *et al.*, 1996c). Normal ρ^+ cells exposed to inhibitors of the respiratory chain die either from necrosis or apoptosis, depending on the cell type (Shimizu *et al.*, 1995; Wolvetang *et al.*, 1994). As to be expected, $\rho°$ cells (which lack a functional respiratory chain) are completely resistant to the apoptosis-inducing effect of respiratory chain inhibitors such as rotenone and antimycin A (Marchetti *et al.*, 1996c; Wolvetang *et al.*, 1994). Moreover, they are resistant to the apoptosis-inducing effect of hyperoxia (Yoneda *et al.*, 1995), presumably because they fail to generate toxic oxygen radicals on the respiratory chain.

The above observations indicate that mtDNA-dependent cellular functions (mainly oxidative phosphorylation and mitochondrial respiration) are not required for the common pathway of apoptosis. Although this interpretation has been clearly formulated in this cautionary form (Jacobson *et al.*, 1993), many investigators over-interpreted the fact that $\rho°$ cells can undergo apoptosis to mean that mitochondria as such are not involved in the apoptotic process. This notion was apparently underscored by the finding that the electron microscopic picture of mitochondria remains essentially normal until late stages of the apoptotic degradation phase. Moreover, in contrast to nuclear DNA, mitochondrial DNA is not degraded during apoptosis (Tepper and Studzinski, 1992; Tepper and Studzinski, 1993), although it may be fragmented during necrosis (Tepper and Studzinski, 1993). Nonetheless, the experiments performed on $\rho°$ cells do not exclude the participation of mitochondria in apoptosis. It would be necessary to produce cells without mitochondria rather than cells without mitochondrial DNA, and such cells are not viable.

The putative involvement of mitochondria in the death process, prompted us to study the mitochondrial function in U937 ρ° cells which undergo apoptosis in response to TNF (Marchetti *et al.*, 1996c). We had previously described that a $\Delta\Psi_m$ disruption was observed early during the apoptotic precess in numerous ρ^+ cells. This $\Delta\Psi_m$ dissipation is meadiated by opening of so called permeability transition (PT) pores (Kroemer, 1997a; Kroemer, 1997b; Kroemer *et al.*, 1997a; Kroemer *et al.*, 1995; Kroemer *et al.*, 1997b; Petit *et al.*, 1996). Importantly enough, we found that ρ° cells manifest an early disruption of the $\Delta\Psi_m$, exactly as this is observed in TNF-treated ρ^+ controls (Marchetti *et al.*, 1996c). In addition, we showed that isolated mitochondria from ρ° cells undergo PT, a phenomenon that we have implicated in apoptosis, in response to a variety of stimuli (atractyloside, calcium, *ter*-butylhydroperoxide) in a fashion that is indistinguishable from control mitochondria (Marchetti *et al.*, 1996c; Zamzami *et al.*, 1996b). This in line with the fact that all putative components of the PT pore are encoded by nuclear rather than mitochondrial genes: the adenine nucleotide translocase, the peripheral benzodiazepine receptor, and the voltage-dependent anion channel (Bernardi *et al.*, 1994; Brandolin *et al.*, 1993; Zoratti and Szabò, 1995). Moreover, ρ° mitochondria liberate the same amount of apoptogenic factors from the intermembrane space as do ρ^+ control organelles (Marchetti *et al.*, 1996c; Zamzami *et al.*, 1996b). Such apoptogenic factors are capable of causing isolated nuclei to undergo hallmarks of nuclear apoptosis (chromatin condensation and DNA fragmentation). Their presence in the supernatant of ρ° mitochondria indicates that they are encoded by nuclear rather than mitochondrial genes.

In conclusion, it appears that the structures responsible for PT and its regulation, as well as the mitochondrial proteins with apoptogenic properties, function normally in ρ° cells. Therfore, the fact that ρ° cells can undergo apoptosis normally cannot be used as an argument against the involvement of mitochondria in apoptosis.

Cell-free system apparently not implying mitochondria

Cell-free systems, that are *in vitro*cultures of disassembled cellular components, have been designed to recapitulate essential steps of the apoptotic process *in vitro*, for instance by combining isolated nuclei with cytosolic extracts and/or isolated organelles, to determine the conditions in which chromatin condensation and endonuclease-mediated DNA fragmentation will occur. A number of cell-free systems involved the use of cytosolic extracts from chicken mitotic cells (Lazebnik *et al.*, 1993; Lazebnik *et al.*, 1995b), from *Xenopus laevis* oocytes undergoing follicular atresia (Cosulich *et al.*, 1996; Evans *et al.*, 1997; Newmeyer *et al.*, 1994), or from various human and murine transformed cell lines including U3937, BT-20, HL-60 and 3T3 cells (Enari *et al.*, 1995; Enari *et al.*, 1996; Martin *et al.*, 1995; Shimizu and Pommier, 1996; Wright *et al.*, 1994; Yoshida *et al.*, 1996). Such cytosolic extacts are generated by mechanical or detergent-mediated disruption of cells and cycles of freeze-thawing, followed by ultracentrifugation. Thus, apparently they are organelle-free. They can be employed to induce key features of nuclear apoptosis in vitro: chromatin condensation, internucleosomal DNA fragmentation, or cleavage of nuclear caspase substrates including PARP and lamins. The fact that such extracts are organelle-free and that addition of caspases to extracts from normal cells creates an apoptogenic activity (Enari *et al.*, 1995; Enari *et al.*, 1996; Muzio *et al.*, 1997), has been thought to imply that mitochondria would not be involved in apoptosis and that rather soluble factors including caspases would be decisive for the cell death process.

This interpretation can be criticized at two levels. First, it has to be critically noted that cell-free systems by definition involve a disruption of subcellular compartimentation, and then, do not necessarly reveal the true physiology of the system. Thus, it can not be excluded that components are relased from various organites (mitochondria, lysosomes, endoplasmic reticulum) during the preparation of the cytosolic extracts. Second, it may be questioned as to whether such cell-free systems truely reveal the execution phase of apoptosis. Indeed, the nuclear degradation, which is the principal read-out of cell-free systems of apoptosis, forms part of the degradation phase and even is not necessary for cell death to occur (Jacobson *et al.*, 1994; Schulze-Osthoff *et al.*, 1994).

Cell-free systems based on the use of carefull separated subcellular fractions have recently suggested that mitochondria and/or mitochondrial products are required for the induction of nuclear apoptosis (Liu *et al.*, 1996; Marchetti *et al.*, 1996c; Newmeyer *et al.*, 1994; Susin *et al.*, 1996; Zamzami *et al.*, 1996b). These data will be discussed in detail in the following section.

MITOCHONDRIAL IMPLICATION IN APOPTOSIS : EVIDENCE FROM CELL-FREE SYSTEMS

A two-step model of apoptosis: an initial mitochondrial step followed by a secondary nuclear step

As discussed above, cytosols from cells undergoing apoptosis contain apoptogenic proteins capable of provoking isolated nuclei to undergo chromatin condensation and DNA fragmentation (Lazebnik *et al.*, 1995a). A similar apoptogenic activity is encountered in whole cytoplasmic preparations of normal cells, and this activity is associated with mitochondria (Martin *et al.*, 1995; Newmeyer *et al.*, 1994). Current data are compatible with a two-step-model of apoptosis. At a first step, different effectors act on mitochondria to disrupt the mitochondrial membrane integrity at the level of the inner and/or outer membrane. As a consequence, mitochondria then release apoptogenic factors into the cytosol, when then can act directly or indirectly on nuclei. This model integrates apparently contradictory data from the litterature. It also predicts that the decisive event of the apoptotic cascade (that is the effector stage) would be disruption of mitochondrial membrane structure or function rather than nuclear demise (that is the degradation stage).

What is then occurring during the effector stage of apoptosis? As mentioned above, we have found that early during apoptosis occurring in intact cells, the mitochondrial inner transmembrane potential ($\Delta\Psi_m$) is disrupted and that this $\Delta\Psi_m$ dissipation involves opening of the mitochondrial megachannel, also called the permeability transition (PT) pore. To investigate the relationship between PT pore opening and release of mitochondrial apoptogenic factors, we incubated isolated mitochondria with agents specifically acting on components of the PT pore. Such agents, which include protoporphyrin IX (a ligand of the peripheral benzodiazepin receptor) or atractyloside (a ligand of the adenine nucleotide translocator), cause an immediate release of apoptogenic factors into the mitochondrial supernatant (Marchetti *et al.*, 1996b; Marchetti *et al.*, 1996c; Zamzami *et al.*, 1996b). Such factors include "apoptosis-inducing factor" (AIF) (Susin *et al.*, 1996) and cytochrome *c*(Kantrow and Piantadosi, 1997) (see below). In a further attempt to understant the apoptotic process, we sat out to identify physiologically relevant inducers of PT accumulating in the cytosol of cells undergoing apoptosis. It appears that a number of physiological effectors (Ca^{2+}, reactive oxygen species, and nitric oxide), caspases, and additional yet non-identified second messengers elicted by ceramide can cause PT in isolated mitochondria, thereby triggering $\Delta\Psi_m$ disruption and release of apoptogenic factors (Hortelano *et al.*, 1997; Susin *et al.*, 1997a; Susin *et al.*, 1996; Susin *et al.*, 1997b; Zamzami *et al.*, 1996b). In contrast, to these observations, Newmeyer and co-workers suggest the presence of a yet un-identified cytosolic factor that causes the release of cytochrome *c* from mitochondria without that disruption of the $\Delta\Psi_m$ occurs (Kluck *et al.*, 1997). The exact relationship between PT, release of apoptogenic factors, and $\Delta\Psi_m$ disruption is currently a field of controversy and intense research efforts.

Apoptogenic factors released by mitochondria:cytochrome c and AIF

Supernatants from mitochondria that have undergone PT but not those from control mitochondria induce signs of apoptosis such as nuclear chromatin condensation in a cell-free system (Susin *et al.*, 1996; Zamzami *et al.*, 1996b). Destruction of mitochondrial membranes via sonication, osmotic shock, or digitonin treatment (which specifically lyses the outer but not the inner membrane, yielding the intermembrane fraction of proteins) also releases this activity, that we have baptized "apoptosis-inducing factor" (AIF), indicating that AIF is pre-formed (Susin *et al.*, 1996) (Table 8.1). Proteinase K treatment as well as heat treatment (70°C, 5 min) destroy AIF activity, indicating that AIF is a protein. Anion exchange chromatography, molecular sieve chromatography, and SDS-PAGE identify mouse hepatocyte AIF as a single ~50 kDa proten. AIF is low-abundant (<0.1% of mitochondrial proteins) and labile at room temperature, requiring purification at 4°C (Susin *et al.*, 1996). AIF activity has been detected in mitochondria from several cell types (liver, heart, brain, myelomonocytic cells, lymphoid cells) and species (mouse, human). It appears phylogenetically conserved, since human AIF induces apoptosis in mouse nuclei and vice versa (Zamzami *et al.*, 1996b). AIF is present in mitochondria from cells lacking mtDNA, indicating that it is encoded for by the nuclear rather than by the mitochondrial genome (Marchetti *et al.*, 1996c; Zamzami *et al.*, 1996b).

Purified AIF suffices to induce hallmarks of nuclear apoptosis such as chromatin condensation and oligonucleosomal DNA fragmentation, in the absence of additional cyoplasmic components (Susin *et al.*, 1996). Moreover, AIF induces nuclear apoptosis much more rapidly (<15 min) (Zamzami *et al.*, 1996b). Purified AIF has no intrinsic DNAse activity, indicating that it probably induces DNA fragmentation via activating pre-existing nuclear DNAses (Susin *et al.*, 1996). Although AIF has a proteolytic activity on unidentified nuclear substrates (our unpublished data), it fails to cleave poly (ADP ribose) polymerase (PARP) or lamin in isolated nuclei (Susin *et al.*, 1996). As a consequence, the proteolytic spectrum of AIF differs from that of other caspases previously implicated in apoptosis induction (Lazebnik *et al.*, 1995a).

Table 8.1 Properties of mouse hepatocyte apoptosis-inducing factor (AIF).

AIF	*Properties*
Physicochemical characteristics	one instable protein, one chainM.W ~50 kDa
Localization	constitutively expressed in different cell types including cells lacking mitochondrial DNA associated with mitochondrial intermembrane space not with membranes nor with matrix released upon induction of permeability transition
Enzymatic activities	no DNAse activity selective protease activity
Effects on isolated nuclei	activation of endonucleases causing oligonucleosomal DNA fragmentation chromatin condensation redistribution of nuclear matrix proteolysis of selected proteins (not PARP nor lamins)
Inhibitors of apoptogenic effects	*p*-mercuryphenylsulfonic acid *N*-phenylmaleimide ZVAD-fmk

To gain information on the mode of action of AIF, we determined the inhibitory profile of this factor. The chromatin condensation-inducing activity of AIF is inhibited by the thiol reagents *p*-chloromercuryphenylsulfonic acid and *N*-phenylmaleimide (Susin *et al.*, 1996; Susin *et al.*, 1997b) but not by specific inhibitors of different calcium, serine, or cysteine proteases including specific inhibitors of interleukin-1β converting enzyme (ICE) and caspase 3 (Zamzami *et al.*, 1996b). The only selective protease inhibitor which blocks AIF activity is *N*-benzyloxycarbonyl-Val-Ala-Asp.fluoromethylketone (z-VAD.fmk) (Susin *et al.*, 1996), an inhibitor of ICE-like proteases (Cain *et al.*, 1996; Fearnhead *et al.*, 1995; Jacobson *et al.*, 1996; Pronk *et al.*, 1996; Slee *et al.*, 1996; Zhivotovsky *et al.*, 1995; Zhu *et al.*, 1995). z-VAD.fmk prevents all manifestations of apoptosis induced by AIF: chromatin condensation, oligonucleosomal DNA fragmentation, and DNA loss from nuclei (Susin *et al.*, 1996). Altogether these data suggest that AIF possesses a cysteine-dependent catalytic activity not identical with but distantly related to proteases from the ICE/CPP32/Ced-3 family. Accordingly, AIF's molecular mass, subcellular localization and proteolytic spectrum differ from those of known members of the ICE/CPP32/Ced-3 family (Henkart, 1996; Patel *et al.*, 1996). Experiments performed in intact cells indicate that z-VAD.fmk inhibits apoptosis in both mammalian (Cain *et al.*, 1996; Fearnhead *et al.*, 1995; Jacobson *et al.*, 1996; Slee *et al.*, 1996;

Zhivotovsky *et al.*, 1995; Zhu *et al.*, 1995) and insect cells (Pronk *et al.*, 1996), in response to a wide array of apoptosis triggers including inducers of PT such as protoporphyrin IX and mClCCP (Marchetti *et al.*, 1996b; Susin *et al.*, 1996). Although this is not a formal proof, this observation underscores the probable importance of AIF as a rate-limiting factor of the apoptotic process *in vivo*.

Nonetheless, AIF is not the only apoptogenic factor released by mitochondria. Cytochrome *c*, a mitochondrial intermembrane heme protein, has been found to exert a co-apoptgeinc activity (Liu *et al.*, 1996). Thus, cytochrome *c* itself is inefficient to cause nuclear apoptosis *in vitro*. However, it acts in conjunction with yet to be characterized cytosolic factors to activate caspase-3 in vitro, which then activates another factor, DNA fragmentation factor (DFF) which activates nuclear endonucleases (Liu *et al.*, 1996; Liu *et al.*, 1997). Cytochrome *c* is encoded by a nuclear gene and translated as apocytochrome *c*, which is subsequently translocated into the mitochondrion where a heme group is attached covalently to form holocytochrome *c*. Only holocytochrome *c* but not its immature precursor apocytochrome *c* is apoptogenic (Kluck *et al.*, 1997). In several models of apoptosis, holocytochrome *c* is released from mitochondrial intermembrane space into the cytosol of living cells (Liu *et al.*, 1996; Yang *et al.*, 1997). The exact mechanism by which cytochrome *c* is released from the mitochondrion is still unclear. Although PT causes cytochrome *c* release (Bernardi, 1996; Kantrow and Piantadosi, 1997), cytochrome *c*release can occur before $\Delta\Psi$ disruption (Kluck *et al.*, 1997; Yang *et al.*, 1997). Thus, it is possible that cytochrome *c* released from mitochondria via a mechanism not involving PT. Alternatively, flickering of the PT pore (which can act in a reversible fashiion with a low conductancy on the inner membrane, Ref. (Ichas *et al.*, 1997)) may have differential effects on the inner membrane proton gradient and the outer membrane function.

MITOCHONDRIAL CONTROL OF APOPTOSIS IN INTACT CELLS

Dissipation of the mitochondrial transmembrane potential

The mitochondrial transmembrane potential ($\Delta\Psi_m$) results from the asymmetric distribution of protons on both sides of the inner mitochondrial membrane, giving rise to a chemical (pH) and electric gradient which is essential for mitochondrial function. The inner side of the inner mitochondrial membrane is negatively charged. Consequently, cationic lipophilic fluorochromes such as rhodamine 123, 3,3'dihexyloxacarbocyanine iodide ($DiOC_6(3)$), chloromethyl-X-rosamine (CMXRos) or 5,5',6,6'-tetrachloro-1,1',3,3'-tetraethylbenzimidazolcarbocyanine iodide (JC-1) distribute to the mitochondrial matrix as a function of the Nernst equation, correlating with the $\Delta\Psi_m$. Using a cytofluorometer, these dyes can be employed to measure variations in the $\Delta\Psi_m$ on a per-mitochondrion- or per-cell-basis. We (Castedo *et al.*, 1996; Castedo *et al.*, 1995; Decaudin *et al.*, 1997; Hirsch *et al.*, 1997; Hortelano *et al.*, 1997; Kroemer *et al.*, 1995; Macho *et al.*, 1995; Macho *et al.*, 1997; Marchetti *et al.*, 1996a; Marchetti *et al.*, 1997; Marchetti *et al.*, 1996b; Marchetti *et al.*, 1996c; Marchetti *et al.*, 1996d; Susin *et al.*, 1997a; Susin *et al.*, 1996; Zamzami *et al.*,

1995a; Zamzami *et al.*, 1995b; Zamzami *et al.*, 1996b) and others (Backway *et al.*, 1997; Boise and Thompson, 1997; Cossarizza *et al.*, 1995; Petit *et al.*, 1995; Polla *et al.*, 1996; Vayssière *et al.*, 1994; Xiang *et al.*, 1996) have shown that cells induced to undergo apoptosis manifest an early reduction in the incorporation of $\Delta\Psi_m$-sensitive dyes, indicating a disruption of the $\Delta\Psi_m$ (Table 8.2). This $\Delta\Psi_m$ collapse can be detected in many different cell types, irrespective of the apoptosis-inducing stimulus. $\Delta\Psi_m$ disruption precedes nuclear apoptosis also in cells lacking mitochondrial DNA (Marchetti *et al.*, 1996c; Marchetti *et al.*, 1996d). It becomes manifest before cells exhibit nuclear DNA fragmentation, hyperproduce ROS, or aberrantly expose phosphatidylserine (PS) on the outer cell membrane leaflet (Castedo *et al.*, 1996; Macho *et al.*, 1995; Zamzami *et al.*, 1995a; Zamzami *et al.*, 1995b). Thus, the $\Delta\Psi_m$ collapse constitutes an early common event of the apoptotic cascade. Since an intact $\Delta\Psi_m$ is indispensable for normal mitochondrial function (Attardi and Schatz, 1988), cells undergoing apoptosis manifest a cessation of mitochondrial biogenesis, both at the transcription and translation levels (Osborne *et al.*, 1994; Vayssière *et al.*, 1994).

Table 8.2 Apoptosis-inducing regimes which induce a $\Delta\Psi_m$ disruption preceding nuclear DNA fragmentation and their inhibition in thymocytes.

Inducer of apoptosis	*Inhibitor of apoptosis*	*References*
glucocorticoids	RU38486 actinomycin D cycloheximide TLCK, Z-VAD.fmk N-acetylcysteine, catalase bongkrekic acid monochlorobimane chloromethyl-X-rosamine	Petit *et al.*, 1995 Castedo *et al.*, 1995 Marchetti *et al.*, 1996a Castedo *et al.*, 1996 Zamzami *et al.*, 1995b Zamzami *et al.*, 1995b
anti-Fas antibody	Ac-YVAD.cmk	Castedo *et al.*, 1996
DNA damage (VP16 or irradiation)	p53 null mutation, actinomycin D cycloheximide, TLCK, ZVAD.fmk bongkrekic acid monochlorobimane chloromethyl-X-rosamine	Castedo *et al.*, 1996 Marchetti *et al.*, 1996a
ter-butylhydroperoxide	monochlorobiman chloromethyl-X-rosamine	Castedo *et al.*, 1995 Marchetti *et al.*, 1997
diamide	monochlorobiman, chloromethyl-X-rosamine	Marchetti *et al.*, 1997
protoporphyrine IX	bongkrekic acid	Marchetti *et al.*, 1996b
nitric oxide	bongkrekic acid	Hortelano *et al.*, 1997

Pharmacological data implying mitochondria in apoptosis

To understand the mechanism by which cells undergoing apoptosis lose their $\Delta\Psi_m$, we performed a series of experiments in which cells were first labeled with

$\Delta\Psi_m$-sensitive fluorochromes and then purified in a fluorocytometer, based on their $\Delta\Psi_m$. In appropriate conditions, this procedure allows for the purification of cells with low $\Delta\Psi_m$ values and a still normal DNA content and morphology (=pre-apoptotic cells) or, alternatively, of cells with a still high $\Delta\Psi_m$ that will lose their $\Delta\Psi_m$ upon a short-term (30 to 120 min) culture period (Zamzami *et al.*, 1995a; Zamzami *et al.*, 1996a; Zamzami *et al.*, 1995b). We have used this system to show that $\Delta\Psi_m$low (but not $\Delta\Psi_m$high) cells will undergo oligonucleosomal DNA fragmentation upon short-term culture at 37°C. Moreover, we have found that some drugs inhibit the $\Delta\Psi_m$ loss of $\Delta\Psi_m$high cells, namely cyclosporin A (CsA) and bongkrekic acid (BA) (Marchetti *et al.*, 1996a; Zamzami *et al.*, 1996a; Zamzami *et al.*, 1995b). It should be noted that cyclosporin A is only a transient (<1 hour) inhibitor of PT (Nicolli *et al.*, 1996), whereas bongkrekic acid is a long-term PT inhibitor (Marchetti *et al.*, 1996a; Marchetti *et al.*, 1997; Zamzami *et al.*, 1995a; Zamzami *et al.*, 1996a). This data suggests that the so-called permeability transition (PT), which is inhibited by CsA and BA (Bernardi and Petronilli, 1996; Zoratti and Szabò, 1995), accounts for the $\Delta\Psi_m$ collapse observed during pre-apoptosis.

PT involves the formation of proteaceous pores ('PT pores'or 'megachannels') (Table 8.3), probably by apposition of inner and outer mitochondrial membrane proteins allowing for the diffusion of solutes <1500 Da and *ipso facto* dissipation of the $\Delta\Psi_m$. CsA is one of the best studied inhibitors of PT. Its PT-inhibitory effect is mediated via a conformational change in a mitochondrial CsA receptor, the matrix cyclophilin D (Nicolli *et al.*, 1996). In contrast, its immunosuppressive effect is mediated via an effect on calcineurin-dependent signalling. A CsA derivative that loses its immunosuppressive (calcineurin-mediated) properties, *N*-methyl-Val-4-CsA, still conserves its $\Delta\Psi_m$-stabilizing (cyclophilin-mediated) effect in apoptotic cells (Zamzami *et al.*, 1996a). This observation is again compatible with the implication of PT in apoptotic $\Delta\Psi_m$ disruption.

To demonstrate that PT might indeed be important for the apoptotic process, we have used two different approaches:

First, we have shown that pharmacological induction of PT with agents specifically affecting mitochondria is sufficient to cause full-blown apoptosis (Marchetti *et al.*, 1996b). Inducers of PT that also trigger signs of nuclear apoptosis include protoporphyrin IX (a ligand of the mitochondrial benzodiazepin receptor, one of the putative constituents of the PT pore (Hirsch *et al.*, 1997; Marchetti *et al.*, 1996b), the protonophore carbamoyl cyanide *m*-chlorophenylhydrazone (mClCCP, that causes dissipation of the $\Delta\Psi_m$) (Hirsch *et al.*, 1997; Susin *et al.*, 1996) and diamide (a divalent thiol-substituting agents causing the crossslinking of vicinial thiols in the mitochondrial matrix) (Marchetti *et al.*, 1997). This indicates that triggering of PT is sufficient to cause apoptosis.

Second, we have used bongrekic acid (BA) to evaluate the effect of PT inhibition in cells in long-term experiments (>120 min). BA does not only prevent the mitochondrial manifestations of apoptosis, but it also abolishes all changes of the apoptotic degradation phase concerning the nucleus (DNA condensation and fragmentation), the cytoplasma (vacuolization, glutathione depletion, ROS hypergeneration, NFκB translocation), and the plasma membrane (exposure of phosphatidylserine residues in the outer membrane leaflet) (Hirsch *et al.*, 1997;

Marchetti *et al.*, 1996a; Marchetti *et al.*, 1996b; Zamzami *et al.*, 1996a). Similarly, chloromethyl-X-rosamine (CMXRos), a substance that prevents oxidation of thiols located in the mitochondrial matrix, can be used as an inhibitor of PT and apoptosis (Marchetti *et al.*, 1997). BA and CMXRos prevent apoptosis induced via both p53-dependent and p53- independent pathways (Marchetti *et al.*, 1996a; Marchetti *et al.*, 1997). The fact that pharmacological inhibition of PT can prevent all post-mitochondrial manifestations of apoptosis (Marchetti *et al.*, 1996a; Zamzami *et al.*, 1996b) suggests that PT constitutes a central coordinating event of the apoptotic process.

Table 8.3 Molecules in the PT pore complex as targets for pharmacological apoptosis modulation.

Molecule (topology)	*Function*	*Ligands and role in PT*	*Reference*
Adenine nucleotide translocator (inner membrane)	ATP/ADP antiport	bongkrekic acid: favors m-state and inhibits PT atractyloside: favors c-state and induces PT	Brustovetsky and Klingenberg, 1996 Klingenberg, 1980
Peripheral benzodiazepin receptor (outer membrane)	receptor for endozepin (CoA-binding protein)	protoporphyrin IX: induces PT PK11195: facilitates PT	Pastorino *et al.*, 1994
Porin (outer membrane)	voltage-dependent anion channel, ATP transport		Beutner *et al.*, 1996
Cyclophilin D (matrix)	peptidyl prolyl isomerase (chaperone function)	cyclosporin A: inhibits interaction with inner membrane N-methyl-Val-4-cyclosporin A (non-immuno suppressive)	Nicolli *et al.*, 1996
Hexokinase (cytosol)	phosphorylates hexosaccharides while hydrolysing ATP	facilitates or regulates PT	Beutner *et al.*, 1996
Creatine kinase (intermembrane)	transfers phosphate from ATP to creatine	inhibits PT ?	Beutner *et al.*, 1996
Matrix thiols	thiol sensors for redox potentials, within ANT?	thiol oxidation and disulfide bridge formation favor PT	Costantini *et al.*, 1996 Halestrup *et al.*, 1997
Ca^{2+} sensitive sites	sensors for divalent cations	calcium favors PT	Bernardi and Petronilli, 1996 Zoratti & Szabò, 1995)

Impact of Bcl-2 on mitochondria

Bcl-2 belongs to a growing family of proteins which can either inhibit (Bcl-2, Bcl-X_L, Mcl-1, Bfl-1, A1 etc.) or favour (Bax, Bcl-X_S, Bad, Bak, Bik etc.) apoptosis (Cory, 1995; Kroemer, 1997b; Reed, 1997; Yang and Korsmeyer, 1996). The Bcl-2 p26 protein possesses a transmembrane domain allowing for its incorporation into different intracellular membranes, including the outer mitochondrial membrane, the endoplasmatic reticulum, and the nuclear envelope (Janiak *et al.*, 1994; Krajewski *et al.*, 1993; Lithgow *et al.*, 1994; Riparbelli *et al.*, 1995). In hematopoietic cells, the mitochondrial localization of Bcl-2 is the quantiatively most important one. At least in certain systems, the specific expression of Bcl-2 in the mitochondrion is sufficient and necessary for its apoptosis action (Greenhalf *et al.*, 1996; Hanada *et al.*, 1995; Nguyen *et al.*, 1994; Tanaka *et al.*, 1993; Zhu *et al.*, 1996). Indeed, in various models, Bcl-2 or Bcl-XL homolog hyperexpression prevents both $\Delta\Psi_m$ and subsequent apoptotic manifestations (Table 8.4).

Table 8.4 Apoptosis-inducing regimes in which Bcl-2 or Bcl-X_L hyperexpression prevents both $\Delta\Psi$ collapse and subsequent apoptotic cell death.

Cell type	*Inducer of apoptosis*	*References*
T cells	glucocorticoids ceramide *ter*-butylhydroperoxide protoporphyrine IX mClCCP etoposide (VP16) γ-irradiation doxorubicin cytosine arabinoside	Zamzami *et al.*, 1995a Susin *et al.*, 1996 Marchetti *et al.*, 1996b Decaudin *et al.*, 1997
B cells	surface IgM crosslinking cyclosporin A etoposide (VP16) γ-irradiation doxorubicin cytosine arabinoside *ter*-butylhydroperoxide ceramide serum withdrawal	Decaudin *et al.*, 1997
PC12 cells	cyanide rotenone antimycin A etoposide (VP16) calcium ionophore	Shimizu *et al.*, 1996
Fibroblasts	p53	Guenal *et al.*, 1997

Based on the findings discussed above, we have formulated three alternative hypotheses: (i) Bcl-2 could neutralize mitochondrial apoptogenic factors (AIF or cytochrome *c*) and thus interfere with their apoptogenic action; (ii) Bcl-2 could inhibit the synthesis or the mitochondrial uptake of apoptogenic factors; (iii) Bcl-2 might interfere with the PT-triggered release of apoptogenic factors from mitochondria. These three possibilities are discussed below.

First, does Bcl-2 interfere with the action of AIF or cytochrome *c*? Nuclei that are purified from cells transfected with the human *bcl-2* gene contain detectable amounts of Bcl-2 protein in their envelope. Such nuclei were incubated with purified AIF to determine a possible inhibitory effect on its apoptosis-inducing effect. Bcl-2 expressing nuclei undergo apoptotic changes in response to AIF, exactly as this is the case for control nuclei from vector-only-transfected cells. Moreover, both control and Bcl-2-overexpressing nuclei manifest the same pattern of chromatin condensation and DNA fragmentation (Susin *et al.*, 1996). This indicates that the nuclear expression of Bcl-2 does not affect the action of AIF. Since Bcl-2 exists also in localizations outside of the nuclear and mitochondrial membrane (Janiak *et al.*, 1994; Krajewski *et al.*, 1993; Lithgow *et al.*, 1994), we evaluated the effect of Bcl-2 overexpression on intact cells exposed to AIF. Introduction of AIF into saponin-treated cells causes the same extent of DNA fragmentation in control and in Bcl-2-hyperexpressing cells (Susin *et al.*, 1996). Thus, Bcl-2 has no detectable effect on AIF action, both in cells and in isolated nuclei. Similarly, Bcl-2 does not prevent the activation of caspase-3 by cytosolic extracts containg cytochrome *c*. Moreover, Bcl-2 does not prevent nuclear apoptosis in cell-free systems when it is added late, after release of cytochrome *c*(Kluck *et al.*, 1997; Newmeyer *et al.*, 1994; Yang *et al.*, 1997). Thus, it appears that Bcl-2 does not neutralize apoptogenic factors once they have been released from mitochondria.

Second, does Bcl-2 interfere with the formation and/or mitochondrial uptake of apoptogenic proteins? On theoretical grounds, Bcl-2 could influence genetic programs of gene expression, could interact with the AIF precursor or apoycto-chrome *c* synthesized in the endoplasmatic reticulum, or could influence the export of AIF and apocytrochrome *c* from the cytoplasma to the mitochondrial intermembrane space. We found that lysis of control and Bcl-2-overexpressing mitochondria with detergents or osmotic shock releases equal amounts of AIF activity and AIF protein (Susin *et al.*, 1996). Similarly, Bcl-2 overexpressing mitochondria contain normal amounts of cytochrome *c* (Yang *et al.*, 1997). Therefore, Bcl-2 does not affect the formation and/or mitochondrial uptake of apoptogenic factors.

Third, does Bcl-2 interfere with the the release of apoptogenic factors? Transfection-enforced hyperexpression of Bcl-2 prevents the disruption of the mitochondrial transmembrane potential that normally precedes apoptosis induced by numerous apoptosis inducers (Decaudin *et al.*, 1997; Guenal *et al.*, 1997; Susin *et al.*, 1996; Zamzami *et al.*, 1995a). The Bcl-2-mediated protection against ceramide-induced $\Delta\Psi_m$ dissipation is observed both in intact cells and in anucleate cells (cytoplasts) (Castedo *et al.*, 1996; Zamzami *et al.*, 1995a), in which Bcl-2 conserves its death-inhibitory function (Jacobson *et al.*, 1994), indicating that its nuclear localization is dispensable for its function. More importantly, Bcl-2 overexpressed in the outer mitochondrial membrane inhibits the PT induced by a variety of PT inducers (*ter*-butylhydroperoxide, mClCCP, atractyloside, protoporphyrin IX) but not by diamide and caspase-1 (Marchetti *et al.*, 1996b; Susin *et al.*, 1997a; Susin *et al.*, 1996; Zamzami *et al.*, 1996a). This effect is observed in isolated mitochondria, correlating with the anti-apoptotic spectrum of Bcl-2 effects in cells. For example, the incapacity of Bcl-2 to interfere with diamide-induced PT correlates with its incapacity to inhibit the diamide-triggered $\Delta\Psi_m$ disruption and subsequent nuclear apoptosis in cells (Zamzami *et al.*, 1996b). Similarly, inefficient inhibition of caspase-1 induced PT by Bcl-2 correlates with poor inhibition of Fas-triggered (caspase-1-dependent) apoptosis (Susin *et al.*, 1997a). Thus, Bcl-2 does exert direct PT-inhibitory effects on mitochondria, although with a limited inhibitory spectrum. The spectrum of activity of Bcl-2 resembles most closely that of BA, a ligand of the ANT. When inhibiting PT, Bcl-2 prevents the release of AIF from the intermembrane space (Susin *et al.*, 1996; Zamzami *et al.*, 1996b). Thus, AIF is present in the supernatants of control mitochondria treated with atractyloside, *ter*-butylhydroperoxide, or mClCCP, yet is absent or greatly reduced in the supernatant of Bcl-2 hyperexpressing mitochondria treated with these reagents (Susin *et al.*, 1996). In addition, Bcl-2 prevents the release of cytochrome *c* from mitochondria (Kluck *et al.*, 1997; Yang *et al.*, 1997).

In synthesis, Bcl-2 suppresses apoptotic PT and AIF and cytochrome *c* release from mitochondria, yet does not interfere with the formation or action of these apoptogenic factors. This interpretation is in accord with previous genetic (Greenhalf *et al.*, 1996; Hanada *et al.*, 1995; Nguyen *et al.*, 1994; Tanaka *et al.*, 1993) and functional data (Greenhalf *et al.*, 1996; Jacobson *et al.*, 1994; Newmeyer *et al.*, 1994), suggesting that, at least in some experimental systems, the mitochondrial but not the nuclear localization of Bcl-2 would determine its anti-apoptotic capacity. In the mitochondrion, Bcl-2 demonstrates a patchy distribution to the contact sites between the outer and the inner mitochondrial membrane (Riparbelli *et al.*, 1995), exactly where the PT pore complex forms. Thus, it is conceivable that Bcl-2 or its homologs might exert a direct regulatory effect on the PT pore. In this context, it appears intriguing that proteins form the Bcl-2 family may constitute ion channels when incorporated into artificial membranes (Antonsson *et al.*, 1997; Minn *et al.*, 1997; Schendel *et al.*, 1997). However, the mechanism via which Bcl-2 prevents PT remains to be clarified. This question is presently addressed in our laboratory.

CONCLUSIONS AND PERSPECTIVES

Although mitochondria have for long been neglected by cell death researchers, it becomes increasingly clear that these organelles play a major role in the regulation of cell death. In an approximate fashion, it may assumed that pre-mitochondrial events determine the initiation phase of apoptosis, that the mitochondrial events constitute the effector stage, and that the degradation phase is basically post-mitochondrial (Figure 8.1). Mitochondria appear particularly attractive as cell death regulators because the sense multiple damage and signal transduction pathways, contain a number of potentially lethal proteins, and constitute the site of action of numerous endogenous regulators of apoptosis including members of the Bcl-2 protein family and CED-4. Although the apoptosis-regulatory function of mitochondria becomes increasingly assumed as a general principle of cell death control, numerous incognita remain to explored. Thus, the exact release, function, and targets of apoptogenic proteins including AIF and cytochrome *c* is unknown. Similarly, the exact functional relationship betwen permeability transition and release of apoptogenic proteins awaits clarification. It thus appears crucial to identify the exact molecular composition and structure/function of the permeability transition pore, a multiprotein complex formed at an anatomical site – the inner/outer membrane contract site – that suggests its implication in the regulation of both mitochondrial membranes. The in-depth-knowledge of the molecules involved in the mitochondrial control of apoptosis may facilitate the comprehension of the effector stage of apoptosis and may furnish targets for future pharmacological interventions on apoptosis.

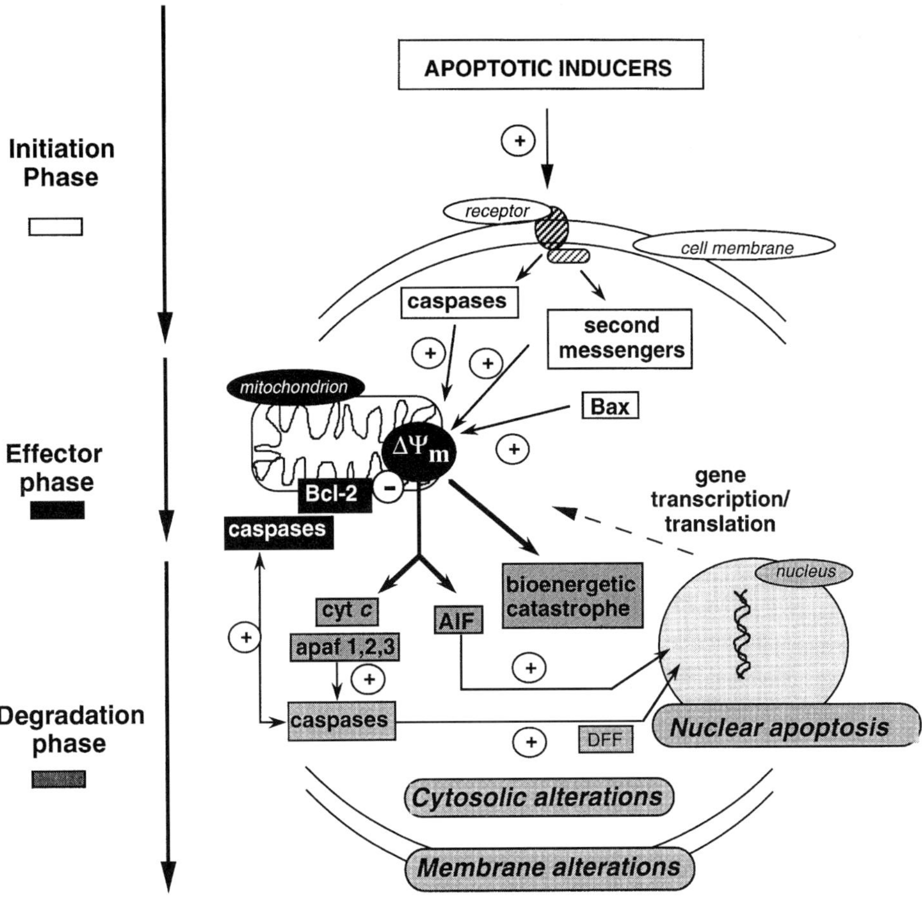

Figure 8.1 The three phases of apoptosis: initiation, effector and degradation phases. cyt *c*, cytochrome *c*; AIF, apoptosis inducing factor; DFF, DNA fragmentation factor.

ABBREVIATIONS

AIF, apoptosis-inducing factor; $\Delta\Psi_m$, mitochondrial transmembrane potential; PT, permeability transition; ROS, reactive oxygen species; z-VAD.fmk, *N*-benzyloxycarbonyl-Val-Ala-Asp-fluoromethyl ketone.

REFERENCES

Antonsson, B., Conti, F., Ciavatta, A., Montessuit, S., Lewis, S., Martinou, I., Bernasconi, M., Bernard, A., Mermod, J.-J., Mazzei, G., Maundrell, K., Gambale, F., Sadoui, R., Martinou, J.-C. (1997) Inhibition of Bax channel-forming activity by Bcl-2. *Science*, **277**, 370–376.

Attardi, G., Schatz, G. (1988) Biogenesis of mitochondria. *Ann. Rev. Cell Biol.*, **4**, 289–333.

Backway, K.L., Mcculloch, E.A., Chow, S., Hedley, D.W. (1997) Relationships between the mitochondrial permeability transition and oxidative stress during ara-C toxicity. *Cancer Res.*, **57**, 2446–2451.

Baixeras, E., Bosca, L., Stauber, C., Gonzalez, A., Carrera, A.C., Gonzalo, J.A., Martinez-A. (1994) From apoptosis to autoimmunity: Insights from the signalling pathways leading to proliferation or to programmed cell death. *Immunol. Rev.*, **142**, 53–91.

Barr, P.J., Tomei, L.D. (1994) Apoptosis and its role in human disease. *Biotechnology*, **12**, 487–493.

Bernardi, P. (1996) The permeability transition pore. Control points of a cyclosporin A-sensitive mitochondrial channel involved in cell death. *Biochim. Biophy. Acta. Bioenergetics*, **1275**, 5–9.

Bernardi, P., Broekemeier, K.M., Pfeiffer, D.R. (1994) Recent progress on regulation of the mitochondrial permeability transition pore; a cyclosporin-sensitive pore in the inner mitochondrial membrane. *J. Bioenergetics Biomembranes*, **26**, 509–517.

Bernardi, P., Petronilli, V. (1996) The permeability transition pore as a mitochondrial calcium release channel; a critical appraisal. *J. Bioenerg. Biomembr.*, **28**, 129–136.

Beutner, G., Rück, A., Riede, B., Welte, W., Brdiczka, D. (1996) Complexes between kinases, mitochondrial porin, and adenylate translocator in rat brain resemble the permeability transition pore. *FEBS. Lett.*, **396**, 189–195.

Boise, L.H., Thompson, C.B. (1997) Bcl-XL can inhibit apoptosis in cells that have undergone Fas-induced protease activation. *Proc. Natl. Acad. Sci. USA*, **94**, 3759–3764.

Borst, P., Grivell, L.A., Groot, G.S.P. (1984) Organelle DNA. *Trends Biochem. Sci.*, **9**, 128–130.

Brandolin, G., Le-Saux, A., Trezeguet, V., Lauquinn, G.J., Vignais, P.V. (1993) Chemical, immunological, enzymatic, and genetic approaches to studying the arrangement of the peptide chain of the ADP/ATP carrier in the mitochondrial membrane. *J. Bioenerg. Biomembr.*, **25**, 493–501.

Brustovetsky, N., Klingenberg, M. (1996) Mitochondrial ADP/ATP carrier can be reversibly converted into a large channel by Ca2+. *Biochemistry*, **35**, 8483–8488.

Cain, K., Inayathussain, S.H., Couet, C., Cohen, G.M. (1996) A cleavage-site-directed inhibitor of interleukin 1 beta-converting enzyme-like proteases inhibits apoptosis in primary cultures of rat hepatocytes. *Biochem. J.*, **314**, 27–32.

Castedo, M., Hirsch, T., Susin, S.A., Zamzami, N., Marchetti, P., Macho, A., Kroemer, G. (1996) Sequential acquisition of mitochondrial and plasma membrane alterations during early lymphocyte apoptosis. *J. Immunol.*, **157**, 512–521.

Castedo, M., Macho, A., Zamzami, N., Hirsch, T., Marchetti, P., Uriel, J., Kroemer, G. (1995) Mitochondrial perturbations define lymphocytes undergoing apoptotic depletion in vivo. *Eur. J. Immunol.*, **25**, 3277–3284.

Cohen, J.J. (1991) Programmed cell death in the immune system. *Adv. Immunol.*, **50**, 55–85.

Cory, S. (1995) Regulation of lymphocyte survival by the Bcl-2 gene family. *Annu. Rev. Immunol.*, **13**, 513–543.

Cossarizza, A., Franceschi, C., Monti, D., Salvioli, S., Bellesia, E., Rivabene, R., Biondo, L., Rainaldi, G., Tinari, A., Malorni, W. (1995) Protective effect of N-acetylcysteine in tumor necrosis factor-alpha-induced apoptosis in U937 cells: the role of mitochondria. *Exp. Cell Res.*, **220**, 232–240.

Costantini, P., Chernyak, B.V., Petronilli, V., Bernardi, P. (1996) Modulation of the mitochondrial permeability transition pore by pyridine nucleotides and dithiol oxidation at two separate sites. *J. Biol. Chem.*, **271**, 6746–6751.

Cosulich, S.C., Green, S., Clarke, P.R. (1996) Bcl-2 regulates activation of apoptotic proteases in a cell-free system. *Curr. Biol.*, **6**, 997–1005.

Decaudin, D., Geley, S., Hirsdch, T., Castedo, M., Marchetti, P., Macho, A., Kofler, R., Kroemer, G. (1997) Bcl-2 and Bcl-XL antagonize the mitochondrial dysfunction preceding nuclear apoptosis induced by chemotherapeutic agents. *Cancer Res.*, **57**, 62–67.

Enari, M., Hase, A., Nagata, S. (1995) Apoptosis by a cytosolic extract from Fas-activated cells. *EMBO J.*, **14**, 5201–5208.

Enari, M.,s Talanian, R.V., Wong, W.W., Nagata, S. (1996) Sequential activation of ICE-like and CPP32-like proteases during Fas-mediated apoptosis. *Nature*, **380**, 723–726.

Evans, E.K., Lu, W., Strum, S.L., Mayer, B.J., Kornbluth, S. (1997) Crk is require for apoptosis in Xenopus egg extracts. *EMBO J.*, **16**, 230–241.

Fearnhead, H.O., Dinsdale, D., Cohen, G.M. (1995) An interleukin-1 beta-converting enzyme-like protease is a common mediator of apoptosis in thymocytes. *FEBS Lett.*, **375**, 283–288.

Gamen, S., Anel, A., Montoya, J., Marzo, I., Piñeiro, A., Naval, J. (1995) mtDNA-depleted U937 cells are sensitive to TNF and Fas-mediated cytotoxicity. *FEBS Lett.*, **376**, 15–18.

Gray, M.W. (1989) Origin and evolution of mitochondrial. DNA *Annu. Rev. Biochem.*, **5**, 25–50.

Greenhalf, W., Stephan, C., Chaudhuri, B. (1996) Role of mitochondria and C-terminal membrane anchor of Bcl-2 in Bax induced growth arrest and mortality in Sacharomyces cerevisiae. *FEBS Lett.*, **380**, 169–175.

Guenal, I., Sidoti-Defraisse, C., Gaumer, S., Mignotte, B. (1997) Bcl-2 and hsp27 act at different levels to suppress programmed cell death. *Oncogene*, **15**, 347–360.

Halestrup, A.P., Woodfield, K.-Y., Connern, C.P. (1997) Oxidative stress, thiol reagent, and membrane potential modulate the mitochondrial permeability transition by affecting nucleotide binding to the adenine nucleotide translocator. *J. Biol. Chem.*, **272**, 3346–3354.

Hanada, M., Aimesempe, C., Sato, T., Reed, J.C. (1995) Structure-function analysis of Bcl-2 protein identification of conserved domains important for homodimerization with Bcl-2 and heterodimerization with Bax. *J. Biol. Chem.*, **270**, 11962–11969.

Henkart, P.A. (1995) Apoptosis: O death, where is thy sting? *J. Immunol.*, **154**, 4905–4908.

Henkart, P.A. (1996) ICE family proteases: Mediators of all apoptotic cell death? *Immunity*, **4**, 195–201.

Hirsch, T., Marchetti, P., Susin, S.A., Dallaporta, B., Zamzami, N., Marzo, I., Geuskens, M., Kroemer, G. (1997) The apoptosis-necrosis paradox. Apoptogenic proteases activated after mitochondrial permeability transition determine the mode of cell death. *Oncogene*, **15**, 1573–1582.

Hortelano, S., Dallaporta, B., Zamzami, N., Hirsch, T., Susin, S.A., Marzo, I., Bosca, L., Kroemer, G. (1997) Nitric oxide induces apoptosis via triggering mitochondrial permeability transition. *FEBS Lett.*, **410**, 373–377.

Ichas, F., Jouavill, L.S., Mazat, J.-P. (1997) Mitochondria are excitable organelles capable of generating and conveying electric and calcium currents. *Cell*, **89**, 1145–1153.

Jacobson, M.D., Burne, J.F., King, M.P., Miyashita, T., Reed, J.C., Raff, M.C. (1993) Bcl-2 blocks apoptosis in cells lacking mitochondrial DNA. *Nature*, **361**, 365–369.

Jacobson, M.D., Burne, J.F., Raff, M.C. (1994) Programmed cell death and Bcl-2 protection in the absence of a nucleus. *EMBO J.*, **13**, 1899–1910.

Jacobson, M.D., Weil, M., Raff, M.C. (1996) Role of Ced-3/ICE-family proteases in staurosporine-induced programmed cell death. *J. Cell Biol.*, **133**, 1041–1051.

Janiak, F., Leber, B., Andrews, D.W. (1994) Assembly of Bcl-2 into microsomal and outer mitochondrial membranes. *J. Biol. Chem.*, **269**, 9842–9849.

Kantrow, S.P., Piantadosi, C.A. (1997) Release of cytochrome c from liver mitochondria during permeability transition. *Biochem. Biophys. Res. Comm.*, **232**, 669–671.

King, M.P., Attardi, G. (1996) Isolation of human cell lines lacking mitochondrial DNA. *Meth. Enzymol.*, **264**, 304–313.

Klingenberg, M. (1980) The ADP-ATP translocation in mitochondria, a membrane potential controlled transport *J. Membrane Biol.*, **56**, 97–105.

Kluck, R.M., Bossy-Wetzel, E., Green, D.R., Newmeyer, D.D. (1997) The release of cytochrome c from mitochondria: a primary site for Bcl-2 regulation of apoptosis. *Science*, **275**, 1132–1136.

Krajewski, S., Tanaka, S., Takayama, S., Schibler, M.J., Fenton, W., Reed, J.C. (1993) Investigation of the subcellular distribution of the bcl-2 oncoprotein: residence in the nuclear envelope, endoplasmic reticulum, and outer mitochondrial membranes. *Cancer Res.*, **53**, 4701–4714.

Kroemer, G. (1995) The pharmacology of T cell apoptosis. *Adv. Immunol.*, **58**, 211–296.

Kroemer, G. (1997a) Mitochondrial implication in apoptosis. Towards an endosymbiotic hypothesis of apoptosis evolution. *Cell Death Differentiation*, **4**, 443–456.

Kroemer, G. (1997b) The proto-oncogene Bcl-2 and its role in regulating apoptosis. *Nature Medicine*, **3**, 614–620.

Kroemer, G., Dallaporta, B., Resche-Rigon, M. (1997a) The mitochondrial death/life rheostat in apoptosis and necrosis. *Annu. Rev. Physiol.*, **60**, in press.

Kroemer, G., Petit, P.X., Zamzami, N., Vayssière, J.-L., Mignotte, B. (1995) The biochemistry of apoptosis. *FASEB J.*, **9**, 1277–1287.

Kroemer, G., Zamzami, N., Susin, S.A. (1997b) Mitochondrial control of apoptosis. *Immunol. Today*, **18**, 44–51.

Lazebnik, Y.A., Cole, S., Cooke, C.A., Nelson, W.G., Earnshaw, W.C. (1993) Nuclear events of apoptosis in vitro in cell-free mitotic extracts: a model system for analysis of the active phase of apoptosis. *J. Cell Biol.*, **123**, 7–22.

Lazebnik, Y.A., Takahashi, A., Poirier, G.G., Kaufman, S.H., Earnshaw, W.C. (1995a) Characterization of the execution phase of apoptosis in vitro using extracts from condemned-phase cells. *J. Cell Sci.*, **S19**, 41–49.

Lazebnik, Y.A., Takayashi, A., Moir, R.D., Goldman, R.D., Poirier, G.D., Kaufmann, S.H., Earnshaw, W.C. (1995b) Studies of the lamin proteinase reveal multiple parallel biochemical pathways during apoptotic execution. *Proc. Natl. Acad. Sci. USA*, **92**, 9042–9046.

Lithgow, T., Vandriel, R., Bertram, J.F., Strasser, A. (1994) The protein product of the oncogene Bcl-2 is a component of the nuclear envelope, the endoplasmic reticulum, and the outer mitochondrial membrane. *Cell Growth Differentiation*, **5**, 411–417.

Liu, X., Kim, C.N., Yang, J., Jemmerson, R., Wang, X. (1996) Induction of apoptic program in cell-free extracts: requirement for dATP and cytochrome. C. *Cell*, **86**, 147–157.

Liu, X., Zou, H., Slaughter, C., Wang, X. (1997) DFF, a heterodimeric protein that functions downstream of caspase 3 to trigger DNA fragmentation during apoptosis. *Cell*, **89**, 175–184.

Macho, A., Castedo, M., Marchetti, P., Aguilar, J.J., Decaudin, D., Zamzami, N., Girard, P.M., Uriel, J., Kroemer, G. (1995) Mitochondrial dysfunctions in circulating T lymphocytes from human immunodeficiency virus-1 carriers. *Blood*, **86**, 2481–2487.

Macho, A., Hirsch, T., Marzo, I., Marchetti, P., Dallaporta, B., Susin, S.A., Zamzami, N., Kroemer, G. (1997) Glutathione depletion is an early and calcium elevation a late event of thymocyte apoptosis *J. Immunol.*, **158**, 4612–4619.

Marchetti, P., Castedo, M., Susin, S.A., Zamzami, N., Hirsch, T., Haeffner, A., Hirsch, F., Geuskens, M., Kroemer, G. (1996a) Mitochondrial permeability transition is a central coordinating event of apoptosis. *J. Exp. Med.*, **184**, 1155–1160.

Marchetti, P., Decaudin, D., Macho, A., Zamzami, N., Hirsch, T., Susin, S.A., Kroemer, G. (1997) Redox regulation of apoptosis: impact of thiol redoxidation on mitochondrial function. *Eur. J. Immunol.*, **27**, 289–296.

Marchetti, P., Hirsch, T., Zamzami, N., Castedo, M., Decaudin, D., Susin, S.A., Masse, B., Kroemer, G. (1996b) Mitochondrial permeability transition triggers lymphocyte apoptosis. *J. Immunol.*, **157**, 4830–4836.

Marchetti, P., Susin, S.A., Decaudin, D., Gamen, S., Castedo, M., Hirsch, T., Zamzami, N., Naval, J., Senik, A., Kroemer, G. (1996c) Apoptosis-associated derangement of mitochondrial function in cells lacking mitochondrial DNA. *Cancer Res.*, **56**, 2033–2038.

Marchetti, P., Zamzami, N., Susin, S.A., Patrice, P.X., Kroemer, G. (1996d) Apoptosis of cells lacking mitochondrial DNA. *Apoptosis*, **1**, 119–125.

Martin, S.J., Green, D.R. (1995) Protease activation during apoptosis: death by a thousand cuts? *Cell*, **82**, 349–352.

Martin, S.J., Newmeyer, D.D., Mathisa, S., Farschon, D.M., Wang, H.G., Reed, J.C., Kolesnick, R.N., Green, D.R. (1995) Cell-free reconstitution of Fas-, UV radiation- and ceramide-induced apoptosis. *EMBO J.*, **14**, 5191–5200.

Minn, A.J., Vélez, P., Schendel, S.L., Liang, H., Muchmore, S.W., Fesik, S.W., Fill, M., Thompson, C.B. (1997) Bcl-XL forms an ion channel in synthetic lipid membranes. *Nature*, **385**, 353–357.

Muzio, M., Salvesen, G.S., Dixit, V.M. (1997) FLICE induced apoptosis in a cell-free system. Cleavage of caspase zymogens. *J. Biol. Chem.*, **272**, 2952–2956.

Nakajima, H., Golstein, P., Henkart, P.A. (1995) The target cell nucleus is not required for cell-mediated granzyme- or Fas-based cytotoxicity. *J. Exp. Med.*, **181**, 1905–1909.

Newmeyer, D.D., Farschon, D.M., Reed, J.C. (1994) Cell-free apoptosis in xenopus egg extracts: inhibition by Bcl-2 and requirement for an organelle fraction enriched in mitochondria. *Cell*, **79**, 353–364.

Nguyen, M., Branton, P.E., Walton, P.A., Oltvai, Z.N., Korsmeyer, S.J., Shore, G.C. (1994) Role of membrane anchor domain of Bcl-2 in suppression of apoptosis caused by E1B-defective adenovirus. *J. Biol. Chem.*, **269**, 16521–16524.

Nicolli, A., Basso, E., Petronilli, V., Wenger, R.M., Bernardi, P. (1996) Interactions of cyclophilin with mitochondrial inner membrane and regulation of the permeability transition pore, a cyclosporin A-sensitive channel. *J. Biol. Chem.*, **271**, 2185–2192.

Oltvai, Z.N., Korsmeyer, S.J. (1994) Checkpoints of dueling dimers foil death wishes *Cell*, **79**, 189–192.

Osborne, B.A., Smith, S.W., Liu, Z.-G., McLaughlin, K.A., Grimm, L., Schwartz, L.M. (1994) Identification of genes induced during apoptosis in T cells. *Immunol. Rev.*, **142**, 301–320.

Pastorino, J.G., Simbula, G., Gilfor, E., Hoek, J.B., Farber, J.L. (1994) Protoporphyrin IX, an endogenous ligand of the peripheral benzodiazepin receptor, potentiates induction of the mitochondrial permeability transition and the killing of culture hepatocytes by rotenone. *J. Biol. Chem.*, **269**, 31041–31046.

Patel, T., Gores, G.J., Kaufman, S.H. (1996) The role of proteases during apoptosis **10**, 587–597.

Petit, P.X., LeCoeur, H., Zorn, E., Dauguet, C., Mignotte, B., Gougeon, M.L. (1995) Alterations of mitochondrial structure and function are early events of dexamethasone-induced thymocyte apoptosis. *J. Cell Biol.*, **130**, 157–167.

Petit, P.X., Susin, S.A., Zamzami, N., Mignotte, B., Kroemer, G. (1996) Apoptosis and mitochondria: back to the future. *FEBS Lett.*, **396**, 7–14.

Polla, B.S., Kantengwa, S., Francois, D., Salvioli, S., Franceschi, C., Marsac, C., Cossarizza, A. (1996) Mitochondria are selective targets for the protective effects of heat shock against oxidative injury. *Proc. Nat. Acad. Sci. USA*, **93**, 6458–6463.

Pronk, G.J., Ramer, K., Amiri, P., Williams, L.T. (1996) Requirement of an ICE-like protease for induction of apoptosis and ceramide generation by REAPER. *Science*, **271**, 808–810.

Reed, J.C. (1997) Double identity for proteins of the Bcl-2 family. *Nature*, **387**, 773–776.

Riparbelli, M.G., Callaini, G., Tripodi, S.A., Cintorino, M., Tosi, P., Dallai, R. (1995) Localization of the Bcl-2 protein to the outer mitochondrial membrane by electron microscopy. *Exp. Cell Res.*, **221**, 363–369.

Schatz, G., Dobberstein, B. (1996) Common principles of protein translocation across membranes. *Science*, **271**, 1519–1526.

Schendel, S., Xie, Z., Montal, M.O., Matsuyama, S., Montal, M., Reed, J.C. (1997) Channel formation by antiapoptotic protein Bcl-2 *Proc. Natl. Acad. Sci. USA*, **94**, 5113–5118.

Schulze-Osthoff, K., Walczak, H., Droge, W., Krammer, P.H. (1994) Cell nucleus and DNA fragmentation are not required for apoptosis. *J. Cell Biol.*, **127**, 15–20.

Shimizu, S., Eguchi, Y., Kamiike, W., Waguri, S., Uchiyama, Y., Matsuda, H., Tsujimoto, Y. (1996) Bcl-2 blocks loss of mitochondrial membrane potential while ICE. inhibitors act at a different step during inhibition of death induced by respiratory chain inhibitors *Oncogene*, **13**, 21–29.

Shimizu, S., Eguchi, Y., Kosaka, H., Kamlike, W., Matsuda, H., Tsujimoto, Y. (1995) Prevention of hypoxia-induced cell death by Bcl-2 and Bcl-xL. *Nature*, **374**, 811–813.

Shimizu, T., Pommier, Y. (1996) DNA fragmentation induced by protease activation in p53-null human leukemia HL60 cells undergoing apoptosis following treatment with the topoisomerase I inhibitor camptothecin: Cell-free system studies. *Exp. Cell Res.*, **226**, 292–301.

Skowronek, P., Haferkamp, O., Rödel, G. (1992) A fluorescence-microscopic and flow-cytometric study of HELA cells with an experimentally induced respiratory deficiency. *Biochem. Biophys. Res. Communications*, **187**, 991–998.

Slee, E.A., Zhu, H.J., Chow, S.C., Macfarlane, M., Nicholson, D.W., Cohen, G.M. (1996) Benzyloxycarbonyl-Val-Ala-Asp (OMe) fluoromethylketone (z-VAD.fmk) inhibits apoptosis by blocking the processing of CPP32. *Biochem. J.*, **315**, 21–24.

Susin, S.A., Zamzami, N., Castedo, M., Daugas, E., Wang, H.-G., Geley, S., Fassy, F., Reed, J., Kroemer, G. (1997a) The central executioner of apoptosis. Multiple links between protease activation and mitochondria in Fas/Apo-1/CD95- and ceramide-induced apoptosis. *J. Exp. Med.*, **186**, 25–37.

Susin, S.A., Zamzami, N., Castedo, M., Hirsch, T., Marchetti, P., Macho, A., Daugas, E., Geuskens, M., Kroemer, G. (1996) Bcl-2 inhibits the mitochondrial release of an apoptogenic protease. *J. Exp. Med.*, **184**, 1331–1342.

Susin, S.A., Zamzami, N., Larochette, N., Dallaporta, B., Marzo, I., Brenner, C., Hirsch, T., Petit, P.X., Geuskents, M., Kroemer, G. (1997b) A cytofluorometric assay of nuclear apoptosis. Application to ceramide-induced apoptosis. *Exp. Cell Res.*, **236**, 387–403.

Tanaka, S., Saito, K., Reed, J.C. (1993) Structure-function analysis of the Bcl-2 oncoprotein. Addition of a heterologous transmembrane domain to portions of the Bcl-2β protein restores function as a regulator of cell survival. *J. Biol. Chem.*, **268**, 10920–10926.

Tepper, C.G., Studzinski, G.P. (1992) Teniposide induces nuclear but not mitochondrial DNA degradation. *Cancer Res.*, **52**, 3384–3390.

Tepper, C.G., Studzinski, G.P. (1993) Resistance of mitochondrial DNA to degradation characterizes the apoptotic but not the necrotic mode of human leukemia cell death. *J. Cell Biochem.*, **52**, 352–361.

Thompson, C.B. (1995) Apoptosis in the pathogenesis and treatment of disease *Science*, **267**, 1456–1462.

Vayssière, J.-L., Petit, P.X., Risler, Y., Mignotte, B. (1994) Commitment to apoptosis is associated with changes in mitochondrial biogenesis and activity in cell lines conditionally immortalized with simian virus 40. *Proc. Natl. Acad. Sci. USA*, **91**, 11752–11756.

Wertz, I.E., Hanley, M.R. (1996) Diverse molecular provocation of programmed cell death. *Trends Biochem. Sci.*, **21**, 359–364.

Wolvetang, E.J., Johnson, K.L., Krauer, K., Ralph, S.J., Linnane, A.W. (1994) Mitochondrial respiratory chain inhibitors induce apoptosis. *FEBS Letters*, **339**, 40–44.

Wright, S.C., Wei, Q.S., Zhong, J., Zheng, H., Kinder, D.H., Larrick, J.W. (1994) Purification of a 24-kD protease from apoptotic tumor cells that activates DNA fragmentation. *J. Exp. Med.*, **180**, 2113–2123.

Xiang, J., Chao, D.T., Korsmeyer, S.J. (1996) Bax-induced cell death may not require interleukin 1beta-converting enzyme-like proteases. *Proc. Natl. Acad. Sci. USA*, **93**, 14559–14563.

Yang, E., Korsmeyer, S.J. (1996) Molecular Thanatopsis: A discourse on the Bcl-2 family and cell death. *Blood*, **88**, 386–401.

Yang, J., Liu, X., Bhalla, K., Kim, C.N., Ibrado, A.M., Cai, J., Peng, T.-I., Jones, D.P., Wang, X. (1997) Prevention of apoptosis by Bcl-2: release of cytochrome c from mitochondria blocked. *Science*, **275**, 1129–1132.

Yoneda, M., Kasumata, K., Hayakawa, M., Tanaka, M., Ozawa, T. (1995) Oxygen stress induces an apoptotic cell death associated with fragmentation of mitochondrial genome. *Biochem. Biophys. Res. Comm.*, **209**, 723–729.

Yoshida, A., Takauji, R., Inuzuka, M., Ueda, T., Nakamura, T. (1996) Role of serine and ICE-like proteases in induction of apoptosis by etoposide in human leukemia HL-60 cells. *Leukemia*, **10**, 821–824.

Zamzami, N., Marchetti, P., Castedo, M., Decaudin, D., Macho, A., Hirsch, T., Susin, S.A., Petit, P.X., Mignotte, B., Kroemer, G. (1995a) Sequential reduction of mitochondrial transmembrane potential and generation of reactive oxygen species in early programmed cell death. *J. Exp. Med.*, **182**, 367–377.

Zamzami, N., Marchetti, P., Castedo, M., Hirsch, T., Susin, S.A., Masse, B., Kroemer, G. (1996a) Inhibitors of permeability transition interfere with the disruption of the mitochondrial transmembrane potential during apoptosis. *FEBS Lett.*, **384**, 53–57.

Zamzami, N., Marchetti, P., Castedo, M., Zanin, C., Vayssière, J.-L., Petit, P.X., Kroemer, G. (1995b) Reduction in mitochondrial potential constitutes an early irreversible step of programmed lymphocyte death in vivo. *J. Exp. Med.*, **181**, 1661–1672.

Zamzami, N., Susin, S.A., Marchetti, P., Hirsch, T., Gómez-Monterrey, I., Castedo, M., Kroemer, G. (1996b) Mitochondrial control of nuclear apoptosis. *J. Exp. Med.*, **183**, 1533–1544.

Zhivotovsky, B., Gahm, A., Ankarcrona, M., Nicotera, P., Orrenius, S. (1995) Multiple proteases are involved in thymocyte apoptosis. *Exp. Cell Res.*, **221**, 404–412.

Zhu, H.J., Fearnhead, H.O., Cohen, G.M. (1995) An ICE-like protease is a common mediator of apoptosis induced by diverse stimuli in human monocytic THP.1 cells. *FEBS. Lett.*, **374**, 303–308.

Zhu, W., Cowie, A., Wasfy, G.W., Penn, L.Z., Leber, B., Andrews, D.W. (1996) Bcl-2 mutants with restricted subcellular localization reveal spatially distinct pathways for apoptosis in different cell types. *EMBO J.*, **15**, 4130–4141.

Zoratti, M., Szabò, I. (1995) The mitochondrial permeability transition. *Biochem. Biophys. Acta. – Rev. Biomembranes*, **1241**, 139–176.

9. CASPASES AND THE COMMITMENT TO DEATH

DEBORAH M. FINUCANE*, THOMAS G. COTTER** AND DOUGLAS R. GREEN*†

*Division of Cellular Immunology, La Jolla Institute for Allergy and Immunology, 10355 Science Center Drive, San Diego, CA 92121, USA
**Tumor Biology Laboratory, Biochemistry Department, University College Cork, Ireland

KEY WORDS: caspase, caspase inhibitors - crmA, IAP, p35, zVAD-fmk, "commitment step", cytotoxic drugs – actinomycin D, dexamethasone, etoposide, staurosporine, developmental death – C.elegans, Drosophila, mitochondria – AIF, cytochrome c, permeability transition (PT), ROS, oncogenes – Bax, Bcl, Bcr-Abl, c-myc

INTRODUCTION

It is now widely accepted that apoptosis, or active cell death, is essential for the development and cellular homeostasis of metazoan animals. The fundamental importance of this form of cell death can be appreciated by its conservation throughout evolution (Ameisen *et al.*, 1995; Ameisen, 1996; Cornillon *et al.*, 1994; Vaux *et al.*, 1994). Apoptosis constitutes an intrinsic suicide mechanism that systematically destroys the cell with characteristic morphological and biochemical changes without initiating an inflammatory response (Kerr *et al.*, 1972; Wyllie, 1980; Arends and Wyllie, 1991; Raff, 1992). This altruistic process is extremely rapid (minutes to hours) and the resulting apoptotic debris is cleared with similar efficiency. Perturbations in this process have been implicated in a range of human pathological disorders including cancer (Green *et al.*, 1994; Thompson, 1995; Williams, 1991; Martin and Green, 1995), acquired immunodeficiency syndrome (AIDS) (Ameisen, 1994; Terai *et al.*, 1991; Meyaard *et al.*, 1992; Martin, 1993; Ameisen and Capron, 1991), ischemic injury (Martinou, 1994) and neurodegenerative disorders (Carson and Ribeiro, 1993; Loo *et al.*, 1993; Holden and Mooney, 1995). Thus investigations into the regulation of apoptosis may have important therapeutic potential.

Apoptosis can be induced in a wide variety of cell types by such diverse stimuli as viral infection, cell-cell interactions, DNA damage, growth factor withdrawal and others (Debbas and White, 1993; Lennon *et al.*, 1991; Henkart, 1994; Berke, 1995; Brunner *et al.*, 1995). Despite such striking heterogeneity in inducing stimuli, cells undergo apoptosis with remarkably similar morphological features. This suggests the existence of a centrally conserved biochemical pathway for this form of programmed cell death. However a major challenge facing researchers into this area over the years has been to distinguish between events critical to activation of the cell death pathway and those that occur as a consequence of such an activation i.e.

† *Corresponding Author*: Tel.: 678 3543. Fax: (619) 558–3525. e-mail: dgreen 5240@aol.com

separating initiation verses orchestration. The rapidity of the apoptosis process further complicates this and so the components of the pathways governing the fate of cells continue to elude us. Recent findings however allow us to re-address the question – what are the critical events in cell death?

WHAT IS CELL DEATH COMMITMENT ?

While apoptosis can be identified by a distinct set of morphological and biochemical features, so far no one event has been definitively shown to be universally required for cell death under all conditions. While it could be speculated that a loss of nuclear function may represent the critical step in cell death, several cases of cell death in the absence of chromatin condensation or nuclear fragmentation have been documented (Schulze-Osthoff *et al.*, 1994; Oberhammer *et al.*, 1993; Zakeri *et al.*, 1993; Martin *et al.*, 1996). Even more compelling is the fact that some cells persist and perform important functions without nuclei: red blood cells are devoid of nuclei yet are functionally viable for around 121 days before dying. Alternatively, the loss of plasma membrane integrity can be a useful indicator of cell death especially in vitro, but in vivo dying apoptotic cells are phagocytosed well before this can occur. Therefore, these and other well documented cellular changes associated with cell death (apoptosis or necrosis), represent only the final stages in the process.

It is still unknown at what biochemical step the cell's fate is irreversibly decided, the point at which the cell is beyond rescue and is destined to die. Past this point, intervention may alter the form of cell death but can not rescue the cell. This "point-of-no-return" where a cell can truly be considered dead, is what we refer to as the cell death "commitment step". The importance of this can be stressed by the fact that most defects in apoptosis identified to date occur at this stage of the cell death pathway. The type of death a cell undergoes can have enormous ramifications on the host system in particular because necrotic cells elicit an inflammatory response while apoptotic cells do not. The mode of cell death has also been linked to tumor prognosis (Arends *et al.*, 1994). Dissociating events that lead to the cells ultimate decision to live or die would allow us the possibility of effectively manipulating cell death with enormous clinical benefits.

HOW CAN WE STUDY CELL DEATH COMMITMENT ?

A major challenge is to ascertain the exact sequence of molecular and cellular events that lead to an irreversible commitment to cell death and the appearance of the apoptotic phenotype. In this regard clonogenic studies have provided a useful approach. For example, Jurkat and HL-60 cells incubated with cytotoxic agents such as actinomycin D, etoposide or staurosporine for a few hours and then washed and re-plated in fresh media appear overtly normal for several hours and even up to several days, but are nonetheless destined to die (Amarante-Mendes *et al.*, 1998). These results demonstrate that commitment to cell death is an early event preceding all the key biological or morphological features of apoptosis reported to date. This strongly argues for separate components regulating cellular commitment to death and the subsequent execution phase characterized as apoptosis.

Studies highlighting caspase involvement in the apoptotic process indicate a role for these unusual cysteine proteases in regulating the execution of many of the events in apoptosis (Martin and Green, 1995; Henkart, 1996; Chinnaiyan and Dixit, 1996; Kumar, 1995; Yuan and Horvitz, 1990; Alnemri *et al.*, 1996; Enari *et al.*, 1995). This has lead to speculation that caspase activation is the commitment point at a molecular level. However what is emerging from recent studies may be far more complex and intriguing. It appears that particular caspases may be essential for some forms of cell death but dispensable for others, depending on the specific insult, cell type and even species type. This leads to a number of issues to resolve concerning this highly conserved group of proteases. What position do the different caspases occupy in the cell death pathway, how are they activated and what are their roles?

CASPASES AND THE COMMITMENT TO CELL DEATH

Several lines of evidence show that proteases are critical mediators of the apoptotic process. Studies on cytotoxic granule-mediated cell death identified granzyme B/fragmentin-2, an enzyme with unusual aspartate cleaving specificity for its substrates, as one of the key components responsible for apoptosis induced by CTLs (Berke 1995; Smyth and Trapani, 1995; Heusel *et al.*, 1994; Shi *et al.*, 1992). Apoptosis induced following microinjection of this enzyme supports such a role (Bleackley, personal communication).

More direct evidence came from elegant genetic studies of two invertebrate models, the nematode Caenorhabditis elegans and the fruit fly, Drosophila melanogaster. Systematic genetic mutations in the nematode C. elegans, have identified 3 genes that are specifically required for the execution of programmed cell death: ced-3, ced-4 and ced-9 (ced is an abbreviation for cell death defective) (Ellis and Horvitz, 1991; Sulston and Horvitz, 1977; Hedgecock *et al.*, 1983; Ellis and Horvitz, 1986; Ellis *et al.*, 1991; Hengartner and Horvitz, 1994; Hengartner *et al.*, 1992). The activities of ced-3 and -4 are required for all somatic cell death in worm development. If either are inactivated all cells that normally die during development will survive. In contrast the third gene, ced-9, is required to protect cells from undergoing programmed cell death: the absence of ced-9 results in widespread ectopic cell death. Double mutants of ced-9 and either ced-3 or -4 lack all cell death, implying that ced-9 functions to prevent apoptosis induced by ced-3 and -4. This suggests that ced-9 acts upstream of these two death promoting genes in this model. Thus there appears to be a delicate balance between the opposing activities of proteins that promote and those that inhibit cell death.

These studies provided important insights into the genes controlling nematode cell death. Homology studies revealed a significant homology between ced-3 and a family of mammalian cysteine proteases referred to as caspases (Yuan *et al.*, 1993; Fernandes-Alnemri *et al.*, 1994; Xue *et al.*, 1996). The existence of mammalian homologues to ced-3 and -9 supports the notion of an evolutionary conserved cell death pathway. To date more than 10 ced-3 mammalian homologues have been isolated and characterized, which are expressed at various stages of development in a broad range of tissues (Yuan *et al.*, 1993; Fernandes-Alnemri *et al.*, 1994; Fernandes-Alnemri *et al.*, 1995a; Kamens *et al.*, 1995; Fernandes-Alnemri *et al.*, 1995b; Whyte, 1996). Interestingly, their unusual substrate specificity for aspartic acid residues (Nicholson *et al.*, 1995; Tewari *et al.*, 1995; Thornberry *et al.*, 1992; Xue and Horvitz, 1995) had only been seen in one other known eukaryotic protease – granzyme B. Ectopic expression of these homologues resulted in apoptotic induction in a variety of cell types (Faucheu *et al.*, 1995; Miura *et al.*, 1993; Kumar *et al.*, 1994), suggesting that caspase proteases are both functionally and structurally homologous to ced-3 and could therefore function as major players in the mammalian cell death process. More importantly studies showed that specific caspase inhibitors dramatically blocked apoptosis initiated by a wide range of cytotoxic stimuli (Loddick *et al.*, 1996; Rodriguez *et al.*, 1996; Jacobsen *et al.*,1996; Milligan *et al.*, 1995; Slee *et al.*, 1996).

With the weight of such evidence implicating caspases as mediators of apoptosis it seems logical that their role could extend to cell death commitment. That is, could caspase inhibitors confer cell survival? Making use of a broad range caspase inhibitor, zVAD-fmk, designed to mimic known target sites of caspases, several studies addressed this question. A range of agents which induced apoptosis with similar characteristics, including the rapid proteolysis of caspase-3 (Nicholson *et al.*, 1995; Duan *et al.*, 1996), were examined in the presence and absence of this caspase inhibitor. While protection was dramatic over a short period irrespective of the death stimulus, it was surprising to discover two very different outcomes after long term monitoring. In one case cell death in general was inhibited by preincubating with caspase inhibitors. This was seen for death induced in response to cell surface CD95 receptor engagement. Similar results were obtained in studies of neuronal survival after growth factor withdrawal in the presence of crmA, a known caspase inhibitor (Gagliardini *et al.*, 1994). It would appear caspases are required for some forms of cell death by acting upstream of a commitment point (Figure 9.1).

In contrast, in some cases caspase inhibitors could efficiently prevent the apoptotic phenotype but not cell death in general. Examples include apoptosis induced by c-myc overexpression, the pro-apoptotic Bcl-2 family members Bak (McCarthy *et al.*, 1997) and Bax (Xiang *et al.*, 1996), growth factor withdrawal (Ohta *et al.*, 1997) and cytotoxic agents such as dexamethasone (Brunet, *et. al.*, 1998) etoposide, actinomycin D and staurosporine (Amarante-Mendes *et al.*, 1998) and by γ-irradiation (McCarthy *et al.*, 1997). In all cases it was found that inhibition of caspase activity only prevented the completion of the apoptotic program but did not rescue cells from cell death in the long term. As seen in Figure 9.2, Jurkat cells treated with staurosporine in the presence of zVAD-fmk displayed impressive inhibition of apoptosis at early time points. However, cell death as assessed by loss of membrane integrity, ultimately followed with only delayed kinetics. Furthermore, treating cells with limiting dilutions of cytotoxic agents such as staurosporine for limiting periods of time and then washing out the insult in the presence or absence of zVAD-fmk produced no difference in clonogenic potential (Amarante-Mendes

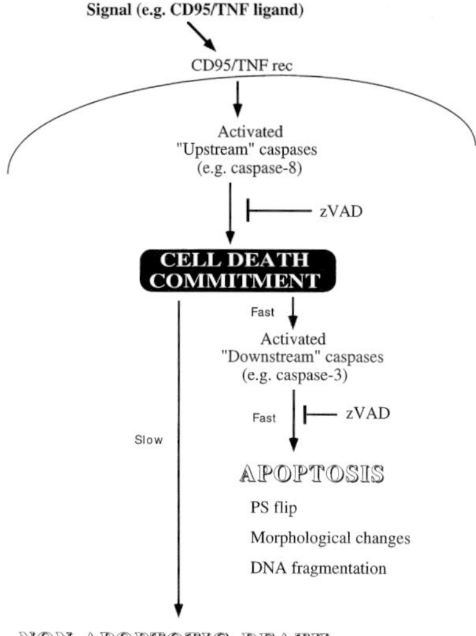

Figure 9.1 Schematic diagram of caspase-dependent cell death.

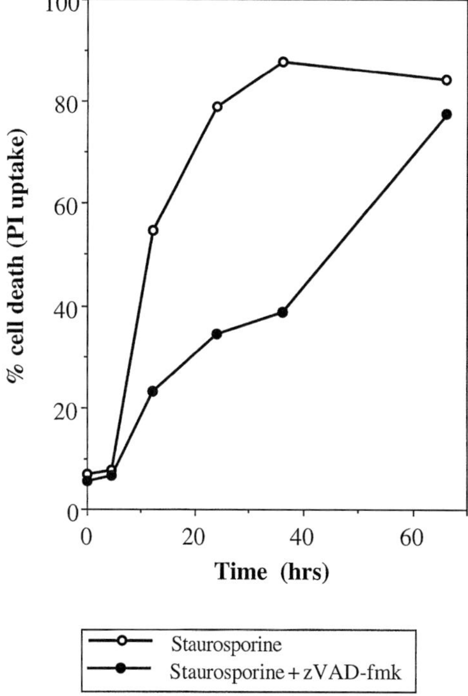

Figure 9.2 Caspase inhibition blocks apoptosis but only delays cell death. Jukat cells were treated with staurosporine (1μM) in the presence or absence of zVAD-fmk (100μM) and cell death was assessed over time by propidium iodide (PI) uptake, indicating loss of membrane integrity.

et al., 1998). Perhaps once a threshold was reached, cells were committed to die regardless of caspase activity. Thus it seems that in certain situations cells commit to die prior to caspase activation. Interestingly, closer examination of the type of death that ensued revealed that rather than apoptosis, extensive cytosolic vacuolization reminiscent of autophagy occurred.

Therefore it appeared a second pathway leading to commitment must exist, one which is caspase-independent. The slower form of cell death observed here does not appear to be merely a type of "necrosis-by-default" caused by irreparable damage to the cell as it can be inhibited by the oncogenes Bcl-2 and Bcr-Abl (Amarante-Mendes *et al.*, 1998; Brunet, *et al.*, 1998). Both Bcr-Abl and Bcl-2 blocked not only apoptosis but cell death and thus maintained cell viability. This indicates that cells are committed to death under molecular control, and that oncogenesis is promoted not by simply interfering with caspase-mediated apoptosis, but by preventing an upstream event which we define as the commitment point to cell death (Figure 9.3). Although caspases may not be required to commit cells to die in this scenario they still play a major role in orchestrating the type of cell death process that ensues.

Attempting to explain these apparent differences led researchers to consider caspase activity in two groups. First in the caspase-dependent death induced by CD95 ligation, the ability of zVAD-fmk to completely block cell death is probably a result of the inhibition of an upstream caspase, caspase-8 (Flice/MACH). This caspase is known to be recruited to the membrane following CD95 ligation, to form part of the receptor complex (Boldin *et al.*, 1996; Muzio *et al.*, 1996), and in the process become activated. Blocking caspase 8 activity would prevent initiation of the cell death pathway and subsequently block both commitment and apoptosis. Second, caspase-independent death as induced by staurosporine, seems to lack the involvement of such upstream caspases in that caspase inhibitors only block the downstream executioner caspases, such as caspase-3 (CPP32), situated after the commitment point. This would suggest that particular caspases can be divided into upstream or downstream molecules depending on their position in the cell death pathway in relation to the point of commitment.

However, the discovery that some caspases can act as both initiators and executioners was a new issue.

CASPASES AND DEVELOPMENTAL CELL DEATH

Genetic studies in C. elegans demonstrated that caspases are essential for the initiation and execution of cell death. This can be seen in ced-3 loss of function mutants by the survival of all cells destined to die during development (Ellis and Horvitz, 1986). However only one caspase has been described in the worm and it is possible that this caspase, ced-3, is the only mechanism for commitment and cell death. It may be that the nematode simply lacks a form of cell death commitment that is present in higher organisms, one that is caspase-independent.

Similarly studies in Drosophila support such a requirement for caspases in cell death commitment. Cell death in Drosophila displays many of the morphological and biochemical hallmarks of mammalian apoptosis (Abrams *et al.*, 1993). Three

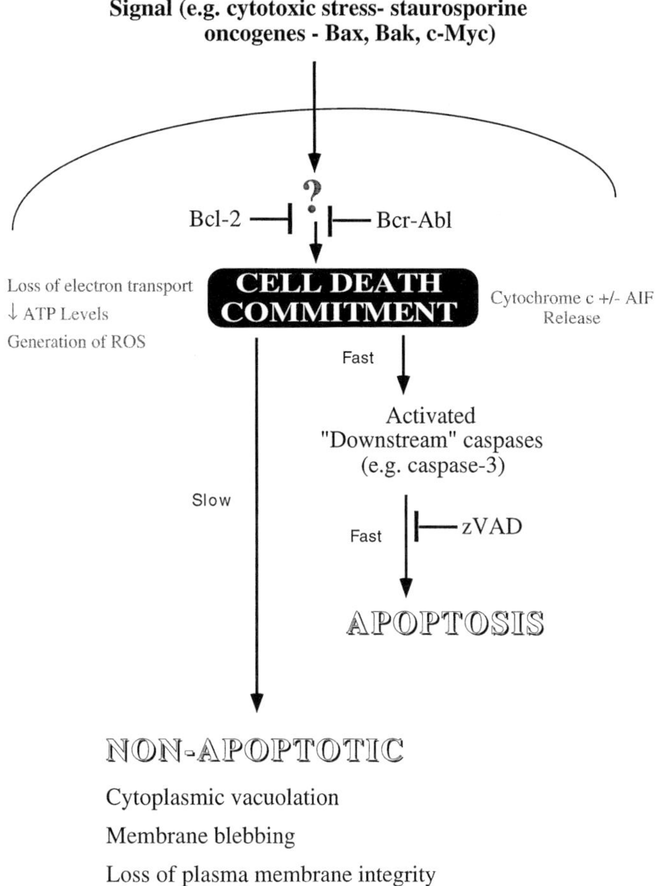

Figure 9.3 Schematic diagram of caspase-independent cell death.

genes have been identified, reaper (rpr) (White *et al.*, 1994), head involution defective (hid) (Grether *et al.*, 1995) and grim (Chen *et al.*, 1996), that appear to act as mediators between different signalling pathways and the cell death program. Transgenic expression of any one of these in the developing embryo or eye is sufficient to kill cells that would normally survive (Grether *et al.*, 1995, Chen *et al.*, 1996, White *et al.*, 1996). Studies indicate that these genes appear to act independently of each other, possibly at the same step in the cell death pathway but their activity is thought to be responsible for activating one or more caspases, as killing is inhibited by the baculovirus p35 protein, a specific inhibitor of caspases (Xue and Horvitz, 1995; Clem and Miller, 1994; Beidler *et al.*, 1995; Hay *et al.*, 1994; Bump *et al.*, 1995). At least one Drosophila caspase, Drosophila caspase-1 (DCP-1), was found to be essential for normal development of some cells (Song *et al.*, 1997). This gene was found to be structurally and biochemically similar to ced-3. Overexpression was sufficient to induce apoptosis, while death was inhibited by the caspase-3 inhibitor DEVD. These studies suggest that caspase activation is necessary for some cell death in Drosophila. This would be in keeping with Figure 9.1 – caspase dependent commitment to death.

The challenge comes with the mammalian model. Caspase-3, a mammalian protease displaying the highest homology to ced-3, is a downstream or executioner caspase with a prominent role in coordinating many of the events in mammalian apoptosis (Kumar and Lavin, 1996). In vitro studies have shown it to be activated once a cell is induced to die in response to a wide variety of cytotoxic agents and although blocking its activation in different cell lines prevents the execution of the apoptotic process it will not prevent death itself (Amarante-Mendes, *et al.*, 1998; Brunet, *et al.*, 1998). Therefore it might be predicted that loss of caspase-3, while inhibiting apoptosis in some cases, would not be capable of blocking cell death in any case, as in Figure 9.3. Despite this, the generation of caspase-3 knockout mice by homologous recombination (Kuida *et al.*, 1996) presented a seemingly complex puzzle. These caspase-3-deficient mice were found to be born at a frequency less than expected by Mendelian genetics, implying a degree of embryonic lethality. Of those mice born however, they appeared smaller than their control littermates and died 1–3 weeks after birth. They presented no discernible histological abnormalities in the tissues of the heart, lung, liver, kidney, spleen and testis but gross defects associated with the nervous system were prominent even before genotyping, with brain development primarily affected. A variety of brain hyperplasias and disorganized cell deployment were observed, indicating a decrease in neuronal cell apoptosis in the absence of caspase-3. Overall it could be seen that the loss of caspase-3 does not affect some tissues but results in dramatic changes in others.

Judging from what is known from *in vitro* studies, the production of mice genetically deficient in a putative downstream caspase would be expected to result in one of two possible outcomes: normal development if an element of redundancy existed or a necrotic-like death in tissues where preventing the primary apoptotic process would instead lead to deployment of a secondary cell death pathway. Redundancy, while possible in light of the fact that caspase-6 (Mch-2) and -7 (Mch-3), the two closest relatives of caspase-3 (Chinnaiyan and Dixit, 1996), are expressed in the nervous system of this knock-out model, has not been proven (or disproven). Neither has inflammation or tissue damage been detected, suggesting the absence of necrosis. It therefore appears that caspase-3 is required not only for apoptosis but also for the commitment to developmental neuronal cell death. This raises the question of how a downstream caspase could occupy such a pivotal role in cell death activation associated with mammalian neuronal development. How can a downstream caspase act in an upstream manner?

COMMITMENT DOWNSTREAM OF DOWNSTREAM CASPASES

To reconcile this it could be proposed that at certain times in the development of some cells a putative downstream caspase can regulate cell death commitment. The brain may represent such a special situation. Here the form of cell death is of profound importance. A necrotic-like death would lead to inflammation and subsequent uncontrolled tissue destruction of irreplacable neuronal cells and a build up of intra-cranial pressure, both of which would be extremely detrimental to the organism. Therefore it might be preferable that only one type of death pathway exist in this developmental cell death – apoptosis. If deregulated it would be more beneficial to the organism to keep the aberrent cells for possible repair than commit to a cell death program that would overall be deleterious to the organism. How this caspase becomes activated remains the question. Since caspase-3 contains only a short prodomain this makes activation via signal transduction events unlikely. Instead it could perhaps be explained by means of expression levels, whereby a certain threshold level must be reached to initiate spontaneous apoptosis. Support for this can be seen in the original papers on caspase-3 showing that transfection of the full length gene can result in processing and activation of this caspase and induce apoptosis (Faucheu *et al.*, 1995). Mutant caspase-3 without protease activity is not processed and subsequently not death-inducing, lending weight to the theory of autoprocessing. Therefore, it remains possible that without neurotropic factors (e.g. NGF) or in the presence of certain neuropoietic cytokines (e.g. LIF), the protein levels of caspase-3 are elevated. This could be due to direct upregulation at the transcription level. Indeed, gene activation has been shown to be required for developmentally programmed cell death (Schwartz *et al.*, 1990). Consistent with this hypothesis, neuronal death in response to NGF withdrawal or LIF has been shown to be blocked in the presence of actinomycin-D or cycloheximide (Martin *et al.*, 1988; Kessler *et al.*, 1993).

However it seems unlikely that commitment in such cases depends solely on transcriptional activation of caspase-3 since this protease is highly expressed in a variety of tissues (Krajewska *et al.*, 1997), and yet these tissues in the knock-out show no discernible abnormalities. Alternatively protein level regulation of caspase-3 regulation may be by a more indirect mechanism. Current data indicate that the sensitivity of each cell to death signals is controlled by a balance between positive and negative modulators of apoptosis, much as in the C.elegans ced-3 and -4 verses -9 proteins (Hengartner and Horvitz, 1994; Hengartner *et al.*, 1992; Steller, 1995). Viewed from this perspective, it may be that certain cell types are set either towards apoptosis by the dominance of positive modulators or away from cell death by the dominance of negative modulators. Where survival factors are limiting, a degradation race may follow between inhibitor and promoter cell death molecules depending on their respective half lives, with caspase-3 winning out and thus the developing neurons die by apoptosis. This may be a possibility in light of the recently reported endogenous X-linked inhibitor of apoptosis protein (XIAP), a member of the IAP family and an direct inhibitor of caspases (Chen and Miller, 1994; Duckett *et al.*, 1996; Deveraux *et al.*, 1997). Therefore if caspase-3 levels increase and/or inhibitor levels decrease the overall result may be spontaneous caspase activation and apoptosis (Figure 9.4). This hypothesis is speculative however, and remains to be formally tested.

In summary there appear to be three ways a cell can commit to die. The first involves upstream caspase-dependent commitment. Here upstream caspases are activated which in turn activate downstream caspases and apoptosis ensues. Inhibiting all caspases will block not only apoptosis but cell death in general (Figure 9.1).

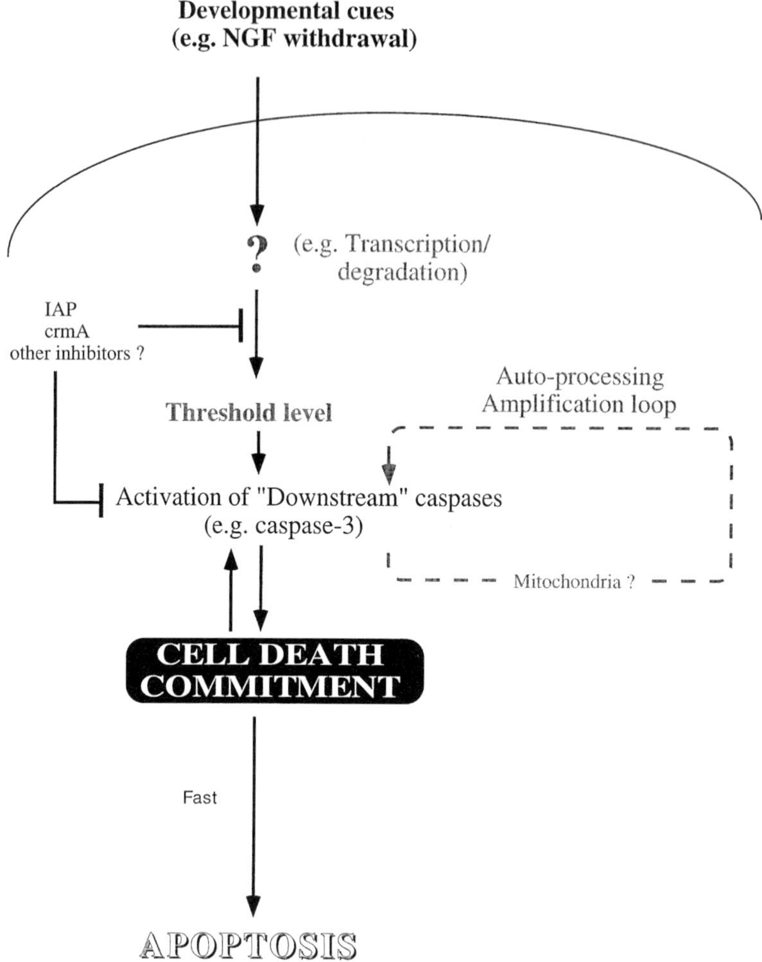

Figure 9.4 A hypothetical model of developmental caspase-dependent cell death.

The second scenario involves caspase-independent commitment. Cellular events independent of caspase activity lead to the commitment step. This is followed by activation of downstream caspases thereby initiating the apoptotic program. Blocking caspase activity will not confer cell survival but only change the form of death to a non-apoptotic form (Figure 9.3).

Finally the third pathway to commitment is one in which downstream caspases are activated directly, leading to commitment and apoptosis (Figure 9.4).

ANTI-APOPTOTIC ONCOGENES AND CELL DEATH COMMITMENT

The importance of studying cell death commitment is underscored by the fact that resistance to apoptosis is now recognized as a crucial factor in human diseases such as tumorgenesis (Green *et al.*, 1994; Williams, 1991; Martin and Green, 1995) and autoimmunity (Carson and Ribeiro, 1993; Carson and Tan, 1995). The type of cell death occurring here is important as it is often indicative of patient prognosis (Arends *et al.*, 1994). Patients with tumors displaying a high necrotic to apoptotic cell death ratio have a less likely chance of survival. This may be due to the fact that necrosis elicits an inflammatory response while apoptotic cells do not.

There is mounting evidence that not only apoptosis but commitment to death in general is regulated at a molecular level. This is substantiated by reports that anti-apoptotic oncogenes are capable of blocking death irrespective of the form (McCarthy *et al.*, 1997; Amarante-Mendes *et al.*, 1998; Brunet, *et al.*, 1998). Caspase-independent death induced by various cytotoxic agents or by expression of c-myc under low serum conditions are inhibitable by anti-apoptotic oncogenes such as Bcl-2 and Bcr-Abl, such that cells not only survive but also go on to proliferate. This indicates that the cell death mechanism has been blocked at a step preceding or at the commitment. It would appear that regardless of the ultimate mode of death, cells initiate some critical event(s) termed the "commitment point" that is inhibitable by survival factors such as oncogenes and possibly growth factors.

WHAT AND WHERE ARE THE COMMITMENT TO CELL DEATH AND APOPTOSIS?

The question still remains however, what are the critical components of the cell death pathway? Furthermore, is the regulation of apoptosis the same as the regulation of cell death? One could envisage a cell death commitment point preceding apoptotic induction or alternatively a commitment point that simultaneously initiates the apoptosis program. It is conceivable that either of these two different mechanisms can be triggered with different outcomes. The pathway(s) involved in cell death commitment are only now beginning to be addressed. Where apoptosis is inappropriately triggered as in cases of autoimmune diseases and AIDS, pharmacological intervention to inhibit caspases (Thornberry *et al.*, 1995; Lewis *et al.*, 1997) would only be successful if caspase activity preceded the commitment to cell death. Hence it will be important to consider the form of death a cell undergoes for future design of therapeutic strategies.

As for the elusive commitment step, the role of mitochondria in cell death is actively being pursued for several reasons (Henkart and Grinstein, 1996; Castedo *et al.*, 1995; Castedo *et al.*, 1996; Susin *et al.*, 1997; Kroemer, 1997; Vayssiere *et al.*, 1994; Kluck *et al.*, 1997; Bossy-Wetzel *et al.*, 1998). Mitochondria have long been thought to impact on cell survival through their ability to produce both ATP and reactive oxygen species (Buttke and Sandstrom, 1994; Richter, 1993; Zamzami *et al.*, 1995; Hockenberry *et al.*, 1993). Studies using cell-free systems have found that mitochondria release at least two molecules that participate in caspase activation and events associated with apoptosis: AIF (apoptosis-inducing factor) and cytochrome c (Kluck *et al.*, 1997; Liu *et al.*, 1996; Susin *et al.*, 1996; Yang *et al.*, 1997). Microinjection of cytochrome c supports the role for this mitochondrial factor in apoptosis (Li *et al.*, 1997). AIF, in contrast, can directly act on the nucleus to cause fragmentation, as well as activating caspases (Susin *et al.*, 1997; Susin *et al.*, 1996). The mitochondrial release of both cytochrome c and AIF is Bcl-2 inhibitable (Kluck *et al.*, 1997; Yang *et al.*, 1997). AIF is currently considered to be released through a mega-pore generated in the mitochondrial membrane as a result of a mitochondrial permeability transition (PT) (Zamzami *et al.*, 1995; Marchetti *et al.*, 1996; Zoratti and Szabo, 1995), although the exact role of PT in cell death commitment is still a matter of debate. Several studies have shown that PT is not universally required to commit cells to death (Ankarcrona *et al.*, 1995; Deckwerth and Johnson, 1993; Tropea *et al.*, 1995). For example, in HL-60 cells, we can induce death with all the characteristics of apoptosis including cytochrome c release, while retaining a permeability transition (Finucane *et al.*, submitted). In addition, in Hela and CEM cells, PT has also been shown to occur several hours after the release of cytochrome c. zVAD-fmk blocked this PT while having no effect on cytochrome c release (Bossy-Wetzel *et al.*, 1998). Together these data suggest that while mitochondrial PT may play some role in cell death it does not appear to represent the commitment point in all cell types.

It remains possible that the mitochondrial release of a factor(s) may represent a critical step in committing a cell to death. Such an event may initiate two cell death programs. One at the cytosolic level where cytochrome c activates caspases and hence the rapid apoptotic arm of cell death, while the second slower death program might be initiated at the mitochondrial level in that the release of a factor(s) would result in mitochondrial failure via disruption of the electron transport chain. With the latter, a resulting gradual drop in ATP levels in conjunction with the simultaneous accumulation of toxic ROS levels might lead to a non-apoptotic death similar to that seen in caspase-independent death. Indeed it was found that Bax-induced death results in an increased production of oxygen radicals levels in the presence or absence of caspase activation despite a difference in the type of cell death that ensued (Xiang *et al.*, 1996). To address the question of whether caspase-independent cell death is a consequence of ROS generation cells were incubated with or without zVAD-fmk and/or N-Acetylcysteine (NAC), an oxygen radical scavenger (Hockenberry *et al.*, 1993; Cossarizza *et al.*, 1995) and then induced to die with staurosporine (Figure 9.5). Dramatic apoptotic protection could be seen by 4.5 hours, as assessed by phosphatidylserine externalization (A). However long-term monitoring of these cells revealed an equal lack of clonogenicity regardless of pre-treatment (B). As a positive control to ensure that this antioxidant was capable of blocking death that proceeded in an oxygen-dependent manner NAC was demonstrated to be capable of inhibiting cell death induced by hydrogen peroxide (A). The question of ATP however, was not addressed. It also remains to be determined whether neutralizing cytosolic cytochrome c might convert apoptosis into necrosis-like caspase-independent cell death. These would be likely to be important areas for future investigation.

Thus, while the precise mechanism of the caspase-independent non-apoptotic death remains unclear, it is nevertheless likely to involve changes in mitochondrial function. This organelle appears to sit at the center of life and death in the cell, and the regulation of these disparate functions is of fundamental importance.

(A)

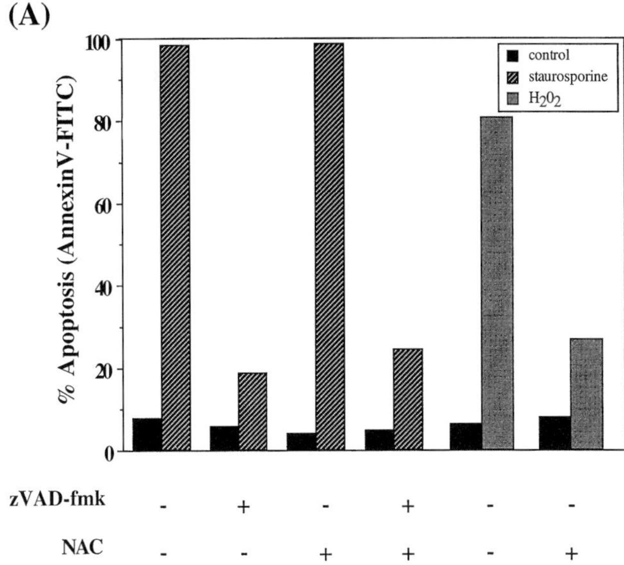

(B)

Treatment	Untreated	Staurosporine
zVAD-fmk	≥ 0.1	< 0.0002
NAC	≥ 0.1	< 0.0002
zVAD-fmk +/ NAC	≥ 0.1	< 0.0002

Figure 9.5 Caspase-independent cell death is not caused solely by ROS. Jurkat cells were pre-treated with zVAD-fmk (100 μM) and/or NAC (20 μM) and then induced to die with staurosporine (1 μM). Hydrogen peroxide (0.5 μM) was used as a positive control for NAC anti-oxidant activity. (A) Percent apoptosis as assessed by Annexin V binding after 4.5 hours for staurosporine and after 11 hours for hydrogen peroxide. (B) Frequency of cells capable of proliferating after incubation with apoptosis-inducing agent determined by clonogenic assay as described in (Amarante-Mendes, *et al.*, 1998).

REFERENCES

Abrams, J.M., White, K., Fessler, L.I. and Steller, H. (1993) Programmed cell death during Drosophila embryogenesis. *Development*, **117**, 29–43.

Alnemri, E.S., Livingston D.J., Nicholson D.W., Salvesen G., Thornberry N.A., Wong W.W., Yuan J. (1996) Human ICE/CED-3 protease nomenclature. *Cell*, **87**, 171.

Amarante-Mendes, G.P., Finucane, D.M., Martin, S.J., Cotter, T.G., Salvesen, G.S. and Green, D.R. (1998) Anti-apoptotic oncogenes prevent caspase-dependent and independent commitment for cell death. *Cell Death and Differentiation*, **5**, 298–306.

Ameisen, J.C. (1994) Programmed cell death (apoptosis) and cell survival regulation: relevance to AIDS and cancer. *AIDS*, **8**, 1197–1213.

Ameisen, J.C. (1996) The origin of programmed cell death. *Science*, **272**, 1278–1279.

Ameisen, J.C. and Capron, A. (1991) Cell dysfunction and depletion in AIDS: the programmed cell death hypothesis. *Immunol. Today*, **12**, 102–105.

Ameisen, J.C., Idziorek, T., Billaut-Mulot, O., Tissier, J.-P., Potentier, A. and Ouaissi, A. (1995) Apoptosis in a unicellular eukaryote (Trypanosoma cruzi): implications for the evolutionary origin and role of programmed cell death in the control of cell proliferation, differentiation and survival. *Cell Death and Differentiation*, **2**, 285–300.

Ankarcrona, M., Dypbukt, J.M., Bonfoco, E., Zhivotovsky, B., Orreneus, S., Lipton, S.A. and Nicotera, P. (1995) Glutamate-induced neuronal death: a succession of necrosis or apoptosis depending on mitochondrial function. *Neuron*, **15**, 961–973.

Arends, M.J. and Wyllie, A.H. (1991) Apoptosis: mechanisms and roles in pathology. *Int. Rev. Exp. Pathol.*, **32**, 223–254.

Arends, M.J., McGregor, A.H. and Wyllie, A.H. (1994) Apoptosis is inversely related to necrosis and determines net growth in tumors bearing constitutively expressed myc, ras, and HPV oncogenes. *Am. J. Pathol.*, **144**, 1045–1057.

Beidler, D.R., Tewari, M., Friesen, P.D., Poiriers, G. and Dixit, V.M. (1995) The baculovirus p35 protein inhibits Fas- and tumor necrosis factor-induced apoptosis. *J. Biol. Chem.*, **270**, 16525–6528.

Berke, G. (1995) The CTL's kiss of death. *Cell*, **81**, 9–12.

Boldin, M.P., Goncharov, T.M., Goltsev, Y.V. and Wallach, D. (1996) Involvement of MACH, a novel MORT1/FADD-interacting protease, in Fas/APO-1- and TNF receptor-induced cell death. *Cell*, **85**, 803–815.

Bossy-Wetzel, E., Newmeyer, D.D. and Green, D.R. (1998) Mitochondrial cytochrome c release in apoptosis occurs upstream of DEVD-specific caspase activation and independently of mitochondrial transmembrane depolarization. *EMBO*, **17**, 37–49.

Brunet, C.J., Gunby, R.H., Benson, R.S.P., Hickman, J.A., Watson, A.J.M. and Brady, G. (1998) Commitment to cell death measured by loss of clonogenicity is separable from the appearance of apoptotic markers. *Cell Death and Differentiation*, **5**, 107–115.

Brunner, T., Mogil, R.J., LaFace, D., Yoo, N.J., Maboubi, A., Echeverri, F., Martin, S.J., Force, W.R., Lynch, D.H., Ware, C.F. and Green, D.R. (1995) Cell-autonomous Fas (CD95)/Fas-ligand interaction mediates activation-induced apoptosis in T-cell hybridomas. *Nature*, **373**, 441–444.

Bump, N.J., Hackett, M., Hugunin, M., Seshagiri, S., Brady, K., Chen, P., Ferenz, C., Franklin, S., Ghayur, T., Li, P., Licari, P., Mankovich, J., Shi, L., Greenberg, A.H., Miller, L.K. and Wong, W.W. (1995) Inhibition of ICE family proteases by baculovirus antiapoptotic protein p35. *Science*, **269**, 1885–1888.

Buttke, T.M. and Sandstrom, P.A. (1994) Oxidative stress as a mediator of apoptosis. *Immunol. Today*, **15**, 7–10.

Carson, D.A. and Ribeiro, J.M. (1993) Apoptosis and disease. *Lancet*, **341**, 1251–1254.

Carson, D.A. and Tan, E.M. (1995) Apoptosis in rheumatic disease. *Bull. Rheum. Dis.*, **44**, 1–3.

Castedo, M., Hirsch, T., Susin, S.A., Zamzami, N., Marchetti, P., Macho, A. and Kroemer, G. (1996) Sequential acquisition of mitochondrial and plasma membrane alterations during early lymphocyte apoptosis *J. Immunol.* **157**, 512–521.

Castedo, M., Macho, A., Zamzami, N., Hirsch T., Marchetti P, Uriel J., Kroemer G. (1995) Mitochondrial perturbations define lymphocytes undergoing apoptotic depletion in vivo. *Eur. J. Immunol.*, **25**, 3277–3284.

Chen, P., Nordstrom, W., Gish, B. and Abrams, J.M. (1996) grim, a novel cell death gene in Drosophila. *Genes & Devel.*, **10**, 1773–1782.

Chinnaiyan, A.M. and Dixit, V.M. (1996) Cytotoxic T-cell-derived granzyme B activates the apoptotic protease ICE-LAP3. *Current Biol.*, **6**, 555–562.

Clem, R.J. and Miller, L.K. (1994) Control of programmed cell death by the baculovirus genes p35 and iap. *Mol. Cell. Biol.*, **14**, 5212–5222.

Cornillon, S., Foa, C., Davoust, J., Buonavista, N., Gross, J.D. and Goldstein, P. (1994) Programmed cell death in Dictyostelium. *J. Cell Sci.*, **107**, 2691–2704.

Cossarizza, A., Franceschi, C., Monti, D., Salviola, S., Bellesia, E., Rivabene, R., Biondo L., Rainaldi, G., Tinari, A. and Malorni, W. (1995) Protective effect of N-acetylcysteine in tumor necrosis factor-alpha-induced apoptosis in U937 cells: the role of mitochondria. *Exp. Cell Res.*, **220**, 232–240.

Debbas, M. and White, E. (1993) Wild-type p53 mediates apoptosis by E1A, which is inhibited by E1B. *Genes Dev.*, **7**, 546–554.

Deckwerth, T.L. and Johnson, E.M.J. (1993) Temporal analysis of events associated with programmed cell death (apoptosis) of sympathetic neurons deprived of nerve growth factor. *J. Cell Biol.*, **123**, 1207–1222.

Deveraux, Q.L., Takahashi, R., Salvesen, G.S. and Reed, J.C. (1997) X-linked IAP is a direct inhibitor of cell-death proteases. *Nature*, **388**, 300–304.

Duan, H., Chinnaiyan, A.M., Hudson, P.L., Wing, J.P., He, W.-W. and Dixit, V.M. (1996) ICE-LAP3, a novel mammalian homologue of the Caenorhabditis elegans cell death protein Ced-3 is activated during Fas- and tumor necrosis factor-induced apoptosis. *J. Biol. Chem.*, **271**, 1621–1625.

Duckett, C.S., Nava V.E., Gedrich R.W., Clem R.J., Van Dongen J.L., Gilfillan M.C, Shiels H., Hardwick J.M., Thompson C.B. (1996) A conserved family of cellular genes related to the baculovirus iap gene and encoding apoptosis inhibitors. *EMBO J.*, **15**, 2685–2694.

Ellis, H.M. and Horvitz, H.R. (1986) Genetic control of programmed cell death in the nematode C. elegans. *Cell*, **44**, 817–829.

Ellis, R.E. and Horvitz, H.R. (1991) Two C. elegans genes control the programmed deaths of specific cells in the pharynx. *Development*, **112**, 591–603.

Ellis, R.E., Jacobson, D.M. and Horvitz, R.H.,(1991) Genes required for the engulfment of cell corpses during programmed cell death in Caenorhabditis elegans. *Genetics*, **129**, 79–94.

Enari, M., Hug, H. and Nagata, S. (1995) Involvement of an ICE-like protease in Fas-mediated apoptosis. *Nature*, **375**, 78–81.

Faucheu, C., Diu, A., Chan, A.W., Blanchet, A.M., Miossec, C., Herve, F., Dutilleul, V.C., Gu, Y., Aldape, R.A., Lippke, J.A., *et al.* (1995) A novel human protease similar to the interleukin-1 beta converting enzyme induces apoptosis in transfected cells. *EMBO J.*, **14**, 1914–1922.

Fernandes-Alnemri, T., Litwack, G. and Alnemri, E.S. (1994) CPP32, a novel human apoptotic protein with homology to Caenorhabditis elegans cell death protein Ced-3 and mammalian interleukin-1 beta-converting enzyme. *J. Biol. Chem.*, **269**, 30761–30764.

Fernandes-Alnemri, T., Litwack, G. and Alnemri, E.S. (1995a) Mch2, a new member of the apoptotic Ced-3/Ice cysteine protease gene family. *Cancer Res.*, **55**, 2737–2742.

Fernandes-Alnemri, T., Takahashi, A., Armstrong, R., Krebs, J., Fritz, L., Tomaselli, K., Wang, L., Yu, Z., Croce, C., Salveson, G., *et al.* (1995b) Mch3, a novel human apoptotic cysteine protease highly related to CPP32. *Cancer Res.*, **55**, 6045–6052.

Gagliardini, V., Fernandez, P.-A., Lee, R.K.K., Drexler, H.C.A., Rotello, R.J., Fishman, M.C. and Yuan, J. (1994) Prevention of vertebrate neuronal death by the crmA gene. *Science* **263**, 826–828.

Green, D.R., Bissonnette, R.P. and Cotter, T.G. (1994) Apoptosis and cancer. *Imp. Adv. Oncol.*, **1**, 37–52.

Grether, M.E., Abrams, J.A., White, K. and Steller, H. (1995) The head involution defective gene of Drosophila melanogaster functions in programmed cell death. *Genes & Development*, **9**, 1694–1708.

Hay, B.A., Wolff, T. and Rubin, G.M. (1994) Expression of baculovirus P35 prevents cell death in Drosophila. *Development*, **120**, 2121–2129.

Hedgecock, E.M., Sulston, J.E. and Thomson, J.N. (1983) Mutations affecting programmed cell deaths in the nematode Caenorhabditis elegans. *Science*, **220**, 1277–1279.

Hengartner, M.O. and Horvitz, R.H. (1994) The ins and outs of programmed cell death during C. elegans development. *Philos. Trans. R. Soc. London Ser. B.*, **345**, 243–246.

Hengartner, M.O., Ellis, R.E. and Horvitz, H.R. (1992) Caenorhabditis elegans gene ced-9 protects cells from programmed cell death. *Nature*, **356**, 494–499.

Henkart, P.A. (1994) Lymphocyte-mediated cytotoxicity: two pathways and multiple effector molecules. *Immunity*, **1**, 343–346.

Henkart, P.A. (1996) ICE family proteases: mediators of all apoptotic cell death? *Immunity*, **4**, 195–201.

Henkart, P.A. and Grinstein, S. (1996) Apoptosis: mitochondria resurrected? *J. Exp. Med.*, **183**, 1293–1295.

Heusel, J.W., Wesselschmidt, R.L., Shresta, S., Russell, J.H. and Ley, T.J. (1994) Cytotoxic lymphocytes require granzyme B for the rapid induction of DNA fragmentation and apoptosis in allogeneic target cells. *Cell*, **76**, 977–987.

Hockenbery, D.M., Oltvari, Z.N., Yin, X.-M., Milliman, C.L. and Korsmeyer, S.J. (1993) Bcl-2 functions in an antioxidant pathway to prevent apoptosis. *Cell*, **75**, 241–251.

Holden, R.J. and Mooney, P.A. (1995) Interleukin-1 beta: a common cause of Alzheimer's disease and diabetes mellitus. *Med. Hypoth.*, **45**, 559–571.

Jacobsen, M.D., Weil, M. and Raff, M.C. (1996) Role of Ced-3/ICE-family proteases in staurosporine-induced programmed cell death. *J. Cell Biol.*, **133**, 1041–1051.

Kamens, J., Paskin, M., Hugunin, M., Talanian, R.V., Allen, H., Banach, D., Bump, N., Hackett, M., Johnston, C.G. and Li, P. (1995) Identification and characterization of ICH-2, a novel member of the interleukin-1 beta-converting enzyme family of cysteine proteases. *J. Biol. Chem.*, **270**, 15250–15256.

Kerr, J.F.R., Wyllie, A.H. and Currie, A.R. (1972) Apoptosis: a basic biological phenomenon with wide-ranging implications in tissue kinetics. *Br. J. Cancer.*, **26**, 239–257.

Kessler, J.A., Ludlam, W.H., Freidin, M., Hall, D.H., Michaelson, M.D., Spray, D.C., Dougherty, M. and Batter, D.K. (1993) Cytokine-induced programmed death of cultured sympathetic neurons. *Neuron*, **11**, 1123–1132.

Kluck, R., Bossy-Wetzel, E., Green, D.R. and Newmeyer, D. (1997) The release of cytochrome c from mitochondria: a primary site for Bcl-2 regulation of apoptosis. *Science*, **275**, 1132–1136.

Krajewska, M., Wang, H.-G., Krajewska, S., Zapata, J.M., Shabaik, A., Gascoyne, R. and Reed, J.C. (1997) Immunohistochemical analysis of in vivo patterns of expression of CPP32 (Caspase-3), a cell death protease. *Cancer Research*, **57**, 1605–1613.

Kroemer, G. (1997) Mitochondrial implication in apoptosis. Towards an endosymbiont hypothesis of apoptosis evolution. *Cell Death and Differentiation*, **4**, 443–456.

Kuida, K., Zheng, T.S., Na, S., Kuan, C.-Y., Yang, D., Karausyama, H., Rakio, P. and Flavell, R.A. (1996) Decreased apoptosis in the brain and premature lethality in CPP32-deficient mice. *Nature*, **384**, 368–372.

Kumar, S. (1995) ICE-like proteases in apoptosis. *TIBS*, **20**, 198–202.

Kumar, S. and Lavin, M.F. (1996) The ICE family of cysteine proteases as effectors of cell death. *Cell Death and Diff.*, **3**, 255–267.

Kumar, S., Kinoshita, M., Noda, M., Copeland, N.G. and Jenkins, N.A. (1994) Induction of apoptosis by the mouse Nedd2 gene, which encodes a protein similar to the product of the Caenorhabditis elegans cell death gene ced-3 and the mammalian IL-1 beta-converting enzyme. *Genes Dev.*, **8**, 1613–1626.

Lennon, S.V., Martin, S.J. and Cotter, T.G. (1991) Dose-dependent induction of apoptosis in human tumour cell lines by widely diverging stimuli. *Cell Prolif.*, **24**, 203–214.

Lewis, J.S., Terriff, C.M., Coulston, D.R. and Garrison, M.W. (1997) Protease inhibitors: a therapeutic breakthrough for the treatment of patients with human immunodeficiency virus *Clin-Ther.*, **19**, 187–214.

Li, F., Srinivasan, A., Wang, Y., Armstrong, R.C., Tomaselli, K.J. and Fritz, L.C. (1997) Cell-specific induction of apoptosis by microinjection of cytochrome c. Bcl-xL has activity independent of cytochrome c release. *J. Biol. Chem.*, **272**, 30299–30305.

Liu, X., Kim, C.N., Yang, I., Jemmerson, R. and Wang, X. (1996) Induction of apoptotic program in cell-free extracts: requirement for dATP and cytochrome c. *Cell*, **86**, 147–157.

Loddick, S.A., MacKenzie, A. and Rothwell, N.J. (1996) An ICE inhibitor, z-VAD-DCB attenuates ischaemic brain damage in the rat. *Neuroreport*, **7**, 1465–1468.

Loo, D.T., Copani, A., Pike, C.J., Whittemore, E.R., Walencewicz, A.J. and Cotman, C.W. (1993) Apoptosis is induced by beta-amyloid in cultured central nervous system neurons. *Proc. Natl. Acad. Sci. USA*, **90**, 7951–7955.

Marchetti, P., Castedo, M., Susin, S.A., Zamzami, N., Hirsch, T., Haeffner, A., Hirsch, F., Geuskens, M. and Kroemer, G. (1996) Mitochondrial permeability transition is a central coordinating event of apoptosis. *J. Exp. Med.*, **184**, 1155–1160.

Martin, D.P., Schmidt, R.E., DiStefano, P.S., Lowry, O.H., Carter, J.G. and Johnson, E.M., Jr. (1988) Inhibitors of protein synthesis and RNA synthesis prevent neuronal death caused by nerve growth factor deprivation. *J. Cell Biol.*, **106**, 829–844.

Martin, S.J. (1993) Programmed cell death and AIDS. *Science*, **262**, 1355–1356.

Martin, S.J. and Green, D.R. (1995a) Apoptosis and cancer: the failure of controls on cell death and cell survival. *Crit. Rev. in Oncol./Hematol.*, **18**, 137–153.

Martin, S.J. and Green, D.R. (1995b) Protease activation during apoptosis: death by a thousand cuts? *Cell*, **82**, 349–352.

Martin, S.J., Finucane, D.M., Amarante-Mendes, G.P., O'Brien, G.A. and Green, D.R. (1996) Phosphatidylserine externalization during CD95-induced apoptosis of cells and cytoplasts requires ICE/CED-3 protease activity. *J. Biol. Chem.*, **271**, 28753–28756.

Martinou, J.C. (1994) Overexpression of BCL-2 in transgenic mice protects neurons from naturally occurring cell death and experimental ischemia. *Neuron*, **13**, 1017–1030.

McCarthy, N J., Whyte, M.K.B., Gilbert, C.S. and Evan, G.I. (1997) Inhibition of Ced-3/ICE-related proteases does not prevent cell death induced by oncogenes, DNA damage, or the Bcl-2 homologue Bak. *J. Cell Biol.*, **136**, 215–227.

Meyaard, L., Otto, S.A., Jonker, R.R., Mijnster, M.J., Keet, R.P.M. and Miedema, F. (1992) Programmed death of T cells in HIV-1 infection. *Science*, **257**, 217–219.

Milligan, C.E., Prevette, D., Yaginuma, H., Homma, S., Cardwell, C., Fritz, L.C., Tomaselli, K.J., Oppenheim, R.W. and Schwartz, L.M. (1995) Peptide inhibitors of the ICE protease family arrest programmed cell death of motoneurons in vivo and in vitro. *Neuron*, **15**, 385–393.

Miura, M., Zhu, H., Rotello, R., Hartwieg, E.A. and Yuan, J. (1993) Induction of apoptosis in fibroblasts by IL-1 beta-converting enzyme, a mammalian homolog of the C. elegans cell death gene ced-3. *Cell*, **75**, 653–660.

Muzio, M., Chinnaiyan, A.M., Kischkel, F.C., O'Rourke, K., Schevchenko, A., Ni, J., Scaffidi, C., Bretz, J.D., Zhang, M., Gentz, R., Mann, M., Krammer, P.H., Peter, M.E. and Dixit, V.M. (1996) FLICE, a novel FADD-homologous ICE/CED-3-like protease, is recruited to the CD95 (Fas/APO-1) death--inducing signalling complex. *Cell*, **85**, 817–827.

Nicholson, D.W., Ali, A., Thornberry, N.A., Vaillancourt, J.P., Ding, C.K., Gallant, M., Gareau, Y., Griffin, P.R., Labelle, M., Lazebnik, Y.A., Munday, N.A., Raju, S.M., Smulson, M.E., Yamin, T.-T., Yu, V.L. and Miller, D.K. (1995) Identification and inhibition of the ICE/CED-3 protease necessary for mammalian apoptosis. *Nature*, **376**, 37–43.

Oberhammer, F., Wilson, J.W. and Dive, C. (1993) Apoptotic death in epithelial cells: cleavage of DNA to 300 and/or 50 kb fragments prior to or in the absence of internucleosomal fragmentation. *EMBO J.*, **12**, 3679–3684.

Ohta, T., Kinoshita, T., Naito, M., Nozaki, T., Masutani, M., Tsuruo, T. and Miyajima, A. (1997) Requirement of the caspase-3/CPP32 protease cascade for apoptotic death following cytokine deprivation in hematopoietic cells. *J. Biol. Chem.*, **272**, 23111–23116.

Raff, M.C. (1992) Social controls on cell survival and cell death. *Nature*, **356**, 397–400.

Richter, C. (1993) Pro-oxidants and mitochondrial Ca2+: their relationship to apoptosis and oncogenesis. *FEBS Lett.*, **325**, 104–107.

Rodriguez, I., Matsuura, K., Ody, C., Nagata, S. and Vassalli, P. (1996) Systemic injection of a tripeptide inhibits the intracellular activation of CPP32-like proteases in vivo and fully protects mice against Fas-mediated fulminant liver destruction and death. *J. Exp. Med.*, **184**, 2067–2072.

Schulze-Osthoff, K., Walczak, H., Droge, W. and Krammer, P.H. (1994) Cell nucleus and DNA fragmentation are not required for apoptosis. *J. Cell Biol.*, **127**, 15–20.

Schwartz, L.M., Kosz, L. and Kay, B.K. (1990) Gene activation is required for developmentally programmed cell death. *Proc. Natl. Acad. Sci. USA.*, **87**, 6594–6598.

Shi, L., Kam, C.M., Powers, J.C., Aebersold, R. and Greenberg, A. (1992) Purification of three cytotoxic lymphocyte granule serine proteases that induce apoptosis through distinct substrate and target cell interactions. *J. Exp. Med.*, **176**, 1521–1529.

Slee, E.A., Zhu, H., Chow, S.C., MacFarlane, M., Nicholson, D.W. and Cohen, G.M. (1996) Benzyloxycarbonyl-Val-Ala-Asp (OMe) fluoromethylketone (Z-VAD.FMK) inhibits apoptosis by blocking the processing of CPP32. *Biochem. J.*, **315**, 21–24.

Smyth, M.J. and Trapani, J.A. (1995) Granzymes: exogenous proteinases that induce target cell apoptosis. *Immunol. Today*, **16**, 202–206.

Song, Z., McCall, K. and Steller, H. (1997) DCP-1, a Drosophila cell death protease essential for development. *Science*, **275**, 536–540.

Steller, H. (1995) Mechanisms and genes of cellular suicide. *Science*, **267**, 1445–1449.

Sulston, J.E. and Horvitz, H.R. (1977) Post-embryonic cell lineages of the nematode, Caenorhabditis elegans. *Dev. Biol.*, **56**, 110–156.

Susin, S.A., Zamzami, N., Castedo, M., Daugas, E., Wang, H.G., Macho, A., Daugas, E., Geuskens, M. and Kroemer, G. (1997) The central executioner of apoptosis: multiple connections between protease activation and mitochondria in Fas/APO-1/CD95- and ceramide-induced apoptosis. *J. Exp. Med.*, **186**, 25–37.

Susin, S.A., Zamzami, N., Castedo, M., Hirsch, T., Marchetti, P., Macho, A., Daugas, E., Geuskens, M. and Kroemer, G. (1996) Bcl-2 inhibits the mitochondrial release of an apoptogenic protease. *J. Exp. Med.*, **184**, 1331–1341.

Terai, C., Kornbluth, R.S., Pauza, C., Richman, D.D. and Carson, D.A. (1991) Apoptosis as a mechanism of cell death in cultured T lymphoblasts acutely infected with HIV-1. *J. Clin. Invest.*, **87**, 1710–1715.

Tewari, M., Quan, L.T., O'Rourke, K., Desnoyers, S., Zeng, Z., Beidler, D.R., Poirier, G.G., Salvesen, G.S. and Dixit, V.M. (1995) Yama/CPP32 beta, a mammalian homolog of CED-3, is a CrmA-inhibitable protease that cleaves the death substrate poly(ADP-ribose) polymerase. *Cell*, **81**, 801–809.

Thompson, C.B. (1995) Apoptosis in the pathogenesis and treatment of disease. *Science*, **267**, 1456–1462.

Thornberry, N.A., Bull, H.G., Calaycay, J.R., Chapman, K.T., Howard, A.D., Kostura, M.J., Miller, D.K., Molineaux, S.M., Weidner, J.R., Aunins, J., *et al.* (1992) A novel heterodimeric cysteine protease is required for interleukin-1 beta processing in monocytes. *Nature (London)*, **356**, 768–774.

Thornberry, N., Miller, D. and Nicholson, D. (1995). *Perspectives in Drug Discovery and Design*, **2**, 389–399.

Tropea, F., Troiano, L., Monti, D., Lovato, E., Malcorni, W., Rainaldi, G., Mattona, P., Viscomi, G., Ingletti, M.C., Portlani, M., Cermelli, C., Cossarizza, A. and Franceschi, C. (1995) Sendai virus and herpes virus type 1 induce apoptosis in human peripheral blood mononuclear cells. *Exp. Cell. Res.*, **218**, 63–7.

Vaux, D.L., Haecker, G. and Strasser, A. (1994) An evolutionary perspective on apoptosis. *Cell*, **76**, 777–779.

Vayssiere, J.L., Petit, P.X., Risler, Y. and Mignotte, B. (1994) Commitment to apoptosis is associated with changes in mitochondrial biogenesis and activity in cell lines conditionally immortalized with simian virus 40. *Proc. Natl. Acad. Sci. USA*, **91**, 11752–11756.

White, K., Grether, M.E., Abrams, J.M., Young, L., Farrell, K. and Steller, H. (1994) Genetic control of programmed cell death in Drosophila. *Science*, **264**, 677–683.

White, K., Tahaoglu, E. and Steller, H. (1996) Cell killing by the Drosophila gene reaper. *Science*, **271**, 805–807.

Whyte, M. (1996) ICE/CED-3 proteases in apoptosis. *Trends Cell Biol.*, **6**, 245–248.

Williams, G.T. (1991) Programmed cell death: apoptosis and oncogenesis. *Cell*, **65**, 1097–1098.

Wyllie, A.H. (1980) Cell death: the significance of apoptosis. *Int. Rev. Cytol.*, **68**, 251–306.

Xiang, J., Chao, D.T. and Korsmeyer, S.T. (1996) BAX-induced cell death may not require interleukin 1 beta-converting enzyme-like proteases. *Proc. Natl. Acad. Sci. USA*, **93**, 14559–14563.

Xue, D. and Horvitz, H.R. (1995) Inhibition of the Caenorhabditis elegans cell-death protease CED-3 by a CED-3 cleavage site in baculovirus p35 protein. *Nature (London)*, **377**, 248–251.

Xue, D., Shaham, S. and Horvitz, H.R. (1996) The Caenorhabditis elegans cell-death protein CED-3 is a cysteine protease with substrate specificities similar to those of the human CPP32 protease. *Genes Dev.*, **10**, 1073–1083.

Yang, J., Liu, X., Bhalla, K., Kim, C.N., Ibrado, A.M., Cai, J., Peng, T.-I., Jones, D.P. and Wang, X. (1997) Prevention of apoptosis by Bcl-2: release of cytochrome c from mitochondria blocked. *Science.*, **275**, 1129–1132.

Yuan, J., Shaham, S., Ledoux, S., Ellis, H.M. and Horvitz, H.R. (1993) The C. elegans cell death gene ced-3 encodes a protein similar to mammalian interleukin-1 beta-converting enzyme. *Cell*, **75**, 641–652.

Yuan, J.-Y. and Horvitz, H.R. (1990) The Caenorhabditis elegans genes ced-3 and ced-4 act cell autonomously to cause programmed cell death. *Dev. Biol.*, **138**, 33–41.

Zakeri, Z.F., Quaglino, D., Latham, T. and Lockshin, R.A. (1993) Delayed internucleosomal DNA fragmentation in programmed cell death. *FASEB J.*, **7**, 470–478.

Zamzami, N., Marchetti, P., Castedo, M., Decaudin, D., Macho, A., Hirsch, T., Susin, S.A., Petit, P.X., Mignotte, B. and Kroemer, G. (1995a) Sequential reduction of mitochondrial transmembrane potential and generation of reactive oxygen species in early programmed cell death. *J. Exp. Med.*, **182**, 367–377.

Zamzami, N., Marchetti, P., Castedo, M., Zamin, C., Vayssiere, J.-L., Petit, P.X. and Kroemer, G. (1995b) Reduction in mitochondrial potential constitutes an early irreversible step of programmed lymphocyte death in vivo. *J. Exp. Med.*, **181**, 1661–1672.

Zoratti, M. and Szabo, I. (1995) The mitochondrial permeability transition. *Biochim. Biophys. Acta.*, **1241**, 139–176.

Part 3
THE EXECUTION OF APOPTOSIS

10. CASPASES: THE MOLECULAR EFFECTORS OF APOPTOSIS

ALISON J. BUTT*[*] AND SHARAD KUMAR[†]

[*]*The Hanson Centre for Cancer Research, Frome Road, Adelaide, Australia 5000*

The caspase family of cysteine proteases were discovered as mammalian homologues of the *Caenorhabditis elegans* death protein CED-3. Caspases are synthesised as precursor molecules which require cleavage at specific aspartate residues to produce the two subunits of the active enzyme. Activation of certain members of this family occurs in response to various apoptotic stimuli and once activated, caspases mediate cleavage of numerous cellular substrates. Recently, caspases have been shown to interact with death receptor complexes, providing a direct link between death receptor signalling and activation of the effector phase of apoptosis. This chapter summarises what is currently known about members of this protease family and how they may play a central role in effecting the apoptotic process.

KEY WORDS: CED-3, Bcl-2, CED-4, protease activation, death receptor.

INTRODUCTION

The detailed cascade of genetic elements involved in the mammalian apoptotic process was first defined from initial studies in *Caenorhabditis elegans*. During the development of this nematode, cells are deleted by an invariant process morphologically and functionally analogous to apoptosis (Ellis *et al.*, 1991). Mutational analysis lead to the discovery of various genes which act as either positive or negative regulators of apoptosis. Of these, *ced-3, ced-4* and *ced-9* appear to be essential for the execution and regulation of cell death (reviewed in Hengartner and Horvitz, 1994a). Both *ced-3* and *ced-4* have pro-apoptotic effects and mutants lacking either of these genes contain additional cells, normally deleted during development (Ellis and Horvitz, 1986). The *ced-9* gene antagonises the function of *ced-3* and *ced-4* to protect cells from apoptosis, as demonstrated in *ced-9* loss-of-function mutants where the majority of cells arrest early in development resulting in embryonic lethality (Hengartner *et al.*, 1992). The first evidence to suggest that the cell death pathways in *C. elegans* and mammalian cells may contain common regulatory elements arose from the discovery that *ced-9* shares both structural and sequence homologies with the mammalian *bcl-2* gene (Hengartner and Horvitz, 1994b). *Bcl-2* has been shown to prevent apoptosis both *in vitro* and *in vivo* in a wide variety of cell types (reviewed in Korsmeyer, 1992) and, when overexpressed, may partially restore functional loss in a *ced-9* mutant (Vaux *et al.*, 1992; Hengartner and Horvitz, 1994b). Further similarities between the *C. elegans* and mammalian death pathways were demonstrated by the discovery that *ced-3* exhibits significant homology to interleukin-1β-converting enzyme (ICE) (caspase-1) (Yuan *et al.*, 1993). ICE was the first identified member of a family of aspartate-specific cysteine pro-

† *Corresponding Author*: (08) 8222 3738. Fax: (08) 8222 3139. e-mail: sharad.kumar@imvs.sa.gov.au

teases (Thornberry *et al.*, 1992; Cerretti *et al.*, 1992) recently designated caspases (Alnemri *et al.*, 1996) which, like CED-3 appear essential for the execution of active cell death. Subsequent studies have revealed several more mammalian ICE/CED-3 homologs: Nedd2/ICH-1 (caspase-2) (Kumar *et al.*, 1994; Wang *et al.*, 1994), CPP32/Yama/apopain (caspase-3) (Fernandes-Alnemri *et al.*, 1994; Nicholson *et al.*, 1995; Tewari *et al.*, 1995b), ICE_{rel}II/TX/ICH2 (caspase-4) (Faucheu *et al.*, 1995; Kamens *et al.*, 1995; Munday *et al.*, 1995), ICE_{rel}III/TY (caspase-5) (Munday *et al.*, 1995; Faucheu *et al.*, 1996); Mch2 (caspase-6) (Fernandes-Alnemri *et al.*, 1995a); ICE-Lap3/Mch3 (caspase-7) (Fernandes-Alnemri *et al.*, 1995b; Duan *et al.*, 1996a), MACH/FLICE/Mch5 (caspase-8) (Boldin *et al.*, 1996; Fernandes-Alnemri *et al.*, 1996; Muzio *et al.*, 1997), ICE-Lap6/Mch6 (caspase-9) (Duan *et al.*, 1996b; Srinivasula *et al.*, 1996b), Mch4 (caspase-10) (Fernandes-Alnemri *et al.*, 1996) and ICH3 (caspase-11) (Wang *et al.*, 1996). In addition, two novel murine caspase homologs have recently been identified and characterised (Van de Craen *et al.*, 1997).

STRUCTURAL FEATURES

Members of the ICE family of proteases can be divided into three distinct sub groups based on their sequence homology: (i) the CED-3-like subfamily consisting of caspase-3, caspase-6, caspase-7, caspase-8 and caspase-10 which are all closely related to CED-3 (~35% sequence identity); (ii) the ICE-like subfamily including caspase-1, caspase-4, caspase-5 and caspase-11 which share just over 50% sequence identity between members and have 26–28% sequence identity with CED-3; (iii) the NEDD2/ICH-1 subfamily of caspase-2 which shares approximately 31% sequence identity with CED-3. All caspases share two unique features that distinguish them from other proteases. They all require an aspartic acid residue in the P_1 position of their substrates and their activation requires cleavage after Asp residues to produce the large and small subunits of the active enzyme. In the caspase family, the residues with the highest degree of sequence identity with CED-3 include residues important for recognition of Asp in P_1 and residues required for catalysis, suggesting functional homology between CED-3 and its mammalian homologs (Wilson *et al.*, 1994). The ability to cleave after Asp residues is shared only with the serine protease granzyme B which has also been implicated in the apoptotic process (Smyth and Trapani, 1995; Greenberg, 1996; Martin *et al.*, 1996).

Much of the present information regarding the structure of caspases is based on studies of caspase-1 and, more recently, caspase-3. The mature functional form of caspase-1 is derived by the proteolytic cleavage of a 404 amino acid precursor polypeptide (p45) at Asp residues 103, 119, 297 and 316 (Thornberry *et al.*, 1992). Active caspase-1 is predicted to be a tetramer of two p20 subunits (residues 120-297) surrounding two adjacent p10 subunits (residues 317-404) (Walker *et al.*, 1994; Wilson *et al.*, 1994). Caspase-1 has the ability to cleave and activate itself as has been demonstrated using a bacterially expressed full-length precursor form of the enzyme (Ramage *et al.*, 1995). However, in monocytic cells caspase-1 is predominantly found in its precursor form (Ayala *et al.*, 1994) suggesting that caspase-1 may be activated by other caspases *in vivo*.

Studies have demonstrated a number of structural similarities between members of the caspase family. More specifically, the residues with the highest degree of sequence identity between members include residues near the cleavage sites of the large and small subunits of caspase-1 (Wilson *et al.*, 1994; Nicholson *et al.*, 1995) suggesting that all members of this family are composed of two subunits. Cloning of caspase-3 (Fernandes-Alnemri *et al.*, 1994) has shown that it consists of p17 and p12 subunits which form the active enzyme complex (Nicholson *et al.*, 1995) in a manner analogous to that seen for caspase-1. Caspase-3 also resembles caspase-1 in overall structure, although differences in the S_4 subsite may explain the variation in their substrate specificities and biological functions (Rotonda *et al.*, 1996). *In vitro* studies of the processing of caspase-2 have similarly shown that the caspase-2 precursor (p51) is cleaved to p19 and p12 subunits by active caspase-1, caspase-3, granzyme B and to a lesser degree by caspase-6 and caspase-2 itself (Harvey *et al.*, 1996; Xue *et al.*, 1996). Recent cloning and characterisation of the other known members of the caspase family have confirmed that activation of these proteases requires proteolytic cleavage of the precursor to the large and small subunits of the active enzyme (Table 10.1).

Table 10.1 Subunits and substrates of the caspase family.

Caspase	*Subunits*	*Substrates*	*References*
caspase-1 (ICE)	p20 + p10	pro-IL-1β, pro-casp1, pro-casp3, IFN-γ-inducing factor	Ramage *et al.*, 1995; Walker *et al.*, 1994; Gu *et al.*, 1997; Ghayur *et al.*, 1997
caspase-2 (Nedd2/ICH1)	p19 + p12		Kumar *et al.*, 1994; Wang *et al.*, 1994; Harvey *et al.*, 1996; Gu *et al.*, 1995
caspase-3 (CPP32/Yama)	p17 + p12	PARP, U1snRNP, DNA-PK, hnRNPs, SREBPs, Rb *etc.*	Nicholson *et al.*, 1995; Fernandes-Alnemri *et al.*, 1994; Tewari *et al.*, 1995b
caspase-4 (ICE_{rel}II/TX/ICH2)	p20 + p10	pro-casp1,	Faucheu *et al.*, 1995; Kamens *et al.*, 1995; Munday *et al.*, 1995
caspase-5 (ICE_{rel}III/TY)	p20 + p10		Munday *et al.*, 1995
caspase-6 (Mch2)	p18 + p11	lamins	Fernandes-Alnemri *et al.*, 1995a; Srinivasula *et al.*, 1996b
caspase-7 (ICE-lap3/Mch3)	p20 + p12	PARP, pro-casp6	Fernandes-Alnemri *et al.*, 1995b; Duan *et al.*, 1996a
caspase-8 (MACH/FLICE/Mch5)	p20 + p10	pro-casp3, pro-casp7, pro-casp4, pro-casp-9	Boldin *et al.*, 1996; Fernandes-Alnemri *et al.*, 1996; Muzio *et al.*, 1996; Srinivasula *et al.*, 1996a
caspase-9 (ICE-lap6/Mch6)	p35 + p10	PARP, pro-casp3	Duan *et al.*, 1996b; Srinivasula *et al.*, 1996b
caspase-10 (Mch4)	p15 + p12	pro-casp3, pro-casp7	Fernandes-Alnemri *et al.*, 1996
caspase-11 (ICH-3)	p20 + p10		Wang *et al.*, 1996

In addition to their sequence homology, members of the caspase family can be subdivided according to the size of their amino-terminal prodomain. Of the caspases discovered thus far, CED-3, caspase-1, caspase-2, caspase-4, caspase-5, caspase-8, caspase-9 and caspase-10 are all characterised by long prodomain regions, whilst caspase-3, caspase-6 and caspase-7 contain short or absent prodomains. During caspase activation the prodomain and in some cases, an internal linker sequence are removed to produce the large and small subunits. Although separate expression of the p20 and p10 subunits of caspase-1 in Sf9 cells results in functional enzyme (Fernandes-Alnemri *et al.*, 1994), time course studies of caspase-1 activation reveal that p10 is released from the p45 precursor before p20 suggesting that the prodomain may have a regulatory role in the activation process (Ramage *et al.*, 1995). Indeed, studies by Van Criekinge *et al.*, (1996) have demonstrated that the prodomain of caspase-1 is absolutely required but not sufficient for dimerisation and subsequent autoactivation. Coexpression of caspase-2 subunits also fails to generate active enzyme (Kumar *et al.*, 1997). A further role for the prodomain region of caspases-2 and -8 has been demonstrated by the recent observation that it mediates the association of these caspases to adaptor molecules which in turn associate with components of the tumour necrosis factor receptor 1 (TNF-R1) (Duan and Dixit, 1997) and Fas/APO1 (CD95) (Boldin *et al.*, 1996; Muzio *et al.*, 1996) signalling pathways. Thus, this link with membrane associated signalling complexes *via* the prodomain suggests that caspases with large prodomains may act upstream as early transducers of diverse apoptotic signals. This is discussed in more depth later on in this chapter.

ACTIVATION DURING APOPTOSIS

In addition to their homology to the *C. elegans* CED-3 protein, there is now accumulating direct evidence to link members of the caspase family with the effector phase of apoptosis. Initial studies with caspase-1 demonstrated that its overexpression in Rat-1 cells induced apoptosis (Miura *et al.*, 1993). Subsequent studies have shown that overexpression of all known members of this family induces apoptosis in a variety of cell systems (Fernandes-Alnemri *et al.*, 1994; Kumar *et al.*, 1994; Wang *et al.*, 1994; Faucheu *et al.*, 1995; Fernandes-Alnemri *et al.*, 1995a; Fernandes-Alnemri *et al.*, 1995b; Kamens *et al.*, 1995; Munday *et al.*, 1995; Duan *et al.*, 1996a). In addition, it has been reported that expression of antisense caspase-2 inhibits apoptosis (Kumar, 1995; Troy *et al.* 1997). Peptidyl inhibitors of caspases including YVAD-CMK, YVAD-CHO, DEVD-CHO and z-VAD-FMK have also been shown to inhibit apoptosis (Lazebnik *et al.*, 1994; Fearnhead *et al.*, 1995; Los *et al.*, 1995; Slee *et al.*, 1996). More directly, specific caspases have been shown to be processed to active subunits following apoptotic stimuli, including caspase-3, caspase-7 (Duan *et al.*, 1996a; Chinnaiyan *et al.*, 1996; Orth *et al.*, 1996; Schlegel *et al.*, 1996) and more recently, caspase-2 (Harvey *et al.*, 1997; MacFarlane *et al.*, 1997) and caspase-6 (Faleiro *et al.*, 1997). However, caspase-1 null mice develop normally, only exhibiting defects in IL-1β processing (Li *et al.*, 1995) and to some extent in Fas-mediated apoptosis (Kuida *et al.*, 1995) suggesting that caspase-1 expression may not be essential for all apoptotic pathways or that there may be some functional redundancy amongst the caspase family. Indeed, Kuida *et al.*, (1996) have reported specific defects in neuronal cell death in caspase-3-deficient mice, whilst thymocytes from these mice retain normal susceptibility to various apoptotic stimuli.

Further evidence for the involvement of caspases in apoptosis was suggested by studies with the cowpox virus encoded protein CrmA (Ray *et al.*, 1992; Komiyama *et al.*, 1994) and the baculovirus protein p35 (Clem and Miller, 1994), both of which are direct inhibitors of certain caspases. CrmA has been shown to inhibit apoptosis induced by a variety of stimuli (Gagliardini *et al.*, 1994; Enari *et al.*, 1995; Los *et al.*, 1995; Miura *et al.*, 1995; Strasser *et al.*, 1995; Tewari and Dixit, 1995; Tewari *et al.*, 1995c) and blocks subsequent activation of caspases-3 and -7 in Jurkat cells following Fas activation (Chinnaiyan *et al.*, 1996). Similarly, p35 blocks apoptosis in insect and mammalian cells (Rabizadeh *et al.*, 1993; Sugimoto *et al.*, 1994; Bump *et al.*, 1995; Martinou *et al.*, 1995). Interestingly, recent work by Datta *et al.*, (1997) has added weight to the suggestion that caspase-specific pathways mediate the induction of apoptosis by demonstrating that ionising radiation induces a CrmA-insensitive, p35-sensitive death pathway in the human myeloid leukemia cell line U-937. In addition, Dorstyn and Kumar (1997) have reported differential inhibitory effects of CrmA and p35 in NIH-3T3 cells following various apoptotic stimuli.

Once activated, caspases have been shown to target a number of cellular substrates (discussed in detail elsewhere in this volume). Kaufmann, (1989) and colleagues (Kaufmann *et al.*, 1993) first demonstrated that the 116 kDa nuclear protein poly (ADP-ribose) polymerase (PARP) is cleaved during apoptosis to produce an 85kDa fragment. Further studies demonstrated that the protease involved in this cleavage activity resembled caspase-1 and was inhibited by CrmA (Lazebnik *et al.*, 1994). This was later confirmed by reports which demonstrated that PARP is cleaved by caspase-3 at a specific P_1-P_4 amino acid sequence ($DEVD^{216}\downarrow G^{217}$) between its DNA binding and catalytic domains (Nicholson *et al.*, 1995; Tewari *et al.*, 1995b). However, in a cell-free system caspase-3 alone did not provoke apoptotic changes, suggesting that additional caspases may be required (Nicholson *et al.*, 1995). Indeed, subsequent studies have shown that other caspases have the ability to cleave PARP with varying efficiencies, including caspase-6 (Fernandes-Alnemri *et al.*, 1995a), caspase-7 (Fernandes-Alnemri *et al.*, 1995b) and caspases-4 and -2 (Gu *et al.*, 1995). Members of the caspase family have been shown to be responsible for the cleavage of other proteins during apoptosis, including sterol regulatory element-binding protein-1 (SREBP-1) and SREBP-2 (Wang *et al.*, 1995; Wang *et al.*, 1996), nuclear lamins (Kaufmann, 1989; Lazebnik *et al.*, 1995; Orth *et al.*, 1996; Takahashi *et al.*, 1996), catalytic subunit of DNA-dependent protein kinase (DNA-PK) (Casciola-Rosen *et al.*, 1995; Song *et al.*, 1996a; Song *et al.*, 1996b), U1 small ribonucleoprotein (U1-70kdal) (Casciola-Rosen *et al.*, 1994; Tewari *et al.*, 1995a), heteronuclear ribonucleoproteins (hnRNP) C1 and C2 (Waterhouse *et al.*, 1996), D4-GDI (Na *et al.*, 1996), huntingtin (Goldberg *et al.*, 1996) and replication factor C (Song *et al.*, 1997).

The importance of such specific proteolysis to the apoptotic process remains to be fully elucidated. However, hnRNPs C1 and C2, U1-70kD, PARP and DNA-PK all play a role in the splicing of mRNA and/or the repair of double-strand DNA breaks (Satoh and Lindahl, 1992; Tazi *et al.*, 1993; Peterson *et al.*, 1995), suggesting that abolition of these repair mechanisms may be essential for the subsequent death of the cell. In addition, PARP has very high activity and, if activated by the chromatin fragmentation that occurs during apoptosis, could deplete cellular ATP stores which are essential for the formation of apoptotic bodies (Zhang *et al.*, 1994). Nuclear lamins are important in maintaining the integrity of the nuclear envelope and chromatin, both of which breakdown during apoptosis. Certain caspases also cleave proteins involved in the formation of the cytoskeleton such as fodrin (Martin *et al.*, 1995; Cryns *et al.*, 1996) and the Gas2 protein which forms part of the microfilament network (Brancolini *et al.*, 1995). Specific proteolysis of these proteins by caspases may be responsible for the ultrastructural changes observed in apoptotic cells. In addition, numerous other proteins have been shown to be degraded during the apoptotic process although the functional significance of these cleavage events is as yet unclear (Browne *et al.*, 1994; Janicke *et al.*, 1996).

In a recent study, Liu *et al.*, (1997) have identified a cytosolic protein that is activated by caspase-3 and induces nuclear DNA fragmentation. Termed DNA fragmentation factor (DFF), this protein consists of two subunits of 40 kDa and 45 kDa, and the latter is cleaved by caspase-3 to generate an active factor that mediates DNA fragmentation without further requirement for active caspases. However, caspase-3 activity is still required for cleavage of other nuclear substrates such as PARP and lamin B1. Thus, this study has provided important evidence for a direct link between caspase activation in the cytosol and induction of nuclear DNA fragmention, a characteristic morphological feature of apoptosis. How DFF induces DNA fragmentation is still unclear, although it has been shown to have no DNase activity and is therefore unlikely to be a nuclease. One possibility is that DFF activates a nuclease situated in the nucleus either by direct activation following translocation to the nucleus or by interaction with an additional protein(s) on the nuclear envelope.

LINK WITH OTHER COMPONENTS OF THE DEATH PATHWAY

Despite numerous studies little is known about how components of the apoptotic pathway interact and, in particular, how apoptotic stimuli lead to caspase activation. However, several recent studies have shed some light on the upstream activation events and how they may interact with caspases. Again, studies in *C. elegans* provided initial clues as to the upstream activators of caspases. These have demonstrated that the apoptotic inhibitor *ced-9* functions upstream of both the death genes *ced-3* and *ced-4* (Ellis *et al.*, 1991) and may exert its protective effects by preventing their activation (Hengartner *et al.*, 1992; Shaham and Horvitz, 1996). Furthermore, apoptosis induced by overexpression of CED-4 requires CED-3 activity suggesting that CED-4 acts upstream of or in parallel to CED-3 (Shaham and Horvitz, 1996). Consequently, members of the Bcl-2 family, the mammalian homologues of CED-9, have been shown to function upstream of two CED-3 homologues, caspases-3 and -7. Overexpression of Bcl-2 or Bcl-x_L can inhibit the activation of caspases and abrogate apoptosis (Chinnaiyan *et al.*, 1996; Erhardt and Cooper, 1996; Estoppey *et al.*, 1997; Perry *et al.*, 1997). Recently, several groups have demonstrated that CED-9 and the mammalian homologue Bcl-x_L can directly interact with and inhibit the function of CED-4 in yeast (James *et al.*, 1997) and mammalian cells (Chinnaiyan *et al.*, 1997; Spector *et al.*, 1997; Wu *et al.*, 1997). In addition, Chinnaiyan *et al.*,

(1997) reported that CED-4 can concurrently complex with CED-3 and caspases-1 and -8 but not caspases-3 and -6. These important findings have suggested that CED-4 may be the link between CED-9 and the Bcl-2 family with CED-3 and caspases. In such a model, Bcl-2 located at the mitochondrial membrane binds and modulates CED-4 which in turn binds CED-3/caspases. Activation of caspases would require that Bcl-2 is removed from the complex. As yet, putative mammalian homologues of CED-4 are unknown but may hold the key to elucidating the mechanisms of caspase activation in mammalian cells.

One possible mechanism for the activation of caspases during apoptosis has been suggested by the observation that translocation of cytochrome c from the mitochondria into the cytosol activates DEVD-specific caspases and triggers apoptosis (Liu *et al.*, 1996). Bcl-2 is normally located in the mitochondrial intermembrane space suggesting that it may exert its antiapoptotic effects by preventing release of cytochrome c from mitochondria following an apoptotic stimulus. Indeed, overexpression of Bcl-2 was demonstrated to block the translocation of cytochrome c into the cytosol (Kluck *et al.*, 1997; Yang *et al.*, 1997). However, the mechanisms by which cytochrome c is released from mitochondria and initiates caspase activation remains to be determined.

Recently, the involvement of caspases in two specific pathways leading to apoptosis appears to be delineated. The cell surface cytokine receptors Fas and TNFR-1 trigger apoptosis by binding of their respective ligands or specific agonist antibodies (reviewed in Cleveland and Ihle, 1995). Caspase activation has been demonstrated to occur during both TNF-R1- and Fas-induced apoptosis (Tewari and Dixit., 1995; Chinnaiyan *et al.*, 1996; Duan *et al.*, 1996a; Enari *et al.*, 1996; Orth *et al.*, 1996). Both Fas and TNFR-1 share a region of homology at their carboxyl terminal designated the 'death domain'. Using the yeast two-hybrid system, three intracellular proteins containing similar death domain regions: TNFR-1-associated death domain (TRADD) (Hsu *et al.*, 1995), Fas-associating protein with death domain (FADD/MORT1) (Boldin *et al.*, 1995; Chinnaiyan *et al.*, 1995) and receptor-interacting protein (RIP) (Stanger *et al.*, 1995), have been isolated and shown to interact with the death domains of Fas and TNFR-1 (Varfolomeev *et al.*, 1996). Subsequently, two groups described a novel caspase designated FLICE (Muzio *et al.*, 1996) or MACH (Boldin *et al.*, 1996) (caspase-8), which has homology to FADD/MORT1 and binds to it *via* a shared sequence motif, the 'death effector domain' (DED) located upstream of the death domain in FADD/MORT1 and in the large prodomain region of caspase-8 (Muzio *et al.*, 1996). This recruitment of caspase-8 to the complex induces a CrmA-inhibitable pathway to apoptosis *via* activation of other downstream caspases (Srinivasula *et al.*, 1996a; Muzio *et al.*, 1997). Caspase-10 also contains a DED motif with homology to that of caspase-8 which mediates its binding to the adaptor molecule FADD and links caspase-10 with both Fas- and TNF-R1-mediated apoptosis (Vincenz and Dixit, 1997).

An analogous situation has been demonstrated following activation of TNFR-1. In this pathway, activation of TNFR-1 results in the formation of a complex consisting of the death domain proteins RIP and TRADD and a death adaptor molecule designated RAIDD (Duan and Dixit, 1997) or CRADD (Ahmad *et al.*, 1997) which associates with the prodomain region of caspase-2 and RIP (Duan and Dixit, 1997) thereby recruiting caspase-2 to the Fas- and TNFR-1-mediated apoptotic pathways (Ahmad *et al.*, 1997). Recently, studies by Mariani *et al.*, (1997) have suggested that caspase activation may occur following activation of an another receptor-ligand pathway to apoptosis, the TRAIL (Apo-2) ligand (Pitti *et al.*, 1996) which interacts with the DR4 death receptor (Pan *et al.*, 1997). Characterisation of these apoptotic pathways has provided the first evidence of a direct link between signalling through death receptors and activation of the caspase cascade. Furthermore, these studies have added weight to the concept that caspases with long prodomain regions such as caspases-8 and -2 may act as upstream transducers of death signals to other members of the caspase family *via* a protease amplification cascade. Consistent with this, Harvey *et al.*, (1997) have recently demonstrated that caspase-2 is activated prior to caspase-3 in mammalian cells following treatment with various apoptotic stimuli.

Interestingly, a novel family of viral inhibitors (v-FLIPS) have recently been isolated which abrogate apoptosis induced by death receptor activation (Thome *et al.*, 1997). v-FLIPs contain two DED which bind to the adaptor protein FADD preventing the recruitment and activation of caspase-8. Consequently, cells expressing v-FLIPs are protected against apoptosis induced by Fas or related death receptors thus enabling viral spread and persistence.

FUTURE PERSPECTIVES

Recent years have seen many important advances in determining the components of and pathways to mammalian cell death and this emerging data strongly points to a central role for caspases as molecular effectors of this process. Clearly, there are components of the death pathway which have been well conserved through evolution and *C. elegans* has provided an invaluable and highly relevant model system in which to study the mechanisms of apoptosis. However, there are many fundamental questions which remain unanswered. The emerging number of mammalian CED-3 homologues suggests either some functional redundancy amongst caspases or that certain caspases may be activated in specific cell types or in response to specific apoptotic stimuli. Alternatively, not all caspases could be involved in mediating apoptosis and the importance of caspase activation solely in the apoptotic process still needs confirmation. It could be envisaged that caspases may have additional cellular roles by processing precursor forms of proteins involved in normal cellular function. For example, the primary function of caspase-1 is processing of inflammatory cytokines such as interleukin-1β (IL-1β) (Thornberry *et al.*, 1992) and interferon-γ-inducing factor (Ghayur *et al.*, 1997; Gu *et al.*, 1997). Caspase-11 has also been shown to promote IL-1β processing by caspase-1 suggesting this caspase may have an additional role in mediating an inflammatory response (Wang *et al.*, 1996). In addition, the *Drosophila* CED-3 homologue DCP-1 has been demonstrated to have an essential role in development which is distinct from its role in apoptosis (Song *et al.*, 1997).

The pathways to Fas- and TNF-R1-induced apoptosis appear to be more fully delineated and have provided evidence for a link between extracellular death signals and caspase activation. However, the intracellular pathways to cell death following other apoptotic signals, for example γ-irradiation or drug-induced DNA damage remain unclear and their elucidation has important therapeutic implications. The isolation of mammalian CED-4 homologues may provide some clues as to the upstream activators of caspases and the negative regulation of this process by members of the Bcl-2 family through their sequestration of CED-4. In addition, the role of cytochrome c in the activation of caspases remains to be fully elucidated but provides another intriguing mechanism by which Bcl-2 could regulate the apoptotic process (Figure 10.1).

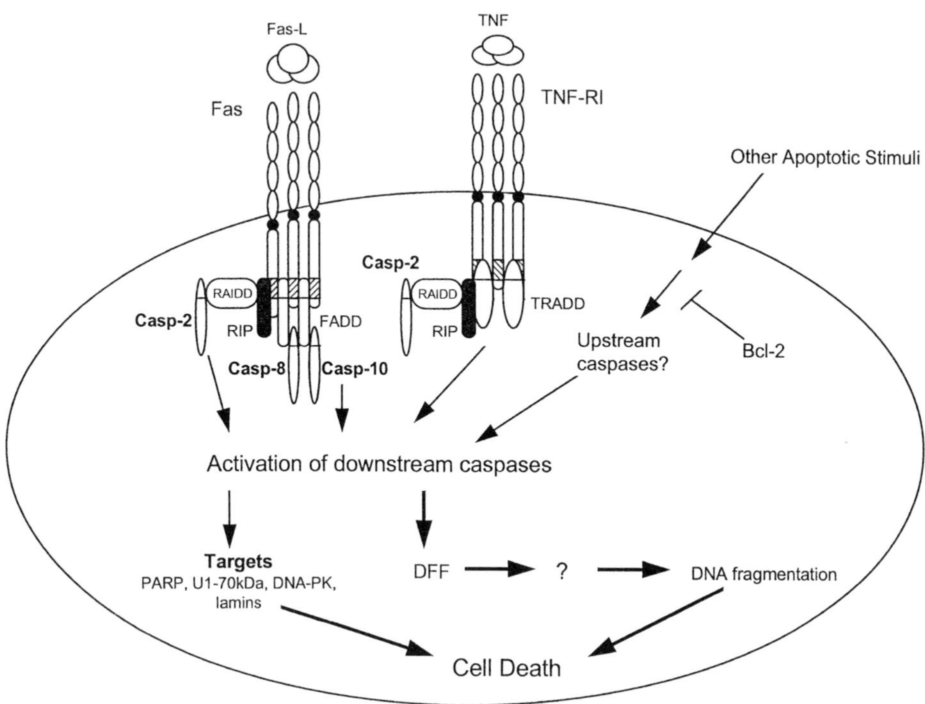

Figure 10.1 A model showing the involvement of caspases as central intracellular effectors of apoptosis. Following activation of the death receptors Fas or TNF-R1 by ligand binding, caspases-2, -8 and -10 are recruited via adaptor proteins such as TRADD, FADD, RIP and RAIDD which interact with the death domains of the receptors and the prodomain regions of the caspase. The death domain of TRADD can also recruit FADD to activate caspase-8. Binding of these caspases may result in self-activation which, in turn results in activation of downstream caspases such as caspase-3. Other apoptotic stimuli such as DNA damaging agents or serum withdrawal can also induce activation of various caspases. This pathway may involve Bcl-2/CED-4/caspase complexes, as discussed in the text. Once activated, caspase-3 cleaves numerous substrates and activate DFF, a protein which mediates DNA fragmentation by a yet unknown mechanism.

REFERENCES

Ahmad, M., Srinivasula, S.M., Wang, L., Talanian, R.V., Litwack, G., Fernandes-Alnemri, T., *et al.* (1997) CRADD, a novel human apoptotic adaptor molecule for caspase-2, and FasL/tumor necrosis factor receptor-interacting protein RIP. *Cancer Res.*, **57**, 615–619.

Alnemri, E.S., Livingston, D.J., Nicholson, D.W., Salveson, G., Thornberry, N.A., Wong, W.W., *et al.* (1996) Human ICE/CED-3 protease nomenclature. *Cell*, **87**, 171

Ayala, J.M., Yamin, T., Egger, L.A., Chin, J., Kostura, M.J. and Miller, D.K. (1994) IL-1β-converting enzyme is present in monocytic cells as an inactive 45-kdal precursor. *J. Immunol.*, **153**, 2592–2599.

Boldin, M.P., Mett, I.L., Varfolomeev, E.E., Chumakov, I., Shemer-Avni, Y., Camonis, J.H., *et al.* (1995) Self-association of the 'death-domains' of the p55 tumor necrosis factor (TNF) receptor and Fas/APO1 prompts signalling for TNF and Fas/APO1 effects. *J. Biol. Chem.*, **270**, 387–391.

Boldin, M.P., Goncharov, T.M., Goltsev, Y.V. and Wallach, D. (1996) Involvement of MACH, a novel MORT1/FADD-interacting protease, in Fas/APO-1-and TNF receptor-induced cell death. *Cell*, **85**, 803–815.

Brancolini, C., Benedetti, M. and Schneider, C. (1995) Microfilament reorganisation during apoptosis: the role of Gas2, a possible substrate for ICE-like proteases. *EMBO J.*, **14**, 5179–5190.

Browne, S.J., Williams, A.C., Hague, A., Butt, A.J. and Paraskeva, C. (1994) Loss of APC protein expressed by human colonic epithelial cells and the appearance of a specific low-molecular-weight form is associated with apoptosis in vitro. *Int. J. Cancer*, **59**, 56–64.

Bump, N.J., Hackett, M., Huguenin, M., Seshagiri, S., Brady, K., Chen, P., *et al.* (1995) Inhibition of ICE family proteases by baculovirus antiapoptotic protein p35. *Science*, **269**, 1885–1888.

Casciola-Rosen, L.A., Miller, K.M., Anhalt, G.J. and Rosen, A. (1994) Specific cleavage of the 70-kDa protein component of the U1 small nuclear ribonucleoprotein is a characteristic biochemical feature of apoptotic cell death. *J. Biol. Chem.*, **269**, 30757–30760.

Casciola-Rosen, L.A., Anhalt, G.J. and Rosen, A. (1995) DNA-dependent protein kinase is one of a subset of autoantigens specifically cleaved early during apoptosis. *J. Exp. Med.*, **182**, 1625–1634.

Cerretti, D.P., Kozlosky, C.J., Mosley, B., Nelson, N., Van Ness, K., Greenstreet, T.A., *et al.* (1992) Molecular cloning of the interleukin-1β converting enzyme. *Science*, **256**, 97–100.

Chinnaiyan, A.M., O'Rourke, K., Tewari, M. and Dixit, V.M. (1995) FADD, a novel death domain-containing protein, interacts with the death domain of Fas and initiates apoptosis. *Cell*, **81**, 505–512.

Chinnaiyan, A.M., Orth, K., O'Rourke, K., Duan, H., Poirier, G.G. and Dixit, V.M. (1996) Molecular ordering of the cell death pathway: Bcl-2 and Bcl-xL function upstream of the ced-3-like apoptotic proteases. *J. Biol. Chem.*, **271**, 4573–4576.

Chinnaiyan, A.M., O'Rourke, K., Lane, B.R. and Dixit, V.M. (1997) Interaction of CED-4 with CED-3 and CED-9: a molecular framework for cell death. *Science*, **275**, 1122–1126.

Clem, R.J. and Miller, L.K. (1994) Control of programmed cell death by the baculovirus genes p35 and iap. *Mol. Cell Biol.*, **14**, 5212–5222.

Cleveland, J.L. and Ihle, J.N. (1995) Contenders in FasL/TNF death signalling. *Cell*, **81**, 479–482.

Cryns, V.L., Bergeron, L., Zhu, H., Li, H. and Yuan, J. (1996) Specific cleavage of alpha-fodrin during Fas- and tumor necrosis factor-induced apoptosis is mediated by an interleukin-1β-converting enzyme/ced3 protease distinct from the poly(ADP-ribose) polymerase protease. *J. Biol. Chem.*, **271**, 31277–31282.

Datta, R., Kojima, H., Banach, D., Bump, N.J., Talanian, R.V., Alnemri, E.S., *et al.* (1997) Activation of CrmA-insensitive, p35-sensitive pathway in ionizing radiation-induced apoptosis. *J. Biol. Chem.*, **272**, 1965–1969.

Dorstyn, L. and Kumar, S. (1997) Differential inhibitory effects of CrmA, p35, IAP and three mammalian IAP homologues on apoptosis in NIH-3T3 cells following various death stimuli. *Cell Death Diff.*, **4**, 570–579.

Duan, H., Chinnaiyan, A.M., Hudson, P.L., Wing, J.P., He, W.W. and Dixit, V.M. (1996a) ICE-LAP3, a novel mammalian homologue of the Caenorhabditis elegans cell death protein Ced-3 is activated during Fas- and tumor necrosis factor-induced apoptosis. *J. Biol. Chem.*, **271**, 1621–1625.

Duan, H., Orth, K., Chinnaiyan, A.M., Poirier, G.G., Froelich, C.J., He, W.W., *et al.* (1996b) ICE-LAP6, a novel member of the ICE/Ced-3 gene family, is activated by the cytotoxic T cell protease granzyme B. *J. Biol. Chem.*, **271**, 16720–16724.

Duan, H. and Dixit, V.M. (1997) RAIDD is a new 'death' adaptor molecule. *Nature*, **385**, 86–89.

Ellis, H.M. and Horvitz, H.R. (1986) Genetic control of programmed cell death in the nematode C. elegans. *Cell*, **44**, 817–829.

Ellis, R., Yuan, J. and Horvitz, H.R. (1991) Mechanisms and functions of cell death. *Annu. Rev. Cell Biol.*, **7**, 663–669.

Enari, M., Hug, H. and Nagata, S. (1995) Involvement of an ICE-like protease in Fas-mediated apoptosis. *Nature*, **375**, 78–81.

Enari, M., Talanian, R.V., Wong, W.W. and Nagata, S. (1996) Sequential activation of ICE-like and CPP32-like proteases during Fas-mediated apoptosis. *Nature*, **380**, 723–726.

Erhardt, P. and Cooper, G.M. (1996) Activation of the CPP32 apoptotic protease distinct signalling pathways with differential sensitivity to Bcl-x_L. *J. Biol. Chem.*, **271**, 17601–17604.

Estoppey, S., Rodriguez, I., Sadoul, R. and Martinou, J-C. (1997) BCL-2 prevents activation of CPP-32 cysteine protease and cleavage of poly(ADP-ribose) polymerase and U1-70 kD proteins in staurosporine-mediated apoptosis. *Cell Death Diff.*, **4**, 34–38.

Faleiro, L., Kobayashi, R., Fearnhead, H. and Lazebnik, Y.A. (1997) Multiple species of CPP32 and Mch2 are the major active caspases present in apoptotic cells. *EMBO J.*, **16**, 2271–2281.

Faucheu, C., Diu, A., Chan, A.W.E., Blanchet, A., Miossec, C., Herve, F., *et al.* (1995) A novel human protease similar to the interleukin-1β converting enzyme induces apoptosis in transfected cells. *EMBO J.*, **14**, 1914–1922.

Faucheu, C., Blanchet, A., Collard-Dutilleul, V., Lalanne, J. and Diu-Hercend, A. (1996) Identification of a cysteine protease closely related to interleukin-1β converting enzyme. *Eur. J. Biochem.*, **236**, 207–213.

Fearnhead, H.O., Dinsdale, D. and Cohen, G.M. (1995) An interleukin-1β-converting enzyme-like protease is a common mediator of apoptosis in thymocytes. *FEBS Lett.*, **375**, 283–288.

Fernandes-Alnemri, T., Litwack, G. and Alnemri, E.S. (1994) CPP32, a novel human apoptotic protein with homology to Caenorhabditis elegans cell death protein Ced-3 and mammalian interleukin-1-β-converting enzyme. *J. Biol. Chem.*, **269**, 30761–30764.

Fernandes-Alnemri, T., Litwack, G. and Alnemri, E.S. (1995a) Mch2, a new member of the apoptotic Ced-3/ICE cysteine protease gene family. *Cancer Res.*, **55**, 2737–2742.

Fernandes-Alnemri, T., Takahashi, A., Armstrong, R., Krebs, J., Fritz, L., Tomaselli, K.J., *et al.* (1995b) Mch3, a novel human apoptotic cysteine protease is highly related to CPP32. *Cancer Res.*, **55**, 6045–6052.

Fernandes-Alnemri, T., Armstrong, R.C., Krebs, J., Srinivasula, S.M., Wang, L., Bullrich, F., *et al.* (1996) In vitro activation of CPP32 and Mch3 by Mch4, a novel human apoptotic cysteine protease containing two FADD-like domains. *Proc. Natl. Acad. Sci. USA*, **93**, 7464–7469.

Gagliardini, V., Fernandez, P.A., Lee, R.K., Drexler, H.C., Rotello, R.J. and Fishman, M.C. *et al.* (1994) Prevention of vertebrate neuronal death by the crmA gene. *Science*, **263**, 826–828.

Ghayur, T., Banerjee, S., Hugunin, M., Butler, D., Herzog, L., Carter, A., Quintal, L., Sekut, L., Talanian, R., Paskind, M., Wong, W., Kamen, R., Tracey, D. and Allen, H. (1997) Caspase-1 processes IFN-γ-inducing factor and regulates LPS-induced IFN-γ production. *Nature*, **386**, 619–623.

Goldberg, Y.P., Nicholson, D.W., Rasper, D.M., Kalchman, M.A., Koide, N.A., Vaillancourt, J.P. and Hayden, M.R. (1996) Cleavage of huntingtin by apopain, a proapoptotic cysteine protease, is modulated by the polyglutamine tract. *Nat. Genet.*, **13**, 442–449.

Greenberg, A.H. (1996) Activation of apoptosis pathways by granzyme B. *Cell Death Diff.*, **3**, 269–274.

Gu, Y., Sarnecki, C., Aldape, R.A., Livingston, D.J. and Su, M.S. (1995) Cleavage of poly(ADP-ribose) polymerase by interleukin-1 beta converting enzyme and its homologs TX and Nedd-2. *J. Biol. Chem.*, **270**, 18715–18718.

Gu, Y., Kuida, K., Tsutsui, H., Ku, G., Hsiao, K., Fleming, M.A., Hayashi, N., Higashino, K., Okamura, H., Nakanishi, K., Kurimoto, M., Tanimoto, T., Flavell, R.A., Sato, V., Harding, M.W., Livingston, D.J. and Su, M.S.-S. (1997) Activation of interferon-γ inducing factor mediated by interleukin-1β converting enzyme. *Science*, **275**, 206–209.

Harvey, N.L., Trapani, J.A., Fernandes-Alnemri, T., Litwack, G., Alnemri, E.S. and Kumar, S. (1996) Processing of the Nedd2 precursor by ICE-like proteases and granzyme B. *Genes to cells*, **1**, 673–685.

Harvey, N.L., Butt, A.J. and Kumar, S. (1997) Functional activation of Nedd2/ICH1 (caspase-2) is an early process in apoptosis. *J. Biol. Chem.*, **272**, 13134–13140.

Hengartner, M.O., Ellis, R.E. and Horvitz, H.R. (1992) Caenorhabditis elegans gene ced-9 protects cells from programmed cell death. *Nature*, **356**, 494–499.

Hengartner, M.O. and Horvitz, H.R. (1994a) Programmed cell death in Caenorhabditis elegans. *Curr. Opi. Genet. Dev.*, **4**, 581–586.

Hengartner, M.O. and Horvitz, H.R. (1994b) C. elegans cell survival gene ced-9 encodes a functional homolog of the mammalian proto-oncogene bcl-2. *Cell*, **76**, 665–676.

Hsu, H., Xiong, J. and Goeddel, D.V. (1995) The TNF receptor 1-associated protein TRADD signals cell death and NF-kappaB activation. *Cell*, **81**, 495–504.

James, C., Gschmeissner, S., Fraser, A. and Evan, G.I. (1997) CED-4 induces chromatin condensation in Schizosaccharomyces pombe and is inhibited by direct physical association with CED-9. *Curr. Biol.*, **7**, 246–252.

Janicke, R.U., Walker, P.A., Lin, X.Y. and Porter, A.G. (1996) Specific cleavage of the retinoblastoma protein by an ICE-like protease in apoptosis. *EMBO J.*, **15**, 101–110.

Kamens, J., Paskind, M., Hugunin, M., Talanian, R.V., Allen, H., Banach, D., *et al.* (1995) Identification and characterisation of ICH-2, a novel member of the interleukin-1β-converting enzyme family of cysteine proteases. *J. Biol. Chem.*, **270**, 15250–15256.

Kaufmann, S.H. (1989) Induction of endonucleolytic DNA cleavage in human acute myelogenous leukemia cells by etoposide, camptothecin and other cytotoxic anticancer drugs: a cautionary note. *Cancer Res.*, **49**, 5870–5878.

Kaufmann, S.H., Desnoyers, S., Ottaviano, Y., Davidson, N.E. and Poirier, G.G. (1993) Specific proteolytic cleavage of poly(ADP-ribose) polymerase: an early marker of chemotherapy-induced apoptosis. *Cancer Res.*, **53**, 3976–3985.

Kluck, R.M., Bossy-Wetzel, E., Green, D.R. and Newmeyer, D.D. (1997) The release of cytochrome c from mitochondria: a primary site for bcl-2 regulation of apoptosis. *Science*, **275**, 1132–1136.

Komiyama, T., Ray, C.A., Pickup, D.J., Howard, A.D., Thornberry, N.A. and Peterson, E.P. (1994) Inhibition of interleukin-1 β converting enzyme by the cowpox virus serpin CrmA. An example of cross-class inhibition. *J. Biol. Chem.*, **269**, 19331–19337.

Korsmeyer, S.J. (1992) Bcl-2 initiates a new category of oncogenes. *Blood*, **80**, 879–886.

Kuida, K., Lippke, J.A., Ku, G., Harding, M.W., Livingston, D.J., Su, M.S., *et al.* (1995) Altered cytokine export and apoptosis in mice deficient in interleukin-1β converting enzyme. *Science*, **267**, 2000–2003.

Kuida, K., Zheng, T.S., Na, S., Kuan, C., Yang, D., Karasuyama, H., *et al.* (1996) Decreased apoptosis in the brain and premature lethality in CPP32-deficient mice. *Nature*, **284**, 368–372.

Kumar, S., Kinoshita, M., Noda, M., Copeland, N.G. and Jenkins, N.A. (1994) Induction of apoptosis by the mouse Nedd2 gene, which encodes a protein similar to the product of the Caenorhabditis elegans cell death gene ced-3 and the mammalian IL-1β-converting enzyme. *Genes Dev.*, **8**, 1613–1626.

Kumar, S. (1995) Inhibition of apoptosis by the expression of antisense Nedd2. *FEBS Letters*, **368**, 69–72.

Kumar, S., Kinoshita, M., Dorstyn, L. and Noda, M. (1997) Origin, expression and possible functions of the two alternatively spliced forms of the mouse Nedd2 mRNA. *Cell Death Diff.*, 378–387.

Lazebnik, Y.A., Kaufmann, S.H., Desnoyers, S., Poirier, G.G. and Earnshaw, W.C. (1994) Cleavage of poly(ADP-ribose) polymerase by a proteinase with properties like ICE. *Nature*, **371**, 346–347.

Lazebnik, Y.A., Takahashi, A., Moir, R.D., Goldman, R.D., Poirier, G.G. and Kaufmann, S.H. (1995) Studies of the lamin proteinase reveal multiple parallel biochemical pathways during apoptotic execution. *Proc. Natl. Acad. Sci. USA*, **92**, 9042–9046.

Li, P., Allen, H., Banerjee, S., Franklin, S., Herzog, L., Johnston, C., *et al.* (1995) Mice deficient in IL-1 beta-converting enzyme are defective in production of mature IL-1 beta and resistant to endotoxic shock. *Cell*, **80**, 401–411.

Liu, X., Kim, C.N., Yang, J., Jemmerson, R. and Wang, X. (1996) Induction of apoptosis program in cell-free extracts: requirement for dATP and cytochrome c. *Cell*, **86**, 147–157.

Liu, X., Zou, H., Slaughter, C. and Wang, X. (1997) DFF, a heterodimeric protein that functions downstream of caspase-3 to trigger DNA fragmentation during apoptosis. *Cell*, **89**, 175–184.

Los, M., Van de Craen, M., Penning, L.C., Schenk, H., Westendorp, M., Baeuerle, P.A., *et al.* (1995) Requirement of an ICE/CED-3 protease for Fas/APO-1-mediated apoptosis. *Nature*, **375**, 81–83.

MacFarlane, M., Cain, K., Sun, X.M., Alnemri, E.S. and Cohen, G.M. (1997) Processing/activation of at least four interleukin-1 beta converting enzyme-like proteases occurs during the execution phase of apoptosis in human monocytic tumor cells. *J. Cell Biol.*, **137**, 469–479.

Mariani, S.M., Matiba, B., Armandola, E.A. and Krammer, P.H. (1997) Interleukin 1β-converting enzyme related proteases/caspases are involved in TRAIL-induced apoptosis of myeloma and leukemia cells. *J. Cell Biol.*, **137**, 221–229.

Martin, S.J., O'Brien, G.A., Nishioka, W.K., McGahon, A.J., Mahboubi, A., Saido, T.C., *et al.* (1995) Proteolysis of fodrin (non-erythroid spectrin) during apoptosis. *J. Biol. Chem.*, **270**, 6425–6428.

Martin, S.J., Amarante Mendes, G.P., Shi, L., Chuang, T.H., Casiano, C.A., O'Brien, G.A., *et al.* (1996) The cytotoxic cell protease granzyme B initiates apoptosis in a cell-free system by proteolytic processing and activation of the ICE/CED-3 family protease, CPP32, via a novel two-step mechanism. *EMBO J.*, **15**, 2407–2416.

Martinou, I., Fernandez, P.A., Missotten, M., White, E., Allet, B., Sadoul, A., *et al.* (1995) Viral proteins E1B19K and p35 protect sympathetic neurons from cell death induced by NGF deprivation. *J. Cell Biol.*, **128**, 201–208.

Miura, M., Zhu, H., Rotello, R., Hartwieg, E.A. and Yuan, J. (1993) Induction of apoptosis in fibroblasts by IL-1 β-converting enzyme, a mammalian homolog of the C. elegans cell death gene ced-3. *Cell*, **75**, 653–660.

Miura, M., Friedlander, R.M. and Yuan, J. (1995) Tumor necrosis factor-induced apoptosis is mediated by a CrmA- sensitive cell death pathway. *Proc. Natl. Acad. Sci. USA*, **92**, 8318–8322.

Munday, N.A., Vaillancourt, J.P., Ali, A., Casano, F.J., Miller, D.K., Molineaux, S., *et al.* (1995) Molecular cloning and pro-apoptotic activity of ICE_{rel}II and ICE_{rel}III, members of the ICE/CED3 family of cysteine proteases. *J. Biol. Chem.*, **270**, 15870–15876.

Muzio, M., Chinnaiyan, A.M., Kischkel, F.C., O'Rourke, K., Shevchenko, A., Ni, J., *et al.* (1996) FLICE, a novel FADD-homologous ICE/CED-3-like protease, is recruited to the CD95 (Fas/APO-1) death-inducing signalling complex. *Cell*, **85**, 817–827.

Muzio, M., Salvesen, G.S. and Dixit, V.M. (1997) FLICE induced apoptosis in a cell-free system. *J. Biol. Chem.*, **272**, 2952–2956.

Na, S., Chuang, T.H., Cunningham, A., Turi, T.G., Hanke, J.H., Bokoch, G.M. and Danley, D.E. (1996) D4-GDI, a substrate of CPP32, is proteolysed during Fas-induced apoptosis. *J. Biol. Chem.*, **271**, 11209–11213.

Nicholson, D.W., Ali, A., Thornberry, N.A., Vaillancourt, J.P., Ding, C.K., Gallant, M., *et al.* (1995) Identification and inhibition of the ICE/CED-3 protease necessary for mammalian apoptosis. *Nature*, **376**, 37–43.

Orth, K., Chinnaiyan, A.M., Garg, M., Froelich, C.J. and Dixit, V.M. (1996) The CED-3/ICE-like protease Mch2 is activated during apoptosis and cleaves the death substrate lamin A. *J. Biol. Chem.*, **271**, 16443–16446.

Pan, G., O'Rourke, K., Chinnaiyan, A.M., Gentz, R., Ebner, R., Ni, J., *et al.* (1997) The receptor for the cytotoxic ligand TRAIL. *Science*, **276**, 111–113.

Perry, D.K., Smyth, M.J., Wang, H-G., Reed, J.C., Poirier, G.G., Obeid, L.M. and Hannun, Y.A. (1997) Bcl-2 acts upstream of the PARP protease and prevents its activation. *Cell Death Diff.*, **4**, 29–33.

Peterson, S.R., Kurimasa, A., Oshimura, M., Dynan, W.S., Bradbury, E.M. and Chen, D.J. (1995) Loss of the catalytic subunit of the DNA-dependent protein kinase in DNA double-strand-break-repair mutant mammalian cells. *Proc. Natl. Acad. Sci. USA*, **92**, 3171–3174.

Pitti, R.M., Marsters, S.A., Ruppert, S., Donahue, C.J., Moore, A. and Ashkenazi, A. (1996) Induction of apoptosis by Apo-2 ligand, a new member of the tumor necrosis factor cytokine family. *J. Biol. Chem.*, **271**, 12687–12690.

Rabizadeh, S., LaCount, D.J., Friesen, P.D. and Bredsen, D.E. (1993) Expression of the baculovirus p35 gene inhibits mammalian neural cell death. *J. Neurochem.*, **61**, 2318–2321.

Ramage, P., Cheneval, D., Chvei, M., Graff, P., Hemmig, R., Heng, R., *et al.* (1995) Expression, refolding and autocatalytic proteolytic processing of the interleukin-1-β converting enzyme precursor. *J. Biol. Chem.*, **270**, 9378–9383.

Ray, C.A., Black, R.A., Kronheim, S.R., Greenstreet, T.A., Sleath, P.R. and Salvesen, G.S. (1992) Viral inhibition of inflammation: cowpox virus encodes an inhibitor of the interleukin-1 β converting enzyme. *Cell*, **69**, 597–604.

Rotonda, J., Nicholson, D.W., Fazil, K.M., Gallant, M., Gareau, Y., Labelle, M., *et al.* (1996) The three-dimensional structure of apopain/CPP32, a key mediator of apoptosis. *Nat. Struct. Biol.*, **3**, 619–625.

Satoh, M.S. and Lindahl, T. (1992) Role of poly(ADP-ribose) formation in DNA repair. *Nature*, **356**, 356–358.

Schlegel, J., Peters, I., Orrenius, S., Miller, D.K., Thornberry, N.A. and Yamin, T.T. (1996) CPP32/apopain is a key interleukin 1β converting enzyme-like protease involved in Fas-mediated apoptosis. *J. Biol. Chem.*, **271**, 1841–1844.

Shaham, S. and Horvitz, H.R. (1996) Developing Caernorhabditis elegans neurons may contain both cell-death protective and killer activities. *Genes Dev.*, **10**, 578–591.

Slee, E.A., Zhu, H., Chow, S.C., MacFarlane, M., Nicholson, D.W. and Cohen, G.M. (1996) Benzyloxycarbonyl-Val-Ala-Asp (OMe) fluoromethylketone (z-VAD.FMK) inhibits apoptosis by blocking the processing of CPP32. *Biochem. J.*, **315**, 21–24.

Smyth, M.J. and Trapani, J.A. (1995) Granzymes: exogenous proteinases that induce target cell apoptosis. *Immunol. Today*, **16**, 202–206.

Song, Q., Lees-Miller, S.P., Kumar, S., Zhang, N., Chan, D.W., Smith G.C.M. *et al.* (1996a) Interleukin-1 β-converting enzyme-like protease cleaves DNA- dependent protein kinase in cytotoxic T cell killing. *J. Exp. Med.*, **184**, 619–626.

Song, Q., Burrows, S.R., Smith, G., Lees-Miller, S.P., Kumar, S., Chan, D.W., *et al.* (1996b) DNA-dependent protein kinase catalytic subunit: a target for an ICE-like protease in apoptosis. *EMBO J.*, **15**, 3238–3246.

Song, Q., Lu, H., Zhang, N., Luckow, B., Shah, G., Poirier, G., *et al.* (1997) Specific cleavage of the large subunit of replication factor C in apoptosis is mediated by CPP32-like protease. *Biochem. Biophys. Res. Commun.*, **233**, 343–348.

Song, Z., McCall, K. and Steller, H. (1997) DCP-1, a Drosophila cell death protease essential for development. *Science*, **275**, 536–540.

Spector, M.S., Desnoyers, S., Hoeppner, D.J. and Hengartner, M.O. (1997) Interaction between the C. elegans cell-death regulators CED-9 and CED-4. *Nature*, **385**, 653–656.

Srinivasula, S.M., Ahmad, M., Fernandes-Alnemri, T., Litwack, G. and Alnemri, E.S. (1996a) Molecular ordering of the fas-apoptotic pathway: the fas/APO-1 protease Mch 5 is a CrmA-inhibitable protease that activates multiple Ced-3/ICE-like cysteine proteases. *Proc. Natl. Acad. Sci. USA*, **93**, 14486–14491.

Srinivasula, S.M., Fernandes-Alnemri, T., Zangrilli, J., Robertson, N., Armstrong, R.C., Wang, L., *et al.* (1996b) The ced-3/interleukin 1 beta converting enzyme-like homolog Mch6 and the lamin-cleaving enzyme Mch2 alpha are substrates for the apoptotic mediator CPP32. *J. Biol. Chem.*, **271**, 27099–27106.

Stanger, B.Z., Leder, P., Lee, T., Kim, E. and Seed, B. (1995) RIP: a novel protein containing a death domain that interacts with Fas/APO-1 (CD95) in yeast and causes cell death. *Cell*, **81**, 513–523.

Strasser, A., Harris, A.W., Huang, D.C.S., Krammer, P.H. and Cory, S. (1995) Bcl-2 and Fas/APO-1 regulate distinct pathways to lymphocyte apoptosis. *EMBO J.*, **14**, 6136–6147.

Sugimoto, A., Friesen, P.D. and Rothman, J.H. (1994) Baculovirus p35 prevents developmentally programmed cell death and rescues a ced-9 mutant in the nematode C. elegans. *EMBO J.*, **13**, 2023–2028.

Takahashi, A., Alnemri, E.S., Lazebnik, Y.A., Fernandes-Alnemri, T., Litwack, G., Goldman, R.D., *et al.* (1996) Cleavage of lamin A by Mch2 alpha but not CPP32: multiple interleukin 1β-converting enzyme-related proteases with distinct substrate recognition properties are active in apoptosis. *Proc. Natl. Acad. Sci. USA*, **93**, 8395–8400.

Tazi, J., Kornstadt, U., Rossi, F., Jeanteur, P., Cathala, G., Brunel, C., *et al.* (1993) Thiophosphorylation of U1-70k protein inhibits pre-mRNA splicing. *Nature*, **363**, 283–286.

Tewari, M., Beidler, D.R. and Dixit, V.M. (1995a) CrmA-inhibitable cleavage of the 70-kDa protein component of the U1 small nuclear ribonucleoprotein during Fas- and TNF-induced apoptosis. *J. Biol. Chem.*, **270**, 18738–18741.

Tewari, M., Quan, L.T., O'Rourke, K., Desnoyers, S., Zeng, Z., Beidler, D.R., *et al.* (1995b) Yama/CPP32 beta, a mammalian homolog of CED-3, is a CrmA- inhibitable protease that cleaves the death substrate poly(ADP- ribose) polymerase. *Cell*, **81**, 801–809.

Tewari, M., Telford, W.G., Miller, R.A. and Dixit, V.M. (1995c) CrmA, a poxvirus-encoded serpin, inhibits cytotoxic T-lymphocyte- mediated apoptosis. *J. Biol. Chem.*, **270**, 22705–22708.

Tewari, M. and Dixit, V.M. (1995) Fas- and tumor necrosis factor-induced apoptosis is inhibited by the poxvirus crmA gene product. *J. Biol. Chem.*, **270**, 3255–3260.

Thome, M., Schneider, P., Hofmann, K., Fickenscher, H., Meini, E., Neipel, F., *et al.* (1997) Viral FLICE-inhibitory proteins (FLIPS) prevent apoptosis induced by death receptors. *Nature*, **386**, 517–521.

Thornberry, N.A., Bull, H.G., Calaycay, J.R., Chapman, K.T., Howard, A.D., Kostura, M.J., *et al.* (1992) A novel heterodimeric cysteine protease is required for interleukin-1 β processing in monocytes. *Nature*, **356**, 768–774.

Troy, C.M., Stefanis, L., Greene, L.A. and Shelanski, M.L. (1997) Nedd2 is required for apoptosis after trophic factor withdrawal, but not superoxide dismutase (SOD1) downregulation, in sympathetic neurons and PC12 cells. *J. Neurosci.*, **17**, 1911–1918.

Van Criekinge, W., Beyaert, R., Van de Craen, M., Vandenabeele, P., Schotte, P., De Valck, D., *et al.* (1996) Functional characterisation of the prodomain of interleukin-1β-converting enzyme. *J. Biol. Chem.*, **271**, 27245–27248.

Van de Craen, M., Vandenabeele, P., Declercq, W., Van den Brande, I., Van Loo, G., Molemans, F., *et al.* (1997) Characterisation of seven murine caspase family members. *FEBS Lett.*, **403**, 61–69.

Varfolomeev, E.E., Boldin, M.P., Goncharov, T.M. and Wallach, D. (1996) A potential mechanism of "cross-talk" between the p55 tumor necrosis factor receptor and Fas/APO1: proteins binding to the death domains of the two receptors also bind to each other. *J. Exp. Med.*, **183**, 1271–1275

Vaux, D.L., Weissman, I.L. and Kim, S.K. (1992) Prevention of programmed cell death in C. elegans by human bcl-2. *Science*, **258**, 1955–1957.

Vincenz, C. and Dixit, V.M. (1997) Fas-associated death domain protein interleukin-1β-converting enzyme 2 (FLICE2), an ICE/Ced-3 homologue, is proximally involved in CD95- and p55-mediated death signalling. *J. Biol. Chem.*, **272**, 6578–6583.

Walker, N.P., Talanian, R.V., Brady, K.D., Dang, L.C., Bump, N.J., Ferenz, C.R., *et al.* (1994) Crystal structure of the cysteine protease interleukin-1β-converting enzyme: a $(p20/p10)_2$ homodimer. *Cell*, **78**, 343–352.

Wang, L., Miura, M., Bergeron, L., Zhu, H. and Yuan, J. (1994) Ich-1, an ICE/ced-3-related gene, encodes both positive and negative regulators of programmed cell death. *Cell*, **78**, 739–750.

Wang, S., Miura, M., Jung, Y., Zhu, H., Gagliardini, V., Shi, L., *et al.* (1996) Identification and characterisation of Ich-3, a member of the interleukin-1β converting enzyme (ICE)/ced 3 family and an upstream regulator of ICE. *J. Biol. Chem.*, **271**, 20580–20587.

Wang, X., Pai, J.T., Wiedenfeld, E.A., Medina, J.C., Slaughter, C.A. and Goldstein, J.L. (1995) Purification of an interleukin-1 beta converting enzyme-related cysteine protease that cleaves sterol regulatory element-binding proteins between the leucine zipper and transmembrane domains. *J. Biol. Chem.*, **270**, 18044–18050.

Wang, X., Zelenski, N.G., Yang, J., Sakai, J., Brown, M.S. and Goldstein, J.L. (1996) Cleavage of sterol regulatory element binding proteins (SREBPs) by CPP32 during apoptosis. *EMBO J.*, **15**, 1012–1020.

Waterhouse, N., Kumar, S., Song, Q., Strike, P., Sparrow, L., Dreyfuss, G., Alnemri, E.S., Litwack, G., Lavin, M. and Watters, D. (1996) Heteronuclear ribonucleo proteins C1 and C2 components of the spliceosome, are specific targets of interleukin 1β-converting enzyme-like proteases in apoptosis. *J. Biol. Chem.*, **271**, 29335–29341.

Wilson, K.P., Black, J.F., Thomson, J.A., Kim, E.E., Griffith, J.P., Navia, M.A., *et al.* (1994) Structure and mechanism of interleukin-1β converting enzyme. *Nature*, **370**, 270–275.

Wu, D., Wallen, H.D. and Nunez, G. (1997) Interaction and regulation of subcellular localisation of CED-4 by CED-9. *Science*, **275**, 1126–1129.

Xue, D., Shaham, S. and Horvitz, H.R. (1996) The Caenorhabditis elegans cell-death protein CED-3 is a cysteine protease with substrate specificities similar to those of the human CPP32 protease. *Genes Dev.*, **10**, 1073–1083.

Yang, J., Liu, X., Bhalla, K., Kim, C.N., Ibrado, A.M., Cai, J., *et al.* (1997) Prevention of apoptosis by bcl-2: release of cytochrome c from mitochondria blocked. *Science*, **275**, 1129–1132.

Yuan, J., Shaham, S., Ledoux, S., Ellis, H. and Horvitz, H.R. (1993) The C. elegans cell death gene ced-3 encodes a protein similar to mammalian interleukin-1β-converting enzyme. *Cell*, **75**, 641–652.

Zhang, J., Dawson, V.L., Dawson, T.M. and Snyder, S.H. (1994) Nitric oxide activation of poly (ADP-ribose) synthetase in neurotoxicity. *Science*, **263**, 687–689.

11. KILLER CELLS – DELIVERERS OF EXOGENOUS DEATH PROTEASES

MARK J. SMYTH, VIVIEN R. SUTTON AND JOSEPH A. TRAPANI[*†]

[*]*Austin Research Institute, Studley Road, Heidelberg 3084, Australia*

Cytotoxic T lymphocytes and natural killer cells together constitute the means by which the immune system detects and clears higher organisms of virus-infected or transformed cells. These killer cells use the same methods for inducing target cell death, despite differing significantly in the way they detect antigens. Effector lymphocytes use two distinct contact-dependent cytolytic mechanisms. The first, granule-exocytosis, depends on the transfer of effector granule proteins into the target cell and in particular, the synergistic action of a calcium-dependent pore-forming protein, perforin, and a family of granule serine proteases (granzymes). The second mechanism requires binding of effector ligand trimer [Fas ligand or tumor necrosis factor] with trimeric Fas or tumor necrosis factor receptor molecules on sensitive target cells. By contrast, this mechanism is calcium-independent and is triggered by the generation of a death signal at the target cell plasma membrane. Recent progress has indicated that both effector pathways impinge on an endogenous signalling cascade that is strongly conserved in species as diverse as helminths and humans and this pathway dictates the death or survival of all cells.

INTRODUCTION

This chapter will discuss the mechanisms by which cytolytic lymphocytes (CL) induce death in pathogen-infected or mutated cells. In the context of this review, CL will primarily constitute cytotoxic T lymphocytes (CTL) and natural killer (NK) cells. CTL are generally $CD3^+CD8^+$ T cells, and express cytocidal molecules in an inducible manner following T-cell receptor ligation by antigen. The nature of the receptors used by T lymphocytes to specifically detect foreign (and some self) antigens has been elucidated, however T cell receptor structure and activation are beyond the scope of this review. By comparison, NK cells recognise non-self antigen by the use of a very different array of facilitatory and inhibitory receptors, which have become the topic of intense recent investigation (for reviews see Karre, 1997; Lanier *et al.*, 1997; Selvakumar *et al.*, 1997). NK cells display a broader spectrum of antigenic specificity. In keeping with their role in the early response to virus infection, NK cells express lytic mediators and important cytokines, such as interferon-γ (IFN-γ), thus either eradicating or at least confining the infection until a specific T cell response can be mounted. The granule exocytosis mechanism (see below) appears to be preferentially used by the $CD8^+$ CTL and NK cells, and this is consistent with the central role of these cells in eliminating pathogen-infected cells. This review will concentrate on the granule exocytosis mechanism rather than Fas ligand (FasL)-/ tumor necrosis factor (TNF)-mediated death which is primarily responsible for the maintenance of T cell populations (homeostasis

† *Corresponding Author*: Tel.: 61-3-92870651. Fax: 61-3-9287 0600. e-mail: ja-trapani @ muwayf. unimelb. edu. au

after activation and during selection), and has been covered in an earlier chapter on death receptors. There is clearly some overlap in the events occurring in cell death induced by both pathways, and thus some general discussion will be devoted to the cell death process itself and how granule exocytosis impinges upon it.

TWO BASIC MECHANISMS OF CL-INDUCED DEATH

A brief history

The granule-exocytosis mechanism was the first CL mechanism to be characterised principally on ultrastructural studies (Zagury *et al.*, 1975; Bykovskaja *et al.*, 1978; Geiger *et al.*, 1982; Yannelli *et al.*, 1986; Young and Cohn, 1986; Kupfer, 1991). Transient direct contact between the effector and the target (conjugate formation) was demonstrated and cytotoxic granule reorientation preceded death of the target cell. The CTL was shown to be refractory to death and indeed could recycle to additional target cells. Discrete lesions were observed appearing specifically on the target cell membrane following an encounter with a CTL (Doumarshkin *et al.*, 1980). Podack et al. then confirmed these observations, identified the target membrane lesions as being remarkably similar to those induced by complement (Podack and Dennert, 1983), and demonstrated that a purified protein, present in high concentration within the presynaptic granules, was by itself capable of inducing changes in target cell membrane permeability. That protein was termed 'perforin' and this work embodied the hypothesis that following antigenic recognition and stabilisation of binding by adhesion structures, prestored lytic mediators could be released by the CL in a vectorial fashion toward the target cell surface (Dennert and Podack, 1983). A larger body of evidence supporting the granule-exocytosis hypothesis rapidly accumulated (detailed below) and the simplicity of the mechanism was inherently appealing.

A number of studies questioned the granule exocytosis hypothesis (see for example Berke, 1991; Krahenbuhl and Tschopp, 1991). Calcium ions are known to be indispensable both for exocytosis and the pore-forming activity of perforin, yet peritoneal exudate lymphocytes (PEL) lysed target cells in the absence of calcium. Some very cytotoxic CTL produced little if any perforin, were non-granulated, and could kill in the absence of exocytosis (Dennert *et al.*, 1987; Berke and Rosen, 1988; Berke, 1989; Allbritton *et al.*, 1988; Ostergaard *et al.*, 1987; Ostergaard and Clark, 1989; Trenn *et al.*, 1987). Thirdly, target cells frequently died by "internal disintegration" (apoptosis), involving nuclear collapse and DNA fragmentation (Russell, 1983; see also Schmid *et al.*, 1986), yet purified perforin alone did not cause these events (Duke *et al.*, 1989). We now know that many of these paradoxes arose because of a previously unidentified second, granule-independent, mechanism mediated by Fas-ligation.

Although not yet complete, the resolution of these mechanisms has been aided by advances in molecular technologies that have now provided indisputable evidence for these two independent cytolytic pathways. The relative biological significance of each of these mechanisms is now under intense study using genetic techniques, particularly gene knock-out mice. At least *in vitro*, the sum total of cytotoxicity seen in killer cell assays is accounted for by the total effects of CL perforin- and FasL/TNF-mediated mechanisms (Lowin *et al.*, 1994a, Table 11.1). The two mechanisms can operate independently of one another, as the FasL mechanism appears to function normally in perforin-deficient CL (accounting for the residual cytotoxicity), while the CTL and NK cells of mice that possess natural mutations of the FasL/Fas mechanism have apparently normal cytolytic granules and express perforin and granzymes in normal amounts (see below). *In vivo*, studies using perforin gene knockout mice, have settled beyond doubt that granule-mediated cytolysis is the predominant mechanism for protection against some non-cytopathic viruses and intracellular bacteria, the principal means for eliminating alloreactive cells, and the dominant mechanism used by NK cells in tumor surveillance (Kagi *et al.*, 1994a; Kagi *et al.*, 1994b; Kagi *et al.*, 1995; Lowin *et al.*, 1994b; Kojima *et al.*, 1994; Walsh *et al.*, 1994; van den Broek *et al.*, 1995; van den Broek *et al.*, 1996). By contrast, natural mutants and gene knockout mice deficient in FasL or TNF-mediated death have demonstrated that this form of CL-mediated death is primarily responsible for negative selection of T cells in the thymus (Castro *et al.*, 1996) and T cell homeostasis following activation by foreign antigen (Singer and Abbas, 1994; Dhein *et al.*, 1995).

Table 11.1 Two distinct mechanisms of CTL / NK-mediated cytolysis.

	granule exocytosis	*FasL /Fas*
Lymphocyte subsets	CD8+ T CD4+ Th2 NK	CD8+ CD4+ Th1
Functions *in vivo*	viral clearance tumor & allograft rejection NK function	T cell homeostasis in development or following infection
Ca^{2+} dependent?	YES	NO
Kinetics of cytolysis	rapid (<4 hours)	rapid (<4 hours)
Natural mutation	not described	autoimmune lymphoproliferative syndrome (ALPS) in humans *gld* (FasL deficiency in mice) *lpr* (Fas deficiency in mice)
Caspase dependence	nuclear-yes; non-nuclear-no	completely dependent
Mechanism of death	apoptosis	apoptosis

GRANULE-EXOCYTOSIS MEDIATED CELL DEATH

Cytotoxic Granules

Lymphocyte cytotoxic granules harbour all the major components that upon exocytosis (Table 11.2), together synergise to inflict target cell death. These granules contain proteins of many types, some whose function remains to be defined (Smyth *et al.*, 1996c). This review will focus on perforin and granzymes which are the key components localised both within the dense core and the peripheral regions of virtually all CL cytotoxic granules (Peters *et al.*, 1991). With only a few exceptions, perforin and granzymes are synthesised specifically by CL. Perforin and granzymes may be liberated from cells both as soluble molecules, and additionally in a membrane-encapsulated form. Granzymes, proteoglycans and perforin are all secreted into the intercellular cleft (Schmidt *et al.*, 1985; Takayama *et al.*, 1987; Jenne and Tschopp, 1988), and their binding to the distal membrane can be observed by cinemicrography and electron microscopy (Yannelli *et al.*, 1986). Ring-like tubular lesions measuring up to 18 nm in diameter and apparently identical to those formed by purified perforin are seen to form in the membrane following attack by CTL clones *in vitro* (Dennert and Podack, 1983). Purified granules can elicit the changes of both membranolysis and apoptosis in a dose-dependent manner and with no particular specificity (Tschopp and Nabholz, 1990; Smyth and Trapani, 1995).

Table 11.2 Molecules isolated from cytolytic granules.

Molecule	*Identified function/s*
Membrane disruptive agents	
Perforin	osmotic lysis; apoptosis induction, in synergy with granzymes; cellular and nuclear granzyme uptake
Granulysin	pore-formation; likely anti-bacterial agent
Granzymes (CL-specific proteases)	
granzymes A-H, J, M,-3, tryptase 2	many and varied: granzyme B is especially implicated in apoptosis induction; other granzymes, especially granzyme A have had a plethora of non-apoptotic functions described (see text)
Other molecules	
proteoglycans	complex granzymes and perforin until release
calreticulin	bind free calcium
DPP1 (cathepsin C)	granzyme activation (cleavage of activation pro-dipeptide)
TIA-1, TIA-R	mRNA binding during cell stress and apoptosis
Lysosomal components	
mannose-6-phosphate receptor	lysosomal targeting
lamp-1, lamp-2, CD63	lysosomal constituents
arylsulfatase cathepsin D β-glucuronidase β-hexosamidase	lysosomal enzymes

Perforin

Discovery

Perforin was purified from CL cytolytic granules and demonstrated to be capable of forming barrel-shaped transmembrane lesions in a calcium-dependent manner (Podack and Dennert, 1983; Dennert and Podack, 1983; Masson and Tschopp, 1985; Podack *et al.*, 1985; Young and Cohn, 1986). The fact that perforin was exocytosed from CTL and could attach to target cell membranes (Podack and Dennert, 1983) argued strongly for the validity of the lytic model of cell death. The cloning of perforin cDNAs (Lowrey *et al.*, 1989; Kwon *et al.*, 1989; Shinkai *et al.*, 1988; Lichtenheld *et al.*, 1988; Shinkai *et al.*, 1989; Ishikawa *et al.*, 1989) predicted an overall similarity between perforin and the final component of the membrane attack complex (MAC) of complement, C9.

A predicted structure

The 100 amino terminal and 150 carboxy terminal residues of perforin are quite unique, however a 300 amino acid stretch in the centre of the perforin sequence (~ residues 100 to 400) show a degree of overall identity with the terminal complement components C6-C9 (about 22%). In particular, two smaller regions within this large central domain have considerably higher similarity to the MAC proteins (eg. between residues 190 to 220), which were thought to subserve the function of membrane insertion, as do the corresponding residues of the complement proteins (Lowrey *et al.*, 1989; Kwon *et al.*, 1989; Liu *et al.*, 1995; Podack, 1989; Podack, 1992). Interestingly however, strong evidence has recently emerged that the carboxy terminal domain is the site of calcium ion binding and initiation of insertion into the lipid bilayer (see below). The second strongly conserved domain is the region between residues 355 and 388 of perforin, which has similarity to the epidermal growth factor (EGF)-like repeat domains also found in the MAC proteins. It should be stressed that no functional role of any of the domains of perforin has been clearly demonstrated experimentally, and proposed roles for various regions rest very much on analogy with the complement proteins.

Insertion into the target membrane lipid bilayer

In the presence of ~1 mM Ca^{2+}, perforin can bind to various lipid molecules directly, provided a phosphorylcholine headgroup is present (Tschopp *et al.*, 1989) and binding is inhibitable by various lipid moieties (Yue *et al.*, 1987). No distinct receptor molecule has ever been described for perforin, and if such a molecule does exist, one would predict it would have to be both abundant in the cell membrane and ubiquitously expressed. Following binding to the plasma membrane, it is likely that a marked conformational change occurs in the perforin molecule, resulting in the exposure of amphipathic alpha-helical regions (possibly residues 190-220). On the basis of molecular modelling, this region is believed to insert into the lipid bilayer by virtue of its contralateral hydrophilic and hydrophobic surfaces. Thus perforin monomers are probably nascent "barrel-stave" structures, that form perforin lesions of homopolymers from as few as 3 to 4 coalesced monomers (Liu *et al.*, 1995). The region that subserves polymerisation function is considered likely to be the EGF-like cysteine-rich domain (residues 355-388). In light of more recent data , it is paradoxical that the amino terminal domain of perforin (the amino-terminal 34 residues of perforin) was shown to cause cell lysis of nucleated and non-nucleated cells and liposomes (Ojcius *et al.*, 1991a). It remains to be seen whether this region is of physiological significance, since baculovirus deletion mutants lacking this region were still able to lyse red cells (Liu *et al.*, 1996).

Interest in the likely function of the carboxy-terminus of perforin has recently been heightened (Uellner *et al.*, 1997). It has not been clear why the membranes of the endoplasmic reticulum and Golgi were not injured by premature activation of perforin during its biosynthesis. Griffiths and coworkers postulate that perforin is synthesized as an inactive precursor molecule, and that final processing to an active form takes place at the carboxy-terminus only under the acidic conditions found in the lytic granules. Unmodified perforin polypeptide (~60 kDa, the likely form in the endoplasmic reticulum) incurs the addition of complex glycans in the Golgi at two positions (one very close to the carboxy-terminus), resulting in a molecule of ~70 kDa. A smaller 'mature' form of ~65 kDa is then created by a proteolytic cleavage (or trimming) close to the carboxy-terminus. Although the exact site of this cleavage is unclear, it appears that a fragment of about 20 amino acids is removed, together with a large carbohydrate moiety attached to it (Uellner *et al.*, 1997). An acidic granule pH is required for this processing and this processing event may also explain how perforin can bind calcium ions. One region has been shown to have significant homology to C2-like calcium binding domains (of which the prototype is synaptotagmin), thought to subserve calcium dependent lipid binding (Shao *et al.*, 1996; Nalefski and Falke, 1996). Following cleavage of the carboxy-terminal region with its attached carbohydrate, the C2-like domain folds into two, 8-stranded β-sheets, bringing together at a single point a number of aspartate residues which, by virtue of their negative charges, can bind a positively charged calcium ion. Having bound calcium, this region of the molecule is now highly reactive, and in the presence of the appropriate lipid moieties can commence the process of attachment and intercalation into the plasma membrane (Uellner *et al.*, 1997). Polymerised perforin is unable to insert into lipid bilayers, so pre-activation by calcium results in rapid and irreversible loss of activity. It is therefore key that target cell membranes are exposed to perforin in the absence of serum, and the membrane-limited space between the effector and target cells probably provides a tightly-regulated "non-inhibitory" molecular environment prior to degranulation. It is still unclear how the membranes of CL are specifically protected from the lytic effects of perforin.

Granzymes

A large family of lymphocyte serine proteases

A family of lymphocyte-specific serine proteases termed "granule enzymes" or "granzymes" (Tschopp and Jongeneel, 1988) comprises about 90% of the protein within the cytolytic granules (Tschopp and Jongeneel, 1988; Henkart, 1987; Ojcius *et al.*, 1991b). The granzymes are related to the chymotrypsin family of serine proteases, and demonstrate structural similarities and genetic linkage to other leucocyte serine proteases, especially those expressed in mast cells and monocytes. The key features of serine proteases is their dependence for catalytic activity on an active site serine residue in a catalytic triad, an oxyanion hole to stabilise transition states of the enzyme-substrate complex, and a substrate-binding pocket, the configuration of which determines the specificity the individual serine protease (Kraut *et al.*, 1977; Smyth *et al.*, 1996a; Smyth *et al.*, 1996b). A total of eight granzymes (A-G, and M) have been identified in the mouse, however only five are identified in humans (A, B, H, M and protease-3). The rat shares granzymes A, B, C, F, M and tryptase-2 in common with humans and mice, but in addition expresses granzyme J (Ewoldt *et al.*, 1997a), and two other proteases of uncertain specificity designated RNK-P4 (most like C) and RNK-P7 (Ewoldt *et al.*, 1997b). No human equivalents of mouse granzymes C-G have been identified, while granzyme H appears to be specifically expressed by human cells (Smyth and Trapani, 1995; Trapani and Smyth, 1993).

Processing

Granzymes are produced as inactive precursor molecules (zymogens) and are fully processed only at the time of packaging into the lytic granules. The nascent granzymes polypeptide is a pre-pro-protein that is equipped with a typical leader sequences to enable transport through the endoplasmic reticulum and Golgi. Leader cleavage generally leaves two amino acids attached to the mature amino terminus (Caputo *et al.*, 1993; Smyth *et al.*, 1995b). It is likely that the final "activation pro-dipeptide" is normally clipped from the remainder of the polypeptide by dipeptidyl peptidase I (DPPI, also known as Cathepsin C), an enzyme expressed by myeloid cells and lymphocytes that also express granzyme-like serine proteases (McGuire *et al.*, 1993). Most granzymes are glycosylated (although heterogeneously) with mannose rich carbohydrates containing mannose-6-phosphate moieties which are important for accurate packaging into granules through the mannose-6-phosphate receptor pathway (Griffiths and Isaaz, 1993). However a mannose-6-phosphate-independent pathway also exists, and accounts for a minority of the granzyme protein ultimately secreted. The pH optimum of granzymes is ~7.5, so optimal activity is not reached until release from the secretory granules.

A variety of protease specificities

Granzymes have several defining characteristics (Smyth *et al.*, 1996a), including; a conserved N-terminus, a unique propeptide sequence, and conserved disulfide bridges. The granzymes also have unusually specific substrate preferences. Serine proteases such as trypsin, chymotrypsin and elastase generally have a very broad range of physiological substrates (ie., the surrounding amino acid context of the residue at the P_1 position) when compared with granzymes (Odake *et al.*, 1991). For example, granzyme B was found to cleave the substrates Boc-ala-ala-**asp**-Sbzl, Boc-ala-phe-**asp**-Sbzl and Boc-phe-ala-**asp**-Sbzl with moderate efficiency, however the substrate Z-**asp**-Sbzl was not cleaved (Odake *et al.*, 1991). This precision of proteolytic cleavage is consistent with a role for granzymes in processing rather than degrading target proteins. Clear peptide substrate specificities have been identified for granzymes A and 3 (tryptases), granzyme B (an "asp-ase", cleaving at asp and possibly glu) and granzyme M (a "met-ase", cleaving at met, but not phe) (Odake *et al.*, 1991; Sayers *et al.*, 1992; Smyth *et al.*, 1992; Smyth *et al.*, 1996a). Granzymes C-G are predicted to be chymases, and preferentially cleave synthetic substrates with phe, leu or asn in the P_1 position (Odake *et al.*, 1991). Although no crystal structure of a granzyme has been reported, molecular modelling based on the known structure of chymotrypsin, elastase (Bode *et al.*, 1989) and rat mast cell protease II (Remington *et al.*, 1988) has permitted a detailed analysis of the substrate binding pockets of granzymes B and M, and these predictions have been validated by mutational analysis (Smyth *et al.*, 1996a; Caputo *et al.*, 1995).

Granzyme function

Many functions have been postulated for granzyme-family members including: matrix degradation and lymphocyte trafficking; a direct role in controlling viral infection by cleaving proteins that are essential for viral replication or infectivity (Simon *et al.*, 1987a; Simon and Kramer, 1994), in particular, granzyme A-deficient mice are profoundly more susceptible to infection with the cytopathic orthopox virus, ectromelia (Mullbacher *et al.*, 1996); and a role for granzymes in the induction and activation of cytokines to potentially amplifying a local inflammatory reaction (Suidan *et al.*, 1996; Irmler *et al.*, 1995). However, by far, the most compelling evidence indicates that granzymes (in particular granzyme B) play an important role in lymphocyte-mediated cell death. Granzyme B is the only known mammalian serine protease with preference for acidic sidechains (Poe *et al.*, 1991), and the significance of this observation for its role in apoptosis will be addressed below.

How Does The Killer Cell Deliver The Lethal Hit?

Perforin – undoubtedly the key mediator

An indispensable role for perforin in target cell death was finally settled only when perforin gene knock-out mice were reported, initially by Kagi *et al.*, (1994a), and independently corroborated by three other groups (Lowin *et al.*, 1994b; Kojima *et al.*, 1994; Walsh *et al.*, 1994). Perforin-deficient animals have apparently normal T cell development and their $CD8^+$T cells were activated normally following infection with lymphocytic choriomeningitis virus (LCMV). However *in vitro* cytotoxic activity was markedly deficient against LCMV-infected targets when compared to the T cells of perforin-expressing littermates. This deficiency in cytolysis was also observed in primary *in vitro*-stimulated alloreactive T cells, and to a substantial extent against alloreactive tumor cell lines (Kagi *et al.*, 1994a). NK cell cytotoxicity was also abolished in these mice. These perforin-deficient mice incidentally provided an excellent model for studying alternative cytolytic pathways in the absence of the granule exocytosis mechanism.

Granzymes trigger apoptosis

The observation that purified perforin alone could not induce DNA fragmentation was important since it hinted that molecules other than perforin were also required for apoptosis induction. A role for granzymes in CTL-mediated cytolysis had been proposed years earlier, based on data demonstrating that protease inhibitors could abrogate cytotoxicity (Chang and Eisen, 1980; Hudig *et al.*, 1991). More recently, specific synthetic oligopeptide chloromethylketone or isocoumarin derivatives such as Boc-ala-ala-**asp**-CH_2Cl, a specific inhibitor of granzyme B (Poe *et al.*, 1991), have been shown to inhibit target cell apoptosis (Shi *et al.*, 1992a; Sun *et al.*, 1996) *in vitro*, indicating that granzyme B proteolytic activity is essential for cytolysis.

Hayes *et al.* first demonstrated that granzyme A could induce DNA fragmentation in membrane-disrupted cells (Hayes *et al.*, 1989). These findings were confirmed and extended, when a DNA-fragmenting activity termed "fragmentin-2" was isolated from a rat natural killer cell line and shown to be identical with RNK-P1, the rat equivalent of granzyme B (Shi *et al.*, 1992a). Granzyme A and B had no apoptosis-inducing or DNA-fragmenting activity when used in isolation *in vitro* but were active when perforin was also present. The quantities of perforin required for this synergy with granzyme B were very low, and exposure to the same quantity of perforin alone produced no DNA damage and only barely detectable (usually <5%) specific ^{51}Cr release from the target cells. Apoptosis of the target cells was rapid provided both perforin and granzyme B were added. Two other "fragmentins" (corresponding to rat granzyme A and tryptase-2) which induced far slower DNA fragmentation in combination with perforin were also identified (Shi *et al.*, 1992b). Shiver *et al.* (1991) using gene transfection into a non-cytotoxic mast cell line also provided evidence that the combination of perforin and granzymes could induce target cell DNA fragmentation (Shiver *et al.*, 1991; Shiver *et al.*, 1992; Nakajima and Henkart, 1994). More recently, the primary *in vitro*-activated alloreactive CTL of mice deficient in granzyme B have been shown to induce slow but reproducible target cell DNA fragmentation (Heusel *et al.*, 1994). In contrast to perforin-deficient mice, granzyme B-deficient mice were not apparently immunocompromised, suggesting that other granule proteases may provide apoptotic function in synergy with perforin. By contrast, the function of perforin is clearly unique and essential.

Perforin – More Than A Passage

Several hypotheses

Intuitively it is most simple to postulate that perforin-induced membrane pores provide access into the cytoplasm for granzymes. Yet, confocal microscopy of cells exposed to perforin and fluoresceinated granzyme B has shown that while perforin does accelerate the uptake of granzyme B into the cell, much of the uptake of granzyme B occurs independently of perforin (Trapani *et al.*, 1996, 1997; Shi *et al.*, 1997; J.A. Trapani and D.A. Jans, unpublished observations). Furthermore, inert and freely diffusible molecules are not taken up appreciably in the presence of the same quantities of perforin that accelerate the uptake of granzyme B (J.A. Trapani and D.A. Jans, unpublished observations). This data disputes both the free diffusion of molecules though perforin pores and the random uptake of extracellular molecules through endocytic repair of perforin-induced lesions. Thus, the collaboration between perforin and granzyme B would appear highly specific. The view that perforin is not necessary for internalization of granzyme B is supported by other studies (Shi *et al.*, 1997; Trapani *et al.*, 1997; Froelich *et al.*, 1996b), but may be target cell-dependent. Importantly however, granzyme B does not induce apoptosis when cells are exposed to it in the absence of perforin (Shi *et al.*, 1991; Shi *et al.*, 1992; Jans *et al.*, 1996; Trapani *et al.*, 1997; Sutton *et al.*, 1997).

An obvious redistribution of granzymes A or B from the cytoplasm into the nucleus has also been demonstrated when perforin was present, and nuclear accumulation correlated precisely with apoptosis (Trapani *et al.*, 1997). Other non-granzyme serine proteases such as chymotrypsin do not accumulate in the nucleus (Trapani *et al.*, 1996). The kinetics of nuclear localization were considerably slower for granzyme A than granzyme B, consistent with the reduced DNA fragmentation observed with tryptase- rather than aspase-induced cell death (Shi *et al.*, 1992b; Shiver *et al.*, 1992). The onset of DNA fragmentation and annexin V binding lagged well behind the extremely rapid (<2 minutes) nuclear targeting of granzyme B. This suggested that penetration into the nucleus preceded, and was not simply a consequence of nuclear membrane disruption during apoptosis. The mechanism of granzyme B entry into the nucleus is not completely understood, but uptake is energy independent, requires a carrier molecule presumed to be a cytosolic protein, and is not dependent on proteolysis by granzyme B (Jans *et al.*, 1996; Trapani *et al.*, 1994).

While perforin is essential for apoptosis, other hypotheses may explain its synergy with granzymes. Firstly, it may be possible for perforin to generate a signalling cascade in its own right, but at the present time there is no report of membrane signalling events being generated following the binding of perforin to a target cell. A further possibility is that perforin provides release of granzymes into the cytoplasm from a sealed compartment such as an endosome. This possibility has been advanced recently by Froelich *et al.* (1996b), when they showed evidence of a saturable granzyme B cell surface receptor that enabled it to enter the cell. The cells exposed to granzyme B alone remained viable, however if granzyme B was introduced with a non-cytopathic replication deficient adenovirus, they underwent rapid apoptosis. A key mechanism of adenovirus pathogenicity is that following its endocytosis, it can escape endosomes into the cytoplasm (Seth *et al.*, 1994). Froelich hypothesises that escape of granzyme B into the cytoplasm due to adenovirus allows it to access key substrates, and that this mimics the role normally played by perforin. This hypothesis would be greatly strengthened by the isolation of a granzyme receptor and demonstration that perforin actually enters the target cell, a finding not well supported by past electron microscopy studies.

Another less well recognised hypothesis regarding the actions of perforin and granzymes has focussed on the possibility that granzyme B can disrupt normal progression through the cell cycle, leading to an untimely entry into mitosis (Krek and Nigg, 1991; Heald *et al.*, 1993). Cells undergoing apoptosis in response to perforin and granzyme B required cdc2 kinase activity to cause chromatin condensation and DNA fragmentation (Shi *et al.*, 1994). Furthermore, Wee 1 kinase overexpression was able to rescue cells from granzyme B / perforin-induced apoptosis (Chen *et al.*, 1995). This hypothesis has remained controversial because there are instances where cdc2 kinase is not required in other forms of apoptosis (Freeman *et al.*, 1994) and target cells continue to be partially sensitive to granzyme B, even when cdc2 kinase activity is reduced to very low levels (Shi *et al.*, 1994). Either this mechanism may be peculiar to CL-mediated apoptosis or at least cdc2-independent pathways of granzyme B / perforin mediated killing must also exist.

An Evolutionarily Conserved Cell Death Cascade

Worms provide the first clues

An endogenous pathway of cell death conserved in all multicellular organisms was first described in the nematode *Caenorhabditis elegans (C. elegans)*. Three genes, ced-3, ced-4 and ced-9 regulate apoptosis in *C. elegans* (Yuan and Horvitz, 1992). Ced-3 and ced-4 are permissive for apoptosis, and inactivation of either gene abolishes apoptosis during nematode development, while ced-9 is inhibitory for apoptosis and functions by blocking cell death in cells that are required in the adult worm. It is now clear that a family of cysteine proteases that share homology with the product of the cell death gene ced-3 are potent inducers of mammalian apoptosis (for review, see Kumar and Lavin, 1996). Ced-9 is similar structurally and functionally to the mammalian Bcl-2 protein (Vaux *et al.*, 1992a), while a mammalian equivalent of the ced-4 gene product has recently been identified (Zou *et al.*, 1997).

Mammalian Ced-3-like proteases

The ced-3 gene encodes a cysteine protease that cleaves target proteins at specific asp residues (Yuan *et al.*, 1993). The cloning of ced-3 suggested it had sequence identity with mammalian cysteine protease interleukin-1β converting enzyme (ICE) (Kostura *et al.*, 1989; Thornberry *et al.*, 1992). A role for ICE in apoptosis was initially unsuspected, however over the past several years multiple mammalian ICE-like proteases have been identified, cloned and characterised as playing an integral role in apoptosis induction (Kumar and Lavin, 1996). The expanding size of the family has required a new nomenclature, and each enzyme is now termed a "caspase" and given a numerical suffix (Alnemri *et al.*, 1996). The mammalian ced-3-like proteases have been categorised into either the ICE-like, CPP32-like or Nedd-2-like groups on the basis of sequence similarity. This chapter will not deal with the caspase family in great detail since others in this volume consider these proteases. All caspases are constitutively expressed as single chain pro-proteins that require autocatalysis or cleavage and activation by another protease. Peptidyl inhibitors of the caspases are potent inhibitors of many types of apoptosis, suggesting that caspase activation is a common feature of diverse forms of apoptosis (Zhivotovsky *et al.*, 1995; Sarin *et al.*, 1997a; Sarin *et al.*, 1997b).

Activated caspases cleave specific target cell proteins to cause apoptosis

The majority of cellular proteins remain intact until late in the apoptotic process (Kumar and Lavin, 1996). Nevertheless, specific structural and catalytic proteins are cleaved early in apoptosis. The DNA repair enzyme, poly(ADP-ribose) polymerase (PARP) (Kaufmann, 1989; Kaufmann *et al.*, 1993; Lindahl *et al.*, 1995), can be cleaved by several of the caspases, including Mch2, Mch3 and CPP32 (Tewari *et al.*, 1995a; Lazebnik *et al.*, 1994; Nicholson *et al.*, 1995; Fernandes-Alnemri *et al.*, 1995b). PARP cleavage is not requisite for apoptosis, since PARP-deficient mice display normal apoptosis (Wang *et al.*, 1995). Other nuclear structures cleaved by activated caspases include; the catalytic subunit of DNA-dependent protein kinase (DNA-PKcs, Casciola-Rosen *et al.*, 1995; Song *et al.*, 1996), the sterol regulatory element-binding proteins 1 and 2 (Wang *et al.*, 1996) and nuclear lamins (Lazebnik *et al.*, 1995). ICE-like proteases also cleave the cell cycle regulatory and anti-apoptotic protein pRb (Janicke *et al.*, 1996). ICE also cleaves actin (thereby reducing its DNAse I binding activity, Kayalar *et al.*, 1996), U1 associated 70 kDa protein (Casciola-Rosen *et al.*, 1996) and D4-GDI (Na *et al.*, 1996). Importantly, some caspases are also able to activate others in a hierarchical manner, thereby amplifying the apoptotic cascade (Figure 11.1). Some caspases including ICE are autocatalytic and others such as CPP32 contribute to their own processing (Fernandes-Alnemri *et al.*, 1996a; Harvey *et al.*, 1996; Fernandes-Alnemri *et al.*, 1995a). There is recent evidence that CPP32 may activate a heterodimeric cytoplasmic molecule, termed DFF (for DNA fragmentation factor), responsible for DNA fragmentation (Liu *et al.*, 1997), suggesting that most of the nuclear events associated with CPP32 activation may be indirect.

Granzyme B is an exogenous aspase that can activate and augment the caspase cascade

Granzyme B is the only mammalian serine protease that cleaves C-terminal to residues with acidic side chains (Poe *et al.*, 1991). Vaux *et al.* (1994) first suggested that granzyme B cleavage of key substrates at asp residues may be a means by which CTL activate the death cascade in target cells. Indeed, killer cells of mice deficient in granzyme B were shown to induce target cell apoptosis with slow kinetics (Heusel *et al.*, 1994), and there is now strong evidence that granzyme B can activate many of the pro-caspases (Figure 11.1). *In vitro*, granzyme B has been shown to activate pro-CPP32 (Darmon *et al.*, 1995; Fernandes-Alnemri *et al.*, 1996; Martin *et al.*, 1996) and pro-Nedd-2 (Harvey *et al.*, 1996), but it has not been demonstrated definitively whether granzyme B acts directly on pro-CPP32 *in vivo*. Killer cell activation of target cell CPP32 appears to be granzyme B-dependent, but the effect might be indirect, since granzyme B can activate upstream caspases, including pro-Fas-like ICE (FLICE, Mch5) and pro-Mch4, both of which can also cleave CPP32 (Fernandes-Alnemri *et al.*, 1996). In theory, the capacity of granzyme B to activate FLICE should be sufficient to activate the whole cascade, but *in vitro* granzyme B can cleave multiple caspases, thus possibly amplifying the cascade both proximally through FLICE and Mch4 and more distally through Mch3, CPP32 (Fernandes-Alnemri *et al.*, 1996), Mch6 and Mch2 (Srinivasula *et al.*, 1996a) and CMH-1 / ICE-LAP3 (Gu *et al.*, 1996). The final result is multiple active caspases that can cleave nuclear and cytoplasmic substrates. Granzyme B does not appear to activate Mch2 directly, although this protease is thought to act on lamins directly (Fernandes-Alnemri, 1995b). Granzyme B may also be capable of accessing (Trapani *et al.*,

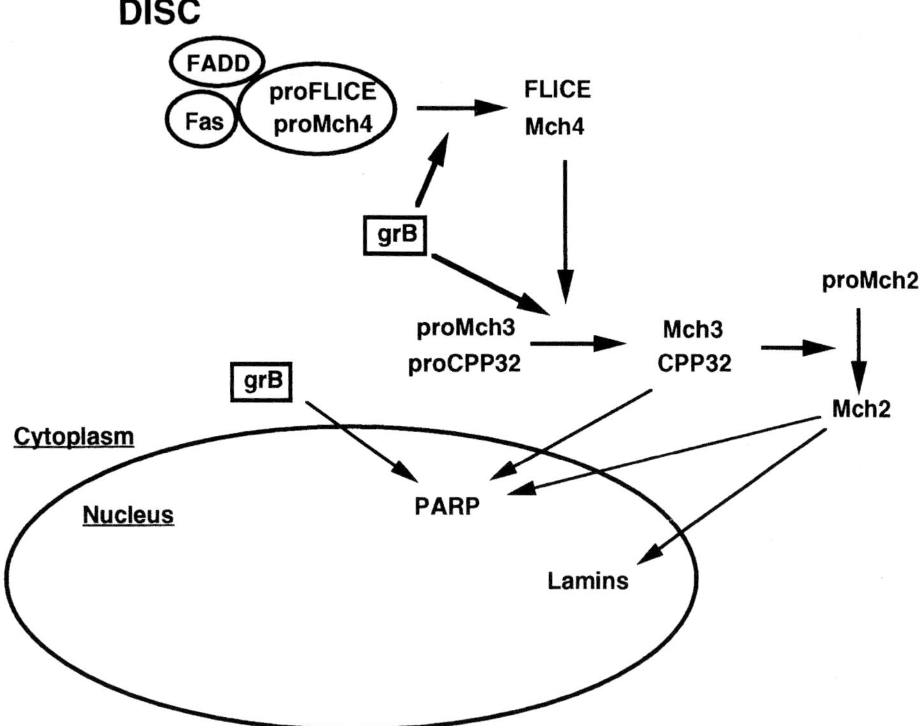

Figure 11.1 A hierarchical scheme of caspase activation, and amplification by granzyme B. Fas ligation and recruitment of FADD results in the formation of a death inducing signalling complex (DISC) at the inner leaflet of the cell membrane, with subsequent production of active FLICE and Mch4. FLICE activation generates at least three caspases that cleave intranuclear target molecules such as PARP and structural lamins, CPP32, Mch3 and Mch 2. Granzyme B can cleave at several points in the pathway, including at the apex of the cascade at pro-FLICE and pro-Mch4, and by cleaving pro-CPP32. In the absence of caspase inhibitors, granzyme B can target the nucleus directly, where it is likely to contribute to apoptotic morphology. The caspases and granzyme B also cleave substrates in the cytoplasm and can kill the cell without the requirement for nuclear collapse.

1996; Jans *et al.*, 1996; Pinkoski *et al.*, 1996) and cleaving nuclear target proteins such as PARP (Froelich *et al.*, 1996) and DNA-PKcs (Song *et al.*, 1996) directly, but in each case the cleavage sites are different from those used by the caspases.

Granzyme A-deficient mice have no obvious defect of apoptosis induction (Ebnet *et al.*, 1995), yet the ability of granzyme A to induce apoptosis in several independent experimental systems in the absence of granzyme B (eg., Shiver *et al.*, 1992; Shi *et al.*, 1992a; Shi *et al.*, 1992b) argues for its ability to complement the actions of granzyme B, or even overcome its absence. However granzyme A has not been demonstrated to activate any of the caspases, and the mechanism by which apoptosis is brought about by granzyme A is uncertain. Indeed, a caspase inhibitor that blocked granzyme B-mediated apoptosis had no effect on cell death caused by granzyme A (Anel *et al.*, 1997). In contrast, granzyme A has been shown to have ICE-like activity, in that it can produce active IL-1β by cleaving pro-IL-1 at a tryptase site adjacent to the asp residue usually cleaved by ICE (Irmler *et al.*, 1995), but the *in vivo* significance of this observation is not clear.

Do granzymes have additional intracellular targets?

Since both the granzyme B/perforin and FasL-mediated apoptotic signalling pathways intersect in the target cell cytoplasm at caspase substrates as far upstream as FLICE, what then distinguishes the two pathways? On the surface, it would seem a threat to the host should both mechanisms used by CL converge so closely in the same biochemical pathway. Evidence is now emerging that both mechanisms are different biochemically, in that, unlike the Fas pathway, the granule exocytosis mechanism can lyse cells independently of caspase-induced proteolysis.

Tetrapeptide inhibitors that are relatively specific for caspase subfamilies have been used to dissect the roles of the different proteases in DNA fragmentation and cell membrane damage (Darmon *et al.*, 1995). Inhibition of CPP32 and related protease Mch3 (Nicholson *et al.*, 1995) had no effect on ^{51}Cr release from target cells killed by a granule-dependent mechanism, but dramatically reduced DNA fragmentation in the same cells. Thus the CPP32 family of proteases are only instrumental in eliciting nuclear damage caused by granzyme B. Similarly Sarin *et al.* (1997a), suggested that the known caspases do not play a significant role in cell membrane damage resulting from granule-induced target cell apoptosis. By contrast, both the nuclear and cytoplasmic phenomena of apoptosis mediated through Fas were dependent on FLICE activation. In another study, serine protease activity was required for both cell membrane damage and DNA fragmentation, however addition of the serine protease inhibitor DCI 15 minutes after the initiation of the lytic cycle resulted in abolition of DNA fragmentation without an effect on membrane damage (Helgason *et al.*, 1995). These caspase-independent events were not due simply to the unopposed actions of perforin acting alone on the cell membrane (Sarin *et al.*, 1997a; Sarin *et al.*, 1997b; Spielman *et al.*, 1997). Attention has now turned to the cytoplasmic events that may distinguish caspase-dependent and caspase-independent pathways of cell death.

More Clues to Cell Death – The Plot Thickens

Attention turns to the cytoplasm/mitochondria interface

Although many of the identified nuclear targets of caspases and granzyme B explain aspects of apoptotic morphology, the cytoplasmic processes that underpin the entire mechanism need to be defined. Some exciting recent biochemical data utilising the key nematode death proteins, indicate that Ced-3 and Ced-4 can physically interact (Chinnaiyan *et al.*, 1997; Irmler *et al.*, 1997), by virtue of their N-terminal domains, which both contain a caspase recruitment domain (Hofmann *et al.*, 1997). Overexpression of Ced-4 in mammalian cells induced cell death and the addition of caspase inhibitors blocked it (Chinnaiyan *et al.*, 1997). Ced-4 also interacts with mammalian ICE and FLICE, which have large prodomains, but not with CPP32 or Mch2a, which have small prodomains. The anti-apoptotic protein Ced-9 can negatively regulate apoptosis by binding to Ced-4 (Wu *et al.*, 1997; Chinnaiyan *et al.*, 1997; Spector *et al.*, 1997, James *et al.*, 1997). This modification of Ced-4 prevents it from activating Ced-3 and dislocates upstream signalling events from caspase activation. Normally, Ced-4 is expressed in the cytoplasm, however when a Ced-9-like mammalian protein, Bcl-X_L, is also expressed, the Ced-4 / Bcl-X_L complex is relocated to mitochondrial membranes (Wu *et al.*, 1997). Conversely, in the absence of Ced-9 / Bcl-2, Ced-4 is free to activate the protease cascade. Therefore, Ced-4 plays a central role in regulating cell death by directly interacting with both the pro-apoptotic proteases of the caspase family and the inhibitory proteins of the Bcl-2 family.

By analogy with both the Ced-3-like and Ced-9-like families, it is indeed possible that multiple Ced-4-like proteins may exist in mammals. At least one mammalian equivalent of Ced-4, apoptotic protease activating factor-1 (Apaf-1), has recently been isolated (Zou *et al.*, 1997). Using purified components, Zou *et al.* (1997) have shown that the Ced-4-homolog Apaf-1, cytochrome c (Apaf-2), Apaf-3 (as yet uncloned), and dATP are sufficient to activate pro-caspase 3. Cytochrome c binds to Apaf-1 in the absence or presence of dATP. Cytochrome c release from mitochondria is also required for caspase activation, and bcl-2 appears to regulate the the loss of mitochondrial membrane potential and release of cytochrome c during apoptosis (Kluck *et al.*, 1994; Yang *et al.*, 1997; Marchetti *et al.*, 1997). Loss of permeability leads to the release of apoptosis-inducing factor (AIF) from mitochondria into the cytosol (Susin *et al.*, 1996). AIF is itself a caspase-inhibitor-sensitive protease which can cause caspase activation and cell death (Kroemer *et al.*, 1997). The relationship between AIF and Apafs 1–3 remains unclear. What implications these latest studies have for granule-mediated death awaits greater dissection of caspase-independent events in mammalian systems.

Natural Inhibitors of CTL-Mediated Apoptosis

Bcl-2- / CED-9-like inhibitors

The mammalian family of bcl-2-like molecules includes both anti- (Bcl-2, Bcl-X_L) and pro- (Bax, Bik, Bcl-X_S) apoptotic members (for review, see Cory *et al.*, 1995; Vaux *et al.*, 1992a). Bax forms heterodimers with Bcl-2 and opposes its activity, and the relative quantities of Bcl-2 and Bax within a cell can determine whether a cell will undergo apoptosis in response to a given stimulus. Bcl-2 can block some, but not all forms of apoptosis reliant on caspase activation. Reports on the ability of Bcl-2 to block CL-mediated apoptosis have been contradictory and may be a function of the target cell type employed (Sutton *et al.*, 1997). Some reports have indicated that Bcl-2 can protect against allogeneic CTL attack (Schroter *et al.*, 1995), while others have suggested otherwise (Vaux *et al.*, 1992b; Sutton *et al.*, 1997). Chiu *et al.* (1995) found protection against granule- but not Fas-mediated attack. Interestingly, although Bcl-2 was unable to block apoptosis induced by intact CL or isolated granules, it completely blocked cell death induced by purified granzyme B and perforin (Sutton *et al.*, 1997). Thus cytolytic granules may contain components that can bypass the Bcl-2-mediated block of granzyme B.

Granzyme B inhibitor (GBI, PI-9), a naturally occurring inhibitor of granzyme B

Novel intracellular serine protease inhibitors (serpins) that can inhibit granzyme B (GBI) were recently described (Sprecher *et al.*, 1995; Sun *et al.*, 1996). These GBI are expressed at high levels in cells with cytolytic capacity (Sun *et al.*, 1997). Their inhibitory loop is strongly related to that of the viral serpin crmA, however the asp at the P_1 position of crmA was replaced with another acidic residue, glu (Sun *et al.*, 1996). GBI and granzyme B stably complex with an association constant within the range for physiologically significant serpin-protease interactions, and this interaction is sufficient to abrogate apoptosis. Transfected GBI also affords protection of cell lines against granzyme/perforin-mediated cell death (V.R. Sutton, J.A. Trapani and P.I. Bird, manuscript in preparation). GBI is a cytosolic protein that is not secreted and is absent from cytolytic granules (Sun *et al.*, 1996). It is hypothesised that GBI might inactivate free granzyme B molecules in the effector cell following packaging or degranulation, thereby protecting the CL from inadvertent autolysis. It is not clear why GBI utilises glu rather than asp at the P1 position, since granzyme B cleaves synthetic substrates at asp far more efficiently than at glu (Poe *et al.*, 1991). Indeed, substitution of the P1 glu with asp does not cause increased binding to granzyme B (P.I. Bird, personal communication). It is not yet known whether GBI can interact with any of the caspase proteins or block Fas-mediated cytolysis.

Viruses Subvert Cell Death Pathways Mediated By Killer Cells

It is not surprising that through the course of evolution, many pathogens have devised ways of delaying apoptosis. Many viruses have developed ways of subverting apoptotic pathways, especially those involving the caspase cascade. Given the exchange of genetic information between viruses and their hosts, it would seem possible that a specific means of blocking perforin or granzyme function might have also evolved. Other than for the inhibitory effects of crmA on granzyme B and the observation that parainfluenza virus can downregulate the expression of granzyme B mRNA in a selective manner in infected T cells (Sieg *et al.*, 1995), there is little information to support this possibility.

Viral inhibitors – CrmA and p35

CrmA is one of a number of poxvirus proteins that interfere with the host's ability to eliminate virus-infected cells (Ray *et al.*, 1992). CrmA is an intracellular serpin whose inhibitory loop (an asp residue in CrmA), mimics the natural substrate preference of the protease. CrmA can inhibit proteases of more than one subclass, including ICE (Ray *et al.*, 1992; Komiyama *et al.*, 1994), CPP32 (Tewari *et al*, 1995) and FLICE (Srinivasula *et al.*, 1996b), in addition to the serine protease granzyme B. CrmA is capable of blocking granzyme B asp-ase activity *in vitro* (Quan *et al.*, 1995), but CrmA binds to granzyme B at least 10-fold less than to ICE (J. Sun, P.I. Bird, J.A. Trapani, unpublished results). CrmA can block Fas-mediated CL-induced apoptosis (Tewari *et al.*, 1995b; Macen *et al.*, 1996) and perforin / granzyme B-mediated apoptosis (Tewari *et al.*, 1995b; Macen *et al.*, 1996), however the effect is far greater on the Fas pathway. The physiological relevance of this observation is unclear, since target cells from CrmA transgenic mice are as susceptible to attack by allogeneic CTL or CTL granules as those of non-transgenic littermates (V.R. Sutton and J.A. Trapani, unpublished observations).

p35 is a baculovirus protein that exerts a strong inhibitory effect on caspase-dependent apoptosis in infected insect cells (Clem *et al.*, 1991; Clem and Miller, 1993) and inhibits a broad variety of caspases (Xue and Horwitz, 1995; Bump *et al.*, 1995). Cleavage of p35 at asp87 is essential for its inhibitory activity, both *in vitro* and in *C. elegans*. Sarin *et al.* (1997a) have demonstrated that p35 can suppress both the nuclear and cytoplasmic events during CL-mediated apoptosis via Fas, but it only inhibits the nuclear consequences of granule-mediated cytolysis.

Other inhibitors

There is a rapidly growing list of viral inhibitors of apoptosis. Indeed it is clear that the co-evolution of viruses and their host cells has resulted in every facet of recognition and apoptosis induction being suppressed by various virus proteins. These include: IAP (inhibitor of apoptosis, Clem and Miller, 1994) – a broadly active serpin with a RING finger domain (Uren *et al.*, 1996; Liston *et al.*, 1996; Roy *et al.*, 1995); the Epstein-Barr encoded protein BHRF, African swine fever virus LMW5-H1, and herpesvirus samrai ORF16, that are Bcl-2-like in structure; the adenovirus E1B protein (White *et al.*, 1991) that can heterodimerise with Bcl-2-like molecules and inhibit their activity (Farrow *et al.*, 1995); and FLICE inhibitory proteins (FLIPs) that encode two death effector domains that can interact with Fas-associated death domain (FADD) to prevent FLICE recruitment and activation (Thome *et al.*, 1997, Bertin *et al.*, 1997; Hu *et al.*, 1997). Most of these proteins have been shown to inhibit Fas- and/or TNFR-mediated cell death, however none have been demonstrated to specifically inhibit granule exocytosis mediated cell death.

Killer Cells Have The Final Say – Or Do They?

Applying these observations hypothetically to the best recognised function of cytotoxic lymphocytes (ie. the elimination to virus-infected cells) – the evolutionary battle between the mammalian host and the virus becomes clearer. In the normal course of an infection with a "benign" (ie., one incapable of blocking the caspase cascade) virus, an infected cell could be killed by either CTL/NK cell – mediated granule exocytosis – and/or endogenous Fas/TNFR – pathways, and dies a rapid death as a consequence of both nuclear and cytoplasmic apoptotic changes. Nuclear degradation would be accomplished by the joint actions of both granzyme B and certain activated caspases, probably acting on different cleavage sites on targeted proteins (for example, as shown in Figure 11.1). In an alternative scenario, a "less benign" virus might be able to completely block the caspase pathway (and therefore Fas-mediated apoptosis) to prolong its own survival, delaying apoptosis, and thus facilitating its spread to uninfected cells. This virus may survive in a dormant or cryptic form until conditions are ideal for further infection of neighbouring cells. However, in periods of serious widespread and life-threatening infection, killer cells must be

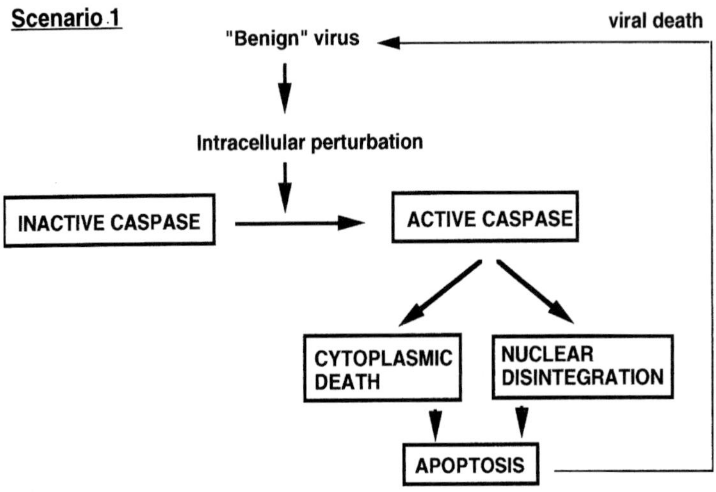

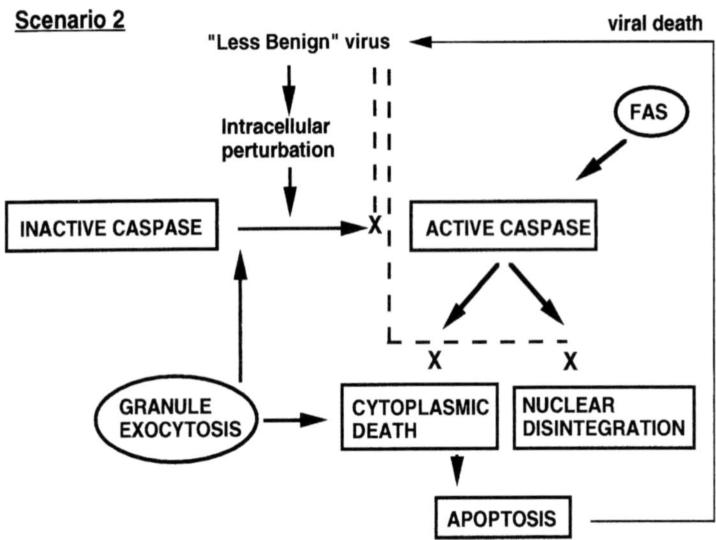

Figure 11.2 Putative mechanisms for clearing two types of viral infection. In *scenario 1*, a cell is infected with a virus that creates sufficient intracellular perturbation to trigger the endogenous caspase cascade, and lacks inhibitory molecules that can block cell suicide. Such a cell dies by a combination of nuclear and cytoplasmic events, and an efficient CTL response may not be necessary to kill the virus. When CTL are raised to such a virus, cell death can proceed through either the Fas or the granule pathway. In this event, both caspases and granzymes are likely to contribute to nuclear apoptotic changes. In *scenario 2*, a different, 'less benign' virus carries genes coding for inhibitors of the caspase cascade, which delay or even prevent altruistic suicide. Killing by CTL through Fas may also be blocked by the same inhibitory molecules, as both the cytoplasmic and nuclear consequences of Fas-induced death are caspase-dependent. In this event, the CTL can still utilize the granule pathway to induce caspase-independent cytolysis in the cytoplasm (see text for more detailed explanation).

capable of confronting infected cells expressing foreign viral antigen, and must have the final say on the survival of an infected cell. Presumably, this final say is delivered by the granule exocytosis mechanism and the plethora of cell death pathways that it can induce in the infected cell. Thus, if the caspase cascade is blocked in a cell, it still dies because of the cytoplasmic consequences of cell death. The means by which cytotoxic granules achieve caspase-independent apoptosis is an important unresolved issue, one possibility being that granzymes can initiate apoptosis by targeting their own unique substrates in the cytoplasm (Figure 11.2). Certainly there remains a myriad of granzyme specificities with no defined function. Perhaps it is not surprising that viral proteins specifically inhibiting granule-mediated cell death have not been discovered, since the virus would threaten the host's and therefore its own viability by eliminating the host's last mainstay of antiviral defense.

ABBREVIATIONS

AIF, apoptosis inducing factor; Apaf, apoptotic protease activating factor; CL, cytotoxic lymphocyte; CTL, cytotoxic T lymphocyte; DPPI, dipeptidyl peptidase I; EGF, epidermal growth factor; FADD, Fas-associated death domain; FLIP, FLICE inhibitory protein; GBI, granzyme B inhibitor; ICE, interleukin-1β-converting enzyme; IFN, interferon; LCMV, lymphocytic choriomeningitis virus; MAC, membrane attack complex; NK, natural killer; PARP, poly (ADP-ribose) polymerase; PBL, peripheral blood lymphocyte; PEL, peritoneal exudate cell; SBzl, thiobenzylester; TNF, tumor necrosis factor

ACKNOWLEDGEMENTS

The authors wish to thank the staff of the Cellular Cytotoxicity Laboratory, Austin Research Institute for their support over many years. We also thank our collaborators, particularly David Jans, Arnold Greenberg, Sharad Kumar, David Vaux, Thomas Sayers and Philip Bird and members of their laboratories. At various times, our laboratories have received generous support from The Wellcome Trust, the National Health and Medical Research Council of Australia and the Anti-Cancer Council of Victoria.

REFERENCES

Allbritton, N.L., Verret, C.R., Wolley, R.C. and Eisen, H.N. (1988) Caslcium ion concentrations and DNA fragmentation in target cell destruction by murine cloned cytotoxic T lymphocytes. *J. Exp. Med.*, **167**, 514–27.

Alnemri, E.S., Livingston, D.J., Nicholson, D.W., Salvesen, G., Thornberry, N.A., Wong, W.W. and Yuan, J. (1996) Human ICE/CED-3 protease nomenclature [letter]. *Cell*, **87**, 171.

Anel, A., Gamen, S., Alava, M.A., Schmitt-Verhulst, A.M, Peneiro, A. and Naval, J. (1997) Inhibition of CPP32-like proteases prevents granzyme B- and Fas-, but not granzyme A-based cytotoxicity exerted by CTL clones. *J. Immunol.*, **158**, 1999–2006.

Berke, G. (1989) The cytolytic T lymphocyte and its mode of action. *Immunol., Lett.*, **20**, 169–78.

Berke, G. (1991) Debate: the mechanism of lymphocyte-mediated killing. Lymphocyte-triggered internal target disintegration. *Immunol. Today*, **12**, 396–9; discussion 402.

Berke, G. and Rosen, D. (1988) Highly lytic in vivo primed cytolytic T lymphocytes devoid of lytic granules and BLT-esterase activity acquire these constituents in the presence of T cell growth factors upon blast transformation in vitro. *J. Immunol.*, **141**, 1429–36.

Bertin, J., Armstrong, R.C., Ottilie, S., *et al.* (1997) A novel family of viral death effector domain containing molecules that inhibit both CD95- and tumor necrosis factor receptor-1-induced apoptosis. *J. Biol. Chem.*, **272**, 9621–9624.

Bode, W., Meyer, E., Jr. and Powers, J.C. (1989) Human leukocyte and porcine pancreatic elastase: X-ray crystal structures, mechanism, substrate specificity and mechanism-based inhibitors. *Biochemistrys*, **28**, 1951–63.

Bump, N.J., Hackett, M., Hugunin, M., Seshagiri, S., Brady, K., Chen, P., Ferenz, C., Franklin, S., Ghayur, T., Li, P. and *et al.* (1995) Inhibition of ICE family proteases by baculovirus antiapoptotic protein p35. *Science*, **269**, 1885–8.

Bykovskaja, S.N., Rytenko, A.N., Rauschenbach, M.O. and Bykovsky, A.F. (1978) Ultrastructural alteration of cytolytic T lymphocytes following their interaction with target cells. II. Morphogenesis of secretory granules and intracellular vacuoles. *Cell Immunol.*, **40**, 175–85.

Caputo, A., Garner, R.S., Winkler, U., Hudig, D. and Bleackley, R.C. (1993) Activation of recombinant murine cytotoxic cell proteinase-1 requires deletion of an amino-terminal dipeptide. *J. Biol. Chem.*, **268**, 17672–5.

Caputo, A., James, M.N., Powers, J.C., Hudig, D. and Bleackley, R.C. (1995) Conversion of the substrate specificity of mouse proteinase granzyme B. *Nat. Struct. Biol.*, **1**, 364–7.

Casciola-Rosen, L., Nicholson, D.W., Chong, T., Rowan, K.R., Thornberry, N.A., Miller, D.K. and Rosen, A. (1996) Apopain/CPP32 cleaves proteins that are essential for cellular repair: a fundamental principle of apoptotic death [see comments]. *J. Exp. Med.*, **183**, 1957–64.

Casciola-Rosen, L.A., Anhalt, G.J. and Rosen, A. (1995) DNA-dependent protein kinase is one of a subset of autoantigens specifically cleaved early during apoptosis. *J. Exp. Med.*, **182**, 1625–34.

Castro, J.E., Listman, J.A., Jacobson, B.A., Wang, Y., Lopez, P.A., Ju, S., Finn, P.W. and Perkins, D. L. (1996) Fas modulation of apoptosis during negative selection of thymocytes. *Immunity*, **5**, 617–27

Chang, T.W. and Eisen, H.N. (1980) Effects of N alpha-tosyl-L-lysyl-chloromethylketone on the activity of cytotoxic T lymphocytes. *J. Immunol.*, **124**, 1028–33.

Chen, G., Shi, L., Litchfield, D.W. and Greenberg, A.H. (1995) Rescue from granzyme B-induced apoptosis by Wee1 kinase. *J. Exp. Med.*, **181**, 2295–300.

Chinnaiyan, A.M., O'Rourke, K., Lane, B.R. and Dixit, V.M. (1997) Interaction of CED-4 with CED-3 and CED-9: a molecular framework for cell death. *Science*, **275**, 1122–6.

Chiu, V.K., Walsh, C.M., Liu, C.C., Reed, J.C. and Clark, W.R. (1995) Bcl-2 blocks degranulation but not fas-based cell-mediated cytotoxicity. *J. Immunol.*, **154**, 2023–32.

Clem, R.J., Fechheimer, M. and Miller, L.K. (1991) Prevention of apoptosis by a baculovirus gene during infection of insect cells. *Science*, **254**, 1388–90.

Clem, R.J. and Miller, L.K. (1993) Apoptosis reduces both the in vitro replication and the in vivo infectivity of a baculovirus. *J. Virol*, **67**, 3730–8.

Clem, R.J. and Miller, L.K. (1994) Control of programmed cell death by the baculovirus genes p35 and iap. *Mol. Cell Biol.*, **14**, 5212–22.

Cory, S. (1995) Regulation of lymphocyte survival by the bcl-2 gene family. *Annu. Rev. Immunol.*, **13**, 513–43.

Darmon, A.J., Nicholson, D.W. and Bleackley, R.C. (1995) Activation of the apoptotic protease CPP32 by cytotoxic T-cell-derived granzyme B. *Nature*, **377**, 446–8.

Dennert, G. anderson, C.G. and Prochazka, G. (1987) High activity of N-alpha-benzyloxycarbonyl-L-lysine thiobenzyl ester serine esterase and cytolytic perforin in cloned cell lines is not demonstrable in in-vivo-induced cytotoxic effector cells. *Proc. Natl. Acad. Sci. USA*, **84**, 5004–8.

Dennert, G. and Podack, E.R. (1983) Cytolysis by H-2-specific T killer cells. Assembly of tubular complexes on target membranes. *J. Exp. Med.*, **157**, 1483–95.

Dhein, J., Walczak, H., Baumler, C., Michael-Debatin, K. and Krammer, P.H. (1995) Autocrine T-cell suicide mediated by APO-1/(Fas/CD95) *Nature*, **373**, 438–41.

Dourmarshkin, R.R., Deteix, P., Simone, C.B. and Henkart, P. (1980) Electron microscopic demonstration of lesions in target cell membranes associated with antibody-dependent cellular cytotoxicity. *Clin. Exp. Immunol.*, **42**, 554–60.

Duke, R.C., Persechini, P.M., Chang, S., Liu, C.C., Cohen, J.J. and Young, J.D. (1989) Purified perforin induces target cell lysis but not DNA fragmentation. *J. Exp. Med.*, **170**, 1451–6.

Ebnet, K., Hausmann, M., Lehmann-Grube, F., Mullbacher, A., Kopf, M., Lamers, M. and Simon, M.M. (1995) Granzyme A-deficient mice retain potent cell-mediated cytotoxicity. *EMBO J.* **14**, 4230–9.

Ewoldt, G.R., Smyth, M.J., Darcy, P.K., Harris, J.L., Craik, C.S., Horowitz, B. and Hudid, D. (1997a) RNKP-4 and RNKP-7, new granzyme serine proteases expressed in rat activated lymphocytes. *J. Immunol.*, **158**, 4574–4583.

Ewoldt, Darcy, P.K., Snook, M.B., Hudig, D. and Smyth, M.J. (1997b) cDNA cloning of granzyme. *Immunogenetics*, **45**, 452–454.

Farrow, S.N., White, J.H., Martinou, I., Raven, T., Pun, K.T., Grinham, C.J., Martinou, J.C. and Brown, R. (1995) Cloning of a bcl-2 homologue by interaction with adenovirus E1B 19K [published erratum appears in Nature 1995 Jun 1;375(6530):431]. *Nature*, **374**, 731–3.

Fernandes-Alnemri, T., Armstrong, R.C., Krebs, J., Srinivasula, S.M., Wang, L., Bullrich, F., Fritz, L.C., Trapani, J.A., Tomaselli, K.J., Litwack, G. and Alnemri, E.S. (1996) In vitro activation of CPP32 and Mch3 by Mch4, a novel human apoptotic cysteine protease containing two FADD-like domains. *Proc. Natl. Acad. Sci. USA*, **93**, 7464–9.

Fernandes-Alnemri, T., Litwack, G. and Alnemri, E.S. (1995a) Mch2, a new member of the apoptotic Ced-3/Ice cysteine protease gene family. *Cancer Res.*, **55**, 2737–42.

Fernandes-Alnemri, T., Takahashi, A., Armstrong, R., Krebs, J., Fritz, L., Tomaselli, K.J., Wang, L., Yu, Z., Croce, C.M., Salveson, G. and *et al.* (1995b) Mch3, a novel human apoptotic cysteine protease highly related to CPP32. *Cancer Res.*, **55**, 6045–52.

Freeman, R.S., Estus, S. and Johnson, E.M., Jr. (1994) Analysis of cell cycle-related gene expression in postmitotic neurons: selective induction of Cyclin D1 during programmed cell death. *Neuron*, **12**, 343–55.

Froelich, C.J., Orth, K., Turbov, J., Seth, P., Gottlieb, R., Babior, B., Shah, G.M., Bleackley, R.C., Dixit, V.M. and Hanna, W. (1996a) New paradigm for lymphocyte granule-mediated cytotoxicity. Target cells bind and internalize granzyme B, but an endosomolytic agent is necessary for cytosolic delivery and subsequent apoptosis. *J. Biol. Chem.*, **271**, 29073–9.

Froelich, C.J., Hanna, W.L., Poirier, G.G., Duriez, D.J., D'Amours, D., Salvesan, G.S., Alnemri, E.S., Earnshaw, W.C. and Shah, G.M. (1996b) Granzyme B/perforin-mediated apoptosis of Jurkat cells results in cleavage of poly(ADP-ribose) polymerase to the 89-kDa apoptotic fragment and less abundant 64-kDa fragment. *Biochem. Biophys. Res. Commun.*, **227**, 658–65.

Garcia-Sanz, J.A., MacDonald, H.R., Jenne, D.E., Tschopp, J. and Nabholz, M. (1990) Cell specificity of granzyme gene expression. *J. Immunol.*, **145**, 3111–8.

Geiger, B., Rosen, D. and Berke, G. (1982) Spatial relationships of microtubule-organizing centers and the contact area of cytotoxic T lymphocytes and target cells. *J. Cell Biol.*, **95**, 137–43.

Griffiths, G.M. and Isaaz, S. (1993) Granzymes A and B are targeted to the lytic granules of lymphocytes by the mannose-6-phosphate receptor. *J. Cell .Biol.*, **120**, 885–96.

Gu, Y., Sarnecki, C., Fleming, M.A., Lippke, J.A., Bleackley, R.C. and Su, M. (1996) Processing and activation of CMH-1 by granzyme B. *J. Biol. Chem.*, **271**, 10816–10820.

Harvey, N.L., Trapani, J.A., Fernandes-Alnemri, T., Litwack, G., Alnemri, E.S. and Kumar, S. (1996) Processing of the Nedd2 precursor by ICE-like proteases and granzyme B. *Genes Cells*, **1**, 673–85.

Hayes, M.P., Berrebi, G.A. and Henkart, P.A. (1989) Induction of target cell DNA release by the cytotoxic T lymphocyte granule protease granzyme A. *J. Exp. Med.*, **170**, 933–46.

Heald, R., McLoughlin, M. and McKeon, F. (1993) Human wee1 maintains mitotic timing by protecting the nucleus from cytoplasmically activated Cdc2 kinase. *Cell*, **74**, 463–74.

Helgason, C.D., Atkinson, E.A., Pinkoski, M.J. and Bleackley, R.C. (1995) Proteinases are involved in both DNA fragmentation and membrane damage during CTL-mediated target cell killing. *Exp. Cell. Res.*, **218**, 50–6.

Henkart, P.A., Berrebi, G.A., Takayama, H., Munger, W.E. and Sitkovsky, M.V. (1987) Biochemical and functional properties of serine esterases in acidic cytoplasmic granules of cytotoxic T lymphocytes. *J. Immunol.*, **139**, 2398–405.

Heusel, J.W., Wesselschmidt, R.L., Shresta, S., Russell, J.H. and Ley, T.J. (1994) Cytotoxic lymphocytes require granzyme B for the rapid induction of DNA fragmentation and apoptosis in allogeneic target cells. *Cell*, **76**, 977–87.

Hu, S., Vincenz, C., Buller, M. and Dixit, V.M. (1997) A novel family of viral death effector domain-containing molecules that inhibit both CD-95-and tumor necrosis factor receptor-1-induced apoptosis. *J. Biol. Chem.*, **272**, 9621–4.

Hudig, D., Allison, N.J., Pickett, T.M., Winkler, U., Kam, C.M. and Powers, J.C. (1991) The function of lymphocyte proteases. Inhibition and restoration of granule-mediated lysis with isocoumarin serine protease inhibitors. *J. Immunol.*, **147**, 1360–8.

Irmler, M., Hertig, S., MacDonald, H.R., Sadoul, R., Becherer, J.D., Proudfoot, A., Solari, R. and Tschopp, J. (1995) Granzyme A is an interleukin 1 beta-converting enzyme. *J. Exp. Med.*, **181**, 1917–22.

Irmler, M., Hofmann, K., Vaux, D. and Tschopp, J. (1997) Direct physical interaction between the Caenorhabditis elegans 'death proteins' CED-3 and CED-4. *FEBS Lett.*, **406**, 189–90.

Ishikawa, H., Shinkai, Y., Yagita, H., Yue, C.C., Henkart, P.A., Sawada, S., Young, H.A., Reynolds, C.W. and Okumura, K. (1989) Molecular cloning of rat cytolysin. *J. Immunol.*, **143**, 3069–73.

Janicke, R.U., Walker, P.A., Lin, X.Y. and Porter, A.G. (1996) Specific cleavage of the retinoblastoma protein by an ICE-like protease in apoptosis. *Embo. J.*, **15**, 6969–6978.

Jans, D.A., Jans, P., Briggs, L.J., Sutton, V. and Trapani, J.A. (1996) Nuclear transport of granzyme B (fragmentin-2) Dependence of perforin in vivo and cytosolic factors in vitro. *J. Biol. Chem.*, **271**, 30781–9.

Jenne, D.E. and Tschopp, J. (1988) Granzymes, a family of serine proteases released from granules of cytolytic T lymphocytes upon T cell receptor stimultion. *Immunol. Rev.*, **103**, 53–71.

Kagi, D., Ledermann, B., Burki, K., Seiler, P., Odermatt, B., Olsen, K.J., Podack, E.R., Zinkernagel, R.M. and Hengartner, H. (1994a) Cytotoxicity mediated by T cells and natural killer cells is greatly impaired in perforin-deficient mice. *Nature*, **369**, 31–7.

Kagi, D., Ledermann, B., Burki, K., Hengartner, H. and Zinkernagel, R.M. (1994b) CD8+ T cell-mediated protection against an intracellular bacterium by perforin-dependent cytotoxicity. *Eur. J. Immunol.*, **24**, 3068–72.

Kagi, D., Seiler, P., Pavlovic, J., Ledermann, B., Burki, K., Zinkernagel, R.M. and Hengartner, H. (1995) The roles of perforin- and Fas-dependent cytotoxicity in protection against cytopathic and noncytopathic viruses. *Eur. J. Immunol.*, **25**, 3256–62.

Karre, K. (1997) How to recognize a foreign submarine. *Immunol. Rev.*, **155**, 5–9.

Kaufmann, S.H. (1989) Induction of endonucleolytic DNA cleavage in human acute myelogenous leukemia cells by etoposide, camptothecin and other cytotoxic anticancer drugs: a cautionary note. *Cancer Res.*, **49**, 5870–8.

Kaufmann, S.H., Desnoyers, S., Ottaviano, Y., Davidson, N.E. and Poirier, G.G. (1993) Specific proteolytic cleavage of poly(ADP-ribose) polymerase: an early marker of chemotherapy-induced apoptosis. *Cancer Res.*, **53**, 3976–85.

Kayalar, C., Ord, T., Testa, M.P., Zhong, L.T. and Bredesen, D.E. (1996) Cleavage of actin by interleukin 1 beta-converting enzyme to reverse DNase I inhibition. *Proc. Natl. Acad. Sci. USA*, **93**, 2234–8.

Kluck, R.M., Bossy-Wetzel, E., Green, D.R. and Newmeyer, D.D. (1997) The release of cytochrome c from mitochondria: a primary site for Bcl-2 regulation of apoptosis [see comments]. *Science*, **275**, 1132–6.

Kojima, H., Shinohara, N., Hanaoka, S., Someya-Shirota, Y., Takagaki, Y., Ohno, H., Saito, T., Katayama, T., Yagita, H., Okumura, K. and *et al.* (1994) Two distinct pathways of specific killing revealed by perforin mutant cytotoxic T lymphocytes. *Immunity* **1**, 357–64.

Kostura, M.J., Tocci, M.J., Limjuco, G., Chin, J., Cameron, P., Hillman, A.G., Chartrain, N.A. and Schmidt, J.A. (1989) Identification of a monocyte specific pre-interleukin 1 beta convertase activity. *Proc. Natl. Acad. Sci., USA* **86**, 5227–31.

Krahenbuhl, O. and Tschopp, J. (1991) Debate: the mechanism of lymphocyte-mediated killing. Perforin-induced pore formation. *Immunol. Today*, **12**, 399–402; discussion 403.

Kraut, J. (1977) Serine proteases: structure and mechanism of catalysis. *Annu. Rev. Biochem.* **46**, 331–58.

Krek, W. and Nigg, E.A. (1991) Mutations of p34cdc2 phosphorylation sites induce premature mitotic events in HeLa cells: evidence for a double block to p34cdc2 kinase activation in vertebrates. *EMBO J.*, **10**, 3331–41.

Kroemer, G., Zamzami, N. and Susin, S.A. (1997) Mitochondrial control of apoptosis. *Immunol. Today*, **18**, 44–51.

Kumar, S. and Lavin, M.F. (1996) The ICE family of cysteine proteases as effectors of cell death. *Cell Death and Differentiation*, **3**, 155–267.

Kupfer, A. (1991) T cell effector functions: mechanisms for delivery of cytotoxicity or help. *Annu. Rev. Cell Biol.*, **7**, 479–504.

Kwon, B.S., Wakulchik, M., Liu, C.C., Persechini, P.M., Trapani, J.A., Haq, A.K., Kim, Y. and Young, J.D. (1989) The structure of the mouse lymphocyte pore-forming protein perforin. *Biochem. Biophys. Res. Commun.*, **158**, 1–10.

Lanier, L.L., Corliss, B. and Phillips, J.H. (1997) Arousal and inhibition of human NK cells. *Immunol. Rev.*, **155**, 145–54.

Lazebnik, Y.A., Cole, S., Cooke, C.A., Nelson, W.G. and Earnshaw, W.C. (1993) Nuclear events of apoptosis in vitro in cell-free mitotic extracts: a model system for analysis of the active phase of apoptosis. *J. Cell. Biol.*, **123**, 7–22.

Lazebnik, Y.A., Kaufmann, S.H., Desnoyers, S., Poirier, G.G. and Earnshaw, W.C. (1994) Cleavage of poly(ADP-ribose) polymerase by a proteinase with properties like ICE. *Nature*, **371**, 346–7.

Lazebnik, Y.A., Takahashi, A., Moir, R.D., Goldman, R.D., Poirier, G.G., Kaufmann, S.H. and Earnshaw, W.C. (1995) Studies of the lamin proteinase reveal multiple parallel biochemical pathways during apoptotic execution. *Proc. Natl. Acad. Sci. USA*, **92**, 9042–6.

Lichtenheld, M.G., Olsen, K.J., Lu, P., Lowrey, D.M., Hameed, A., Hengartner, H. and Podack, E.R. (1988) Structure and function of human perforin. *Nature*, **335**, 448–51.

Liston, P., Roy, N., Tamai, K., Lefebvre, C., Baird, S., Cherton-Horvat, G., Farahani, R., McLean, M., Ikeda, J.E., MacKenzie, A. and Korneluk, R.G. (1996) Suppression of apoptosis in mammalian cells by NAIP and a related family of IAP genes. *Nature*, **379**, 349–53.

Liu, C.C., Walsh, C.M. and Young, J.D. (1995) Perforin: structure and function. *Immunol. Today*, **16**, 194–201.

Liu, C.C., Persechini, P.M. and Young, J.D. (1996) Expression and characterization of functionally active recombinant perforin produced in insect cells. *J. Immunol.*, **156**, 3292–300.

Liu, X., Zou, H., Slaughter, C. and Wang, X. (1997) DFF, a heterodimeric protein that functions downstream of caspase-3 to trigger DNA fragmentation during apoptosis. *Cell*, **89**, 175–84.

Lowin, B., Hahne, M., Mattmann, C. and Tschopp, J. (1994a) Cytolytic T-cell cytotoxicity is mediated through perforin and Fas lytic pathways. *Nature*, **370**, 650–2.

Lowin, B., Beermann, F., Schmidt, A. and Tschopp, J. (1994b) A null mutation in the perforin gene impairs cytolytic T lymphocyte-and natural killer cell-mediated cytotoxicity. *Proc. Natl. Acad. Sci. USA*, **91**, 11571–5.

Lowrey, D.M., Aebischer, T., Olsen, K., Lichtenheld, M., Rupp, F., Hengartner, H. and Podack, E.R. (1989) Cloning, analysis and expression of murine perforin 1 cDNA, a component of cytolytic T-cell granules with homology to complement component C9. *Proc. Natl. Acad. Sci. USA*, **86**, 247–51.

Macen, J.L., Garner, R.S., Musy, P.Y., Brooks, M.A., Turner, P.C., Moyer, R.W., McFadden, G. and Bleackley, R.C. (1996) Differential inhibition of the Fas and granule-mediated cytolysis pathways by the orthopoxvirus cytokine response modifier A/SPI-2 and SPI-1 protein. *Proc. Natl. Acad. Sci. USA*, **93**, 9108–13.

Marchetti, P., Hirsch, T., Zamzami, N., Castedo, M., Decaudin, D., Susin, S.A., Masse, B. and Kroemer, G. (1996) Mitochondrial permeability transition triggers lymphocyte apoptosis. *J. Immunol.*, **157**, 4830–6.

Martin, S.J., Amarante-Mendes, G.P., Shi, L., Chuang, T.H., Casiano, C.A., O'Brien, G.A., Fitzgerald, P., Tan, E.M., Bokoch, G.M., Greenberg, A.H. and Green, D.R. (1996) The cytotoxic cell protease granzyme B initiates apoptosis in a cell-free system by proteolytic processing and activation of the ICE/CED-3 family protease, CPP32, via a novel two-step mechanism. *EMBO J.*, **15**, 2407–16.

Masson, D., Corthesy, P., Nabholz, M. and Tschopp, J. (1985) Appearance of cytolytic granules upon induction of cytolytic activity in CTL-hybrids. *EMBO J.*, **4**, 2533–8.

Masson, D., Peters, P.J., Geuze, H.J., Borst, J. and Tschopp, J. (1990) Interaction of chondroitin sulfate with perforin and granzymes of cytolytic T-cells is dependent on pH. *Biochemistry*, **29**, 11229–35.

Masson, D. and Tschopp, J. (1987) A family of serine esterases in lytic granules of cytolytic T lymphocytes. *Cell*, **49**, 679–85.

Masson, D. and Tschopp, J. (1988) Inhibition of lymphocyte protease granzyme A by antithrombin III. *Mol. Immunol.*, **25**, 1283–9.

Masson, D. and Tschopp, J. (1985) Isolation of a lytic, pore-forming protein (perforin) from cytolytic T-lymphocytes. *J. Biol. Chem.*, **260**, 9069–72.

McGuire, M.J., Lipsky, P.E. and Thiele, D.L. (1993) Generation of active myeloid and lymphoid granule serine proteases requires processing by the granule thiol protease dipeptidyl peptidase I. *J. Biol. Chem.*, **268**, 2458–67.

Mullbacher, A., Ebnet, K., Blanden, R.V., Hla, R.T., Stehle, T., Museteanu, C. and Simon, M.M. (1996) Granzyme A is critical for recovery of mice from infection with the natural cytopathic viral pathogen, ectromelia. *Proc. Natl. Acad. Sci. USA*, **93**, 5783–7.

Muller, C., Kagi, D., Aebischer, T., Odermatt, B., Held, W., Podack, E.R., Zinkernagel, R.M. and Hengartner, H. (1989) Detection of perforin and granzyme A mRNA in infiltrating cells during infection of mice with lymphocytic choriomeningitis virus. *Eur. J. Immunol.*, **19**, 1253–9.

Na, S., Chuang, T.H., Cunningham, A., Turi, T.G., Hanke, J.H., Bokoch, G.M. and Danley, D.E. (1996) D4-GDI, a substrate of CPP32, is proteolyzed during Fas-induced apoptosis. *J. Biol. Chem.*, **271**, 11209–13.

Nakajima, H. and Henkart, P.A. (1994) Cytotoxic lymphocyte granzymes trigger a target cell internal disintegration pathway leading to cytolysis and DNA breakdown. *J. Immunol.*, **152**, 1057–63.

Nalefski, E.A. and Falke, J.J. (1996) The C2 domain calcium-binding motif. Structural and functional diversity. *Protein Sci.*, **5**, 2375–2390.

Nicholson, D.W., Ali, A., Thornberry, N.A., Vaillancourt, J.P., Ding, C.K., Gallant, M., Gareau, Y., Griffin, P.R., Labelle, M., Lazebnik, Y.A. and *et al.* (1995) Identification and inhibition of the ICE/CED-3 protease necessary for mammalian apoptosis. *Nature*, **376**, 37–43.

Odake, S., Kam, C.M., Narasimhan, L., Poe, M., Blake, J.T., Krahenbuhl, O., Tschopp, J. and Powers, J.C. (1991) Human and murine cytotoxic T lymphocyte serine proteases: subsite mapping with peptide thioester substrates and inhibition of enzyme activity and cytolysis by isocoumarins. *Biochemistry*, **30**, 2217–27.

Ojcius, D.M., Persechini, P.M., Zheng, L.M., Notaroberto, P.C., Adeodato, S.C. and Young, J.D. (1991a) Cytolytic and ion channel-forming properties of the N terminus of lymphocyte perforin. *Proc. Natl. Acad. Sci. USA*, **88**, 4621–5.

Ojcius, D.M., Zheng, L.M., Sphicas, E.C., Zychlinsky, A. and Young, J.D. (1991b) Subcellular localization of perforin and serine esterase in lymphokine- activated killer cells and cytotoxic T cells by immunogold labeling. *J. Immunol.*, **146**, 4427–32.

Ostergaard, H.L. and Clark, W.R. (1989) Evidence for multiple lytic pathways used by cytotoxic T lymphocytes. *J. Immunol.*, **143**, 2120–6.

Ostergaard, H.L., Kane, K.P., Mescher, M.F. and Clark, W.R. (1987) Cytotoxic T lymphocyte mediated lysis without release of serine esterase. *Nature*, **330**, 71–2.

Peters, P.J., Borst, J., Oorschot, V., Fukuda, M., Krahenbuhl, O., Tschopp, J., Slot, J.W. and Geuze, H.J. (1991) Cytotoxic T lymphocyte granules are secretory lysosomes, containing both perforin and granzymes. *J. Exp. Med.*, **173**, 1099–109.

Pinkoski, M.J., Winkler, U., Hudig, D. and Bleackley, R.C. (1996) Binding of granzyme B in the nucleus of target cells. Recognition of an 80-kilodalton protein. *J. Biol. Chem.*, **271**, 10225–9.

Podack, E.R. (1989) Granule-mediated cytolysis of target cells. *Curr. Top Microbiol. Immunol.*, **140**, 1–9.

Podack, E.R. (1992) Perforin: structure, function and regulation. *Curr. Top Microbiol. Immunol.*, **178**, 175–84.

Podack, E.R. and Dennert, G. (1983) Assembly of two types of tubules with putative cytolytic function by cloned natural killer cells. *Nature*, **302**, 442–5.

Podack, E.R., Young, J.D. and Cohn, Z.A. (1985) Isolation and biochemical and functional characterization of perforin 1 from cytolytic T-cell granules. *Proc. Natl. Acad. Sci. USA*, **82**, 8629–33.

Poe, M., Blake, J.T., Boulton, D.A., Gammon, M., Sigal, N.H., Wu, J.K. and Zweerink, H.J. (1991) Human cytotoxic lymphocyte granzyme B. Its purification from granules and the characterization of substrate and inhibitor specificity. *J. Biol. Chem.*, **266**, 98–103.

Quan, L.T., Caputo, A., Bleackley, R.C., Pickup, D.J. and Salvesen, G.S. (1995) Granzyme B is inhibited by the cowpox virus serpin cytokine response modifier A. *J. Biol. Chem.*, **270**, 10377–9.

Ray, C.A., Black, R.A., Kronheim, S.R., Greenstreet, T.A., Sleath, P.R., Salvesen, G.S. and Pickup, D.J. (1992) Viral inhibition of inflammation: cowpox virus encodes an inhibitor of the interleukin-1 beta converting enzyme. *Cell*, **69**, 597–604.

Remington, S.J., Woodbury, R.G., Reynolds, R.A., Matthews, B.W. and Neurath, H. (1988) The structure of rat mast cell protease II at 1.9-A resolution. *Biochemistry*, **27**, 8097–105.

Roy, N., Mahadevan, M.S., McLean, M., Shutler, G., Yaraghi, Z., Farahani, R., Baird, S., Besner-Johnston, A., Lefebvre, C., Kang, X. and *et al.* (1995) The gene for neuronal apoptosis inhibitory protein is partially deleted in individuals with spinal muscular atrophy [see comments]. *Cell*, **80**, 167–78.

Russell, J.H. (1983) Internal disintegration model of cytotoxic lymphocyte-induced target damage. *Immunol Rev.*, **72**, 97–118.

Sarin, A., Williams, M.S., Alexander-Miller, M.A., Berzofsky, J.A., Zacharchuk, C.M. and Henkart, P.A. (1997a) CTL lysis via the two CTL pathways differs with respect to utilization of target caspases: lysis via Fas requires caspases while granule exocytosis does not. In Proceedings of the Sixth EMBO Workshop on Cell-mediated Cytotoxicity (Kerkrade, The Netherlands, pp. 28 (abstract)

Sarin, A., Williams, M.S., Alexander-Miller, M.A., Berzofsky, J.A., Zacharchuk, C.M. and Henkart, P.A. (1997b) Target cell lysis by CTL granule exocytosis is independent of ICE/Ced-3 family proteases. *Immunity*, **6**, 209–15.

Sayers, T.J., Wiltrout, T.A., Sowder, R., Munger, W.L., Smyth, M.J. and Henderson, L.E. (1992) Purification of a factor from the granules of a rat natural killer cell line (RNK) that reduces tumor cell growth and changes tumor morphology. Molecular identity with a granule serine protease (RNKP-1) *J. Immunol*, **148**, 292–300.

Schmid, D.S., Tite, J.P. and Ruddle, N.H. (1986) DNA fragmentation: manifestation of target cell destruction mediated by cytotoxic T-cell lines, lymphotoxin-secreting helper T-cell clones and cell-free lymphotoxin-containing supernatant. *Proc. Natl. Acad. Sci. USA*, **83**, 1881–5.

Schmidt, R.E., MacDermott, R.P., Bartley, G., Bertovich, M., Amato, D.A., Austen, K.F., Schlossman, S.F., Stevens, R.L. and Ritz, J. (1985) Specific release of proteoglycans from human natural killer cells during target lysis. *Nature*, **318**, 289–91.

Schroter, M., Lowin, B., Borner, C. and Tschopp, J. (1995) Regulation of Fas(Apo-1/CD95)-and perforin-mediated lytic pathways of primary cytotoxic T lymphocytes by the protooncogene bcl-2. *Eur. J. Immunol.*, **25**, 3509–13.

Selvakumar, A., Steffens, U. and Dupont, B. (1997) Polymorphism and domain variability of human killer cell inhibitory receptors. *Immunol. Rev.*, **155**, 183–96.

Seth, P. (1994) A simple and efficient method of protein delivery into cells using adenovirus. *Biochem. Biophys. Res. Commun.*, **203**, 582–7.

Shi, L., Kam, C.M., Powers, J.C., Aebersold, R. and Greenberg, A.H. (1992b) Purification of three cytotoxic lymphocyte granule serine proteases that induce apoptosis through distinct substrate and target cell interactions. *J. Exp. Med.*, **176**, 1521–9.

Shi, L., Kraut, R.P., Aebersold, R. and Greenberg, A.H. (1992a) A natural killer cell granule protein that induces DNA fragmentation and apoptosis. *J. Exp. Med.*, **175**, 553–66.

Shi, L., Mai, S., Israels, S., Browne, K., Trapani, J.A. and Greenberg, A.H. (1997) Granzyme B (GraB) autonomously crosses the cell membrane and perforin initiates apoptosis and GraB nuclear localization. *J. Exp. Med.*, **185**, 855–66.

Shi, L., Nishioka, W.K., Th'ng, J., Bradbury, E.M., Litchfield, D.W. and Greenberg, A.H. (1994) Premature p34cdc2 activation required for apoptosis [see comments]. *Science*, **263**, 1143–5.

Shinkai, Y., Takio, K. and Okumura, K. (1988) Homology of perforin to the ninth component of complement (C9) *Nature*, **334**, 525–7.

Shinkai, Y., Yoshida, M.C., Maeda, K., Kobata, T., Maruyama, K., Yodoi, J., Yagita, H. and Okumura, K. (1989) Molecular cloning and chromosomal assignment of a human perforin (PFP) gene. *Immunogenetics*, **30**, 452–7.

Shiver, J.W. and Henkart, P.A. (1991) A noncytotoxic mast cell tumor line exhibits potent IgE-dependent cytotoxicity after transfection with the cytolysin/perforin gene. *Cell*, **64**, 1175–81.

Shiver, J.W., Su, L. and Henkart, P.A. (1992) Cytotoxicity with target DNA breakdown by rat basophilic leukemia cells expressing both cytolysin and granzyme A. *Cell*, **71**, 315–22.

Sieg, S., Xia, L., Huang, Y. and Kaplan, D. (1995) Specific inhibition of granzyme B by parainfluenza virus type 3. *J. Virol.*, **69**, 3538–41.

Simon, M.M. and Kramer, M.D. (1994) Granzyme A. *Methods Enzymol.*, **244**, 68–79.

Simon, M.M., Simon, H.G., Fruth, U., Epplen, J., Muller-Hermelink, H.K. and Kramer, M.D. (1987a) Cloned cytolytic T-effector cells and their malignant variants produce an extracellular matrix degrading trypsin-like serine proteinase. *Immunology*, **60**, 219–30.

Singer, G.G. and Abbas, A.K. (1994) The Fas antigen is involved in peripheral but not thymic deletion of T lymphocytes in T cell receptor transgenic mice. *Immunity*, **1**, 365–71.

Smyth, M.J., McGuire, M.J. and Thia, K.Y. (1995b) Expression of recombinant human granzyme B.A processing and activation role for dipeptidyl peptidase I. *J. Immunol.*, **154**, 6299–305.

Smyth, M.J., O'Connor, M.D. and Trapani, J.A. (1996c) Granzymes: a variety of serine protease specificities encoded by genetically distinct subfamilies. *J. Leukoc. Biol.*, **60**, 555–62.

Smyth, M.J., O'Connor, M.D., Trapani, J.A., Kershaw, M.H. and Brinkworth, R.I. (1996a) A novel substrate-binding pocket interaction restricts the specificity of the human NK cell-specific serine protease, Met-ase-1. *J. Immunol.*, **156**, 4174–81.

Smyth, M.J., Sutton, V.R., Kershaw, M.H. and Trapani, J.A. (1996b) Xenospecific cytotoxic T lymphocytes use perforin-and Fas-mediated lytic pathways. *Transplantation*, **62**, 1529–32.

Smyth, M.J. and Trapani, J.A. (1995) Granzymes: exogenous proteinases that induce target cell apoptosis. *Immunol. Today*, **16**, 202–6.

Smyth, M.J., Wiltrout, T., Trapani, J.A., Ottaway, K.S., Sowder, R., Henderson, L.E., Kam, C.M., Powers, J.C., Young, H.A. and Sayers, T.J. (1992) Purification and cloning of a novel serine protease, RNK-Met-1, from the granules of a rat natural killer cell leukemia. *J. Biol. Chem.*, **267**, 24418–25.

Song, Q., Lees-Miller, S.P., Kumar, S., Zhang, Z., Chan, D.W., Smith, G.C., Jackson, S.P., Alnemri, E.S., Litwack, G., Khanna, K.K. and Lavin, M.F. (1996) DNA-dependent protein kinase catalytic subunit: a target for an ICE-like protease in apoptosis. *EMBO J.*, **15**, 3238–46.

Spielman, J., Baker, M.B., Levy, R.B., Lee, R.K. and Podack, E.R. (1997) Perforin, antigen presentation and T cell homeostasis. In Proceedings of the Sixth EMBO Workshop on Cell-mediated Cytotoxicity (Kerkrade, The Netherlands, 51–52 (abstract).

Sprecher, C.A., Morgenstern, K.A., Mathewes, S., Dahlen, J.R., Schrader, S.K., Foster, D.C. and Kisiel, W. (1995) Molecular cloning, expression and partial characterization of two novel members of the ovalbumin family of serine proteinase inhibitors. *J. Biol. Chem.*, **270**, 29854–61.

Srinivasula, S.M., Fernandes-Alnemri, T., Zangrilli, J., Robertson, N., Armstrong, R.C., Wang, L., Trapani, J.A., Tomaselli, K.J., Litwack, G. and Alnemri, E.S. (1996a) The Ced-3/interleukin 1beta converting enzyme-like homolog Mch6 and the lamin-cleaving enzyme Mch2alpha are substrates for the apoptotic mediator CPP32. *J. Biol. Chem.*, **271**, 27099–106.

Srinivasula, S.M., Ahmad, M., Fernandes-Alnemri, T., Litwack, G. and Alnemri, E. (1996b) Molecular ordering of the Fas-apoptotic pathway: the Fas / APO-1 protease Mch5 is a crmA-inhibitable protease that activates multiple Ced-3/ICE-like cysteine proteases. *Proc. Natl. Acad. Sci. USA*, **93**, 14486–14491.

Suidan, H.S., Bouvier, J., Schraer, S., Stone, S.R., Monard, D. and Tschopp, J. (1994) Granzyme A release upon stimulation of cytotoxic T lymphocytes activates the thrombin receptor on neuronal cells and astrocytes. *Proc. Natl. Acad. Sci. USA*, **91**, 812–8116.

Sun, J., Bird, C.H., Sutton, V., McDonald, L., Coughlin, P.B., De Jong, T.A., Trapani, J.A. and Bird, P.I. (1996) A cytosolic granzyme B inhibitor related to the viral apoptotic regulator cytokine response modifier A is present in cytotoxic lymphocytes. *J. Biol. Chem.*, **271**, 27802–9.

Sun, J., Ooms, L., Bird, C.H., Sutton, V.R., Trapani, J.A. and Bird, P.I. (1997) A new family of 10 murine ovalbumin serpins includes two homologs of proteinase inhibitor 8 and two homologs of the granzyme B inhibitor (Proteinase inhibitor 9) [In Process Citation]. *J. Biol. Chem.*, **272**, 15434–41.

Susin, S.A., Zamzami, N., Castedo, M., Hirsch, T., Marchetti, P., Macho, A., Daugas, E., Geuskens, M. and Kroemer, G. (1996) Bcl-2 inhibits the mitochondrial release of an apoptogenic protease. *J. Exp. Med.*, **184**, 1331–41.

Sutton, V.R., Vaux, D.L. and Trapani, J.A. (1997) Bcl-2 prevents apoptosis induced by perforin and granzyme B, but not that mediated by whole cytotoxic lymphocytes. *J. Immunol.*, **158**, 5783–90.

Takayama, H., Trenn, G., Humphrey, W., Jr., Bluestone, J.A., Henkart, P.A. and Sitkovsky, M.V. (1987) Antigen receptor-triggered secretion of a trypsin-type esterase from cytotoxic T lymphocytes. *J. Immunol.*, **138**, 566–9.

Tewari, M. and Dixit, V.M. (1995) Fas- and tumor necrosis factor-induced apoptosis is inhibited by the poxvirus crmA gene product. *J. Biol. Chem.*, **270**, 3255–60.

Tewari, M., Quan, L.T., O'Rourke, K., Desnoyers, S., Zeng, Z., Beidler, D.R., Poirier, G.G., Salvesen, G.S. and Dixit, V.M. (1995a) Yama/CPP32 beta, a mammalian homolog of CED-3, is a CrmA-inhibitable protease that cleaves the death substrate poly(ADP-ribose) polymerase. *Cell*, **81**, 801–9.

Tewari, M., Telford, W.G., Miller, R.A. and Dixit, V.M. (1995b) CrmA, a poxvirus-encoded serpin, inhibits cytotoxic T-lymphocyte-mediated apoptosis. *J. Biol. Chem.*, **270**, 22705–8.

Thome, M., Schneider, P., Hofmann, K., Fickenscher, H., Meinl, E., Neipel, F., Mattmann, C., Burns, K., Bodmer, J.L., Schroter, M., Scaffidi, C., Krammer, P.H., Peter, M.E. and Tschopp, J. (1997) Viral FLICE-inhibitory proteins (FLIPs) prevent apoptosis induced by death receptors. *Nature*, **386**, 517–21.

Thornberry, N.A., Bull, H.G., Calaycay, J.R., Chapman, K.T., Howard, A.D., Kostura, M.J., Miller, D.K., Molineaux, S.M., Weidner, J.R., Aunins, J. and *et al.* (1992) A novel heterodimeric cysteine protease is required for interleukin-1 beta processing in monocytes. *Nature*, **356**, 768–74.

Trapani, J.A., Browne, K.A., Dawson, M. and Smyth, M.J. (1993) Immunopurification of functional Asp-ase (natural killer cell granzyme B) using a monoclonal antibody. *Biochem. Biophys. Res. Commun.*, **195**, 910–20.

Trapani, J.A., Browne, K.A., Smyth, M.J. and Jans, D.A. (1996) Localization of granzyme B in the nucleus. A putative role in the mechanism of cytotoxic lymphocyte-mediated apoptosis. *J. Biol. Chem.*, **271**, 4127–33.

Trapani, J.A., Jans, P., Froelich, C.J., Smyth, M.J., Sutton, V.R. and Jans, D. (1997) Cytoplasmic entry of granzyme B is perforin independent, however its nuclear accumulation requires perforin and indicates imminent apoptosis. (in press)

Trapani, J.A. and Smyth, M.J. (1993) Killing by cytotoxic T cells and natural killer cells: multiple granule serine proteases as initiators of DNA fragmentation. *Immunol. Cell Biol.*, **71**, 201–8.

Trapani, J.A., Smyth, M.J., Apostolidis, V.A., Dawson, M. and Browne, K.A. (1994) Granule serine proteases are normal nuclear constituents of natural killer cells. *J. Biol. Chem.*, **269**, 18359–65.

Trenn, G., Takayama, H. and Sitkovsky, M.V. (1987) Exocytosis of cytolytic granules may not be required for target cell lysis by cytotoxic T-lymphocytes. *Nature*, **330**, 72–4.

Tschopp, J. and Jongeneel, C.V. (1988) Cytotoxic T lymphocyte mediated cytolysis. *Biochemistry*, **27**, 2641–6.

Tschopp, J., Schafer, S., Masson, D., Peitsch, M.C. and Heusser, C. (1989) Phosphorylcholine acts as a Ca2+-dependent receptor molecule for lymphocyte perforin. *Nature*, **337**, 272–4.

Tschopp, J. and Nabholtz, M. (1990) Perforin-mediated target cell lysis by cytotoxic T lymphocytes. *Annu. Rev. Immunol.*, **8**, 279–302.

Uellner, R., Jones, J. and Griffiths, G.M. (1997) Perforin is activated by proteolytic cleavage during biosynthesis which reveals a phospholipid binding domain. In Proceedings of the Sixth EMBO Workshop on Cell-mediated Cytotoxicity (Kerkrade, The Netherlands, pp. 25 (abstract)

Uren, A.G., Pakusch, M., Hawkins, C.J., Puls, K.L. and Vaux, D.L. (1996) Cloning and expression of apoptosis inhibitory protein homologs that function to inhibit apoptosis and/or bind tumor necrosis factor receptor-associated factors. *Proc. Natl. Acad. Sci. USA*, **93**, 4974–8.

van den Broek, M.E., Kagi, D., Ossendorp, F., Toes, R., Vamvakas, S., Lutz, W.K., Melief, C.J., Zinkernagel, R.M. and Hengartner, H. (1996) Decreased tumor surveillance in perforin-deficient mice. *J. Exp. Med.*, **184**, 1781–90.

van den Broek, M.F., Kagi, D., Zinkernagel, R.M. and Hengartner, H. (1995) Perforin dependence of natural killer cell-mediated tumor control in vivo. *Eur. J. Immunol.*, **25**, 3514–6.

Vaux, D.L., Aguila, H.L. and Weissman, I.L. (1992b) Bcl-2 prevents death of factor-deprived cells but fails to prevent apoptosis in targets of cell mediated killing. *Int. Immunol.*, **4**, 821–4.

Vaux, D.L., Haecker, G. and Strasser, A. (1994) An evolutionary perspective on apoptosis. *Cell*, **76**, 777–9.

Vaux, D.L., Weissman, I.L. and Kim, S.K. (1992a) Prevention of programmed cell death in Caenorhabditis elegans by human bcl-2. *Science*, **258**, 1955–7.

Walsh, C.M., Matloubian, M., Liu, C.C., Ueda, R., Kurahara, C.G., Christensen, J.L., Huang, M.T., Young, J.D., Ahmed, R. and Clark, W.R. (1994) Immune function in mice lacking the perforin gene. *Proc. Natl. Acad. Sci. USA*, **91**, 10854–8.

Wang, X., Zelenski, N.G., Yang, J., Sakai, J., Brown, M.S. and Goldstein, J.L. (1996) Cleavage of sterol regulatory element binding proteins (SREBPs) by CPP32 during apoptosis. *EMBO J.*, **15**, 1012–20.

Wang, Z.Q., Auer, B., Stingl, L., Berghammer, H., Haidacher, D., Schweiger, M. and Wagner, E.F. (1995) Mice lacking ADPRT and poly(ADP-ribosyl)ation develop normally but are susceptible to skin disease. *Genes Dev.*, **9**, 509–20.

White, E., Cipriani, R., Sabbatini, P. and Denton, A. (1991) Adenovirus E1B 19-kilodalton protein overcomes the cytotoxicity of E1A proteins. *J. Virol.*, **65**, 2968–78.

Wu, D., Wallen, H.D. and Nunez, G. (1997) Interaction and regulation of subcellular localization of CED-4 by CED-9 [see comments]. *Science*, **275**, 1126–9.

Xue, D. and Horvitz, H.R. (1995) Inhibition of the Caenorhabditis elegans cell-death protease CED-3 by a CED-3 cleavage site in baculovirus p35 protein. *Nature*, **377**, 248–51.

Yang, J., Liu, X., Bhalla, K., Kim, C.N., Ibrado, A.M., Cai, J., Peng, T.I., Jones, D.P. and Wang, X. (1997) Prevention of apoptosis by Bcl-2: release of cytochrome c from mitochondria blocked [see comments]. *Science*, **275**, 1129–32.

Yannelli, J.R., Sullivan, J.A., Mandell, G.L. and Engelhard, V.H. (1986) Reorientation and fusion of cytotoxic T lymphocyte granules after interaction with target cells as determined by high resolution cinemicrography. *J. Immunol.*, **136**, 377–82.

Young, J.D. and Cohn, Z.A. (1986) Cell-mediated killing: a common mechanism? *Cell*, **46**, 641–2.

Yuan, J. and Horvitz, H.R. (1992) The Caenorhabditis elegans cell death gene ced-4 encodes a novel protein and is expressed during the period of extensive programmed cell death. *Development.*, **116**, 309–20.

Yuan, J., Shaham, S., Ledoux, S., Ellis, H.M. and Horvitz, H.R. (1993) The C. elegans cell death gene ced-3 encodes a protein similar to mammalian interleukin-1 beta-converting enzyme. *Cell*, **75**, 641–52.

Yue, C.C., Reynolds, C.W. and Henkart, P.A. (1987) Inhibition of cytolysin activity in large granular lymphocyte granules by lipids: evidence for a membrane insertion mechanism of lysis. *Mol. Immunol.*, **24**, 647–53.

Zagury, D., Bernard, J., Thierness, N., Feldman, M and Berke, G. (1975) Isolation and characterization of individual functionally reactive cytotoxic T lymphocytes: conjugation, killing and recycling at the single cell level. *Eur. J. Immunol.*, **5**, 818–822.

Zhivotovsky, B., Gahm, A., Ankarcrona, M., Nicotera, P. and Orrenius, S. (1995) Multiple proteases are involved in thymocyte apoptosis. *Exp. Cell. Res.*, **221**, 404–12.

Zou, H., Henzel, W.J., Liu, X., Lutschg, A. and Wang, X. (1997) Apaf-1, a human protein homologous to C.elegans CED-4, participates in cytochrome c-dependent activation of caspase-3. *Cell*, **90**, 405–13.

12. SUBSTRATES OF CELL DEATH PROTEASES AND THEIR ROLE IN APOPTOSIS

DIANNE WATTERS*† AND NIGEL WATERHOUSE

**Queensland Cancer Fund Research Unit, Queensland Institute of Medical Research, P.O. Royal Brisbane Hospital, 4029, Australia*

KEY WORDS: caspases, proteasome, calpain

INTRODUCTION

Apoptosis or programmed cell death is a morphologically distinct form of cell death, highly conserved in multicellular organisms. It is characterised by a precisely orchestrated sequence of morphological changes which result in the engulfment of the dying cell by macrophages in the absence of inflammation. These changes include cell shrinkage, plasma membrane blebbing, chromatin condensation, nuclear segmentation and formation of apoptotic bodies. A variety of death signals including DNA damage, ionizing or UV radiation, cytotoxic agents, growth factor withdrawal, and cytokines can induce apoptosis. The cellular signalling pathways involved in controlling apoptosis remain poorly defined and little is known about the biochemical mechanisms underlying the dramatic changes that accompany cell death. The importance of proteases in the execution of apoptosis has become increasingly apparent, in particular the cysteine proteases of the interleukin 1β-converting enzyme (ICE)-like family (caspases) (Martin and Green, 1995; Kumar and Lavin, 1996; Cohen, 1997) and the serine proteases of the granzyme B family (Greenberg, 1996). A role for calpain (Squier and Cohen, 1996) and the proteasome has also been described (Grimm *et al.*, 1996).

Caspases display an absolute requirement for aspartic acid at the P1 position of their substrate (Alnemri *et al.*, 1996). The existence of multiple caspases with potentially overlapping specificity, differential tissue expression and potential for participation in a protease cascade suggests that each enzyme has a defined role to play in the execution of apoptosis although there is also possibly some redundancy built into the system. Caspase-1 (ICE) plays a critical role in inflammation by the regulation of multiple proinflammatory cytokines and also in some forms of apoptosis (Li *et al.*, 1995; Kuida *et al.*, 1995; Gu *et al.*, 1996; Ghayur *et al.*, 1997; Friedlander *et al.*, 1997) and caspase-3 is required for normal development of the mouse brain (Kuida *et al.*, 1996).

As detailed elsewhere in this book, caspases are synthesised as inactive precursors and activated by other caspases/proteases in activation cascades. Several studies have attempted to define substrate specifities of the caspases in order to

† *Corresponding Author*: Tel.: 61–7–33620335. Fax: 61–7–33620106. e-mail: dianneW@qimr.edu.au

determine their individual roles (Talanian *et al.*, 1997; Thornberry *et al.*, 1997). The results divide the caspases into three distinct groups and suggest that several have redundant functions: group I (caspases -1,4,5) have a preference for the sequence WEHD; group II (CED-3 and caspases -2,3,7, prefer the DEXD sequence, which is similar or identical to the cleavage sites in known macromolecular substrates cleaved during apoptosis; and group III (caspases -6,8,9, and granzyme B) have an optimal sequence which resembles activation sites in caspase proenzymes, consistent with a role for these enzymes as upstream components in the proteolytic cascade.

Since the majority of cellular proteins remain intact during apoptosis (Kaufmann *et al.*, 1989; Baxter *et al.*, 1989), the proteins which are targeted for cleavage must have a crucial role in the cell, or be involved in certain cellular functions which are shut down in order to conserve energy for the apoptotic process. Structural proteins must also be targeted to enable the morphological changes associated with apoptosis to occur. Some substrates are activated by cleavage and this may commit the cell to undergo apoptosis. Others are inactivated by cleavage which may accelerate the apoptotic process or conserve energy. The identification of the specific substrates for individual caspases and elucidating the consequences of their cleavage remains a major challenge but is essential for our understanding of the molecular mechanisms of cell death. This chapter will review the current knowledge of apoptosis substrates excluding the proteolytic activation of the caspases themselves.

PROTEOLYTIC TARGETS WHICH ARE ACTIVATED AFTER CLEAVAGE

The classical hallmark of apoptosis is the cleavage of chromatin into nucleosomal fragments (Wyllie *et al.*, 1980). A DNA Fragmentation Factor (DFF) was first purified and characterised from HeLa cytosol (Liu *et al.*, 1997). It is the first caspase substrate with a proven function in apoptosis. DFF was shown to consist of two previously uncharacterised protein subunits of 45 kDa and 40 kDa in size. Caspase-3 cleaves the 45 kDa subunit of DFF into an intermediate fragment of 30 kDa and an 11 kDa fragment, while the 40 kDa subunit remains intact. N-terminal sequence analysis and mass spectrometry of the cleavage products revealed that caspase-3 cleaves DFF at Asp 117 and Asp 224. When activated by caspase-3 DFF initiates the fragmentation of DNA into oligonucleosomal sized fragments. Once DFF is activated, caspase-3 activity is no longer required for DNA fragmentation although it is required for cleavage of other substrates. Interestingly, DFF and caspase-3 showed no detectable nuclease activity when incubated with naked DNA, thus it is unlikely that DFF is a nuclease that directly cleaves DNA. Liu *et al.* (1997) suggest that DFF might either translocate into the nuclei or interact with protein(s) on the outer surface of the nuclear envelope to trigger a signal transduction pathway which activates a nuclease.

Many of the substrates which are activated after cleavage, like DFF are components of signal transduction pathways. Radiation-induced apoptosis in U937 cells results in the proteolytic cleavage of PKCδ to an activated form (Emoto *et al.*,

1995). The cleavage occurs next to an aspartic acid supporting a role for caspases and is inhibited by Bcl-2. The caspase responsible for PKCδ cleavage was later identified as caspase-3 (Ghayur *et al.*, 1996). PKCδ proteolysis results in a sixfold increase in kinase activity. Overexpression of the PKCδ catalytic fragment in HeLa cells resulted in the phenotypic changes of apoptosis, detachment from the culture dish and loss of viability. Transfection of a kinase inactive mutant produced no changes in morphology.

Another member of the novel PKC family, PKCθ has recently been shown to be cleaved in the third variable region (V3) in apoptosis induced by a variety of signals (Datta *et al.*, 1997). Cleavage is mediated by caspase-3 at $DEVD_{354}{\downarrow}K$. Overexpression of the cleaved kinase-active fragment, but not the full-length or kinase-inactive fragment, results in induction of apoptosis. The functional role of these novel PKCs in cells is currently unknown and the substrates phosphorylated by them are also unknown, hence their role in apoptosis is unclear.

PITSLRE kinases are a superfamily of protein kinases related to the mitotic kinase Cdc2. Ectopic expression of the smallest member of this superfamily induces apoptosis (Bunnell *et al.*, 1990). Induction of apoptosis by Fas has been correlated with proteolysis and increased activity of these kinases (Lahti *et al.*, 1995). Beyaert *et al.* (1997) showed the cleavage of the p110 α2 form of PITSLRE kinase to 60 kDa and 43 kDa fragments and the p170 isoform to a 130 kDa fragment, while the p90 isoform remained unaffected, during apoptosis in a rodent T cell hybridoma treated with TNF. *In vitro* translated p110 PITSLRE kinase could be cleaved to 60 kDa and 43 kDa fragments only by caspase-1 and caspase-3 however *in vivo*, it is likely that caspase-3 mediates PITSLRE kinase cleavage since the cleavage was similar in fibroblasts from mice deficient in caspase-1. Caspase cleavage separates the C-terminal kinase domain from the N-terminal portion of the molecule which contains two nuclear localisation signals as well as an SH2 binding domain. Thus cleavage might be a mechanism to modulate the localisation of these kinases or their interaction with other proteins. To date no specific substrates for PITSLRE kinases have been identified.

Epithelial, endothelial and muscle cells undergo apoptosis when they lose contact with the extracellular matrix, a process which has been termed "anoikis" (Ruoshlati and Reed, 1994). The Jun-N-terminal kinase (JNK) pathway is activated in and promotes anoikis (Frisch *et al.*, 1996). Cardone *et al.* (1997) have now shown that MEKK-1, an upstream activator of JNK, is cleaved by caspases releasing the inhibitory N-terminal domain. When overexpressed the MEKK-1 cleavage product stimulates apoptosis and a cleavage-resistant mutant of MEKK-1 partially protects cells from anoikis. In addition the cleavage resistant and kinase inactive mutants also inhibited the caspase proteolytic cascade. The activation of JNKK and in turn JNK by MEKKs is likely to be one important downstream effector of MEKK activation. A major substrate of JNK, c-Jun was shown to promote apoptosis (Bossy-Wetzel *et al.*, 1997) although the role of the JNK pathway in other systems has been variable.

p21-activated kinases (PAKs) are serine-threonine kinases whose activity is regulated by the small GTPases, Rac and Cdc42, and the effects of Rac and Cdc42 on the cytoskeleton are probably mediated by PAKs (Sells *et al.*, 1997). Fas-induced apoptosis of Jurkat T cells and TNF-α and C2 ceramide induced apoptosis of

MCF-7 cells, results in early cleavage of the 62 kDa PAK2 to fragments of 34 kDa and 28 kDa. (Rudel and Bokoch, 1997). The cleavage site was determined to be at Asp 212 and most likely to be mediated by caspase-3. This site is also present in PAK1, which is not cleaved during apoptosis. In PAK1, Asp 212 is surrounded by a number of features (including continuous proline residues) which could disrupt caspase recognition. PAK3, which also remains intact, lacks the relevant Asp residue. Cleavage of PAK2 would result in removal of the regulatory domain and subsequent activation of PAK2. Cells expressing dominant-negative PAK2 did not form apoptotic bodies during Fas-mediated cell death, but remained as intact rounded cells in which DNA fragmentation could still be detected by the TUNEL assay. Blocking PAK2 function also accelerated the externalisation of phosphatidylserine on the plasma membrane. PAK2 is a ubiquitously expressed protein in mammalian tissues and is thus likely to play a crucial role in the morphological changes associated with apoptosis.

Wissing *et al.* (1997) have recently shown that cytosolic phospholipase A2 (cPLA2) is cleaved by a caspase-3-like enzyme in TNF-induced apoptosis resulting in activation of the enzyme. The 100 kDa cPLA2 was cleaved to a 70 kDa fragment. Inhibition of caspase activity by DEVD-CHO or CrmA inhibited both the induced cleavage and activation of cPLA2. Arachidonic acid generated by cPLA2 activation has been implicated in a signal transduction pathway resulting in cell death (Hayakawa *et al.*, 1993; Voelkel-Johnson *et al.*, 1996). These results show that activation of cPLA2 requires caspases and that cPLA2 acts as a death mediator in TNF-induced apoptosis, consistent with earlier results from the same group showing that inhibition of TNF-induced apoptosis is always accompanied by inhibition of TNF-induced activation of cPLA2.

When cells are depleted of cholesterol, a protease cleaves the sterol regulatory element binding proteins (SREBPs) to release amino terminal fragments of approx. 500 amino acids which enter the nucleus and activate transcription of genes encoding the low density lipoprotein receptor and enzymes of cholesterol biosynthesis. Wang, Pai *et al.* (1995) purified the hamster equivalent of caspase-3 (named SCA-1) which cleaves SREBP-1/2 at a conserved Asp in the consensus DEPD↓S. Cleavage of SREBPs by SCA-1 is completely distinct from the sterol-regulated cleavage process and occurs in response to various apoptotic stimuli (Wang *et al.*, 1996). A second SREBP cleaving activity (SCA-2), which is the hamster homolog of caspase-7, has now been purified and it cleaves at the same site as SCA-1 (Pai *et al.*, 1996). The significance of the cleavage of SREBPs during apoptosis is not yet established and it is not known whether the liberated N-terminal fragments can activate transcription in apoptosis.

SUBSTRATES WHICH ARE INACTIVATED BY CLEAVAGE

Poly (ADP-ribose) polymerase (PARP) is an abundant nuclear enzyme which recognises and binds to DNA strand-breaks and synthesises poly (ADP-ribose) quantitatively according to the number of breaks (Masutani *et al.*, 1995). The cleavage of PARP to a 85 kDa fragment has been regarded as a hallmark of apoptosis

(Kaufmann *et al.*, 1993) and every caspase examined has been shown to have the ability to cleave this protein with varying catalytic efficiencies although the most likely mediator of PARP cleavage *in vivo* is caspase-3 (Takahashi & Earnshaw, 1996; Margolin *et al.*, 1997). Proteolytic cleavage of PARP results in separation of the two zinc finger DNA-binding motifs from the C-terminal catalytic domain (Nicholson *et al.*, 1995). PARP knockout mice develop normally and their fibroblasts efficiently repair DNA damage caused by UV and alkylating agents. These mice do however develop skin hyperplasia with age (Wang, Auer *et al.*, 1995) thus PARP may function in response to environmental stress. Proliferation of mutant fibroblasts and thymocytes following ionising radiation was impaired consistent with the proposed role of PARP in cell cycle checkpoint mechanisms following γ-irradiation (Masutani *et al.*, 1995). The physiological role of this enzyme in apoptosis is unclear. Cleavage of PARP may decrease consumption of NAD+ and its precursor ATP, conserving energy for the completion of apoptosis (Lazebnik *et al.*, 1994) or it may permit deribosylation and consequent activation of endonucleases.

Another DNA repair enzyme which is cleaved in apoptosis is the catalytic subunit of DNA-dependent protein kinase (DNA-PKcs) which belongs to a subset of autoantigens found to be specifically cleaved early during apoptosis (Casciola-Rosen *et al.*, 1995). The complex of the 460 kDa catalytic subunit of DNA-PK and the heterodimeric Ku protein has been shown to be involved in double-strand break repair (DSB) (Anderson *et al.*, 1996; Jackson, 1996) and mutations in DNA-PKcs give rise to the severe combined immunodeficiency (scid) phenotype, which includes defective DNA DSB repair (Weaver *et al.*, 1996a). DNA-PKcs is selectively cleaved in cells undergoing apoptosis and also during CTL-mediated cytolysis, but not in lines resistant to apoptosis (Song *et al.*, 1996a; Han *et al.*, 1996; Song *et al.*, 1996b). DNA-PKcs activity progressively decreased with time in apoptotic cells in parallel with the cleavage (Song *et al.*, 1996a). Interestingly the p70/p86 Ku proteins which associate with DNA-PKcs remained intact.

Replication factor C (RFC) is a multisubunit DNA polymerase accessory complex required for DNA replication *in vitro* and consists of five polypeptides of masses 140, 40, 38, 37, and 36 kDa in humans (Cullman *et al.*, 1995). RFC possesses a primer/template DNA-binding activity that has been localised to the 140 kDa subunit. Two groups have shown that RFC 140 is specifically cleaved in cells undergoing apoptosis to an 87 kDa fragment (Song *et al.*, 1997a; Ubeda and Habener, 1997). The protease responsible was identified as caspase-3 or a closely related enzyme, and the cleavage site as $DEVD_{706}{\downarrow}G$. The cleavage of this substrate would result in the separation of its DNA binding from its association domain, required for replication complex formation, therefore resulting in the impairment of DNA replication.

Another autoantigen U1-70 kDa (70 kDa protein component of the U1 small ribonucleoprotein) is redistributed in apoptotic cells appearing first as a rim around condensing chromatin and subsequently around apoptotic bodies at the cell surface (Casciola-Rosen *et al.*, 1994a). It is specifically cleaved in apoptotic cells to a 40 kDa fragment (Casciola Rosen *et al.*, 1994b; Tewari *et al.*, 1995). Purified ICE was unable to cleave this protein suggesting that another caspase, possibly caspase-3, is responsible. Cleavage of U1-70 kDa separates the N-terminal RNA recognition motif from the C-terminal RSD-rich domain which potentially has a dominant negative effect on mRNA splicing (Romac and Keene, 1995).

We have identified the heteronuclear ribonucleoproteins (hnRNP) C1 and C2 as specific targets for proteolytic cleavage in apoptosis induced by a variety of stimuli including ionising radiation (Waterhouse *et al.*, 1996). HnRNP C1/C2 (alternate splice products differing by a stretch of 13 amino acids) are abundant nuclear proteins, which bind to nascent pre-messenger RNA and to poly (U) (Dreyfuss *et al.*, 1993). The C proteins are highly conserved in evolution and are thought to play a role in pre-mRNA splicing since antibodies to hnRNP C have been shown to inhibit splicing reactions *in vitro*. The carboxyterminal region of hnRNP is rich in aspartic acid residues with five potential cleavage sites for caspases. The effect of cleaving off a small segment of the C-terminus on the activity of hnRNP is not known, since the RNA binding occurs through an N-terminal domain. It is interesting that two components of the splicing complex (U1-70 kDa and hnRNP C) have now been identified as proteolytic targets in apoptosis. This strategy would ensure that energy is not wasted in the dying cell by processing messenger RNA that is no longer required for cellular functions.

GTPases of the Rho family regulate oxidant production in leukocytes, cell adhesion, motility and stress responses. The **GDP d**issociated **i**nhibitor (D4-GDI) is a highly abundant regulator of Rho GTPases in haemopoietic cells and is highly homologous to Rho GDI. Rho GDIs form a complex with members of the Rho GTPase family (Rho, Rac, Cdc42) and thereby maintain an inactive cytosolic form of the GTPase (Chuang *et al.*, 1993). Two truncated forms of D4-GDI copurified with caspase activity from THP-1 cells (Na *et al.*, 1996). Sequence analysis revealed two potential caspase cleavage sites in the protein, at Asp 19 and Asp 55. Neither site is present in Rho GDI. D4-GDI is cleaved to a 23 kDa fragment during Fas and staurosporine-induced apoptosis of Jurkat T cells consistent with cleavage at $DELD_{19}{\downarrow}S$. Mutation of Asp 19 to Asn abolished the ability of recombinant caspase-3 to cleave ^{35}S-labelled *in vitro* transcribed and translated D4-GDI. Truncated D4-GDI is unable to effectively bind and regulate GTPases of the Rho family (Danley *et al.*, 1996). The resulting deregulation of RhoGTPase activity could affect the cytoskeleton and contribute to membrane blebbing. It could also affect the activation of stress activated protein kinases, which have been implicated in apoptosis (as described elsewhere in this book).

The product of the retinoblastoma susceptibility gene, Rb, is a negative regulator of cell proliferation in its hypophosphorylated form (Weinberg *et al.*, 1995). A protein serine/threonine phosphatase is responsible for the anticancer drug-induced Rb hypophosphorylation, G1 arrest and apoptosis in the p53-null human leukemic cell lines, HL60 and U937 (Dou *et al.*, 1995). An and Dou (1996) showed that the hypophosphorylated Rb is immediately cleaved into at least two fragments p68 and p48, by a caspase, probably caspase-3. The time-course of cleavage coincided with DNA fragmentation and addition of a phosphatase inhibitor, calyculin A, to AraC-treated HL60 cells prevented hypophosphorylation, subsequent cleavage of Rb, and apoptosis. However, three groups have now determined the site of caspase cleavage to be Asp 886 at the C-terminal end of the molecule releasing a 42 amino-acid fragment: Janicke *et al.* (1996) in a study of TNF- and staurosporine-induced apoptosis in tumour cell lines; Chen *et al.* (1997) using HL-60 cells treated with etoposide or araC and MCF10 breast epithelial cells treated with anti-Fas; and Tan *et al.* (1997) in a study of TNF- and Fas- induced apoptosis. The cleaved Rb still bound cyclin D3 and had enhanced binding affinity to E2F, but failed to bind the regulatory protein Mdm2 which has been implicated in apoptosis and is itself a proteolytic target in apoptosis. Expression of a non-degradable mutant of Rb was found to attenuate the death response toward TNF but not Fas ligation (Tan *et al.*, 1997). Overexpression of Rb has also been shown to inhibit p53-mediated apoptosis (Hass-Kogan *et al.*, 1995; Haupt *et al.*, 1995) presumably by competition for the cleavage enzyme.

Mdm2 is a negative regulator of the tumour suppressor p53 which induces both apoptosis and cell cycle arrest. It is amplified and overexpressed in a variety of tumours containing normal p53, and overexpression of Mdm2 inhibits both p53-induced apoptosis and cell cycle arrest (Haupt *et al.*, 1996; Chen *et al.*, 1996). Mdm2 is cleaved during apoptosis by caspase-3 both *in vitro* and *in vivo* (Erhardt *et al.*, 1997). Although there are six potential DXXD sites in Mdm2, only one of these sites (Asp 359) is preferentially cleaved. Induction of Mdm2 by p53 forms a negative feedback loop, which is critical for regulation of the growth suppressive and apoptotic activities of p53. Cleavage of Mdm2 in apoptosis may break this loop allowing apoptosis to proceed.

CYTOSKELETAL PROTEINS

The lamins are intranuclear intermediate filament proteins which play a role in maintaining nuclear shape and in mediating chromatin-nuclear membrane associations (Georgatos *et al.*, 1994). Lamins have also been shown to form a part of a diffuse skeleton that ramifies throughout the interior of the nucleus (Hozak *et al.*, 1995). Mitosis is accompanied by p34cdc2-mediated phosphorylation of lamins, resulting in their depolymerisation. In contrast apoptosis involves proteolytic degradation of lamin subunits (Kaufmann *et al.*, 1989; Oberhammer *et al.*, 1994). Lamins A and C are alternate splice products of the same gene whereas the B-type lamins are related but distinct. Proteolysis of lamin A into a 45 kDa fragment by a caspase was demonstrated in a cell-free system (Lazebnik *et al.*, 1995) and lamins A and B were shown to be cleaved to a 45 kDa fragment coinciding with chromatin condensation (Oberhammer *et al.*, 1994). The contribution of lamin degradation to the process of apoptosis is not clear but several suggestions have been put forward. Lamin B1 degradation during apoptosis was reported to precede the onset of DNA fragmentation and could lead to collapse of chromatin due to loss of attachment points on the nuclear matrix (Neamati *et al.*, 1995). Lazebnik *et al.* (1995) reported that inhibition of lamin cleavage activity does not interfere with chromatin margination or DNA degradation but does prevent collapse of chromatin. Rao *et al.* (1996) using an uncleavable mutant of lamin A or B showed that the onset of apoptosis was delayed and mutant lamin expressing cells failed to show chromatin condensation and nuclear shrinkage. The nuclear envelope collapsed and the lamina remained intact however the later stages of apoptosis and the formation of apoptotic bodies was unaltered. In addition lamin cleavage was shown to be insufficient to drive the changes in nuclear morphology characteristic of apoptosis (Takahashi *et al.*, 1997a). One of five caspases active in a cell free system of apoptosis was inhibited by the rabbitpox crmA/SPI-2 and by a peptide spanning the lamin A cleavage site, suggesting that the same enzyme is responsible for lamin A cleavage and nuclear disintegration (Takahashi *et al.*, 1996b).

Treatment of HeLa cells with anti-CD95 mAb resulted in preferential degradation of lamin B compared with lamins A and C (Mandal *et al.*, 1996). This raises the possibility that different proteases are responsible for lamin B and lamin A/C cleavage. Based on sensitivity to YVAD-cmk, the lamin B protease appeared to be distinct from the PARP protease (caspase-3). Previous studies by Orth *et al.* (1996a) and Takahashi *et al.* (1996a) indicated that lamin A is cleaved to its characteristic apoptotic fragment by caspase-6 but not by caspases -3 or -7. Zhivotovsky *et al.* (1997) also found that the PARP cleaving enzyme was different to the lamin cleaving enzyme and that there are two different proteases involved in lamin proteolysis, one of which is cytoplasmic and translocated to the nucleus, and the other nuclear and activated by calcium.

Another intermediate filament protein, keratin 18 (K18) is present in epithelial like cells. During apoptosis, K18 intermediate filaments reorganize into granular structures enriched for K18 phosphorylated on Ser53 and K18, but not K8, is proteolytically cleaved (Caulin *et al.*, 1997). Similar fragments are produced *in vitro* by caspase-6, -3 and -7. Caspase-6 produces a C-terminal 22 kDa fragment which is further cleaved by caspase-3 and -7 to a 19 kDa fragment. Mutation studies showed that the phosphorylation of Ser 53 was not required for cleavage. A non-cleavable mutant was shown to cause keratin filament reorganisation in stably transfected clones.

Nuclear Matrix Protein (NuMA) is a ubiquitous 236 kDa nuclear matrix protein which relocates to the spindle poles during mitosis and is required for organisation of the mitotic spindle. NuMA is soluble in mitotic extracts but forms extensive insoluble structures on dephosphorylation (Saredi *et al.*, 1997). It is also proteolytically processed during apoptosis (Weaver *et al.* 1996b). The cleavage site has been located to between residues 1701 and 1725 (Gueth-Hallonet *et al.*, 1997), however the protease responsible is unknown since caspase inhibitors did not prevent the cleavage of NuMA, the only effective inhibitor being TPCK. Cleavage of NuMA may result in interference with the binding to matrix attachment regions and contribute to the collapse of the matrix, or it might induce chromatin reorganisation.

Fodrin (non-erythroid spectrin), a major component of the cortical cytoskeleton, consists of heterodimers of the α (240 kDa) and β (235 kDa) subunits, aligned side-by-side. α-Fodrin binds actin (Bennett and Gilligan, 1993) and ankyrin, which contains a "cell death" domain similar to those in the TNF and Fas receptors (Feinstein *et al.*, 1995). Fodrin is thought to be responsible for coupling transmembrane proteins to the cytoplasm. Martin *et al.* (1995) first demonstrated the proteolytic cleavage of fodrin to fragments of 150 kDa and 120 kDa during apoptosis and hypothesised that this may lead to membrane blebbing. Both calpain and caspases are capable of cleaving fodrin but it is not yet clear which enzyme(s) is responsible for fodrin cleavage during apoptosis. Calpain has been reported to play a role in membrane blebbing (Cotter *et al.*, 1992), and phosphatidylserine (PS) exposure (Comfurius *et al.*, 1990). Several recent studies have examined the events associated with the cleavage of fodrin. Fodrin cleavage appears not to be involved in PS exposure in platelets, since the cleavage is not inhibited by chelation of extracellular calcium whereas PS exposure is (Hampton *et al.*, 1996). Vanags *et al.* (1996) investigated the kinetics of PS exposure, membrane blebbing, nuclear fragmentation, caspase activation and fodrin cleavage in TNF-induced apoptosis of U937 cells and in Fas-induced apoptosis of Jurkat cells. They were unable to discern which extranuclear events occurred first however their results suggested that activation of calpain occurs after activation of caspase-3. It was concluded that in the TNF/U937 system calpain was responsible for cleavage of fodrin to the 150 kDa fragment, however calpain did not seem to be involved in the Fas/Jurkat model. Cryns *et al.* (1996) using the Fas/Jurkat and TNF/HeLa models concluded that the fodrin protease acts upstream of the PARP protease (caspase-3). The caspase-3 inhibitor DEVD-CHO could protect cells from Fas-induced apoptosis but did not prevent fodrin proteolysis and cell permeable calpain inhibitors had no effect on fodrin cleavage suggesting that a caspase only partially sensitive to DEVD-CHO was responsible *in vivo* for fodrin cleavage. In neuronal apoptosis a role for both calpain and caspases in fodrin cleavage is indicated (Nath *et al.*, 1996; Jordan *et al.*, 1997). The calpain cleavage site in fodrin, VY↓GMMP, is located just N-terminal to the calmodulin-binding domain. The sequence DETD↓SK was proposed as the cleavage site yielding the 150 kDa fragment but this is yet to be confirmed by sequence analysis.

Two groups reported that actin is cleaved *in vitro* in extracts of apoptotic cells, Mashima *et al.* (1995) using etoposide-treated U937 cells and Kayalar *et al.* (1996) using purified actin was incubated with caspase-1 *in vitro*. The 43 kDa actin band was cleaved into fragments of 41 kDa, 30 kDa and 14 kDa, however during apoptosis of PC12 cells induced by serum withdrawal, the pattern of fragments produced was different to that seen *in vitro*. Mashima *et al.* (1997) have since reported that a 15 kDa fragment was detected in apoptotic U937 cells, however a time course of cleavage was not presented nor was evidence of the extent of cleavage occurring during apoptosis. Song *et al.* (1997a) carried out an extensive investigation of the fate of actin in different cell types in response to a variety of apoptosis-inducing stimuli but failed to observe degradation of actin although cleavage of DNA-PKcs was observed. Actin cleavage in neutrophils, which spontaneously undergo apoptosis, was found to occur at a site near the N-terminus (between Val 43 and Met 44) and was inhibited by the calpain inhibitor ALLN but not by zVAD-fmk (Brown *et al.*, 1997). The 38 kDa fragment generated is of similar size to that reported to be cleaved by caspases in PC12 cells (41 kDa).

It is unlikely that actin in its polymerised form in microfilaments is accessible to caspases *in vivo* and the functional relevance of actin cleavage in some forms of apoptosis remains to determined. Actin has recently been identified as a major glutaminyl substrate for tissue transglutaminase in HL-60 and U937 cells undergoing apoptosis (Nemes *et al.*, 1997).

Changes in actin organisation occur after overexpression of some actin binding proteins, the small GTP binding proteins Rho and Rac, and various C-terminal deleted derivatives of Gas2. The cell shape changes induced by C-terminal deletion mutants of Gas2 resemble those seen in apoptosis (Brancolini *et al.*, 1995). With Gas2 mutants Δ276-314 and Δ236-314, cell shrinkage and collapse of the cell body occurred however wild type Gas 2 did not induce cell shape changes. In NIH3T3 cells undergoing apoptosis upon serum-withdrawal, Brancolini *et al.* (1995) showed that Gas2 is processed to a 31 kDa form exclusively in the non-adherent (apoptotic) population. When Asp 279 was mutated to alanine (Gas 2D279A) no proteolytic cleavage was observed, whereas degradation of PARP proceeded normally. Gas2 shows a tissue-specific pattern of expression thus, it is unlikely that it represents a universal effector for the microfilament-associated transitions during apoptosis.

Another actin-regulating protein, gelsolin, has been identified as a substrate for caspase-3 (Kothakota *et al.*, 1997). The cleaved gelsolin severed actin filaments *in vitro* in a calcium independent manner and expression of the cleavage product in multiple cell types caused the cells to round up, detach and undergo nuclear fragmentation. These results suggest that gelsolin may be another physiological effector of the morphological changes in apoptosis. In addition Ohtsu *et al.* (1997) have shown that gelsolin inhibits apoptosis induced by Fas when overexpressed in Jurkat cells. Caspase-3 activation was strongly suppressed in the gelsolin transfectants indicating that gelsolin blocks upstream of this protease.

The APC (adenomatous polyposis coli) protein, the product of the gene mutated in familial adenomatous polyposis, is implicated in colonic maturation. Its expression in the normal colonic crypt gradually increases as cells migrate up the crypt. Expression of APC in an inducible system inhibits the growth of human colorectal cancer cells lacking APC by induction of apoptosis (Morin *et al.*, 1996). APC has been shown to bind to β-catenin (Rubinfeld *et al.*, 1993; Su *et al.*, 1993) and to microtubules (Munemitsu *et al.*, 1994; Smith *et al.*, 1994). APC is cleaved in apoptosis to a 90 kDa fragment (Browne *et al.*, 1994) however the function of this protein and its role in apoptosis are not known.

PROTEOLYTIC TARGETS INVOLVED IN NEURODEGENERATIVE DISORDERS

Huntingtin is a ubiquitously expressed 350 kDa protein of unknown function which contains a polyglutamine repeat region in its N-terminal domain. Mutation of this protein gives rise to the human genetic disorder Huntington's disease (HD), a progressive neurodegenerative disorder characterised by choreic movements and intellectual impairment.

The polyglutamine stretch is encoded by a polymorphic stretch of CAG repeats, the length of which correlates with disease phenotype. Expression of the CAG repeat beyond 36, is associated with HD. Goldberg *et al.* (1996) have shown that huntingtin is a substrate for caspase-3 and that the rate of cleavage increases with the length of the polyglutamine tract. Huntingtin cleavage occurs coincident with the onset of apoptosis at a cluster of 4 DXXD sites within the N-terminal region to yield an 80 kDa product. Huntingtin may normally play an important role in cell survival and prevention of apoptosis, as suggested by the embryonic lethality in HD–/– mice (Nasir *et al.*, 1995; Zeitlin *et al.*, 1995; Duyao *et al.*, 1995). Although huntingtin is ubiquitously expressed, inappropriate apoptosis occurs only in specific areas of the brain, suggesting that another protein expressed only in these regions co-operates with huntingtin to lead to apoptosis. Further understanding of the role of huntingtin cleavage in neuronal cell death in HD could lead to specific therapeutic targets for delaying the progression of diseases such as HD characterised by inappropriate apoptosis.

Familial Alzheimer's disease is caused by mutations in the genes encoding the presenilin 1 and 2 proteins (PS1 and PS2). Presenilins are localised to the nuclear membrane, its associated kinetochores and the centrosomes suggesting a role in chromosome organisation and segregation (Li *et al.*, 1997). These proteins undergo regulated endoproteolytic processing. Mutations in PS1 and PS2 increase the production of the highly amyloidogenic 42-residue form of amyloid beta-protein (Abeta42). Xia *et al.* (1997) have demonstrated an interaction between presenilins and the beta-amyloid precursor protein (APP). Their results suggest that mutant PS interacts with APP in a way that enhances proteolysis to Abeta42. Apoptotic cell death has been reported to be a pathological feature of Alzheimer's disease and during apoptosis PS1 and PS2 proteins are cleaved at sites distal to their normal proteolysis sites by a caspase-3 like enzyme (Kim *et al.*, 1997; Loetscher *et al.*, 1997). The cleavage site was identified as Asp345 and Asp329 in PS1 and PS2 respectively. In cells expressing mutant PS2 the ratio of alternative to normal PS2 cleavage fragments was increased relative to wild-type PS2-expressing cells, suggesting a potential role for apoptosis-associated cleavage of presenilins in the pathogenesis of Alzheimer's disease. $PS1^{-/-}$ mice die shortly after birth and exhibit skeletal defects and neuronal loss indicating that PS1 is required for proper formation of the axial skeleton, normal neurogenesis and neuronal survival (Shen *et al.*, 1997).

TARGETS OF THE PROTEASOME

Most of the death substrates identified to date are cleaved by the caspases, however there is a role for other proteases in apoptosis. Calpain has been shown to be important in thymocyte apoptosis (Squier and Cohen, 1996) and in hepatocytes (Gressner *et al.*, 1997). At least one substrate of calpain has been identified – fodrin. Work on the involvement of calpain in other apoptosis systems has been hampered by the limitations of the so-called calpain specific inhibitors. In many cases these actually cause apoptosis themselves and they have since been shown to inhibit the proteasome which has recently emerged as an important regulator of apoptosis. The proteasome is a high molecular weight (26S) multicatalytic protease which is involved in the rapid degradation of proteins covalently linked to chains of ubiquitin. It consists of a 20S catalytic core and regulatory subunits and its disruption has a lethal effect. The importance of the proteasome in the regulation of apoptosis has recently been demonstrated by several groups. Inhibition of the proteasome can induce apoptosis in several cell types, for example MOLT-4 (Shinohara *et al.*, 1996), U937 (Imajoh-Ohmi *et al.*, 1995; Fujita *et al.*, 1996), HL-60

(Drexler , 1997), RVC (Tanimoto *et al.*, 1997), and PC12 (Lopes *et al.*, 1997). The induction of apoptosis by the proteasome inhibitors is dependent on p53, since dominant negative p53 inhibited this process. Levels of p53 increased rapidly in cells treated with proteasome inhibitors as did the p53-inducible gene products p21 and Mdm-2 (Lopes *et al.*, 1997). On the other hand, in thymocytes, proteasome inhibitors blocked cell death induced by ionizing radiation, glucocorticoids or phorbol ester, and also the degradation of PARP. (Grimm *et al.*, 1996). The induction of apoptosis in Ewing's sarcoma cells by ionizing radiation is accompanied by accumulation of ubiquinated proteins (Soldatenkov and Dritischilo, 1997). Sadoul *et al.* (1996) showed that nanomolar concentrations of several proteasome inhibitors, including the highly specific lactacystin, prevented apoptosis and PARP degradation in NGF-deprived sympathetic neurones. These results place the proteasome upstream of the caspases which is supported by the fact that in macrophages lactacystin prevented the proteolytic activation of caspase-1 and subsequent generation of IL-1β. Activation-induced T-cell death is also inhibited by lactacystin (Cui *et al.*, 1997) .

Several proteins are degraded into small fragments during apoptosis, for example histones H1 and H2B and Topoisomerase II. The latter has recently been shown to be degraded by the ubiquitin proteolysis pathway (Nakajima *et al.*, 1996). Another important target of the proteasome which may have a role in apoptosis is IκB. Cui *et al.* (1997) provided evidence for a link between the proteasome-dependent degradation of Iκ-B and the AICD that occurs through activation of the FasL gene and up-regulation of the Fas gene.

CONCLUSIONS

Several studies have now demonstrated the activation of multiple caspases during the execution of apoptosis in several cell types. In the study by MacFarlane *et al.* (1997), the activation of caspases -2,3,6,7 but not caspase-1 was observed in apoptotic THP.1 cells. Martins *et al.* (1997) observed the activation of caspases -3, and -6, plus other unidentified species, but not caspases -1 or -2 in etoposide-treated HL60 cells. Takahashi *et al.* (1997b) demonstrated the stepwise appearance of different caspases and their distinct substrate preferences by competition experiments.

Others have also attempted to elucidate the temporal sequence of apoptotic events with respect to cleavage of various substrates, Greidinger *et al.* (1996) identified three tiers of caspase activity after Fas ligation of Jurkat cells. The earliest detected cleavage was of fodrin to the 150 kDa fragment, the second phase involved the cleavage of PARP , U1-70 kDa and DNA-PKcs, all substrates of caspase-3, and the third phase involved cleavage of lamin B. Lazebnik *et al.* (1995) also concluded that lamin cleavage was a relatively late event compared to PARP cleavage. However these results are quite disparate from those of Mandal *et al.* (1996) who reported that degradation of lamin B is an early event in Fas-induced apoptosis in HeLa cells whereas the cleavage of lamins A and C occurred much later implying that different proteases are responsible for the cleavage of the different lamins. Caspase-6 has been shown to cleave lamin A (Orth *et al.*, 1996a; Takahashi *et al.*, 1996a) with caspases -3 and -7 being unable to cleave this substrate. Orth *et al.* (1996b) have also placed caspase-6 upstream of caspases -3 and -7 in the temporal sequence of events indicating that caspase-6 activates the other two caspases. The reason for the discrepancy of these results is not clear but may be related to different cell types under study or different proteolytic cascades induced by different apoptotic stimuli. Differential sensitivity of caspases to phenylarsine oxide revealed that two or more caspases are required for the initiation and completion of nuclear apoptotic changes (Takahashi *et al.*, 1997a).

These results demonstrate that apoptotic execution involves the coordinate action of multiple proteases each with a distinctive set of targets. Cleavage of some of these targets may help in initiating the process of apoptosis (e.g. PKCδ) while the cleavage of others leads to the characteristic nuclear and morphological changes seen in apoptotic cells.

Increasing numbers of substrates are now being identified as summarised in Table 12.1, but certain critical questions still remain to be answered. For example, what are the specific targets of individual caspases/proteases? as most substrates so far defined are cleaved by caspase-3. What is the effect of cleavage on substrate function? and what is the role of the cleaved fragments which are activated? The expression of caspases and targets may vary between different cell types as do the signalling pathways inducing activation of the caspases. In addition the substrates of proteases other than the caspases need to be identified to define the role of these proteases in apoptosis. In many cases the cellular function of the proteolytic targets are unknown thus the study of proteolytic targets in apoptosis is important not only for understanding the molecular mechanisms of cell death but may also lead to a better understanding of the role of these critical proteins in cellular function.

Table 12.1 Proteolytic targets in apoptosis.

Substrate	*Fragment Size kDa#*	*Cellular location*	*Cleavage Site*	*Protease*	*Reference*
DFF	30/11	Cytoplasm	$DETD_{117}$↓S; $DAVD_{224}$↓T	Caspase-3	Liu *et al.*, 1997
PKCδ	40	Cytoplasm	$DMQD_{330}$↓N	Caspase-3	Emoto *et al.*, 1995
PKCθ	40	Cytoplasm	$DEVD_{354}$↓K	Caspase-3	Datta *et al.*, 1997
PITSLRE α2-1	60+43	Nucleus	$YVPD_{393}$↓S	Caspase-3/-1	Beyaert *et al.*, 1997
PAK2	34+28	Cytoplasm	EVD_{212}↓G	Caspase-3	Rudel and Bokoch, 1997
PLA2	70	Cytoplasm	$DELD_{522}$↓A	Caspase-3	Wissing *et al.*, 1997
SREBP-1	70	Cytoplasm	$SEPD_{460}$↓S	Caspase-3	Wang *et al.*, 1995
SREBP-2	70	Cytoplasm	$DEPD_{468}$↓S	Caspase-3	Wang *et al.*, 1995
PARP	85	Nucleus	$DEVD_{213}$↓G	Caspase-3	Kaufmann *et al.*, 1993
DNA-PKcs	240/120	Nucleus	$DEVD_{2712}$↓N	Caspase-3	Song *et al.*, 1996a
RFC	87	Nucleus	$DEVD_{706}$↓G	Caspase-3	Song *et al.*, 1997a; Ubeda and Habener, 1997
U1-70kDa	40	Nucleus	$DGPD_{359}$↓G	Caspase-3	Casciola-Rosen *et al.*, 1996
HnRNP C1/C2		Nucleus	5 DXXD potential cleavage sites	Caspase-3/-7	Waterhouse *et al.*, 1996

Table 12.1 (*contd.*)

Substrate	*Fragment Size kDa#*	*Cellular location*	*Cleavage Site*	*Protease*	*Reference*
D4-GDI	23	Cytoplasm	$DELD_{19}\downarrow S$	Caspase-3	Na *et al.*, 1996
Rb	4	Nucleus	$DEAD_{886}\downarrow G$	Caspase-3	Janicke *et al.*, 1996; Chen *et al.*, 1997
Mdm2	60+30	Nucleus	$DVPD_{359}\downarrow G$	Caspase-3	Erhardt *et al.*, 1997
Lamin A/C	42/33	Nucleus	$VEID_{230}\downarrow N$	Caspase-6	Lazebnik *et al.*, 1995; Takahashi *et al.*, 1996
Lamin B1	45	Nucleus			Kaufmann *et al.*, 1989; Oberhammer *et al.*, 1994
Fodrin	150/120	Cortical cytoskeleton		Calpain/ Caspase	Martin *et al.*, 1995
Actin	41/30/15 38	Micro-filaments	Asp 244 $HQGV_{43}V\downarrow M$	Caspase Calpain	Mashima *et al.*, 199;5 Brown *et al.*, 1997
Keratin-18	26/22	Intermediate filaments		Caspase-6, -3,-7	Caulin *et al.*, 1997
NuMA	190	Nucleus			Gueth-Hallonet *et al.*, 1997
Gelsolin		Micro-filaments		Caspase-3	Kothakota *et al.*, 1997
Gas2	31	Micro-filaments	$SRVD_{279}\downarrow G$	Caspase	Brancolini *et al.*, 1995
APC	90	Cytoskeletal			Browne *et al.*, 1994
Huntingtin	80	Cytoplasm	4 DXXD potential cleavage sites		Goldberg *et al.*, 1997
Presenilins PS1 PS2	14 20	Membrane	$AQRD_{345}\downarrow S$ $DSYD_{329}\downarrow S$	Caspase-3 like	Kim *et al.*, 1997; Loetscher *et al.*, 1997
Terminin	30				Hebert *et al.*, 1994
Topoisomerase 1	70	Nucleus			Kaufmann *et al.*, 1989
Topoisomerase IIα	Small fragments	Nucleus		Proteasome	Nakajima *et al.*, 1996
Histone H1	Small fragments	Nucleus			Kaufmann *et al.*, 1989
Histone H2B	Small fragments	Nucleus			Baxter *et al.*, 1989

x/y denotes fragment y is produced by further cleavage of fragment x; x+y denotes the N- and C-terminal fragments.

ACKNOWLEDGEMENTS

This work was supported by grants from the National Health and Medical Research Council of Australia, the Queensland Cancer Fund and the University of Queensland Cancer Research Fund.

REFERENCES

Alnemri, E.S., Livingston, D.J., Nicholson, D.W., Salveson, G., Thornberry, N.A., Wong, W.W. and Yuan, J. (1996) Human caspase-1/CED-3 protease nomenclature. *Cell*, **87**, 171.

An, B. and. Dou, Q.P. (1996) Cleavage of retinoblastoma protein during apoptosis: an interleukin 1β-converting enzyme-like protease as candidate. *Cancer Res.*, **56**, 438–442.

Anderson, C.W. and Carter, T.H. (1996) The DNA activated protein kinase- DNA-PK. *Current Topics in Microbiol. Immunol.*, **217**, 91–111.

Baxter, G.D., Smith, P.J. and Lavin, M.F. (1989) Molecular changes associated with induction of cell death in a human T-cell leukemia line: putative nucleases identified as histones. *Biochem. Biophys. Res. Commun.*, **162**, 30–37.

Bennett, V. and Gilligan, D.M. (1993) The spectrin-based membrane skeleton and micron-scale organization of the plasma membrane. *Annu. Rev. Cell Biol.*, **9**, 27–66.

Beyaert, R., Kidd, V.J., Cornelis, S., Van de Craen, M., Denecker, G., Lahti, J.M., Gururajan, R., Vandenabeele, P. and Fiers, W. (1997) Cleavage of PITSLRE kinases by ICE/CASP-1 and CPP32/CASP-3 during apoptosis induced by tumor necrosis factor. *J. Biol. Chem.*, **272**, 11694–11697.

Bossy-Wetzel, E., Bakiri, L. and Yaniv, M. (1997) Induction of apoptosis by the transcription factor c-jun. *EMBO J.*, **16**, 1695–1709.

Brancolini, C., Benedetti, M. and Schneider, C. (1995) Microfilament reorganisation during apoptosis: the role of Gas2, a possible substrate for ICE-like proteases. *EMBO J.*, **14**, 5179–5190.

Brown, S.B., Bailey, K. and Savill, J. (1997) Actin is cleaved during constitutive apoptosis. *Biochem. J.*, **323**, 233–237.

Browne, S.J., Williams, A.C., Hague, A., Butt, A.J. and Paraskeva, C. (1994) Loss of APC protein expressed by human colonic epithelial cells and the appearance of a specific low-molecular weight form is associated with apoptosis *in vitro*. *Int. J. Cancer*, **59**, 56–64.

Bunnell, B.A., Heath, L.S., Adams, D.E., Lahti, J.M. and Kidd, V.J. (1990) Increased expression of a 58-kDa protein kinase leads to changes in the CHO cell cycle. *Proc. Natl. Acad. Sci. USA*, **87**, 7467–7471.

Cardone, M.H., Salveson, G.S., Widman, C., Johnson, G. and Frisch, S.M. (1997) The regulation of anoikis: MEKK-1 activation requires cleavage by caspases. *Cell*, **90**, 315–323.

Casciola-Rosen, L.A., Anhalt, G.J. and Rosen, A. (1994a) Autoantigens targeted in systemic lupus erythematosus are clustered in two populations of surface structures on apoptotic keratinocytes. *J. Exp. Med.*, **179**, 1317–1330.

Casciola-Rosen, L.A., Miller, D.K., Anhalt, G.J. and Rosen, A. (1994b) Specific cleavage of the 70-kDa protein component of the U1 small nuclear ribonucleoprotein is a characteristic biochemical feature of apoptotic cell death. *J. Biol. Chem.*, **269**, 30757–30760.

Casciola-Rosen, L.A., Anhalt, G.J. and Rosen, A. (1995) DNA-dependent protein kinase is one of a subset of autoantigens specifically cleaved early during apoptosis. *J. Exp. Med.*, **182**, 1625–1634.

Caulin C., Salvesen, G.S. and Oshima, R.G. (1997) Caspase cleavage of keratin 18 and reorganization of intermediate filaments during epithelial cell apoptosis. *J. Cell Biol.*, **138**, 1379–1394.

Chen, J., Wu, X., Lin, J. and Levine, A.J. (1996) Mdm-2 inhibits the G1 arrest and apoptosis functions of the p53 tumor suppressor protein. *Mol. Cell. Biol.*, **16**, 2445–2452.

Chen, W.-D., Otterson, G.A., Lipkowitz, S., Khleif, S.N., Coxon, A.B. and Kaye, F.J. (1997) Apoptosis is associated with cleavage of a 5 kDa fragment from RB which mimics dephosphorylation and modulates E2F binding. *Oncogene*, **14**, 1243–1248.

Chuang,T.H., Bohl, B.P. and Bokoch, G.M. (1993) Biologically active lipids are regulators of Rac.GDI complexation. *J. Biol. Chem.*, **268**, 26206–26211.

Cohen, G.M. (1997) Caspases: the executioners of apoptosis. *Biochem. J.*, **326**, 1–16.

Comfurius, P., Senden, J.M.G., Tilly, R.H.J., Schroit, A.J., Bevers, E.M. and Zwaal, R.F.A. (1990) Loss of membrane phospholipid asymmetry in platelets and red cells may be associated with calcium-induced shedding of plasma membrane and inhibition of aminophospholipid translocase. *Biochim. Biophys. Acta.*, **1026**, 153–160.

Cotter, T.G., Lennon, S.V., Glynn, J.M. and Green, D.R. (1992) Microfilament-disrupting agents prevent the formation of apoptotic bodies in tumor cells undergoing apoptosis. *Cancer Res.*, **52**, 997–1005.

Cryns, V., Bergeron, L., Zhu, H., Li, H. and Yuan, J. (1996) Specific cleavage of α-fodrin during Fas- and tumor necrosis factor-induced apoptosis is mediated by an interleukin-1β-converting enzyme/Ced-3 protease distinct from the poly(ADP-ribose) polymerase protease. *J. Biol. Chem.*, **271**, 31277–31282.

Cui, H., Matsui, K., Omura, S., Schauer, S.L., Matulka, R.A., Sonenshein, G.E. and Ju, S.T. (1997) Proteasome regulation of activation-induced T cell death. *Proc. Natl. Acad. USA*, **94**, 7515–7520.

Cullman, G., Fien, K., Kobayashi, R. and Stillman, B. (1995) Characterisation of the five replicon factor C genes of *Saccharomyces cerevisiae*. *Mol. Cell. Biol.*, **14**, 1626–1634.

Danley, D.E., Chuang, T.-H. and Bokoch, G.M. (1996) Defective Rho GTPase regulation by IL-1β-converting enzyme-mediated cleavage of D4 GDP dissociation inhibitor. *J. Immunol.*, **157**, 500–503.

Datta, R., Kojima, H., Yoshida, K. and Kufe, D. (1997) Caspase-3-mediated cleavage of protein kinase C theta in induction of apoptosis. *J. Biol. Chem.*, **272**, 20317–20320.

Dou, Q.P., An, B. and Will, P.L. (1995) Induction of retinoblastoma phosphatase activity by anticancer drugs accompanies p53-independent G1 arrest and apoptosis. *Proc. Natl. Acad. Sci. USA*, **92**, 9019–9023.

Drexler, H.C. (1997) Activation of the cell death program by inhibition of proteasome function. *Proc. Natl. Acad. Sci. USA*, **94**, 855–860.

Dreyfuss, G., Matunis, M.J., Pinol-Roma, S. and Burd, C.G. (1993) hnRNP proteins and the biogenesis of mRNA. *Annu. Rev. Biochem.*, **62**, 289–321.

Duyao, M.P., Auerbach, A.B., Ryan, A., Persichetti, F., Barnes, G.T., McNeil, S.M., Ge, P.,Vonsattel, J.P., Gusella, J.F., Joyner, A.L. and MacDonald, N.E. (1995) Inactivation of the mouse Huntington's disease gene homolog Hdh. *Science*, **269**, 407–410.

Emoto, Y., Manome, Y., Meinhardt, G., Kisaki, H., Kharbanda, S., Robertson, M., Ghayur, T., Wong, W.W., Kamen, R., Weichselbaum, R. and Kufe, D. (1995) Proteolytic activation of protein kinase C δ by an ICE-like protease in apoptotic cells. *EMBO J.*, **14**, 6148–6156.

Erhardt, P., Tomaselli, K.J., Cooper, G.M. (1997) Identification of the Mdm2 oncoprotein as a substrate for CPP32-like apoptotic proteases. *J. Biol. Chem.*, **272**, 15049–15052.

Farschon, D.M., Couture, C., Mustelin, T., Newmeyer, D.D. (1997) Temporal phases in apoptosis defined by the actions of Src homology 2 domains, ceramide, Bcl-2, interleukin-1β converting enzyme family proteases, and a dense membrane fraction. *J. Cell Biol.*, **137**, 1117–1125.

Feinstein, E., Kimuchi, A., Wallach, D., Boldin, M. and Varfolomeev, E. (1995) The death domain: a module shared by proteins with diverse cellular functions. *Trends Biochem. Sci.*, **20**, 342–344.

Friedlander, R.M., Gagliardini, V., Hara, H., Fink, K.B., Li, W., MacDonald, G., Fishman, M.C., Greenberg, A.H., Moskowitz, M.A. and Yuan, J. (1997) Expression of a dominant negative mutant of interleukin-1 beta converting enzyme in transgenic mice prevents neuronal cell death induced by trophic factor withdrawal and ischemic brain injury. *J. Exp. Med.*, **185**, 933–940.

Frisch, S., Vuori, K., Kelaita, D. and Sicks, S. (1996) A role for Jun-N-terminal kinase in anoikis; suppression by bcl-2 and crmA. *J. Cell Biol.*, **135**, 1377–1382.

Fujita, E., Mukasa, T., Tsukahara, T., Arahata, K., Omura, S. and Momoi, T. (1996) Enhancement of CPP32-like activity in the TNF-treated U937 cells by the proteasome inhibitors. *Biochem. Biophys. Res. Commun.*, **224**, 74–79.

Georgatos, S.D., Meier, J. and Simos, G. (1994) Lamins and lamin-associated proteins. *Curr. Opin. Cell Biol.*, **6**, 347–353.

Ghayur, T., Hugunin, M., Talanian, R.V., Ratnofsky, S., Quinlan, C., Emoto,Y., Pandey, P., Datta, R., Huang, Y., Kharbanda, S., Allen, H., Kamen, R., Wong, W. and Kufe, D. (1996) Proteolytic activation of protein kinase C delta by an ICE/CED3-like protease induces characteristics of apoptosis. *J. Exp. Med.*, **184**, 2399–2404.

Ghayur, T., Banerjee, S., Hugunin, M., Butler, D., Herzog, L., Carter, A., Quintal, L., Sekut, L., Talanian, R., Paskind, M., Wong, W., Kamen, R., Tracey, D. and Allen, H. (1997) Caspase-1 processes IFN-γ-inducing factor and regulates LPS-induced IFN-γ production. *Nature*, **386**, 619–623.

Goldberg, Y.P., Nicholson, D.W., Rasper, D.M., Kalchman, M.A., Koide, H.B., Graham, R.K., Bromm, M., Kazemi-Esfarjani, P., Thornberry, N.A., Vaillancourt, J.P. and Hayden, M.R. (1996) Cleavage of huntingtin by apopain, a proapoptotic cysteine protease, is modulated by the polyglutamine tract. *Nature Genetics*, **13**, 442–449.

Greidinger, E.L., Miller, D.K., Yamin, T.-T., Casciola-Rosen, L. and Rosen, A. (1996) Sequential activation of three distinct ICE-like activities in Fas-ligated Jurkat cells. *FEBS Letts.*, **390**, 99–303.

Greenberg, A.H. (1996) Activation of apoptosis pathways by granzyme B. *Cell Death and Differentiation*, **3**, 269–274.

Gressner, A.M., Lahme, B. and Roth, S. (1997) Attenuation of TGF-β-induced apoptosis in primary cultures of hepatocytes by calpain inhibitors. *Biochem. Biophys. Res. Commun.*, **231**, 457–462.

Grimm, L.M., Goldberg, A.L., Poirier, G.G., Schwartz, L.M. and Osborne, B.A. (1996) Proteasomes play an essential role in thymocyte apoptosis. *EMBO J.*, **15**, 3835–3844.

Gu, Y., Kuida, K., Tsutsui, H., Ku, G., Hsiao, K., Fleming, M.A., Hayashi, N., Higashino, K., Okamura, H., Nakanishi, K., Kurimoto, M., Tanimoto, T., Flavell, R.A., Sato, V., Harding, M.W., Livingston, D.J. and Su, M.S. (1996) Activation of interferon-γ inducing factor mediated by interleukin-1β converting enzyme. *Science*, **275**, 206–209.

Gueth-Hallonet, C., Weber, K. and Osborn, M. (1997) Cleavage of the Nuclear Matrix Protein NuMA during apoptosis. *Exp. Cell Res.*, **233**, 21–24.

Hampton, M.B., Vanags, D.M., Pron-Ares, M.I. and Orrenius, S. (1996) Involvement of extracellular calcium in phosphatidylserine exposure during apoptosis. *FEBS Letts.*, **399**, 271–282.

Han, Z., Malik, N., Carter, T., Reeves, W.H., Wyche, J.H. and Hendrickson, E.A. (1996) DNA-dependent protein kinase is a target for a CPP32-like apoptotic protease. *J. Biol. Chem.*, **271**, 25035–25040.

Hass-Kogan, D.A., Kogan, S.C., Levi, D., Dazin, P., T'Ang, A., Fung, Y-K.T. and Israel, M.A. (1995) Inhibition of apoptosis by the retinoblastoma gene product. *EMBO J.*, **14**, 461–472.

Haupt, Y., Rowan, S. and Oren, M. (1995) p53-mediated apoptosis in HeLa cells can be overcome by excess pRB. *Oncogene*, **10**, 1563–1571.

Haupt, Y., Barak, Y. and Oren, M. (1996) Cell type specific inhibition of p53 mediated apoptosis by mdm-2. *EMBO J.*, **15**, 1596–1606.

Hayakawa, M., Ishida, N., Takeuchi, K., Shibamoto, S., Hori, T., Oku, N., Ito, F., Tsujimoto, M. (1993) Arachidonic acid-selective cytosolic phospholipase A2 is crucial in the cytotoxic action of tumor necrosis factor. *J. Biol. Chem.*, **268**, 11290–11295.

Hebert, L., Pandey, S. and Wang, E. (1994) Commitment to cell death is signaled by the appearance of a terminin protein of 30 kDa. *Exp. Cell Res.*, **210**, 10–18.

Hozak, P., Sasseville, A.M-J., Raymond, Y. and Cook, P.R. (1995) Lamin proteins form an internal nucleoskeleton as well as a peripheral lamina in human cells. *J. Cell Sci.*, **108**, 635–644.

Imajoh-Ohmi, S., Kawaguchi, T., Sugiyama, S., Tanaka, K., Omura, S. and Kikuchi, H. (1995) Lactacystin, a specific inhibitor of the proteasome, induces apoptosis in human monoblast U937 cells. *Biochem. Biophys. Res. Commun.*, **217**, 1070–1077.

Jackson, S.P. (1996) The recognition of DNA damage. *Curr. Opin. Genet. Dev.*, **6**, 19–25.

Janicke R, Walker P.A., Lin X.Y., Porter A.G. (1996) Specific cleavage of the retinoblastoma protein by an ICE-like protease in apoptosis. *EMBO J.*, **15**, 6969–6978.

Jordan, J., Galindo, M.F. and Miller, R.J. (1997) Role of calpain- and interleukin-1 beta converting enzyme-like proteases in the beta-amyloid-induced death of rat hippocampal neurons in culture. *J. Neurochem.*, **68**, 1612–1621.

Kaufmann, S.H. (1989) Induction of endonucleocytic DNA cleavage in human acute myelogenous leukemia cells by etoposide, camptothecin and other cytotoxic anticancer drugs: a cautionary note. *Cancer Res.*, **49**, 5870–5878.

Kaufmann, S.H., Desnoyers, S., Ottaviano, Y., Davidson, N.E. and Poirier, G.G. (1993) Specific proteolytic cleavage of poly(ADP-ribose) polymerase: an early marker of chemotherapy-induced apoptosis. *Cancer Res.*, **53**, 3976–3985.

Kayalar, C., Ord, T., Testa, M.P., Zhong, L.-T. and Bredesen, D.E. (1996) Cleavage of actin by interleukin 1β-converting enzyme to reverse Dnase 1 inhibition. *Proc. Natl. Acad. Sci. USA*, **93**, 2234–2238.

Kim, T.W., Pettingell, W.H., Jung, Y.K., Kovacs, D.M. and Tanzi, R.E. (1997) Alternative cleavage of Alzheimer-associated presenilins during apoptosis by a caspase-3 family protease. *Science*, **277**, 373–376.

Kothakota, S., Azuma, T., Reinhard, C., Klippel, A., Tang, J., Chu, K., McGarry, T.J., Kirschner, M.W., Koths, K., Kwiatkowski, D.J. and Williams, L.T. (1997) Caspase-3-generated fragment of gelsolin: effector of morphological change in apoptosis. *Science*, **278**, 294–298.

Kuida, K., Lippke, J.A., Ku, G., Harding, M.W., Livingston, D.J., Su, M.S. and Flavell, R.A. (1995) Altered cytokine export and apoptosis in mice deficient in interleukin-1 beta converting enzyme. *Science*, **267**, 2000–2003.

Kuida, K. Zheng, T.S., Na, S., Kuan, C., Yang, D., Karasuyama, H., Rakic, P. and Flavell, R.A. (1996) Decreased apoptosis in the brain and premature lethality in CPP32-deficient mice. *Nature*, **384**, 368–372.

Kumar, S. and Lavin, M.F. (1996) The ICE family of cysteine proteases as effectors of cell death. *Cell Death and Differentiation*, **3**, 255–267.

Lahti, J.M., Xiang, J., Heath, L.S., Campana, D. and Kidd, V.J. (1995) PITSLRE kinase activity is associated with apoptosis. *Mol. Cell Biol.*, **15**, 1–11.

Lazebnik, Y.A., Kaufmann, S.H., Desnoyers, S., Poirier, G.G. and Earnshaw, W.C. (1994) Cleavage of poly(ADP-ribose)polymerase by a proteinase with properties like ICE. *Nature*, **371**, 346–347.

Lazebnik, Y.A., Takahashi, A., Moir, R.D., Goldman, R.D., Poirier, G.G., Kaufmann, S.H. and Earnshaw, W.C. (1995) *Proc. Natl. Acad. Sci. USA*, **92**, 9042–9046.

Li, P., Allen, H., Banerjee, S., Franklin, S., Herzog, L., Johnston, C., McDowell, J., Paskind, M., Rodman, L., Salfeld, J., *et al.* (1995) Mice deficient in IL-1 beta-converting enzyme are defective in production of mature IL-1 beta and resistant to endotoxic shock. *Cell*, **80**, 401–411.

Li, J., Xu, M., Zhou, H., Ma, J. and Potter, H. (1997) Alzheimer presenilins in the nuclear membrane, interphase kinetochores, and centrosomes suggest a role in chromosome segregation. *Cell*, **90**, 917–927.

Liu, X., Zou, H., Slaughter, C. and Wang, X, (1997) DFF, a heterodimeric protein that functions downstream of caspase-3 to trigger DNA fragmentation during apoptosis. *Cell*, **89**, 175–184.

Loetscher, H., Deuschle, U., Brockhaus, M., Reinhardt, D., Nelboeck, P., Mous, J., Grunberg, J., Haass, C. and Jacobsen, H. (1997) Presenilins are processed by caspase-type proteases. *J. Biol. Chem.*, **272**, 20655–20659.

Lopes, U.G., Erhardt, P., Yao, R. and Cooper, G.M. (1997) p53-dependent induction of apoptosis by proteasome inhibitors. *J. Biol. Chem.*, **272**, 12893–12896.

MacFarlane, M., Cain, K., Sun, X.-M., Alnemri, E.S., Cohen, G.M. (1997) Processing/activation of at least four interleukin-1β converting enzyme-like proteases occurs during the execution phase of apoptosis in human monocytic tumor cells. *J. Cell Biol.*, **137**, 469–479.

Mandal, M., Maggirwar, S.B., Sharma, N., Kaufmann, S.H., Sun, S.-C. and Kumar, R. (1996) Bcl-2 prevents CD95 (Fas/APO-1)-induced degradation of lamin B and poly(ADP-ribose)polymerase and restores the NF-κB signalling pathway. *J Biol. Chem.*, **271**, 30354–30359.

Margolin, N., Raybuck, S.A., Wilson, K.P., Chen, W., Fox, T., Gu, Y. and Livingston, D.J. (1997) Substrate and inhibitor specificity of interleukin-1β-converting enzyme and related caspases. *J. Biol. Chem.*, **272**, 7223–7228.

Martin, S.J. and Green, D.R. (1995) Protease activation in apoptosis: death by a thousand cuts? *Cell*, **82**, 349–352.

Martin, S.J., O'Brien, G.A., Nishioka, W.K., McGahon, A.J., Mahboubi, A., Saido, T.C. and Green, D.R. (1995) Proteolysis of fodrin (non-erythroid spectrin) during apoptosis. *J. Biol. Chem.*, **270**, 6425–6428.

Martins, L.M., Kottke, T., Mesner, P.W., Basi, G.S., Sinha, S., Frigon Jr., N., Tatar, E., Tung, J.S., Bryant, K., Takahashi, A., Svingen, P.A., Madden, B.J., McCormick, D.J., Earnshaw, W.C. and Kaufmann, S.H. (1997) Activation of multiple interleukin-1β converting enzyme homologues in cytosol and nuclei of HL-60 cells during etoposide-induced apoptosis. *J. Biol. Chem.*, **272**, 7421–7430.

Mashima, T., Naito, M., Fujita, N., Noguchi, K. and Tsuruo, T. (1995) Identification of actin as a substrate of ICE and an ICE-like protease and involvement of an ICE-like protease but not ICE in VP-16-induced U937 apoptosis. *Biochem. Biophys. Res. Commun.*, **217**, 1185–1192.

Mashima, T., Naito, M., Noguchi, K., Miller, D.K., Nicholson, D.W. and Tsuruo, T. (1997) Actin cleavage by CPP-32/apopain during the development of apoptosis. *Oncogene*, **14**, 1007–1012.

Masutani, N., Nozaki, T., Wakabayashi, K. and Sugimura, T. (1995) Role of poly(ADP-ribose) polymerase in cell-cycle checkpoint mechanisms following γ-irradiation. *Biochimie.*, **77**, 462–465.

Morin, P.J., Vogelstein, B. and Kinzler, K.W. (1996) Apoptosis and APC in colorectal tumorigenesis. *Proc. Natl. Acad. Sci. USA*, **93**, 7950–7954.

Munemitsu, S., Souza, B., Muller, O., Albert, I., Rubinfeld, B. and Polakis, P. (1994) The APC gene product associates with microtubules *in vivo* and promotes their assembly *in vitro*. *Cancer Res.*, **54**, 3676–3681.

Na, S., Chang, T.-H., Cunningham, A., Turi, T.G., Hanke, J.H., Bokoch, G.M., Danley, D.E. (1996) D4-GDI, a substrate of CPP32, is proteolyzed during Fas-induced apoptosis. *J. Biol. Chem.*, **271**, 11209–11213.

Nakajima, T., Morita, K., Ohi, N., Arai, T., Nozaki, N., Kikuchi, A., Osaka,F., Yamao, F. and Oda, K. (1996) Degradation of Topoisomerase IIα during E1A-induced apoptosis is mediated by the activation of the ubiquitin proteolysis system. *J. Biol. Chem.*, **271**, 24842–24849.

Ohtsu, M., Sakai, N., Fujita, H., Kashiwagi, M., Gasa, S., Shimuzu, S., Eguchi, Y., Tsujimoto, Y., Sakiyama Y., Kobayashi, K. and Kuzumaki, N. (1997) Inhibition of apoptosis by the actin-regulatory protein gelsolin. *EMBO J.*, **16**, 4650–4656.

Nasir, J., Floresco, S.B., O'Kusky, J.R., Diewert, V.M., Richman, J.M., Zeisler, J., Borowski, A., Marth, J.D., Phillips, A.G., Hayden, M.R. (1995) Targeted disruption of the Huntington's disease gene results in embryonic lethality and behavioural and morphological changes in heterozygotes. *Cell*, **81**, 811–823.

Nath, R., Raser, K.J., Stafford, D., Hajimohammadreza, I., Posner, A., Allen, H., Talanian, R.V., Yuen, P.-W., Gilbertsen, R.B. and Wang, K.K.W. (1996) Non-erythroid α-spectrin breakdown by calpain and interleukin 1β-enzyme-like protease(s) in apoptotic cells: contributory roles of both protease families in neuronal apoptosis. *Biochem. J.*, **319**, 683–690.

Neamati, N., Fernandez, A., Wright, S., Kiefer, J. and McConkey, D.J. (1995) Degradation of Lamin B1 precedes oligonucleosomal DNA fragmentation in apoptotic thymocytes and isolated thymocyte nuclei. *J. Immunol.*, **154**, 3788–3795.

Nemes Jr., Z., Adany, R., Balazs, M., Boross, P. and Fesus, L. (1997) Identification of cytoplasmic actin as an abundant glutaminyl substrate for tissue transglutaminase in HL-60 and U937 cells undergoing apoptosis. *J. Biol. Chem.*, **272**, 20577–20583.

Nicholson, D,W., Ali, A., Thornberry, N.A., Vaillancourt, J.P., Ding, C.K., Gallant, M., Gareau, Y., Griffin, P.R., Labelle, M., Lazebnik, Y.A., Munday, N.A., Raju, S.M., Smulson, M.E., Yamin, T., Yu, V.L. and Miller, D.K. (1995) Identification and inhibition of the ICE/CED-3 protease necessary for mammalian apoptosis. *Nature*, **376**, 37–43.

Oberhammer, F.A., Hochegger, K., Froschl, G., Tiefenbacher, R. and Pavelka, M. (1994) Chromatin condensation during apoptosis is accompanied by degradation of lamin A+B without enhanced activation of cdc2 kinase. *J. Cell Biol.*, **126**, 827–837.

Orth, K., Chinnaiyan, A.M., Garg, M., Froelich, C.J. and Dixit, V.M. (1996a) The CED-3/ICE-like protease Mch2 is activated during apoptosis and cleaves the death substrate lamin A. *J. Biol. Chem.*, **271**, 16443–16446.

Orth, K., O'Rourke, K., Salveson, G.S., Dixit, V.M. (1996b) Molecular ordering of apoptotic mammalian CED-3/ICE-like proteases. *J. Biol. Chem.*, **271**, 20977–20980.

Pai, J.-T., Brown, M.S., Goldstein, J.L. (1996) Purification and cDNA cloning of a second apoptosis-related cysteine protease that cleaves and activates sterol regulatory element binding proteins. *Proc. Natl. Acad. Sci. USA*, **93**, 5437–5442.

Porter, A.G., Ng, P. and Janicke, R.U. (1997) Death substrates come alive. *Bioessays*, **19**, 501–507.

Rao, L., Perez, D., White, E. (1996) Lamin proteolysis facilitates nuclear events during apoptosis. *J. Cell Biol.*, **135**, 1441–1455.

Romac, J.M. and Keene, J.D. (1995) Overexpression of the arginine-rich carboxy-terminal region of U1 snRNP 70K inhibits both splicing and nucleocytoplasmic transport of mRNA. *Genes Dev.*, **9**, 1400–1410.

Rosen, A. and Casciola-Rosen, L. (1997) Macromolecular substrates of the ICE-like proteases during apoptosis. *J. Cell Biochem.*, **64**, 50–54.

Rubinfeld, B., Souza, B., Albert, I., Muller, O., Chamberlain, S.H., Masiarz, F.R., Munemitsu, S., Polakis, (1993) Association of the APC gene product with β-catenin. *Science*, **262**, 1731–1734.

Rudel, T. and Bokoch, G.M. (1997) Membrane and morphological changes in apoptotic cells regulated by caspase-mediated activation of PAK2. *Science*, **276**, 1571–1574.

Ruoslahti, E. and Reed, J. (1994) Anchorage independence, integrins and apoptosis. *Cell*, **77**, 477–478.

Sadoul, R., Fernandez, P.A., Quiquerez, A.L., Martinou, I., Maki, M., Schroter, M., Becherer, J.D., Irmler, M., Tschopp, J. and Martinou, J.C. (1996) Involvement of the proteasome in the programmed cell death of NGF-deprived neurons. *EMBO J.*, **15**, 3845–3852.

Saredi, A., Howard, L. and Compton, D.A. (1997) Phosphorylation regulates the assembly of NuMA in a mammalian mitotic extract. *J. Cell Sci.*, **110**, 1287–1297.

Sells, M.A., Knaus, U.G., Bagrodia, S., Ambrose, D.M., Bukoch, G.M. and Chernoff, J. (1997) Human p21-activated kinase (PAK1) regulates actin organisation in mammalian cells. *Curr. Biol.*, **7**, 202–210.

Shen, J., Bronson, R.T., Chen, D.F., Xia, W., Selkoe, D.J. and Tonegawa, S. (1997) Skeletal and CNS defects in Presenilin-1-deficient mice. *Cell*, **89**, 629–639.

Shinohara, K., Tomioka, M., Nakano, H., Tone, S., Ito, H. and Kawashima, S. (1996) Apoptosis induction resulting from proteasome inhibition. *Biochem. J.*, **317**, 385–388.

Smith, K.J., Levy, D.B., Maupin, P., Pollard, T.D., Vogelstein, B. and Kinzler, K.W. (1994) Wild-type but not mutant APC associates with the microtubule cytoskeleton. *Cancer Res.*, **54**, 3672–3675.

Soldatenkov, V.A. and Dritschilo, A. (1997) Apoptosis of Ewing's sarcoma cells is accompanied by accumulation of ubiquitinated proteins. *Cancer Res.*, **57**, 3881–3885.

Song, Q., Lees-Miller, S.P., Kumar, S., Zhang, N., Chan, D.W., Smith, G.C.M., Jackson, S.P., Alnemri, E.S., Litwack, G., Khanna, K.K. and Lavin, M.F. (1996a) DNA-dependent protein kinase catalytic subunit: a target for an ICE-like protease in apoptosis. *EMBO J.*, **15**, 3238–3246.

Song, Q., Burrows, S.R., Smith, G., Lees-Miller, S.P., Kumar, S., Chan, D.W., Trapani, J.A., Alnemri, E., Litwack, G., Lu, H., Moss, D.J., Jackson, S.P. and Lavin, M.F. (1996b) ICE-like protease cleaves DNA-dependent protein kinase in cytotoxic T-cell killing. *J. Exp. Med.*, **184**, 619–626.

Song, Q., Lu, H., Zhang, N., Luckow, B., Shah, G., Poirier, G. and Lavin, M. (1997a) Specific cleavage of the large subunit of replication factor C in apoptosis is mediated by CPP32-like protease. *Biochem. Biophys. Res. Commun.*, **233**, 343–348.

Song, Q., Wei, T., Lees-Miller, S., Alnemri, E., Watters, D. and Lavin, M.F. (1997b) Resistance of actin to cleavage during apoptosis. *Proc. Natl. Acad.Sci. USA*, **94**, 157–162.

Squier, M.K.T. and Cohen, J.J. (1996) Calpain activation in apoptosis. *Cell Death and Differentiation*, **3**, 275–283.

Su, L.-K., Vogelstein, B. and Kinzler, K.W. (1993) Association of the APC tumor suppressor protein with catenins. *Science*, **262**, 1734–1737.

Takahashi, A. and Earnshaw, W.C. (1996) ICE-related proteases in apoptosis. *Curr. Opin. Genet. Dev.*, **6**, 50–55.

Takahashi, A., Alnemri, E.S., Lazebnik, Y.A., Fernandes-Alnemri, T., Litwack, G., Moir, R.D., Goldman, R.D., Poirier, G.G., Kaufmann, S.H. and Earnshaw,W.C. (1996a) Cleavage of lamin A by Mch2α but not CPP32: multiple interleukin 1β-converting enzyme-related proteases with distinct substrate recognition properties are active in apoptosis. *Proc. Natl. Acad. Sci. USA*, **93**, 8395–8400.

Takahashi, A., Musy, P.Y., Martins, L.M., Poirier, G.G., Moyer, R.W. and Earnshaw, W.C. (1996b) CrmA/SPI-2 inhibition of an endogenous ICE-related protease responsible for lamin A cleavage and apoptotic nuclear fragmentation. *J. Biol. Chem.*, **271**, 32487–32490.

Takahashi, A., Goldschmidt-Clermont, P.J., Alnemri, E.S., Fernandes-Alnemri, T., Yoshizawa-Kumagaya, K., Nakajima, K., Sasada, M., Poirier, G.G. and Earnshaw, W.C. (1997a) Inhibition of ICE-related proteases (caspases) and nuclear apoptosis by phenyarsine oxide. *Exp. Cell Res.*, **231**, 123–131.

Takahashi, A., Hirata, H., Yonehara, S., Imai, Y., Lee, K.K., Moyer, R.W., Turner, P.C., Mesner, P.W., Okazaki, T., Sawai, H., Kishi, S., Yamamoto, K., Okuma, M. and Sasada, M. (1997b) Affinity labeling displays the stepwise activation of ICE-related proteases by Fas, staurosporine, and crmA-sensitive caspase-8. *Oncogene*, **14**, 2741–2752.

Talanian, R.V., Quinlan, C., Trautz, S., Hackett, M.C., Mankovich, J.A., Banach, D., Ghayur, T., Brady, K.D. and Wong, W.W. (1997) Substrate specificities of caspase family proteases. *J. Biol. Chem.*, **272**, 9677–9682.

Tan, X., Martin, S.J., Green, D.R. and Wang, J.Y.J. (1997) Degradation of Retinoblastoma Protein in Tumor necrosis factor- and CD95-induced cell death. *J. Biol. Chem.*, **272**, 9613–9616.

Tanimoto, Y., Onishi, Y., Hashimoto, S. and Kizaki, H. (1997) Peptidyl aldehyde inhibitors of proteasome induce apoptosis rapidly in mouse lymphoma RVC cells. *J. Biochem.*, **121**, 542–549.

Tewari, M., Beidler, D.R. and Dixit, V.M. (1995) CrmA-inhitable cleavage of the 70 kDa protein component of the U1 small nuclear ribonucleoprotein during Fas- and tumor necrosis factor-induced apoptosis. *J. Biol. Chem.*, **270**, 18738–18741.

Thornberry, N.A., Rano, T.A., Peterson, E.P., Rasper, D.M., Timkey, T., Garcia-Calvo, M., Houtzager, V.M., Nordstrom, P.A., Roy, S., Vaillancourt, J.P., Chapman, K.T. and Nicholson, D.W. (1997) A combinatorial approach defines specificities of members of the caspase family and granzyme B. *J. Biol. Chem.*, **272**, 17907–17911.

Ubeda, M. and Habener, J.F. (1997) The large subunit of the DNA replication complex C (DSEB/RF-C140) cleaved and inactivated by caspase-3 (CPP32/YAMA) during fas-induced apoptosis. *J. Biol. Chem.*, **272**, 19562–19568.

Vanags, D.M., Porn-Ares, M.I., Coppola, S., Burgess, D.H. and Orrenius, S. (1996) Protease involvement in fodrin cleavage and phosphatidyserine exposure in apoptosis. *J. Biol. Chem.*, **271**, 31075–31085.

Voelkel-Johnson, C., Thorne, T.E. and Laster, S.M. (1996) Susceptibility to TNF in the presence of inhibitors of transcription or translation is dependent on the activity of cytosolic phospholipase A2 in human melanoma tumor cells. *J. Immunol.*, **156**, 201–207.

Wang, X., Pai, J.-T., Wiedenfeld, E.A., Medina, J.C., Slaughter, C.A., Goldstein, J.L. and Brown, M.S. (1995a) Purification of an interleukin-1 beta converting enzyme-related cysteine protease that cleaves sterol regulatory element-binding proteins between the leucine zipper and transmembrane domains. *J. Biol. Chem.*, **270**, 18044–18050.

Wang, X., Zelenski, N.G., Yang, J., Sakai, J., Brown, M.S. and Goldstein, J.L. (1996) Cleavage of sterol regulatory element binding proteins (SREBPs) by CPP32 during apoptosis. *EMBO J.*, **15**, 1012–1020.

Wang, Z.-Q., Auer, B., Stingl, L., Berghammer, H., Haidacher, D., Schweiger, M. and Wagner, E.F. (1995b) Mice lacking ADPRT and poly(ADP-ribosyl)ation develop normally but are susceptible to skin disease. *Genes Devel.*, **9**, 509–520.

Waterhouse, N., Kumar, S., Song, Q., Strike, P., Sparrow, L., Dreyfuss, G., Alnemri, E.S., Litwack, G., Lavin, M. and Watters, D. (1996) Heteronuclear ribonucleoproteins C1 and C2, components of the spliceosome, are specific targets of interleukin 1β-converting enzyme-like proteases in apoptosis. *J. Biol. Chem.*, **271**, 29335–29341.

Weaver, D.T. (1996a) Regulation and repair of double strand DNA breaks. *Crit. Rev. Eukaryot. Gene Expr.*, **6**, 345–375.

Weaver, V.M., Carson, C.E., Walker, P.R., Chaly, N., Lach, B., Raymond, Y., Brown, D.L. and Sikorska, M. (1996b) Degradation of nuclear matrix and DNA cleavage in apoptotic thymocytes. *J. Cell Sci.*, **109**, 45–56.

Weinberg, R.A. (1995) The retinoblastoma protein and cell cycle control. *Cell*, **81**, 323–330.

Wissing, D., Mouritzen, H., Egeblad, M., Poirier, G.G. and Jaattela, M. (1997) Involvement of caspase-dependent activation of cytosolic phospholipase A2 in tumor necrosis factor-induced apoptosis. *Proc. Natl. Acad. Sci. USA*, **94**, 5073–5077.

Wyllie, A.H., Kerr, J.F. and Currie, A.R. (1980) Cell death: the significance of apoptosis. *Int. Rev. Cytol.*, **68**, 251–306.

Xia, W., Zhang, J., Perez, R., Koo, E.H. and Selkoe, D.J. (1997) Interaction between amyloid precursor protein and presenilins in mammalian cells: implications for the pathogenesis of alzheimer disease. *Proc. Natl. Acad. Sci. USA*, **94**, 8208–8213.

Zeitlin, S., Liu, J.-P., Chapman, D.L., Papaioannou, V.E. and Efstratiadis, A. (1995) Increased apoptosis and early embryonic lethality in mice nullizygous for the Huntington's disease gene homologue. *Nature Genet.*, **11**, 155–162.

Zhivotovsky, B., Gahm, A. and Orrenius, S. (1997) Two different proteases are involved in the proteolysis of lamin during apoptosis. *Biochem. Biophys. Res. Commun.*, **233**, 96–101.

INDEX